Springer Hydrogeology

Series Editor

Juan Carlos Santamarta Cerezal, San Cristóbal de la Laguna, Sta. Cruz Tenerife, Spain

The *Springer Hydrogeology* series seeks to publish a broad portfolio of scientific books, aiming at researchers, students, and everyone interested in hydrogeology. The series includes peer-reviewed monographs, edited volumes, textbooks, and conference proceedings. It covers the entire area of hydrogeology including, but not limited to, isotope hydrology, groundwater models, water resources and systems, and related subjects.

Chaitanya B. Pande · Manish Kumar ·
N. L. Kushwaha
Editors

Surface and Groundwater Resources Development and Management in Semi-arid Region

Strategies and Solutions for Sustainable Water Management

Springer

Editors
Chaitanya B. Pande
Indian Institute of Tropical Meteorology
Pune, India

Institute of Energy Infrastructure
Universiti Tenaga Nasional (UNITEN)
Kajang, Malaysia

N. L. Kushwaha
Indian Agricultural Research Institute
New Delhi, India

Manish Kumar
Dr. Rajendra Prasad Central Agricultural University
Bihar, India

ISSN 2364-6454 ISSN 2364-6462 (electronic)
Springer Hydrogeology
ISBN 978-3-031-29393-1 ISBN 978-3-031-29394-8 (eBook)
https://doi.org/10.1007/978-3-031-29394-8

This Springer imprint is published by the registered company Springer Nature Switzerland AG
The registered company address is: Gewerbestrasse 11, 6330 Cham, Switzerland

Preface

Land and water are two main resources on earth for the survival of human habitat, natural plant growth and wildlife habitat. Mismanagement of land resources and scarcity of water resources day by day affects the sustainable development of any region. Therefore, it is necessary to preserve, control and maintain these resources through improved engineering practices. The semi-arid region is the median of arid and humid region and somewhat characteristics of other two regions. The interventions in semi-arid regions can also be an alternate solution to arid and humid region. Semi-arid regions of earth are today suffering from a lack of sustainable water and groundwater. There is a growing demand world wide for sustainable watershed development, management and planning. The demand is more significant in the rain-fed and drought-prone area of earth surface, where watershed management is poorer and groundwater is limited.

In the present text, an attempt has been made to provide comprehensive coverage of surface and groundwater resources development and management in semi-arid region. The book covers the management of surface and groundwater using new engineering applications in two sections. The first section is about sustainable surface and groundwater development and hydrological systems. The sections deal about applications of the remote sensing, machine learning, GIS techniques and modelling used for the surface and groundwater resources and management. The chapters are received from different Scientists, Professors, Researchers, Post-doctoral researchers and Post-graduate students from all over the world.

Pune, India
Bihar, India
New Delhi, India

Chaitanya B. Pande
Manish Kumar
N. L. Kushwaha

Contents

Evapotranspiration Importance in Water Resources Management Through Cutting-Edge Approaches of Remote Sensing and Machine Learning Algorithms

Ali Raza, Yongguang Hu, Siham Acharki, Noman Ali Buttar, Ram L. Ray, Aftab Khaliq, Nadeem Zubair, Muhammad Zubair, Neyha Rubab Syed, and Ahmed Elbeltagi

Abstract Evapotranspiration (ET) is an important component of the water cycle and agricultural water balance. Planning, managing, and regulating agricultural water resources rely on reliable estimates of agricultural water consumption. It leads

---

A. Raza (✉) · Y. Hu (✉) · N. A. Buttar
School of Agricultural Engineering, Jiangsu University, Zhenjiang 212013, P.R. China
e-mail: alir3201@gmail.com

Y. Hu
e-mail: deerhu@ujs.edu.cn

S. Acharki
Department of Earth Sciences, Faculty of Sciences and Techniques, Abdelmalek Essaadi University, 90000 Tangier, Morocco

N. A. Buttar
Department of Agricultural Engineering, Khwaja Fareed University of Engineering and Information Technology, Rahim Yar Khan 64200, Pakistan

R. L. Ray
Department of Agriculture, Nutrition and Human Ecology, College of Agriculture and Human Sciences, Prairie View A&M University, Prairie View, TX 77446, USA
e-mail: raray@pvamu.edu

A. Khaliq · N. Zubair
Department of Agricultural Engineering, Bahauddin Zakariya University 60800, Multan, Pakistan

M. Zubair
School of Transportation, Southeast University, Nanjing 21009, China

N. R. Syed
School of Energy & Environment, Power Engineering & Engineering Thermophysics, Southeast University, Nanjing 21009, China

A. Elbeltagi
Agricultural Engineering Department, Faculty of Agriculture, Mansoura University, Mansoura 35516, Egypt
e-mail: ahmedelbeltagy81@mans.edu.eg

C. B. Pande et al. (eds.), *Surface and Groundwater Resources Development and Management in Semi-arid Region*, Springer Hydrogeology,
https://doi.org/10.1007/978-3-031-29394-8_1

to establishing a sustainable water balance, mitigates the consequences of water scarcity, and prevents the waste and overuse of scarce water resources. An accurate assessment of ET is important to improve water use efficiency and sustainable water management in agriculture because ET is a primary consumer of irrigation water and precipitation in agricultural landscapes. To increase the accuracy of ET calculation in agricultural water management, precision and digital agricultural technologies, cutting-edge approaches such as satellite data, remote sensing technology, and machine learning algorithms would be useful. This chapter examines the effective approaches used in estimating ET for agricultural water management and provides a summary of the technical advances made in the methodologies currently in use. It also identifies potential enhancements that could be made to boost the precision of ET estimation and achieve precise agricultural water management.

Keywords Remote sensing · Water resources management · Evapotranspiration · Machine learning algorithms · Challenges and solutions · Mitigation and adaptation

1 Introduction

Weather, climate, and soil moisture are only a few of the dynamic and complicated environmental elements that greatly impact agricultural production. These circumstances have limited control over the processes and cannot be foreseencompletely (van Mourik et al. 2021). Irrigation can boost agricultural productivity and income compared to rain-fed agriculture (Jaramillo et al. 2020; Vanschoenwinkel and Van Passel 2018). Furthermore, irrigated agriculture is dependable and provides a wider range of crops with higher value (Asmamaw et al. 2021). Furthermore, irrigated agriculture uses 20% of all cultivated lands, produces 40% of the world's crops, and significantly contributes to food security globally (Meybeck and Redfern 2016; Zongo et al. 2022). In 2016, agriculture used around 92% of the freshwater withdrawn globally (World Bank 2022). In 2050, the Food and Agriculture Organization (FAO) predicted an 11% increase in water demand for agriculture (Bruinsma 2009). De Fraiture et al. (2010) concluded in their research that water resources could be sufficient till 2050 if water usage properly manage and mitigation strategies will apply to reduce risk in agriculture sector by controlling green land into urbanization (Liu et al. 2022).

Water is considered the most crucial resource in agriculture (Chartzoulakis and Bertaki 2015). As the world's population is expected to increase, agriculture will require more food and water (Boretti and Rosa 2019; Islam and Karim 2019; OECD 2013). Chartzoulakis (Chartzoulakis and Bertaki 2015) indicated that less than 65 percent of the applied water is actually used by crops, demonstrating irrigation's poor efficiency in the agricultural sector. Additionally, the majority of water extraction worldwide roughly 70% is used for irrigation, which has been emphasized as the main driver for water depletion in most regions (Aryalekshmi et al. 2021). Domestic, industrial, and energy-generating uses of water are further strained for various reasons, such as recreation and cultural enhancement and the protection of ecosystems and the environment (Karimi and Bastiaanssen 2015). Extensive research revealed that poor and inappropriate water management was to blame for the water shortages in several areas (Entezari et al. 2019). Sustainable water resource management in agriculture can be achieved by supporting farmers using water-saving techniques and putting legal restrictions on water allocations in place. However, water management in the agricultural sector is a crucial issue that technological improvements can assist in resolving in comparison to other issues (Pande et al. 2023a).

Currently, enterprises and infrastructure building can now make more money economically from a water investment than from agriculture. The production from irrigated agriculture is anticipated to increase greatly while utilizing less water in the future (de Jong et al. 2021). The escalating demand for water requires more efficient procedures to satisfy it while meeting the finite resources' constraints. In many river basins, water resource management plans prioritize cutting back on water use without sacrificing agricultural productivity (Sun et al. 2017). In this case, it is possible to use water-conserving irrigation methods and tools. For instance, alternate wetting and drying (AWD) and semi-dry culture (SDC) can both conserve up to 50% of water when compared to standard irrigation approaches (Enriquez et al. 2021; Wang et al. 2018). A significant amount of agricultural water is lost as evapotranspiration. It is important to use in-situ plants and soil measurements (e.g., soil moisture, plant moisture, soil and plant physical parameters) to estimate actual crop water requirements based on evapotranspiration (Altobelli et al. 2016). Most of the water from semi-arid cultivated regions is lost through evapotranspiration (ET) (Eliades et al. 2022; Liou and Kar 2014). Understanding how ET affects the water budget is important for managing water resources, forest development and the diversity of species, the production of sustainable crops, the safety of our food supply, and the stability of society (Althoff et al. 2019; de Jong et al. 2021; Karimi and Bastiaanssen 2015; Moiwo and Tao 2015; Stefanidis and Alexandridis 2021; Zongo et al. 2022). It should be mentioned that more effective ET-reduction strategies must be implemented to ensure a sustainable and efficient approach to water use (Reyes-Gonzalez 2017; Zheng et al. 2020). These measures will balance water distribution between industries, households, ecosystems, and agriculture (Afzaal et al. 2020; Bogawski and Bednorz 2014; Moiwo and Tao 2015; Wang et al. 2012).

This chapter aims to investigate the effective methods for estimating ET for agricultural water management; provide a summary of the technical advances made in the methodologies currently in use such as Remote Sensing (RS), Geographical Information System (GIS) and Artificial Intelligence (AI); identify potential improvements that could be implemented to improve the accuracy of ET estimation and achieve precise agricultural water management; demonstrate the potential benefits of technological advancement and future opportunities for sustainable agriculture.

2 Importance of Evapotranspiration in Water Resource Management

Evapotranspiration (ET), which combines evaporation and transpiration, is actually the absorption of water by plants from the zone of the root. Evaporation is the vaporization of water or soil from the surface as shown in Fig. 1. Agricultural decision support tools such as precipitation and ET, a measure of climate, represent regional climates. The ET monitors the amount of water required for efficient water management while also assisting with surface energy balance (Bogawski and Bednorz 2014; Wang et al. 2012). It is becoming more difficult to conserve water in irrigation scheduling while water productivity is increasing, both globally and locally (Krishna 2019). One of the key problems in river basin hydrology is estimating the local ET (Karimi and Bastiaanssen 2015). Li et al. (2009) estimated that 60% or more of the typical precipitation would contain ET from the ground surface. ET may be used to measure plant water tension since, for vegetated fields, the rates at which the vegetation absorbs water are the same (Islam and Karim 2019) A decrease in water resources could have a negative impact on the harvest because there are not enough water allocations, endangering food security. The water management system needs to be optimized in this context and properly estimate evapotranspiration (Reyes-Gonzalez 2017). According to Krishna (Hao et al. 2022), it is essential to estimate ET precisely because knowing and measuring the factors affecting ET is necessary to clarify the ambiguities in the hydrologic cycle's behavior in response to climate change. Well-founded ET estimates are required to control the irrigation system's components; they considerthe size and power of canals, dams, and pumps because ET plays an important role in water balance at all scales, from individual plots to global plots systems (Kharrou et al. 2021).

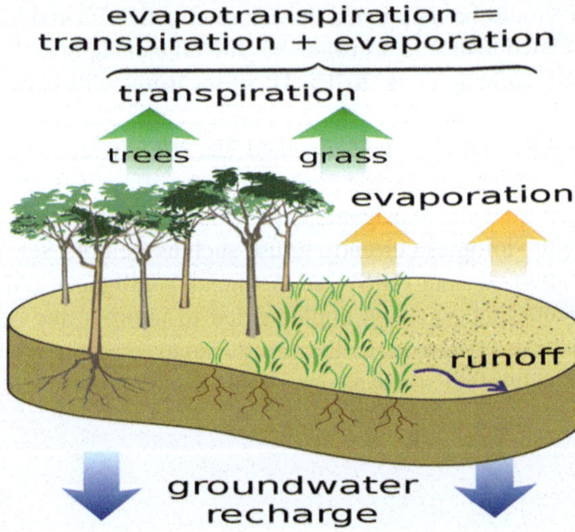

Fig. 1 Conceptual diagram of near-surface hydrology, showing evapotranspiration, evaporation, transpiration, runoff, and recharge processes (*source* Buttar et al. 2018)

The evaporation process allows the hydrosphere, atmosphere, and biosphere to exchange energy continuously (Krishna 2019; Wang et al. 2012). Because it is a dynamic metric, the crop water need should consider water flows, stocks, and variations over time. Since all measurements require enough hardware and sensors for dependable monitoring and data logging, they can all be challenging (Calera et al. 2017). The ET mechanism significantly aids in the return of moisture to the atmosphere (Karimi and Bastiaanssen 2015). Moiwo and Tao (2015) reported that precipitation, which accounts for 39.0% of ET is by far the largest contributor, irrigation (24.7%) and soil water (36.3%) are next, after analyzing the respective contributions of the three water supply sources to ET. Water moves from the seas to the atmosphere to the continents and back to the oceans across and under the land surface, as seen in Fig. 2. The numbers in parentheses after the different water forms (such as ice) indicate their volumes in millions of cubic kilometers, whereas the numbers after the processes (such as precipitation) indicate their fluxes in millions of cubic kilometers of water each year. Moreover, ET is necessary for all aspects of productivity in the environment (Liu and El-Kassaby 2018). In most cases, the vegetation's heterogeneity influences ET estimation, which is made more difficult during periods of dynamic flux following irrigation and precipitation (de Andrade et al. 2021; Geli et al. 2019; Gowda et al. 2007; Wang et al. 2012).

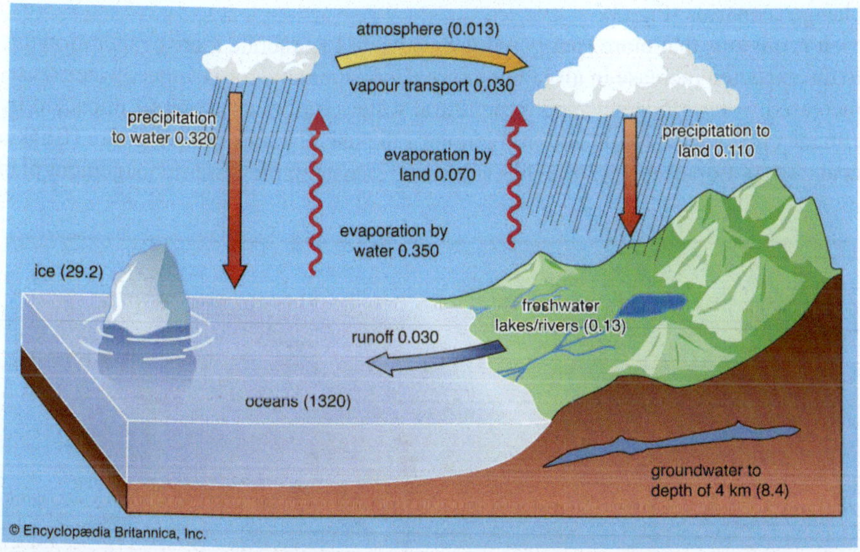

Fig. 2 Representation of surface hydrological cycle (*source* Encyclopædia Britannica)

2.1 *Water Resource Management and Climate Change*

Agriculture is a highly vulnerable and affected sector by climate change and fluctuations. According to research from the Food and Agriculture Organization (FAO) and the Intergovernmental Panel on Climate Change (IPCC), agriculture is one of the most susceptible to climate change, particularly in developing nations. Due to current advanced technologies, drones can now be used as part of an Innovative Agriculture Meteorological Methodology for the Precise and Real-Time Estimation of Crop Water (Alexandris et al. 2021). This has caused worries among the scientific community. Climate change will have a variety of complex implications in agriculture and water resources (Sun et al. 2018). The soil's water balance will be affected by a changing climate, which will change how much water evaporates and transpires. Significant changes in agricultural productivity, water availability and quality variations, frequent and severe floods, and droughts are only a few of the effects (Hardelin and Lankoski 2015). Climate change adaptation and mitigation impact agricultural water consumption, especially when it comes to water reuse, increased water efficiency, and agricultural water conservation (Bhakta et al. 2019). Lopez et al. (2022) has developed a sustainable water management plan to reduce extensive groundwater extraction from irrigated agriculture and guarantee that water management policies comply with sustainable water usage and storage under various climate change scenarios (Fig. 3).

It is obvious that more energy will be available to generate more evaporation due to the expected increase in air temperature. Unfortunately, for instance, water reuse, increased water efficiency, and agricultural water conservation are all impacted by climate change adaptation and mitigation of the use of water in agriculture (Bhakta et al. 2019). Lopez et al. (2022) has suggested a sustainable water management plan

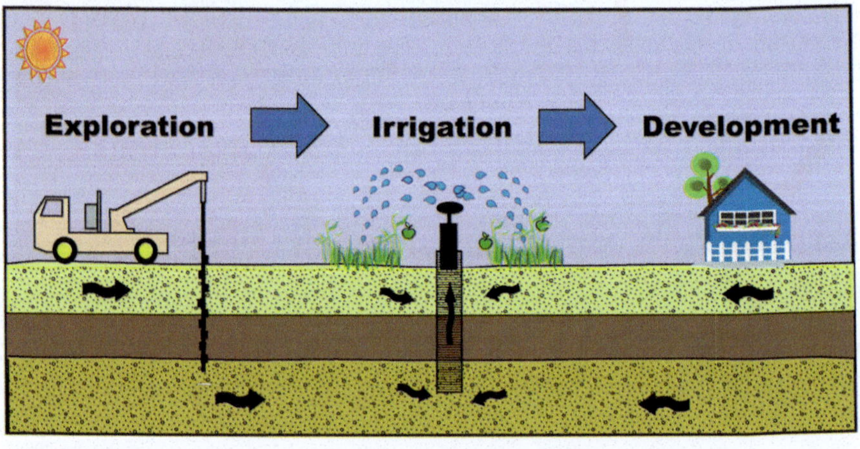

Fig. 3 Sustainable use of the groundwater resource (*source* Gallardo 2019)

to reduce the amount of groundwater significantly removed from irrigated agriculture and ensure that water management policies follow sustainable water usage and storage under various climate change scenarios (Pande et al. 2023b). The decreasingprecipitation and rising temperatures will have a negatively impact on crop yields and outputs. To avoid water shortages caused by climate change, it is essential to understand evapotranspiration's function (Lopez et al. 2022). Entezari et al. (2019) introduced air–water harvesting technology (AWH) to encourage sustainable agriculture. In addition, they examined the possibility of recycling evapotranspiration water within greenhouses to acquire liquid water in arid or desert environments. Based on the Global Climate Model (GCM) composite forecasts, Xing-Guo (Bhatt & Hossain 2019) showed that there had been a substantial change in the climate over the past 60 years in the study region. Across all three scenarios, regional average ET increased by 6–10% more in the 2050s than in the 1990s, and that increase would be more pronounced in the twenty-first century. However, local variations are not sufficiently taken into account by GCMs. The climate change impact on water availability has been studied using Regional Climate Models (RCM) (Mo et al. 2017; Olmos Gimenez and García-Galiano 2018; Stefanidis 2021), and researchers have discussed the model resolution effect on forecast accuracy. According to Salman (Tolika et al. 2016), crop water demand and climatic water availability (CWA) vary spatiotemporally based on long-term rainfall and temperature data. The study findings indicated that rising temperatures contribute to increased evapotranspiration, which raises crop water needs significantly and reduces water availability in a particular climate (Fig. 4).

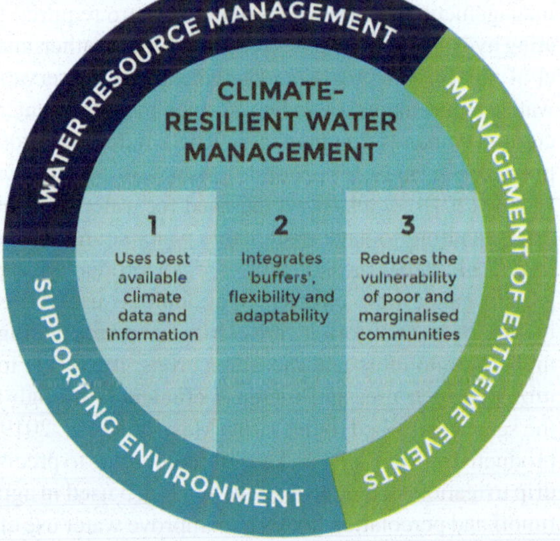

Fig. 4 Climate resilient water management (*source* Will Bugler 2020)

2.2 Role in Water Resource Management of Accurate ET Estimation

Smart farming uses cutting-edge technology to improve plant growth conditions (MacPherson et al. 2022). The smart agriculture concept involves crop monitoring through optimization and automation to some extent. A digital agriculture system incorporates web-based data platforms and analytics combined with quality, efficient, and productive farm processes; this is an application of "Quality and Efficient Agriculture" (Salman et al. 2020). Data management in digital agriculture involves big data analysis. Insufficient mobile cellular infrastructure and facilities are hindered by a lack of key technologies (MacPherson et al. 2022) and the fact that agriculture is not usually practiced in most countries, putting digitization into practice is difficult. Negative effects, such as profits loss, environment contamination, and leaching of nutrient are caused by geomorphology, soil properties, crop development phases, and other agronomic parameters. Nevertheless, traditional farming systems fail to account for the heterogeneity in agricultural areas (Boretti and Rosa 2019). On the other hand, precision agriculture makes use of spatially distributed data, precise data processing, and reliable decision-making tools. The use of harvest monitoring, variable-rate irrigation technology (VRT), GIS, remote sensing, and the Global Navigation Satellite System (GNSS) are just a few of the technologies incorporated into precision agriculture (Kingra et al. 2016).

ET plays a crucial role in precision farming. An accurate ET computation is required to comprehend water balances, hydrological processes, climate change, and ecosystem activities. Evapotranspiration is the biggest obstacle to agricultural water management. Accurate ET calculations are required for drought monitoring, evaluating hydrological models, forecasting the weather, and predicting forest fires (Kingra et al. 2016). For precise management and conservation of agricultural water, crop water requirements are essential since irrigation water cannot completely meet agricultural demands (Djaman et al. 2018). Indeed, evapotranspiration must be measured precisely to make an accurate estimation of crop water requirements. Koech and Langat (2018) emphasized the need for water-efficient technologies and techniques for agriculture to have sustainable water sources. Additionally, Blatchford (Blatchford et al. 2019) measures crop water productivity (CWP) using digital technologies to gauge how effectively agriculture uses water. Since precision agriculture incorporates monitoring, measuring, and responding to variability in field equipment that monitors and measures crops, it expects to lower cultivation costs, optimize resource use, and increase efficiency through real-time data transmitted via the sensors attached to the farm (Mavridou et al. 2019). Irrigated agriculture is more productive in semi-arid and desert regions due to precision agriculture applications. A drip irrigation technique as shown in Fig. 5 used in agriculture field which can reduce runoff and percolation losses and improve water use efficiency (Shanmugapriya et al. 2019).

Fig. 5 Use of drip irrigation system in agricultural field

2.3 Quantitative ET Estimation: Current Perspective

In the twenty-first century, it was widely believed that advances in ET technology were still utilized more frequently for research than for practical applications. In recent years, the use of spatial science tools in agriculture, such as satellite technology and remote sensing, has risen substantially. It offers an effective and inexpensive treatment for strategies for field measurements. Typically, field sensors advise inputs and manage the appropriate amounts of nutrients and water. GPS receivers can capture the spatial variability of these needs (Kingra et al. 2016). Consequently, future automated farm management will extensively use agricultural automation systems and technologies. Technologies such as deep learning and spectral analysis fall into this category (Tian et al. 2020). Systems for automating agriculture can be made more cost-effective, dependable, and stable with computer vision enhanced by artificial intelligence (AI) (Tian et al. 2020). The ET database's availability with open access will help future ET-based agriculture water management research. Numerous organizations are developing this idea, including the Chinese Academy of Sciences, the US Geological Survey, the US Department of Agriculture, and the Commonwealth Science and Industrial Research Organization of Australia. It is difficult to estimate evapotranspiration precisely. Nonetheless, it is necessary to develop and implement irrigation systems and water management (Liakos et al. 2018). Although the unidentified water traverses the water balancing method in this approach, the boundary contributes to an error in estimating evapotranspiration.

Numerous studies have demonstrated the importance of water balance in estimating evapotranspiration. According to Nolz (2016), an enhanced sensor arrangement system would help detect these motions by gaining a deeper understanding of the existence and movement of subsoil and groundwater. Traditional ET estimating procedures utilize external measurements; several factors are considered when calculating the leaf's surface roughness, vapor pressure, wind speed, temperature, and gas concentrations (CO_2, water vapor) (Ghiat et al. 2021; Maina et al.

2014; Subedi and Chávez 2015). On expansive terrain, it is challenging to measure these features; consequently, extrapolation or interpolation must be used to approximate them (Abdollahnejad et al. 2018; Li et al. 2009). Empirical approaches have the advantage of reducing calculation time and the number of ground-based observations across regions with high variability. Still, they have the disadvantage of requiring fewer ground-based observations in homogeneous areas. There is a possibility that it may not work properly depending upon the ground surface characteristics. The Priestley-Taylor model, the Stanghellini model, and the Hargreaves-Samani model are additional models to the Penman–Monteith equation included in Ghiat's list of models (Ghiat et al. 2021). Ghiat et al. (2021) also mentions the Stanghellini equation. Nonetheless, empirical constants integration in empirical models leads to an overestimation of ET. Despite their sometimes high data requirements, physically based analytical approaches can provide ET estimates that are in good agreement with measurements (An 2011; Ghiat et al. 2021). Examples of field scale measurement techniques include scintillometers, Bowen meters, lysimeters, conventional soil water balance, eddy covariance systems, and others (Blatchford et al. 2019; Ghiat et al. 2021; Karimi and Bastiaanssen 2015). In some cases, installing infrastructure across the entire watershed may not be feasible financially. Most experts believe that FAO's Penman–Monteith model is the best technique for determining crop coefficient (K), thereby making ET the most accurate becauseit performs with precise lysimeter observations (Ghiat et al. 2021; Maina et al. 2014). Subedi and Chávez (2015) and Maina et al. (2014) indicated that Penman–Monteith equation is the most frequent approach for determining ET. Despite the unquestionable accuracy of the aerodynamic terms of the Penman–Monteith equation, one of its major challenges remains is the computation of the canopy surface resistance (Meraz-Maldonado and Flores-Magdaleno 2019; Zhao et al. 2020). Therefore, accurate surface resistance estimation requires more attention. Subedi and Chávez (2015) emphasized that the Penman–Monteith equation under an advective state is flawed because it cannot account for the horizontal passage of perfect sensible heat flow. Without good meteorological data, the Penman–Monteith method cannot be employed. A comparison of the Penman–Monteith method with five other temperature-based methods (Hargreaves-Samani, Blaney-Riddle, and Thornthwaite, Hamon) can be found in Lang et al. (2017) and three radiation-based approaches on an annual and seasonal scale (Makkink, Abtew, and Priestley-Taylor). The most important finding of the study was that strategies for PET estimation based on radiation performed better than those based on temperature. The Abtew method is suited for warm, low-latitude locations, whereas the Makkink method is suitable for regions with complicated geographical features. Using parametric PET estimation based on radiation, Tegos et al. (2022) proposesa brand new, extremely precise method. This model's only downside is that it must be calibrated locally to perform for similar watersheds. Although it is also highly accurate it is time consuming and expensive to measure evapotranspiration in the field with a lysimeter experiment. Therefore, climatological data is usually used to predict ET. Numerous experts evaluated ET estimation methods for different case studies based on radiation and temperature. In addition, multiple researchers have successfully predicted the spatial and temporal distribution of ET using cutting-edge machine learning and

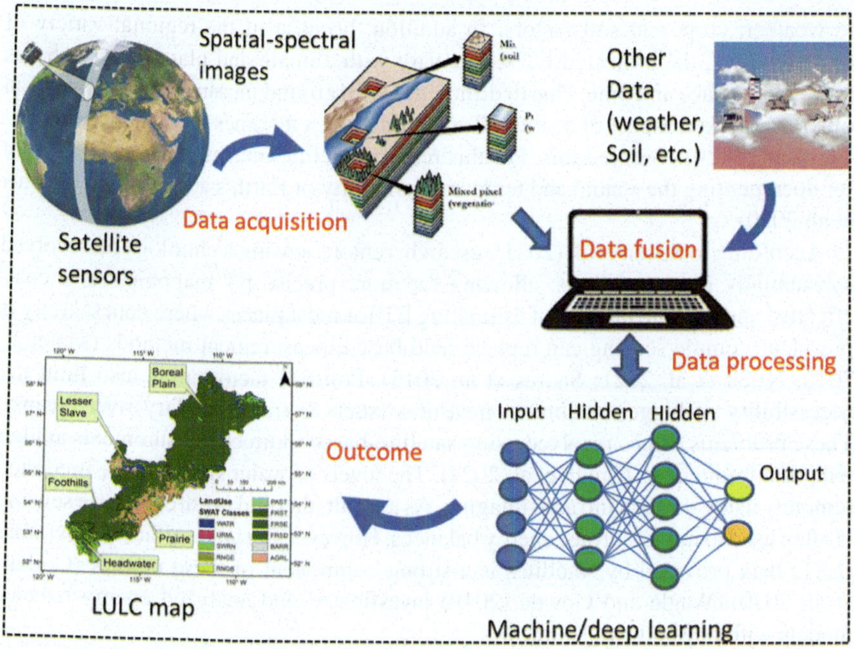

Fig. 6 LULC mapping using machine learning and remote sensing methods (*source* Wang et al. 2022)

remote sensing methods (Bhakta et al. 2019). Wang et al. (2022) explored cutting-edge machine learning and remote sensing methods as shown in Fig. 5 to investigate land use and land changes (LULC) (Fig. 6).

In agricultural research, numerous meteorological elements, physio-chemical variables, and soil features are utilized to make changes (Alvino and Marino 2017; Blatchford et al. 2019; Calera et al. 2017; Shanmugapriya et al. 2019; Zeyliger and Ermolaeva 2021). Due to the expanded availability of EO data and remote sensing technology, global PET datasets can now be developed using monthly remote temperature measurements (Tegos et al. 2022). Agricultural growth monitoring and crop classification can also be performed utilizing crop productivity season remote sensing analyses. As a result, remote sensing technologies coupled with GPS and the effectiveness of agricultural activities can be increased with GIS, such as estimating the size of fields. Assessing soil fertility and moisture levels, determining crop stress, tracking disease and pest spread, and forecasting droughts and floods are overwhelming topics (Chen et al. 2019; Meybeck and Redfern 2016; Nsiah et al. 2021; Pereira et al. 2015; Shanmugapriya et al. 2019).

Reyes-Gonzalez (2017) used ML as an effective tool for assessing evapotranspiration and agricultural water demand, emphasizing satellite-based remote sensing. They have investigated fundamental components that regulate meteorological observations, crop data, and geohydraulic features that will affect ET rates, in addition to weather, crop, and soil factors. In addition, because of the regional variety of these features, the projected ET values vary with climate and plant growth phases throughout space and time. Due to difficulties with ground measurements, there is an often overlooked aspect of spatial ET modeling when it comes to water accounting. Wu et al. (2020) observed this. For this reason, satellite images are a powerful tool for documenting the spatial and temporal variability of Earth's atmosphere (Corbari et al. 2020).

According to Stisen et al. (2021) research, remote sensing technology has evolved substantially in recent years, allowing for more precise ET mapping. As a cost-effective and reliable method of estimating ET for most places where data scarcity is prevalent, remote sensing can replace field-based measurement methods (Ma et al. 2018; Nsiah et al. 2021; Sadras et al. 2015). Political factors may also limit the accessibility of bigger catchment measures (such as transboundary river basins). These problems can be resolved using satellite-based indirect measurements at high temporal frequencies (Singh et al. 2022). The levels of water stress can be predicted remotely using thermal infrared imaging. As a result, thermal infrared remote sensing is often used to model surface energy balances. However, the land surface temperature (LST) data provided by satellites is a strong component of these models (Corbari et al. 2020). Wagle and Gowda (2019) investigated and analyzed energy balance models with one and two sources.

Agricultural and food security are negatively impacted by climate change when ET is not calculated accurately based on ET's geographic spread, plant water status, reflection, thermal radiance, and vegetation index coefficients (Caldwell et al. 2017; Calera et al. 2017). Vegetation index coefficients should be utilized in crop water status studies based on reflectance, thermal radiance, and vegetation index coefficients. According to Alvino, crop-water condition and yield are positively correlated and argued for a larger emphasis on remote sensing research to boost agricultural output. As these techniques grow more user-friendly, remote sensing enables the mapping of extraterrestrial life with unparalleled precision and little effort (Caldwell et al. 2017). Remote sensing data and crop coefficients (Kc) are combined to estimate actual ET (Calera et al. 2017). Reflectance-based models are capable of estimating potential crop transpiration. According to Reyes-González findings (Reyes-González et al. 2018), on regional and field scales, ET maps can be generated using remotely sensed multispectral vegetation indices. Moreover, their results showed seasonal water consumption variations may be reduced by 18% if irrigation programs incorporate ET estimation (Reyes-González et al. 2018). Remote sensing applications use digital infrared thermography to measure the temperature of the canopy for early diagnosis of agricultural water stress and water conservation through site-specific irrigation control. Combining satellite information, mobile devices, and a web-GIS platform, IrriSatSMS is a system developed by the CSIRO of Australia to manage irrigation water (Calera et al. 2017). Remote sensing techniques, such as unmanned aircraft systems (UAS) and wireless soil moisture sensors, are indispensable in precision agriculture (Alvino and Marino 2017), which enables the monitoring of soil moisture levels and plant growth to enhance the effectiveness of irrigation management at specific locations. These methodologies are capable of assessing the state of soil moisture and certain physical properties, as well as harvest prediction, crop water conditions, canopy status, and pest infestations (Neupane and Guo 2019). It is more practical to monitor unmeasured watersheds using remote sensing. According to Andriambeloson r, wind speed, air temperature, and humidity should be measured precisely remotely as part of the energy balance equation. As another challenge, the airspace between the surface of the earth and satellite sensors disturbs satellite radiometers, reducing the precision of measurements (Li et al. 2009). ET measurements are strongly influenced by solar radiation and temperature but have a very low correlation with them (An 2011). ET governs wind velocity and relative humidity. The exchange of materials and energy between plants and the atmosphere. Thus, plant development and yield are highly dependent on ET. Using the most recent remote sensing techniques, it is possible to monitor agricultural performance with excellent spatial and temporal accuracy (Blatchford et al. 2019). Using remote sensing, Karimi and Bastiaanssen (2015) discovered that 95% precision could be achieved in estimating evapotranspiration based on actual rainfall, precipitation, and land use. In addition, they emphasized the necessity for further research into multiple space-borne sensors for geographical land cover and precipitation mapping. Machine learning algorithms have enabled machine vision technologies to be deployed in agriculture that accurately evaluate huge amounts of data (Jing et al. 2019; Mavridou et al. 2019). Assuming an energy balance at the surface, various algorithms and functions for using satellite observations to estimate the actual ET have developed in recent years. Soil moisture information, which can be gathered using thermal or microwave measurements, is essential to determining ET. Microwave measurements have the great advantage of being usable in various circumstances and at varying spatial resolutions (Karimi and Bastiaanssen 2015). Energy is consumed during the evaporation process. A lower canopy surface temperature can be achieved by increasing evapotranspiration (Jones et al. 2018). Using this methodology, evapotranspiration and drought stress have been calculated in agricultural water management using ground-based thermal remote sensing (Singh et al. 2022; Winbourne et al. 2020). On agricultural lands, monitoring canopy surface temperature is difficult because of the varied and distributed canopy cover. To ensure that uniform croplands are monitored, thermal remote sensing sensors are largely used on the ground. Yet modern thermal cameras offer accurate canopy surface temperatures while minimizing background and soil noise, making them useful for various crops. The leaf water potential or leaf stomatal conductance is often estimated in advanced irrigation scheduling to determine plant water status (Spohrer et al. 2010). Infrared thermography (IRT) is a practical non-contact method for calculating the relationship between transpiration and stomata opening and leaf temperature. Therefore, it is possible to glean information about the physiological health of all crops in the field from an increased leaf cluster. Nevertheless, leaf temperature is dependenton additional elements. Thermography-based water status assessments can be impacted by factors including air temperature, radiation, humidity, and wind speed (Spohrer et al. 2010). By using sophisticated methods for ET estimation, such as remote sensing and satellite technology, one can improve agricultural water management significantly and with greater precision.

3 Directions for Future Research

The potential and usefulness of the present evapotranspiration estimating systems, as well as a number of their shortcomings, will be shown by examining the methodologies of Agricultural water management and ET estimations as stated in the literature. Findings from ET modeling have only recently begun to be applied in the field (Howell 1996; Pereira et al. 1996). However, the majority of agricultural regions around the world have adopted the most contemporary scientific discoveries (Bhattarai and Wagle 2021; Sattari et al. 2021). Surface variables, geographical and temporal data scales, land surface fluxes, validations of modeled latent heat fluxes, and proximate meteorological information acquisition are all part of the physical interpretation of surface variables and land surface fluxes are major obstacles to using remote sensing techniques to estimate ET. It is notable that ET should be physically estimated as an energy variable, not immediately observed from space.

For a better understanding of seasonal fluctuations, studies of phenology, plant cover, and water movement from the land are required. Advanced space-borne observational equipment and remote sensing methods are employed to quantify energy flows into the atmosphere at the top and estimate ET. No precise measurements of the energy flow on the earth's surface have been developed. According to Ustin and Middleton (2021), the qualitative data gleaned from optical, thermal, radar, and LiDAR imaging is expected to close this gap over the next couple of decades. Jing et al. (2019) has pointed out that it does not matter how climates and terrains differ in land surface features; no standard model could be utilized to calculate the ET using satellite data. For an accurate ET estimation in future applications, the relationship between distributed hydrological modeling and remote sensing should be strengthened (Mhawej et al. 2020). Higher-resolution input data are also necessary because they must direct satellite-based ET measurements for managing water resources and conducting hydrological research (Karimi and Bastiaanssen 2015). Accuracy is also improved by using machine learning algorithms to analyze massive amounts of observable data. Future research methodologies should modify existing methodologies and consider the possibility of a changing climate (Krishna 2019; Mhawej et al. 2020).

4 Conclusions

An accurate water usage assessment of agricultural areas is necessary for planning, managing, and controlling agricultural water resources. Water shortages affect agricultural growth, harvest, and ultimately, food scarcity. Evapotranspiration (ET) is the main cause of water loss in agricultural areas. Utilizing more efficient ET reduction techniques is essential for efficient and sustainable water management in agricultural settings. The air temperature is also projected to increase due to climate change. Thus, more evaporation will take place when more energy becomes accessible. Consequently, maximizing agricultural water efficiency and putting long-term water management techniques in place should be based on an accurate estimation of ET. Taking into account these specifics, our comprehensive review focuses on the technical development of accurate ET measuring methods and approaches, as well as the identification of more efficient ET reduction strategies which hinder crop growth. These are the two primary approaches that are available in the literature. It is critical that the ET estimation be accurate in order to precisely manage agricultural water. It was also found that more research is required to detect energy interactions and to bridge the knowledge and technological gaps. There are many unknowns regarding our understanding of energy exchange during different crop growth phases. The ET assessment is more accurate due to the use of digital and precision agriculture technologies. The use of innovative techniques, such as the application of machine learning algorithms, remote sensing, and satellite technology to analyze the data used in ET estimation, has improved agricultural water management.

Acknowledgements The authors are grateful to the reviewers for their comments and valuable suggestions.

Author Contributions A. R and Y. H conceptualized the book chapter. A.R. completed the original draft preparation. All the coauthors edited and reviewed the chapter. All authors have read and agreed to the published version of the book chapter.

Funding The research was supported by Key R&D program of Jiangsu Provincial Government (BE2021340) and the Priority Academic Program Development of Jiangsu Higher Education Institutions (PAPD-2018-87).

Ethics Approval and Consent to Participate Not applicable
Institutional Review Board Statement: Not applicable.
Informed Consent Statement: Not applicable.

Data Availability Statement Data available on reasonable request.

Conflicts of Interest The authors declare no conflict of interest.

References

Abdollahnejad A, Panagiotidis D, Surový P (2018) Estimation and extrapolation of tree parameters using spectral correlation between UAV and Pléiades data. Forests 9(2):85
Afzaal H, Farooque AA, Abbas F, Acharya B, Esau T (2020) Precision irrigation strategies for sustainable water budgeting of potato crop in Prince Edward Island. Sustainability 12(6):2419
Alexandris S, Psomiadis E, Proutsos N, Philippopoulos P, Charalampopoulos I, Kakaletris G, Papoutsi E-M, Vassilakis S, Paraskevopoulos A (2021) Integrating drone technology into an innovative agrometeorological methodology for the precise and real-time estimation of crop water requirements. Hydrology 8(3):131
Althoff D, Alvino FCG, Filgueiras R, Aleman CC, da Cunha FF (2019) Evapotranspiration for irrigated agriculture using orbital satellites. Bioscience J 35(3)

Altobelli F, Meybeck A, Gitz V (2016) Accounting for water use in agriculture. Knowl Inform Sustain Food Syst 53

Alvino A, Marino S (2017) Remote sensing for irrigation of horticultural crops. Horticulturae 3(2):40

An Y (2011) Evaluation of evapotranspiration estimation methods and their impacts on crop yield simulations. Carleton University

Aryalekshmi BN, Biradar RC, Chandrasekar K, Ahamed JM (2021) Analysis of various surface energy balance models for evapotranspiration estimation using satellite data. Egypt J Remote Sens Space Sci 24(3):1119–1126

Asmamaw DK, Janssens P, Desse M, Tilahun S, Adgo E, Nyssen J, Walraevens K, Cornelis WM (2021) Deficit irrigation as a sustainable option for improving water productivity in Sub-Saharan Africa: the case of Ethiopia. A critical review. Environ Res Commun

Bhakta I, Phadikar S, Majumder K (2019) State-of-the-art technologies in precision agriculture: a systematic review. J Sci Food Agric 99(11):4878–4888

Bhatt R, Hossain A (2019) Concept and consequence of evapotranspiration for sustainable crop production in the era of climate change. Adv Evapotranspiration Methods Appl 1:1–13

Bhattarai N, Wagle P (2021) Recent advances in remote sensing of evapotranspiration. Remote Sens 13(21):4260

Blatchford ML, Mannaerts CM, Zeng Y, Nouri H, Karimi P (2019) Status of accuracy in remotely sensed and in-situ agricultural water productivity estimates: a review. Remote Sens Environ 234:111413

Bogawski P, Bednorz E (2014) Comparison and validation of selected evapotranspiration models for conditions in Poland (Central Europe). Water Resour Manage 28(14):5021–5038

Boretti A, Rosa L (2019) Reassessing the projections of the world water development report. NPJ Clean Water 2(1):1–6

Encyclopædia Britannica (2022) Surface hydrologic cycle. retrieved from url https://www.britannica.com/science/hydrosphere/The-water-cycle#/media/1/279025/69410. Access 11 Dec 2022

Bruinsma J (2009) The resource outlook to 2050: by how much do land, water and crop yields need to increase by 2050? In: How to feed the World in 2050. Proceedings of a technical meeting of experts, Rome, Italy, 24–26 June 2009. Food and Agriculture Organization of the United Nations (FAO), pp 1–33

Will Bugler (2020) Climate-resilient water management: an operational framework from South Asia. Retrieved from url: https://www.weadapt.org/knowledge-base/national-adaptation-planning/climate-resilient-water-management-an-operational-framework-from-south-asia. Access Date: 10 Dec 2022

Buttar NA, Yongguang H, Shabbir A, Lakhiar IA, Ullah I, Ali A, Yasin MA (2018) Estimation of evapotranspiration using Bowen ratio method. IFAC-PapersOnLine 51(17):807–810

Caldwell T, Huntington J, Scanlon B, Joros A, Howard T (2017) Improving irrigation water use estimates with remote sensing technologies: a feasibility study for Texas. Texas Water Development Board

Calera A, Campos I, Osann A, D'Urso G, Menenti M (2017) Remote sensing for crop water management: from ET modelling to services for the end users. Sensors 17(5):1104

Chartzoulakis K, Bertaki M (2015) Sustainable water management in agriculture under climate change. Agric Agric Sci Procedia 4:88–98

Chen A, Orlov-Levin V, Meron M (2019) Applying high-resolution visible-channel aerial imaging of crop canopy to precision irrigation management. Agric Water Manag 216:196–205

Corbari C, Skokovic Jovanovic D, Nardella L, Sobrino J, Mancini M (2020) Evapotranspiration estimates at high spatial and temporal resolutions from an energy–water balance model and satellite data in the capitanata irrigation consortium. Remote Sensing 12(24):4083

de Andrade BCC, de Andrade Pinto EJ, Ruhoff A, Senay GB (2021) Remote sensing-based actual evapotranspiration assessment in a data-scarce area of Brazil: a case study of the Urucuia Aquifer System. Int J Appl Earth Obs Geoinf 98:102298 de Jong IH, Arif SS, Gollapalli PKR, Neelam P, Nofal ER, Reddy KY, Röttcher K, Zohrabi N (2021) Improving agricultural water productivity with a focus on rural transformation. Irrigation Drainage 70(3):458–469

Djaman K, O'Neill M, Owen CK, Smeal D, Koudahe K, West M, Allen S, Lombard K, Irmak S (2018) Crop evapotranspiration, irrigation water requirement and water productivity of maize from meteorological data under semiarid climate. Water 10(4):405

Eliades M, Bruggeman A, Djuma H, Christofi C, Kuells C (2022) Quantifying evapotranspiration and drainage losses in a semi-arid nectarine (*Prunus persica* var. nucipersica) field with a dynamic crop coefficient (Kc) derived from leaf area index measurements. Water 14(5):734

Enriquez Y, Yadav S, Evangelista GK, Villanueva D, Burac MA, Pede V (2021) Disentangling challenges to scaling alternate wetting and drying technology for rice cultivation: distilling lessons from 20 years of experience in the Philippines. FrontSustain Food Syst 5:675818

Entezari A, Wang RZ, Zhao S, Mahdinia E, Wang JY, Tu YD, Huang DF (2019) Sustainable agriculture for water-stressed regions by air-water-energy management. Energy 181:1121–1128

Gallardo AH (2019) Hydrogeological characterisation and groundwater exploration for the development of irrigated agriculture in the West Kimberley region, Western Australia. Groundw Sustain Dev 8:187–197

Geli HME, González-Piqueras J, Neale CMU, Balbontín C, Campos I, Calera A (2019) Effects of surface heterogeneity due to drip irrigation on scintillometer estimates of sensible, latent heat fluxes and evapotranspiration over vineyards. Water 12(1):81

Ghiat I, Mackey HR, Al-Ansari T (2021) A review of evapotranspiration measurement models, techniques and methods for open and closed agricultural field applications. Water 13(18):2523

Gowda PH, Chavez JL, Colaizzi PD, Evett SR, Howell TA, Tolk JA (2007) Remote sensing based energy balance algorithms for mapping ET: current status and future challenges. Trans ASABE 50(5):1639–1644

Hao P, Di L, Guo L (2022) Estimation of crop evapotranspiration from MODIS data by combining random forest and trapezoidal models. Agric Water Manag 259:107249

Hardelin J, Lankoski J (2015) Climate change, water and agriculture: challenges and adaptation strategies. Euro Choices 14(2):10–15

Howell TA (1996) Irrigation scheduling research and its impact on water use. In: Evapotranspiration and irrigation scheduling, proceedings of the international conference, pp 21–33

Islam SMF, Karim Z (2019) World's demand for food and water: the consequences of climate change. Desalination-Challenges and Opportunities 57–84

Jaramillo S, Graterol E, Pulver E (2020) Sustainable transformation of rainfed to irrigated agriculture through water harvesting and smart crop management practices. Front Sustain Food Syst 4:437086

Jing W, Yaseen ZM, Shahid S, Saggi MK, Tao H, Kisi O, Salih SQ, Al-Ansari N, Chau K-W (2019) Implementation of evolutionary computing models for reference evapotranspiration modeling: short review, assessment and possible future research directions. Eng Appl Comput Fluid Mech 13(1):811–823

Jones HG, Hutchinson PA, May T, Jamali H, Deery DM (2018) A practical method using a network of fixed infrared sensors for estimating crop canopy conductance and evaporation rate. Biosys Eng 165:59–69

Karimi P, Bastiaanssen WGM (2015) Spatial evapotranspiration, rainfall and land use data in water accounting–Part 1: review of the accuracy of the remote sensing data. Hydrol Earth Syst Sci 19(1):507–532

Kharrou MH, Simonneaux V, Er-Raki S, Le Page M, Khabba S, Chehbouni A (2021) Assessing irrigation water use with remote sensing-based soil water balance at an irrigation scheme level in a semi-arid region of Morocco. Remote Sensing 13(6):1133

Kingra PK, Majumder D, Singh SP (2016) Application of remote sensing and GIS in agriculture and natural resource management under changing climatic conditions. Agric Res J 53(3):295–302

Koech R, Langat P (2018) Improving irrigation water use efficiency: a review of advances, challenges and opportunities in the Australian context. Water 10(12):1771

Krishna PR (2019) Evapotranspiration and agriculture—a review. Agric Rev 40:1–11
Lang D, Zheng J, Shi J, Liao F, Ma X, Wang W, Chen X, Zhang M (2017) A comparative study of potential evapotranspiration estimation by eight methods with FAO Penman-Monteith method in southwestern China. Water 9(10):734
Li Z-L, Tang R, Wan Z, Bi Y, Zhou C, Tang B, Yan G, Zhang X (2009) A review of current methodologies for regional evapotranspiration estimation from remotely sensed data. Sensors 9(05):3801–3853
Liakos KG, Busato P, Moshou D, Pearson S, Bochtis D (2018) Machine learning in agriculture: a review. Sensors 18(8):2674
Liou Y-A, Kar SK (2014) Evapotranspiration estimation with remote sensing and various surface energy balance algorithms—a review. Energies 7(5):2821–2849
Liu X, Liu Y, Wang Y, Liu Z (2022) Evaluating potential impacts of land use changes on water supply–demand under multiple development scenarios in dryland region. J Hydrol 610:127811. https://doi.org/10.1016/j.jhydrol.2022.127811
Liu Y, El-Kassaby YA (2018) Evapotranspiration and favorable growing degree-days are key to tree height growth and ecosystem functioning: meta-analyses of Pacific Northwest historical data. Sci Rep 8(1):1–12
Lopez JR, Winter JM, Elliott J, Ruane AC, Porter C, Hoogenboom G, Anderson M, Hain C (2022) Sustainable use of groundwater may dramatically reduce irrigated production of maize, soybean, and wheat. Earth's Future 10(1):e2021EF002018
Ma Y, Liu S, Song L, Xu Z, Liu Y, Xu T, Zhu Z (2018) Estimation of daily evapotranspiration and irrigation water efficiency at a Landsat-like scale for an arid irrigation area using multi-source remote sensing data. Remote Sens Environ 216:715–734
MacPherson J, Voglhuber-Slavinsky A, Olbrisch M, Schöbel P, Dönitz E, Mouratiadou I, Helming K (2022) Future agricultural systems and the role of digitalization for achieving sustainability goals. A review. Agron Sustain Develop 42(4):1–18. https://doi.org/10.1007/s13593-022-00792-6
Maina MM, Amin MSM, Rowshon MK, Aimrun W, Samsuzana AA, Yazid MA (2014) Effects of crop evapotranspiration estimation techniques and weather parameters on rice crop water requirement. Aust J Crop Sci 8(4):495–501
Mavridou E, Vrochidou E, Papakostas GA, Pachidis T, Kaburlasos VG (2019) Machine vision systems in precision agriculture for crop farming. J Imaging 5(12):89
Meraz-Maldonado N, Flores-Magdaleno H (2019) Maize evapotranspiration estimation using penman-monteith equation and modeling the bulk canopy resistance. Water 11(12):2650
Meybeck A, Redfern S (2016) Knowledge and information for sustainable food systems. Joint FAO/UNEP workshop on knowledge and information for sustainable food systems Rome (Italy) 10–11 Sep 2014
Mhawej M, Elias G, Nasrallah A, Faour G (2020) Dynamic calibration for better SEBALI ET estimations: validations and recommendations. Agric Water Manag 230:105955
Mo X-G, Hu S, Lin Z-H, Liu S-X, Xia J (2017) Impacts of climate change on agricultural water resources and adaptation on the North China Plain. Adv Clim Chang Res 8(2):93–98
Moiwo JP, Tao F (2015) Contributions of precipitation, irrigation and soil water to evapotranspiration in (semi)-arid regions. Int J Climatol 35(6):1079–1089
van Mourik S, van der Tol R, Linker R, Reyes-Lastiri D, Kootstra G, Koerkamp PG, van Henten EJ (2021) Introductory overview: Systems and control methods for operational management support in agricultural production systems. Environ Modell Softw 139:105031
Neupane J, Guo W (2019) Agronomic basis and strategies for precision water management: a review. Agronomy 9(2):87
Nolz R (2016) A review on the quantification of soil water balance components as a basis for agricultural water management with a focus on weighing lysimeters and soil water sensors/Ein Überblick über die Ermittlung von Wasserhaushaltsgrößen als Basis für die landeskulturelle Wasserwirtschaft mit Fokus auf Lysimeter und Bodenwassersensoren. Die Bodenkultur: J Land Manage Food Environ 67(3):133–144

Nsiah JJ, Gyamfi C, Anornu GK, Odai SN (2021) Estimating the spatial distribution of evapotranspiration within the Pra River Basin of Ghana. Heliyon 7(4):e06828
OECD (2013) Education at a glance 2016 OECD indicators. OECD Publishing. https://doi.org/10.1787/eag-2012-en
Olmos Gimenez P, García-Galiano SG (2018) Assessing regional climate models (RCMs) ensemble-driven reference evapotranspiration over Spain. Water 10(9):1181
Pande CB, Moharir KN (2023a) Application of hyperspectral remote sensing role in precision farming and sustainable agriculture under climate change: A review. In: Pande CB, Moharir KN, Singh SK, Pham QB, Elbeltagi A (eds) Climate change impacts on natural resources, ecosystems and agricultural systems. Springer Climate. Springer, Cham. https://doi.org/10.1007/978-3-031-19059-9_21
Pande CB, Costache R, Sammen SS et al (2023b) Combination of data-driven models and best subset regression for predicting the standardized precipitation index (SPI) at the Upper Godavari Basin in India. Theor Appl Climatol 152:535–558. https://doi.org/10.1007/s00704-023-04426-z
Pereira LS, Allen RG, Smith M, Raes D (2015) Crop evapotranspiration estimation with FAO56: past and future. Agric Water Manag 147:4–20
Pereira LS, Perrier A, Allen RG, Alves I (1996) Evapotranspiration: review of concepts and future trends. International conference
Reyes-González A, Kjaersgaard J, Trooien T, Hay C, Ahiablame L (2018) Estimation of crop evapotranspiration using satellite remote sensing-based vegetation index. Adv Meteorol
Reyes-Gonzalez A (2017) Using remote sensing to estimate crop water use to improve irrigation water management. South Dakota State University
Sadras VO, Cassman K, Grassini P, Bastiaanssen WGM, Laborte AG, Milne AE, Sileshi G, Steduto P (2015) Yield gap analysis of field crops: methods and case studies
Salman SA, Shahid S, Afan HA, Shiru MS, Al-Ansari N, Yaseen ZM (2020) Changes in climatic water availability and crop water demand for Iraq region. Sustainability 12(8):3437
Sattari MT, Apaydin H, Band SS, Mosavi A, Prasad R (2021) Comparative analysis of kernel-based versus ANN and deep learning methods in monthly reference evapotranspiration estimation. Hydrol Earth Syst Sci 25(2):603–618
Shanmugapriya P, Rathika S, Ramesh T, Janaki P (2019) Applications of remote sensing in agriculture: a review. Int J Curr Microbiol App Sci 8(1):2270–2283
Singh PK, Jain SK, Mishra PK, Goel MK (2022) An assessment of water consumption patterns and land productivity and water productivity using WA+ framework and satellite data inputs. Phys Chem Earth, Parts A/B/C 126:103053
Spohrer K, Merkt N, Wenyong D, Xiongkui H, Joachim M (2010) Non-invasive water status detection in grapevine (Vitis vinifera L.) by thermography. Int J Agric Biol Eng 2(4):46–54
Stefanidis S (2021) Ability of different spatial resolution regional climate model to simulate air temperature in a forest ecosystem of central Greece. J Environ Prot Ecol 22:1488–1495
Stefanidis S, Alexandridis V (2021) Precipitation and potential evapotranspiration temporal variability and their relationship in two forest ecosystems in greece. Hydrology 8(4):160
Stisen S, Soltani M, Mendiguren G, Langkilde H, Garcia M, Koch J (2021) Spatial patterns in actual evapotranspiration climatologies for europe. Remote Sensing 13(12):2410
Subedi A, Chávez JL (2015) Crop evapotranspiration (ET) estimation models: a review and discussion of the applicability and limitations of ET methods. J Agric Sci 7(6):50
Sun T, Huang Q, Wang J (2017) Estimation of irrigation water demand and economic returns of water in Zhangye Basin. Water 10(1):19
Sun SK, Li C, Wu PT, Zhao XN, Wang YB (2018) Evaluation of agricultural water demand under future climate change scenarios in the Loess Plateau of Northern Shaanxi, China. Ecol Ind 84:811–819
Tegos A, Malamos N, Koutsoyiannis D (2022) Raspotion—a new global PET dataset by means of remote monthly temperature data and parametric modelling. Hydrology 9(2):32
Tian H, Wang T, Liu Y, Qiao X, Li Y (2020) Computer vision technology in agricultural automation—a review. Inform Process Agric 7(1):1–19

Tolika K, Anagnostopoulou C, Velikou K, Vagenas C (2016) A comparison of the updated very high resolution model RegCM3_10 km with the previous version RegCM3_25 km over the complex terrain of Greece: present and future projections. Theoret Appl Climatol 126(3):715–726

Ustin SL, Middleton EM (2021) Current and near-term advances in Earth observation for ecological applications. Ecol Process 10(1):1–57

Vanschoenwinkel J, Van Passel S (2018) Climate response of rainfed versus irrigated farms: the bias of farm heterogeneity in irrigation. Clim Change 147(1):225–234

Wagle P, Gowda PH (2019) Editorial for the special issue "Remote Sensing of Evapotranspiration (ET)." Remote Sensing 11(18):2146

Wang P, Song X, Han D, Zhang Y, Zhang B (2012) Determination of evaporation, transpiration and deep percolation of summer corn and winter wheat after irrigation. Agric Water Manag 105:32–37

Wang C, Wang S, Chen H, Wang J, Tao Y, Liu J (2018) Evaluation of water-storage and water-saving potential for paddy fields in Gaoyou, China. Water 10(9):1176

Wang J, Bretz M, Dewan MAA, Delavar MA (2022) Machine learning in modelling land-use and land cover-change (LULCC): current status, challenges and prospects. Sci Total Environ 153559

Winbourne JB, Jones TS, Garvey SM, Harrison JL, Wang L, Li D, Templer PH, Hutyra LR (2020) Tree transpiration and urban temperatures: current understanding, implications, and future research directions. Bioscience 70(7):576–588

World Bank (2022) World bank data: annual freshwater withdrawals, agriculture (% of total freshwater withdrawal). Available online: https://data.worldbank.org/indicator/ER.H2O.FWAG.ZS?end=2016&start=2011&view=chart. Accessed on 01 Oct 2022

Wu G, Hu Z, Keenan TF, Li S, Zhao W, Cao RC, Li Y, Guo Q, Sun X (2020) Incorporating spatial variations in parameters for improvements of an evapotranspiration model. J Geophys Res Biogeosci 125(11):e2019JG005504

Zeyliger AM, Ermolaeva OS (2021) Water stress regime of irrigated crops based on remote sensing and ground-based data. Agronomy 11(6):1117

Zhao WL, Qiu GY, Xiong YJ, Gentine P, Chen BY (2020) Uncertainties caused by resistances in evapotranspiration estimation using high-density eddy covariance measurements. J Hydrometeorol 21(6):1349–1365

Zheng Q, Hao L, Huang X, Sun L, Sun G (2020) Effects of urbanization on watershed evapotranspiration and its components in southern China. Water 12(3):645

Zongo B, Barbier B, Diarra A, Zorom M, Atewamba C, Combary OS, Ouédraogo S, Toé P, Hamma Y, Dogot T (2022) Economic analysis and food security contribution of supplemental irrigation and farm ponds: evidence from northern Burkina Faso. Agric Food Secur 11(1):1–18

Application of Remote Sensing and GIS for Morphometric Analysis: A Case Study of Burhanpur Watershed

Abhishek Patel, Kethavath Ajaykumar, Arvind Dhaloiya, K. V. Ramana Rao, Yogesh Rajwade, and C. K. Saxena

Abstract Morphometric analysis of watershed is quantitative representation of its drainage network based on topography of a landform. Morphometric analysis-based basin attributes provide insights on drainage behavior of watershed and strengthen watershed planning, environmental monitoring and basin management activities. In broad categories these attributes are categorized as linear (length related), areal (area and shape related), and relief (elevation related) aspects of watershed. In this study Burhanpur watershed, upper part of Tapi river basin, has been selected for detailed morphometric analysis using Remote Sensing (RS) data and Geographic Information Systems (GIS) technology. The study examines about 10,585 km^2 drainage area for linear, arial and relief aspects of Burhanpur watershed. The values for: linear aspects (Stream order, Stream length ratio, Length of overland flow etc.), arial aspect (Drainage density, Form factor, Circulatory ratio etc.) and relief aspect (Ruggedness number, Relief ratio etc.) were recorded after analysis. This study improves the understanding of drainage characteristics and stream pattern of selected watershed and reflects potential of RS and GIS for land and water resource conservation and management with in watershed. Further study recommends the way forward in research at study area.

Keywords GIS · RS · Drainage analysis · Geo-morphometry analysis · Tapi basin

A. Patel (✉)
ICAR-Central Arid Zone Research Institute, Regional Research Institute, Bhuj 370105, Gujarat, India
e-mail: abhi.patel.ape121@gmail.com

K. Ajaykumar
ICAR-Research Complex for NEH Region, Umiam 793103, Meghalaya, India

A. Dhaloiya
Department of Soil and Water Engg, Punjab Agricultural University, Ludhiana 141004, Punjab, India

K. V. R. Rao · Y. Rajwade · C. K. Saxena
ICAR-Central Institute of Agriculture Engineering, Bhopal 462038, Madhya Pradesh, India

C. B. Pande et al. (eds.), *Surface and Groundwater Resources Development and Management in Semi-arid Region*, Springer Hydrogeology,
https://doi.org/10.1007/978-3-031-29394-8_2

1 Introduction

A watershed is an area which contributes the total runoff as a response to rainfall event(s) to a single common outlet point for disposal. Often it is considered similar to a drainage basin or a catchment area to a reservoir. As watershed is an integrated form of natural resources i.e., land, water, and vegetation, study of watershed and management of its natural resources is crucial for attaining the environmental sustainability (Sangma and Guru 2020; Patel et al. 2021; Kumar et al. 2021a). Watershed topography and drainage network, being the most delicate components of a landscape, greatly influences both structure and fluvial dynamics of watershed. River's flow pattern and drainage systems are dynamic in nature, changing over time and space as a result of a variety of factors such as regional geology, structural elements, vegetation, and soils (Rekha et al. 2011). Poor watershed conditions like excessive runoff generation, increased soil erosion, poor infiltration conditions, drought and floods can be overcome by a well-planned managemental techniques (Choudhari et al. 2018). Such strategies suitably provide optimal utilization of resources without any degradation and loss in sustainability (Arvind et al. 2018; Bajirao et al. 2019). In this direction the characteristics of watershed (topographical, climatic, and hydrologic) and related processes must be understood qualitatively as well as quantitatively by the decision-makers (Singh et al. 2021).

Geomorphometry is the mathematical analysis of the earth's surface that describes its topographic reliefs and drainage behavior (Pakhmode et al. 2003). According to Rastogi and Sharma (1976), a number of hydrologic phenomena are associated with the physiographic features of watersheds. Several morphometric parameters (belonging to linear, arial and relief aspects) were derived (Horton 1932, 1945; Smith 1950; Miller 1953; Schumm 1956; Hadley and Schumm 1961; Strahler 1964; Faniran 1968) and used by researches in the geomorphometric studied of watershed (Nooka Ratnam et al., 2005; Mesa 2006 and Ozdemir and Bird 2009).

1.1 Role of RS and GIS

Initiation of morphometric research began in the middle of the twentieth century using a traditional methodology based on hand assessments of topographic maps (Schumm 1956). However, the traditional method of assessing river morphology is a labor-intensive and time consuming (Pande and Moharir 2017). On the other hand, with the development of geospatial and computational technology and accessibility of remotely sensed data it is now much easier to undertake exact and precise assessments with lesser time and resources. When topographic surveys are not feasible or area which are not accessible satellite terrain data such as Digital Elevation Models (DEM) are useful for geomorphometric parameterization of a watershed. DEMs give continuous data, and are simple to integrate into Geographical Information Systems (GIS) (Moore et al. 1991; Patel et al. 2022). Senthamizhan et al. (2016)

used satellite data and GIS technologies in morphometrics for constructing watershed management studies. To calculate morphometric parameters in the sub-basin susceptible to both erosion and sedimentation, a combined approach employing toposheet, remotely sensed digital elevation model, and morphometric ArcGIS toolbox has been utilized Singh et al. (2021). Moreover, RS ang GIS alone are the advance tools capable to study and manage natural resources (Kushwaha et al. 2022a; Machiwal et al. 2022; Dhaloiya et al. 2022).

1.2 *Application of Morphometric Analysis*

Geo-morphometric analysis offers the way out to investigate the geometrical and drainage characteristics of watershed based on topography. This information becomes useful to study the ungauged watersheds where information on hydrology, geology, geomorphology, and soil is limited. Forecasting of other basin features such as travel time, time to peak, and severity of erosional processes can be performed using this technique (Romshoo et al. 2012; Puno and Puno 2019). Moreover, information obtained from morphometric study of watersheds could be an important tool for managing water resources, preventing soil erosion, mapping landslide susceptibility, assessing groundwater potential, and prioritizing watersheds (Kumar et al. 2021b; Kushwaha et al. 2022b; Sharma et al. 2022).

The geomorphometric analysis along with other supplemental data such as runoff, flooding, groundwater development, soil erosion, land use and cover, socioeconomic status of the local population etc. can be used to priorities sub-watersheds (Kumar et al. 2019). It is widely used to identify changes in characteristics of drainage basin, understand the several environmental issues at watershed (Mangan et al. 2019; Choudhari et al. 2018). In general, geomorphometric analysis gives suitable information to understand the spatial and temporal behavior of watershed hydrology and related risks (Pande et al. 2018). Which are crucial for sustainable management and planning of natural resources (Karabulut and Özdemır 2019; Nıtheshnırmal et al. 2019). This study is conducted with the objective to examine watershed geomorphometric characteristics using RS data and GIS technology for Burhanpur watershed at upper Tapi river in central India. Study quantifies the hydro-geo-morphic characteristics of watershed to be used in further development of management strategies for the area against several degradation processes. The study discusses in detail the different morphometric parameters, process of their quantification and importance in the context of Burhanpur watershed. Considering the previous studies, this study delivers important insights for the scholars and decision makers of the similar research area.

2 Materials and Methods

2.1 Study Area: Geography and Environment

The Tapi river system is one of the biggest west flowing river systems of peninsular India. The river originates from Multai a place of Madhya Pradesh passes through Maharashtra and falls in to the Arabian Sea near city of Surat, Gujarat. From origin to its mergence into sea it covers approx. 728 km of distance (https://indiawris.gov.in). The Tapi river system drains about total of 65,145 km^2 area (second largest westward draining inter-state river basin) out of which roughly 16, 79 and 5% area falls in Madhya Pradesh, Maharashtra and Gujarat respectively (Fig. 1).

The Part that is drained from Madhya Pradesh is the approximately half of Upper Tapi Basin. Burhanpur, being last district of Madhya Pradesh that is passed by Tapi river, is the outlet of the upper reach of Tapi river and a gauging station for flow measurement. Hance, this study considers Burhanpur as the outlet of watershed and distinguishes the study area as Burhanpur watershed. The river travels near about 340 km from origin to reach the outlet at Burhanpur. The spatial extent of watershed expands from 75° 55′ E to 78° 18′ 14″ E longitude and from 21° 1′ 51″ N to 22° 1′ 52″ N latitude (Fig. 1). Within this extent the watershed covers about 10,585 km^2 area. Topographic elevation of the area varies from 188 to 1171 m (above Mean Sea Level). The area experiences the average annual rainfall of 900 mm. Study area is predominantly having clayey to loamy clayey soils (Chandra et al. 2016) and comprises agriculture, range lands, water bodies, barren land, built-up area and forest as main land uses types.

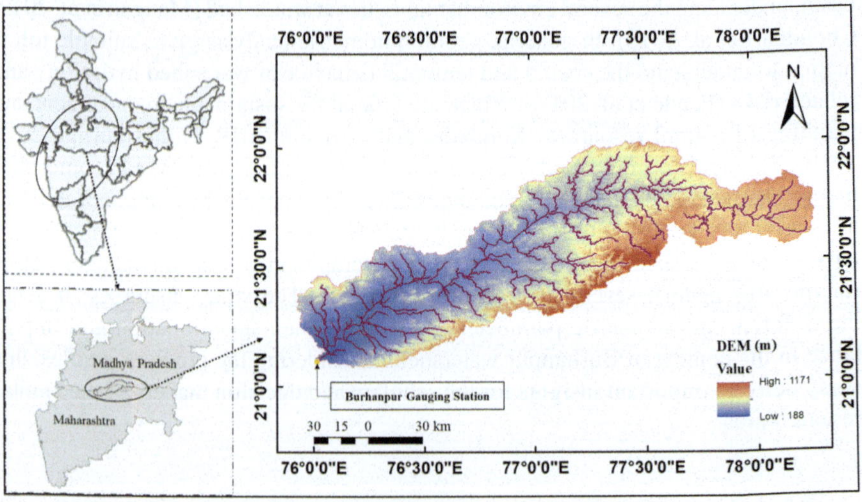

Fig. 1 Study area with DEM map and river reaches

2.2 Geo-morphometric Analysis: Parameter Quantification

The present study quantifies the geo-morphometric parameters of Burhanpur watershed using Remote Sensing (RS) data and Geographical Information System (GIS). The Shuttle Radar Topography Mission (SRTM) 1 Arc-Second Digital Elevation Model (DEM) from USGS Earth Resources Observation and Science (EROS) Center archive was downloaded to get elevation and topography of study area (https://www.usgs.gov/). This RS data was used for delineation of watershed and morphometric analysis through Hydrology tool box and other spatial tools in ArcGIS (version 10.4) (Band 1986; Morris and Heerdegen 1988; Tarboton et al. 1991; Maidment 2002). In order to quantify the geomorphometry of watershed three different aspects: linear (length related), areal (area and shape related), and relief (elevation related) aspects were considered (Strahler 1964). The total 23 basic and derived aspects were analyzed to study the Burhanpur watershed as listed and described in Table 1. Linear aspect of watershed was determined using (i) *Stream order* (u), (ii) *Stream length* (L_u), (iii) *Number of streams* (N_u), (iv) *Mean stream length* (L_{ms}) (v) *Stream length ratio* (R_l), (vi) *Bifurcation ratio* (R_{bf}), (vii) *Mean bifurcation ratio* (R_{bfm}), (viii) *Basin length* (L_b), and (ix) *Basin perimeter* (P). Further areal aspect was analyzed through (i) *Basin area* (A), (ii) *Drainage density* (D_d), (iii) *Stream frequency*, (iv) *Form factor* (F_f), (v) *Shape factor* (F_s), (vi) *Circulatory ratio* (R_c), (vii) *Elongation ratio* (R_e), (viii) *Constant of channel maintenance* (C), (ix) *Drainage texture* (T), (x) *Length of overland flow* (L_g), and (xi) *Texture ratio* (T). Moreover, to study the morphometry through relief aspect the (i) *Basin relief* (H), (ii) *Relief ratio* (Rr), and (iii) *Ruggedness number* (R_n) were studied (Fig. 2 and Table 1).

In this study the linear, aerial, and relief aspects were quantified and analyzed based on the procedure suggested by Horton (1932) Horton (1945), Smith (1950), Miller (1953), Schumm (1956), Hadley and Schumm (1961), Strahler (1964), Faniran (1968), and followed by Nooka Ratnam et al. (2005), Mesa (2006) and Ozdemir and Bird (2009).

3 Results and Discussion

SRTM-DEM from USGS-EROS Center archive was downloaded to get elevation and topography of study at 30 × 30 m spatial resolution and used for the delineation of boundary and stream of the watershed. From the slop analysis of DEM is was found that the slope of DEM varies from 1 to 190% across watershed area (Fig. 3). The area near to the lower (west)-middle boundary of watershed is predominant with higher hill slope (>30%). Whereas the Eastern part of watershed (which is origin of Tapi river) is having lower slope value (<30%).

Figure 4 depicts the flow direction at the spatial resolution of 30 × 30 m derived from DEM. It was observed that flow is being directed in all the directions, however,

Table 1 Aspect wise morphometric parameters

Aspects	Quantification	Detail	References
Linear aspects: describe of arrangement of linear elements			
Stream order (u)	Using GIS utility	Hierarchical rank	Strahler (1964)
Stream length (L_u)	Using GIS utility	Length of uth order stream	Horton (1945)
Number of streams (N_u)	Using GIS utility	Total number of uth order stream	Horton (1945), Strahler (1964)
Mean stream length (L_{ms})	$L_{ms} = L_u/N_u$	Unit length of uth order stream	Strahler (1964)
Stream length ratio (R_l)	$R_l = L_u/L_{u-1}$	Horton postulated the constant values for this ration throughout the successive orders	Horton (1945)
Bifurcation ratio (R_{bf})	$R_{bf} = N_u/N_{u+1}$	Higher values indicate the rockey area with sleep slope with valleys in-between	Horton (1932)
Mean bifurcation ratio (R_{bfm})		The average of bifurcation ratios of all orders	Schumm (1956)
Basin length (L_b)	$L_b = 1.321A^{0.568}$		Nooka Ratnam et al. (2005)
Basin perimeter (P)	Using GIS utility	Outer periphery of basin	Schumn (1956)
Aerial aspect: describe of arrangement of areal elements			
Basin area (A)	Using GIS utility	area bounded within the periphery of basin	Strahler (1964)
Drainage density (D_d)	$D_d = \Sigma L_u/A$	The ratio between the total stream length of all (D_d) orders to the area of the basin	Horton (1945)
Stream frequency (S_f)	$F_s = \Sigma N_u/A$	number of stream segments in unit area of watershed	Horton (1945)
Form factor (F_f)	$F_f = A/L_b^2$	It indicates the flow intensity of a basin of a defined area	Horton (1932, 1945)
Shape factor (F_s)	$F_s = L_b^2/A$	Becomes greater than 1 for basins which are elongated along some characteristic length of the basin and less than 1 for basins which are perpendicular to this characteristic length	Horton (1932)

(continued)

Table 1 (continued)

Aspects	Quantification	Detail	References
Circulatory ratio (R_c)	$Rc = 4\pi A/P^2$	It is influenced by the length and frequency of stream, geological structures, land use/ landcover, climate, relief and slope of the basin	Miller (1953), Strahler (1964)
Elongation ratio (R_e)	$Re = 1.128\sqrt{A}/L_b$	A circular basin is more efficient in the discharge of runoff than an elongated basin	Schumm (1956)
Constant of channel maintenance (C)	$C = 1/D_d$	Drainage area needed to generate a unit length of stream	Schumm (1956)
Infiltration number (I_f)	$I_f = D \times F_s$	It describes about the infiltration physiognomies of the basin area	Faniran (1968)
Length of overland flow (L_g)	$L_g = 1/2D_d$	It relates inversely to the average slope of the channel and is quite synonymous with the length of sheet flow to a large degree	Horton (1945)
Texture ratio (T)	$T = \Sigma N_u/P$	The ratio between the total number of streams of all orders and perimeter of the basin	Smith (1950)
Relief aspect: describe of arrangement of elevation elements			
Basin relief (H)	$H = H_{max}\text{-}H_{min}$	It has significant role to understand landforms development, drainage development, surface and erosional properties of area	Hadley and Schumm (1961)
Relief ratio (Rr)	$R_r = H/L$	It indicates the overall steepness of a drainage basin and is an indicator of intensity of erosion processes operating on the slope of the basin	Schumm (1956)
Ruggedness number (R_n)	$R_n = (H \times D_d)/K$	It is useful for steepness and slope of the drainage network	Schumm (1956)

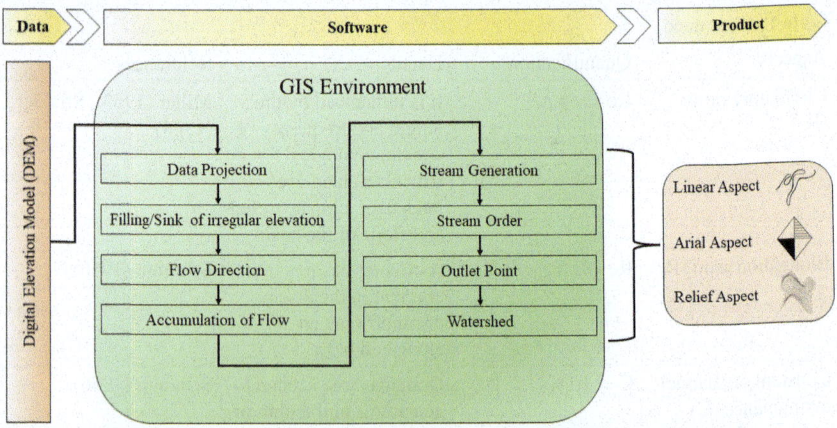

Fig. 2 Conceptual process for watershed delineation and morphometric parameters analysis

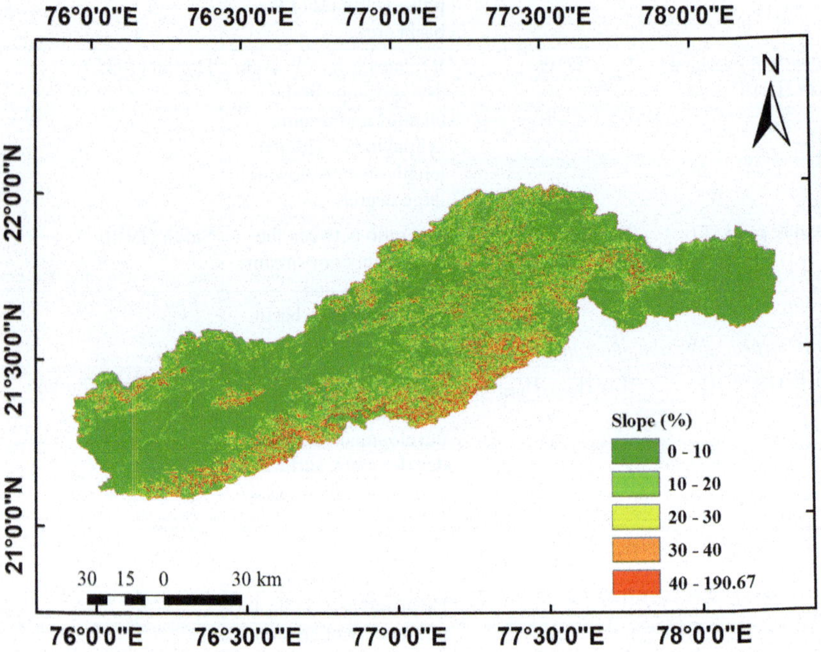

Fig. 3 Slope map of Burhanpur watershed

the predominant direction is the West (16) with the 17.4% cells of DEM. This is followed by South (16.9%), North (15.9%) and East (11.3%).

The area of delineated watershed was 10,585 km^2, which is further sub divided into 18 sub-watersheds (Fig. 5). The maximum and minimum area of sub watershed are

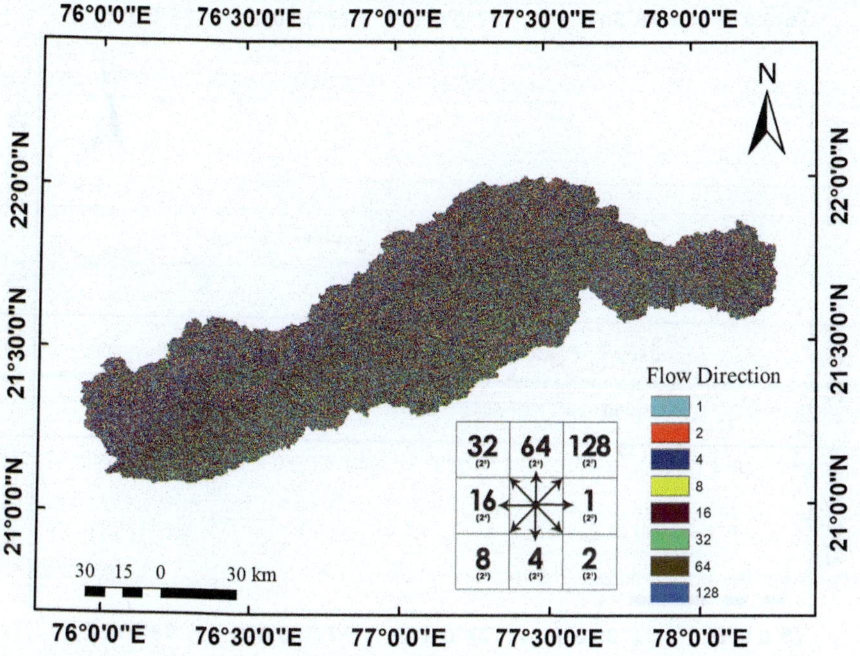

Fig. 4 Flow direction map of Burhanpur watershed

1670.3 km^2 and 146 km^2 for SW_0 and SW_7 respectively. Perimeter of 272.1 and 79.8 km respectively.

The study area comes under the dendritic type drainage system which has widespread pattern of drainage. The network of tributes found to have highest order stream of 4th order (Table 2). Table 2 suitably depicts that delineated watershed consists 4 numbers of stream order starting from 1st order to 4th order. Total length of stream in the watershed was 2203.7 km for 281 numbers of streams. The largest sub-watershed (WS_0) covers total 43 streams (21—1st order, 10—2nd order, 11—3rd order, and 1—4th order) with stream length of 338.2 km (174.1 km—1st order, 54.5 km—2nd order, 99.0 km—3rd order, and 10.6 km—4th order).

The perimeter of sub watersheds found to be of range between 78.6 and 272.1 km. The stream orders reflect the slope of watershed, usually 1st and 2nd order streams lie on steep slopes and delivers the flow to the next order stream. The length of stream characterizes the size of different components of drainage system. It was observed from the Table 2 that the watershed has the basin length of 253.5 km whereas, among sub-watersheds, WS_0 and WS_7 have the longest and smallest sub-watershed's lengths respectively. The total length of streams of Burhanpur watershed is about 2203.7 km (accounting 1108.3, 551.9, 213.4, and 330.1 for first, second, third and fourth order of streams respectively). Whereas maximum mean stream length of 9.70 km is recorded for 3rd order stream, as 3rd order stream is not present in all sub watersheds. The highest (1.55) and lowest (0.39) stream length ratio is recorded

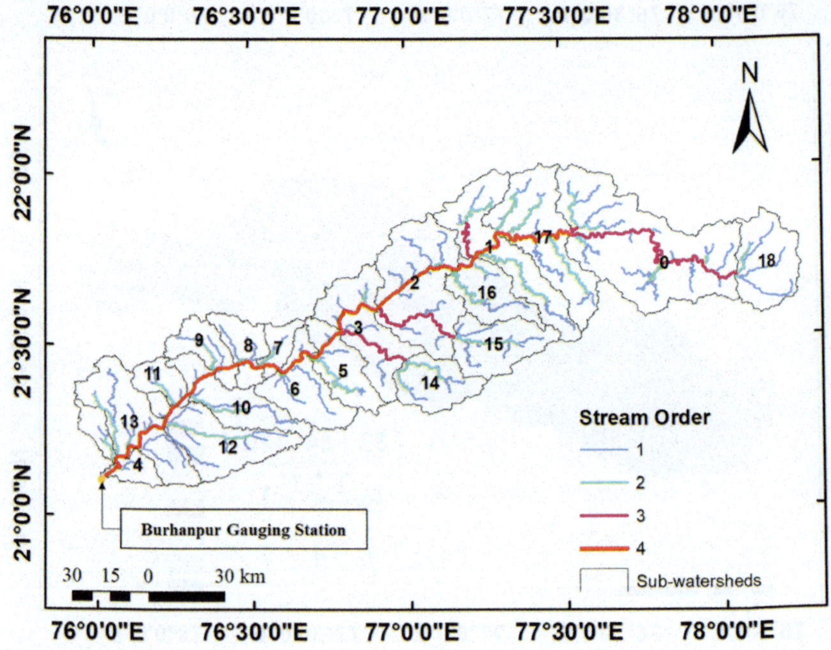

Fig. 5 Sub-watershed wise stream order of Burhanpur watershed

between 4th and 3rd order stream and between 3nd order and 2nd order streams respectively. The stream length ratio for sub-watershed varies from 0.04 to 9.71 with in the sub watershed. Sub-watershed SW_2 has highest stream length ratio for 2nd, 3rd and 4th order streams. This variation among the stream length ratio between the successive stream orders shows the corresponding variation in the slope and topographic conditions of watershed.

The bifurcation ratio of Burhrapur watershed varies between 0.34 and 2.10 with mean bifurcation ratio of 1.77. Among sub-watersheds the highest bifurcation ratio is observed to be 4.67 for SW_0 whereas least bifurcation ratio of 0.33 for SW_3. These values are common in the basin where the drainage pattern is not influenced by the geological structure of watershed (Sreedevi et al. 2009). For bifurcation ratio less than 3 the watershed would be structurally less disturbed with less distortion in drainage pattern and geological formation (Patel et al. 2013). Higher value of bifurcation ratio also shows that watershed has steeply dipping rock strata and severe over land flow (Chandniha and Kansal 2017). The bifurcation ratio more than 3 indicates severe over land flow and low discharge to the sub-watersheds. Further, the values of bifurcation ratio also display shape of watersheds. Here the delineated sub watersheds SW_0, SW_2, SW_7, SW_10, SW_11 and SW_15 were elongated in shape whereas remaining all the sub watersheds bifurcation ratio less than 3 are normal and approximately circular in shape. The higher value of bifurcation ratio

Table 2 Sub-watershed wise computation of morphometric parameters for linear aspect

Sub watershed	Area (km^2)	Perimeter (km)	Total length of uth order streams (km)				Total number of uth order streams				Total
			1	2	3	4	1	2	3	4	
SW_0	1670.3	272.1	174.1	54.5	99.0	10.6	21	10	11	1	43
SW_1	750.6	248.3	43.9	71.9	21.7	17.8	8	5	1	3	17
SW_2	996.1	193.9	100.1	4.2	40.9	38.9	11	1	5	8	25
SW_3	644.3	187.7	50.8	7.6	32.8	27.5	7	2	2	6	17
SW_4	264.9	142.2	22.1	14.8		11.9	5	3		3	11
SW_5	394.5	137.1	33.2	36.3		17.5	6	3		4	13
SW_6	480.8	153.0	59.5	2.7		18.9	5	2		4	11
SW_7	146.0	79.8	10.3	9.2		15.9	3	1		3	7
SW_8	266.1	103.6	34.0	5.9		15.7	4	2		3	9
SW_9	278.3	108.1	30.6	17.9		15.6	5	2		4	11
SW_10	474.0	167.7	55.3	25.7		15.2	6	2		5	13
SW_11	188.6	78.6	13.3	12.4		12.2	3	1		3	7
SW_12	647.0	182.4	92.7	38.8		17.4	8	6		3	17
SW_13	696.2	189.0	105.9	38.7		27.6	9	5		5	19
SW_14	390.3	103.7	29.5	38.9	8.3	0.0	5	3	1		9
SW_15	492.5	133.4	67.9	29.2	3.9	0.0	8	6	1		15
SW_16	414.9	137.8	36.4	44.2		18.3	4	2		3	9
SW_17	827.6	211.5	72.6	84.5		49.2	9	4		6	19
SW_18	562.2	131.4	76.2	14.6	6.8	0.0	5	3	1	0	9
Burhanpur watershed	**10585.0**	**1145.5**	**1108.3**	**551.9**	**213.4**	**330.1**	**132**	**63**	**22**	**64**	**281**

Sub watershed (SW)	SW length (km)	Mean stream length (km)				Stream length ratio			Bifurcation ratios			Mean R_{bf}
		1	2	3	4	2	3	4	1	2	3	
SW_0	88.8	8.29	5.45	9.00	10.65	0.31	1.82	0.11	2.10	0.91	11.00	4.67
SW_1	56.4	5.49	14.38	21.75	5.92	1.64	0.30	0.82	1.60	5.00	0.33	2.31
SW_2	66.2	9.10	4.21	8.17	4.86	0.04	9.71	0.95	11.00	0.20	0.63	3.94
SW_3	51.7	7.25	3.82	16.39	4.59	0.15	4.29	0.84	3.50	1.00	0.33	1.61
SW_4	31.2	4.41	4.93		3.97	0.67			1.67			1.67
SW_5	39.1	5.53	12.10		4.37	1.09			2.00			2.00
SW_6	43.8	11.91	1.35		4.73	0.05			2.50			2.50
SW_7	22.2	3.42	9.21		5.29	0.90			3.00			3.00
SW_8	31.3	8.49	2.95		5.22	0.17			2.00			2.00
SW_9	32.1	6.11	8.93		3.89	0.58			2.50			2.50
SW_10	43.4	9.22	12.85		3.03	0.46			3.00			3.00

(continued)

Table 2 (continued)

Sub watershed (SW)	SW length (km)	Mean stream length (km)				Stream length ratio			Bifurcation ratios			Mean R_{bf}
		1	2	3	4	2	3	4	1	2	3	
SW_11	25.7	4.43	12.43		4.08	0.93			3.00			3.00
SW_12	51.8	11.59	6.47		5.81	0.42			1.33			1.33
SW_13	54.0	11.77	7.73		5.52	0.37			1.80			1.80
SW_14	38.9	5.91	12.96	8.29		1.32	0.21		1.67	3.00		2.33
SW_15	44.4	8.49	4.86	3.91		0.43	0.13		1.33	6.00		3.67
SW_16	40.3	9.09	22.09		6.09	1.22			2.00			2.00
SW_17	59.6	8.07	21.12		8.20	1.16			2.25			2.25
SW_18	47.9	15.25	4.87	6.80		0.19	0.47	0.00	1.67	3.00		2.33
Burhanpur Watershed	**253.5 (L_b)**	**8.40**	**8.76**	**9.70**	**5.16**	**0.50**	**0.39**	**1.55**	**2.10**	**2.86**	**0.34**	**1.77**

spectacles the mature topography and large variation between successive streams order (Sreedevi et al. 2009).

Drainage density illustrate the runoff potential, infiltration capacity, climatic conditions and vegetative cover of watershed (Horton 1932). Table 3 shows the different aerial parameters of Burhanpur watershed along with all sub-watersheds.

The drainage density of Burhanpur watershed is 0.21 which indicates that overland flow is predominate and having coarse drainage structure in the basin. Lower values of drainage density also characterize the watershed highly resistance and permeable sub soil materials with low relief and covers dense vegetation (Sreedevi et al. 2009). The drainage density of sub watershed varies between 0.17 and 0.25. For the watersheds the higher drainage density implies the weak impermeable surface, sparsely vegetated and high relief conditions for corresponding sub-watersheds. From Table 3 it was observed that stream frequency of Burhanpur watershed is 0.03 and varies between 0.02 and 0.05 for sub watershed. The lower values of stream frequency show that the area has gentle ground slope and more permeable rocks. The lower values of drainage density and stream frequency implies that Burhanpur watershed as whole has less surface runoff and flooding is likely to be less. It also gives the inference of more percolation within the watershed which shows greater ground water potentials.

Form factor is the ratio of area of the basin to the square of basin length. Generally it values less than 0.79 (0.79 is for perfectly circular watershed) (Chandniha and Kansal 2017; Patel et al. 2013). The low value of form factor confirms that the basin is elongated in shape. The form factor of Burhanpur watershed is 0.16 and for the sub watershed it varies from 0.21 to 0.29. This depicts that the sub-watersheds with high F_f (circular shape) will have peak flow of short duration, whereas those have low F_f (elongated shape) will be having less peak of longer duration. Hence flood can easily be managed in elongated sub-watersheds than in those which are of circular in shape. As per the Horton (1932) the shape factor of a basin always remains greater than 1 for basin which are elongated in some characteristic length of the basin Shape

Table 3 Sub-watershed wise computation of morphometric parameters for arial aspect

Sub watershed	L	N	D_d	S_f	F_f	F_S	R_c	R_e	C	If	L_g	T
SW_0	338.29	43	0.20	0.03	0.21	4.72	0.28	0.52	4.94	0.01	2.47	0.16
SW_1	155.31	17	0.21	0.02	0.24	4.24	0.15	0.55	4.83	0.00	2.42	0.07
SW_2	184.05	25	0.18	0.03	0.23	4.40	0.33	0.54	5.41	0.00	2.71	0.13
SW_3	118.70	17	0.18	0.03	0.24	4.15	0.23	0.55	5.43	0.00	2.71	0.09
SW_4	48.76	11	0.18	0.04	0.27	3.68	0.16	0.59	5.43	0.01	2.72	0.08
SW_5	86.99	13	0.22	0.03	0.26	3.88	0.26	0.57	4.53	0.01	2.27	0.09
SW_6	81.14	11	0.17	0.02	0.25	3.99	0.26	0.56	5.93	0.00	2.96	0.07
SW_7	35.33	7	0.24	0.05	0.29	3.39	0.29	0.61	4.13	0.01	2.07	0.09
SW_8	55.51	9	0.21	0.03	0.27	3.68	0.31	0.59	4.79	0.01	2.40	0.09
SW_9	63.98	11	0.23	0.04	0.27	3.70	0.30	0.59	4.35	0.01	2.17	0.10
SW_10	96.18	13	0.20	0.03	0.25	3.98	0.21	0.57	4.93	0.01	2.46	0.08
SW_11	37.94	7	0.20	0.04	0.28	3.51	0.38	0.60	4.97	0.01	2.48	0.09
SW_12	148.97	17	0.23	0.03	0.24	4.15	0.24	0.55	4.34	0.01	2.17	0.09
SW_13	172.18	19	0.25	0.03	0.24	4.19	0.24	0.55	4.04	0.01	2.02	0.10
SW_14	76.69	9	0.20	0.02	0.26	3.88	0.46	0.57	5.09	0.00	2.54	0.09
SW_15	100.95	15	0.20	0.03	0.25	4.00	0.35	0.56	4.88	0.01	2.44	0.11
SW_16	98.81	9	0.24	0.02	0.26	3.91	0.27	0.57	4.20	0.01	2.10	0.07
SW_17	206.29	19	0.25	0.02	0.23	4.29	0.23	0.54	4.01	0.01	2.01	0.09
SW_18	97.62	9	0.17	0.02	0.25	4.07	0.41	0.56	5.76	0.00	2.88	0.07
Burhanpur Watershed	2203.68	281	0.21	0.03	0.16	6.07	0.10	0.46	4.80	0.01	2.40	0.25

L = Total stream length, N = Total stream numbers, D_d = Drainage density (km/km^2), S_f = Stream frequency (number/km^2), F_f = Farm factor, F_S = shape factor, R_c = Circularity ration, R_e = Elongation ration, C = Constant of channel maintenance (km^2/km), I_f = Infiltration number, L_g = Length of overland flow (km), T = Texture ration

factor of Burhapur watershed is 6.07 which shows that the watershed is more or less elongated in shape.

The value of circularity ratio ranges from 0.2 to 0.8 or less than 1. The value more than 0.5 indicates that the watershed is more homogeneous in geological formation and more circular in shape whereas less than 0.5 shows that watershed is elongated in shape (Miller 1953). In circular watershed as the high flow could accumulated from whole area of watershed simultaneously circularity ratio of watershed is used for assessment of flood hazard. The circularity ratio of Burhanpur watershed is 0.10 depicting it as elongated in shape. Further emphasizes that the discharge from the watershed would be less. Whereas the circularity ration vary between 0.15 and 0.46 for the sub-watersheds. Higher value of 0.46 is for sub watershed SW_14 (relatively circular in shape) shows that it could be flood hazard zone with in the watershed. Similarly, the zones with higher circularity ration are prone to flood hazard.

Constant of channel maintenance is the inverse of drainage density of watershed. It indicates the magnitude of Sq. km of watershed surface area required to sustain the one km liner length of stream segment. The value of constant of channel maintenance for Burhanpur watershed is 4.80. Its value varies from 4.01 to 5.93 with in the sub watersheds. Overland flow is flow of precipitated water which move over the land surface leading to the stream channels which is differ from surface runoff. Overland flow is predominant in small watershed than larger watersheds. Burhanpur watershed has length of overland flow of 2.40 km, whereas its value varies between 2.01 and 2.96 km for sub-watersheds. SW_6 is more dominated with overland flow because of more relief than other sub-watersheds.

Texture ratio is also called as drainage texture which is the ratio of total no of stream segments of all order to the perimeter of the watershed. It depends on the lithological properties of basin, infiltration of the soil as well as relief aspects of the terrain (Chandniha and Kansal 2017; Shelar et al. 2022). The texture ratio of Burhanpur watershed is 0.25. The value varies within the sub-watershed between 0.07 and 0.16. Lower values of texture ratio indicate that watershed is plain with lower degree of slope. Relief ratio and Ruggedness number (R_n) of Burhanpur watershed observed as 3.88 and 0.2 respectively. The higher value of relief ratio characterizes the Burhanpur watershed with hilly regions (Fig. 3). Relief of the Burhanpur watershed is 983 m further indicates the mountainous area of watershed.

4 Conclusion

The Hydrological study of a watershed greatly relay on its geo-morphometric characteristics. No doubt, watershed delineation along with its stream network may be performed with traditional observational survey and using the maps. However, advance remote sensing and GIS technologies play a great role to provide quick yet authentic delineation of study area. Further this analysis not only helps to study hydrologic behavior but also to prioritize susceptible areas under erosion, management and utilization of land and water resources, status of landform etc. this study reviled that the Burhanpur watershed, as a whole, being elongated in shape (low form factor of 0.16 and shape factor of 6.07) and having dominant channel flow (low bifurcation ratio of 1.77) shows lesser susceptibility to erosion. Further it was found that the watershed is having low values of drainage density, stream frequency, and circularity ration of 0.21, 0.03, and 0.10 respectively, indicating that watershed, as a whole at outlet, do not contribute to instance flood hazards.

Nevertheless, when the sub-watersheds are considered, the results show the other face of a coin. It was observed that the part of watershed having sever overland flow based on bifurcation ration indicating the susceptibility under erosion problem. The fact of having low drainage density again confirms the threat of high overland flow for some sub-watersheds and their susceptibility to soil erosion. Moreover, comparative higher values of form factor and elongation ration of some sub watershed (SW_4, SW_7, SW_8, SW_9, SW_14, and SW_16) confirms their circular shapes.

In turn this indicates their capability to generate concentrated peak flow and hance susceptibility under flash flood situation. From this study it is concluded that the Burhanpur watershed is having some areas which may be under critical threat of soil erosion, sedimentation, and flash flood hazards.

The geomorphometric study being first step to understand the watershed's hydrological processes and their potential risk on resources. However, the microscopic characterization and prioritization of sub-watersheds within the watershed area is further essential for comprehensive management, detailed planning and effective implementation. Hance this study also highlights the need of further research to prioritize the sub-watershed for strategies building to promote soil conservation via control of soil erosion and on-site water harvesting via control of flash floods. Which will help the decision makers to allocate the investments to critical sub-watersheds in technically efficient and economically effective way.

References

Arvind, Sharma P, Sangwan K (2018) Performance assessment of integrated watershed management programme (IWMP) of Rohtak District, Haryana. J Rural Agric Res 18(1):50–53

Bajirao TS, Kumar P, Kumar A (2019) Application of remote sensing and GIS for morphometric analysis of watershed: a review. Int J Chem Stud 7(2):709–713

Band LE (1986) Topographic partition of watersheds with digital elevation models. Water Resour Res 22(1):15–24

Chandniha SK, Kansal ML (2017) Prioritization of sub-watersheds based on morphometric analysis using geospatial technique in Piperiya watershed, India. Appl Water Sci 7(1):329–338. https://doi.org/10.1007/s13201-014-0248-9

Chandra P, Patel PL, Porey PD (2016) Prediction of sediment erosion pattern in Upper Tapi Basin, India. Curr Sci 110(6):1038–1049

Choudhari PP, Nigam GK, Singh SK, Thakur S (2018) Morphometric based prioritization of watershed for groundwater potential of Mula river basin, Maharashtra, India. Geol Ecol Landscapes 2(4):256–267. https://doi.org/10.1080/24749508

Dhaloiya A, Hooda RS, Kumar D, Malik A, Kumar A (2022) Geoinformatics-based assessment of gross irrigation requirement of different crops grown in the south-western region of Haryana, India. In: Current directions in water scarcity research, vol 7. Elsevier, pp 299–316. https://doi.org/10.1016/B978-0-323-91910-4.00018-2

Faniran A (1968) The index of drainage intensity: a provisional new drainage factor. Aust J Sci 31(9):326–330

Hadley RF, Schumm SA (1961) Sediment sources and drainage basin characteristics in upper Cheyenne River basin. US Geological Survey Water-Supply Paper 1531:198

Horton RE (1932) Drainage-basin characteristics. EOS Trans Am Geophys Union 13(1):350–361. https://doi.org/10.1029/TR013i001p00350

Horton RE (1945) Erosional development of streams and their drainage basins hydrophysical approach to quantitative morphology. GSA Bull 56(3):275–370. https://doi.org/10.1130/0016-7606(1945)56[275:EDOSAT]2.0.CO;2

Karabulut MS, Özdemir H (2019) Comparison of basin morphometry analyses derived from different DEMs on two drainage basins in Turkey. Environ Earth Sci. https://doi.org/10.1007/s12665-019-8585-5

Kumar D, Arvind, Nain AS, Darshana, Arya S, Bhardwaj S, Abhilash (2019) Soil loss estimation using geo-spatial technology in north western trai region of India. J Agrometeorol 21:182–188

Kumar D, Arvind, Nain AS, Singh A, Mor A, Bhardwaj S (2021a) Geo-spatial technology application for prioritization of land resources in Udham Singh Nagar District of Uttarakhand, India. Indian J Trad Knowl (IJTK) 20(2):595–603

Kumar D, Dhaloiya A, Nain AS, Sharma MP, Singh A (2021b) Prioritization of watershed using remote sensing and geographic information system. Sustainability 13(16):9456

Kushwaha NL, Elbeltagi A, Patel A, Zakwan M, Rajput J, Sharma P (2022a) Assessment of water resources using remote sensing and GIS techniques. In: Current directions in water scarcity research, vol 7. Elsevier, pp 85–98

Kushwaha NL, Elbeltagi A, Mehan S, Malik A, Yousuf A (2022b) Comparative study on morphometric analysis and RUSLE-based approaches for micro-watershed prioritization using remote sensing and GIS. Arab J Geosci 15(7):1–18. https://doi.org/10.1007/s12517-022-09837-2

Machiwal D, Patel A, Kumar S, Naorem A (2022) Status and challenges of monitoring soil erosion in croplands of arid regions. In: Shit PK, Adhikary PP, Bhunia GS, Sengupta D (eds) Soil health and environmental sustainability. Environmental science and engineering. Springer, Cham. https://doi.org/10.1007/978-3-031-09270-1_8

Maidment DR, Morehouse S (2002) Arc hydro: GIS for water resources. ESRI, Inc.

Mangan P, Haq MA, Baral P (2019) Morphometric analysis of watershed using remote sensing and GIS—a case study of Nanganji River Basin in Tamil Nadu, India. Arab J Geosci 12(6):202. https://doi.org/10.1007/s12517-019-4382-4

Mesa LM (2006) Morphometric analysis of a subtropical Andean basin (Tucumán, Argentina). Environ Geol 50(8):1235–1242. https://doi.org/10.1007/s00254-006-0297-y

Miller VC (1953) A quantitative study of drainage basin characteristics in the mountain area. Virginia and Tennessee. Technical Report. Office of Naval Research, Department of Geology, Columbia University, New York

Moore ID, Grayson RB, Ladson AR (1991) Digital terrain modelling: a review of hydrological, geomorphological and biological applications. Hydrol Process 5:3–30

Morris DG, Heerdegen RG (1988) Automatically derived catchment boundaries and channel networks and their hydrological applications. Geomorphology 1(2):131–141. https://doi.org/10.1016/0169-555X(88)90011-6

Nitheshnirmal S, Bhardwaj A, Dineshkumar C, Rahaman SA (2019) Prioritization of erosion prone micro-watersheds using morphometric analysis coupled with multi-criteria decision making. Proceedings 24(1):11. https://doi.org/10.3390/iecg2019-06207

Nooka Ratnam K, Srivastava YK, Venkateswara Rao V, Amminedu E, Murthy KSR (2005) Check dam positioning by prioritization of micro-watersheds using SYI model and morphometric analysis—remote sensing and GIS perspective. J Indian Soc Remote Sens 33(1):25–38. https://doi.org/10.1007/BF02989988

Ozdemir H, Bird D (2009) Evaluation of morphometric parameters of drainage networks derived from topographic maps and DEM in point of floods. Environ Geol 56(7):1405–1415. https://doi.org/10.1007/s00254-008-1235-y

Pakhmode V, Kulkarni H, Deolankar SB (2003) Hydrological drainage analysis in watershed-programme planning: a case from the Deccan basalt. India Hydrogeol J 11(5):595–604

Pande CB, Moharir K (2017) GIS based quantitative morphometric analysis and its consequences: a case study from Shanur River Basin, Maharashtra India. Appl Water Sci 7:861–871. https://doi.org/10.1007/s13201-015-0298-7

Pande C, Moharir K, Pande R (2018) Assessment of morphometric and hypsometric study for watershed development using spatial technology—a case study of Wardha River Basin in Maharashtra, India. Int J River Basin Manag 19:43–53

Patel DP, Gajjar CA, Srivastava PK (2013) Prioritization of Malesari mini-watersheds through morphometric analysis: a remote sensing and GIS perspective. Environ Earth Sci 69(8):2643–2656. https://doi.org/10.1007/s12665-012-2086-0

Patel A, Kethavath A, Kumar M, Rao K, Srinivasrao C (2021) Sustainable land and water management for reducing soil erosion in tropical India. In: Srinivasrao C, Balakrishnan M, Krishnan P, Sumanthkumar V (eds) Agricultural research, technology and policy: innovations and advances.

National Academy of Agricultural Research Management, pp 333–347. https://www.researchgate.net/publication/354170929

Patel A, Jena PP, Khatun A et al (2022) Improved Cartosat-1 based DEM for flood inundation modeling in the delta region of Mahanadi River Basin, India. J Indian Soc Remote Sens 50:1227–1241. https://doi.org/10.1007/s12524-022-01525-8

Puno GR, Puno RCC (2019) Watershed conservation prioritization using geomorphometric and land use-land cover parameters. Glob J Environ Sci Manag 5(3):279–294

Rastogi RA, Sharma TC (1976) Quantitative analysis of drainage basin characteristics. J Soil Water Conserv India 26(1&4):18–25

Rekha VB, George AV, Rita M (2011) Morphometric analysis and micro-watershed prioritization of Peruvanthanam Sub-watershed, the Manimala River Basin, Kerala, South India. Environ Res Eng Manag 3(57):6–14

Romshoo SA, Bhat SA, Rashid I (2012) Geoinformatics for assessing the morphometric control on hydrological response at watershed scale in the upper Indus basin. J Earth Syst Sci 121(3):659–686

Sangma F, Guru B (2020) Watersheds characteristics and prioritization using morphometric parameters and fuzzy analytical hierarchal process (FAHP): a part of lower Subansiri sub-basin. J Indian Soc Remote Sens. https://doi.org/10.1007/s12524-019-01091-6

Schumm SA (1956) Evaluation of drainage system and slopes in badlands at Perth Amboy, New Jersey. Geol Soc Am Bull 67(5):597–646. https://doi.org/10.1130/0016-7606(1956)67[597:EODSAS]2.0.CO;2

Sharma N, Kaushal A, Yousuf A, Sood A, Kaur S, Sharda R (2022) Geospatial technology for assessment of soil erosion and prioritization of watersheds using RUSLE model for lower Sutlej sub-basin of Punjab, India. Environ Sci Pollut Res 1–17. https://doi.org/10.1007/s11356-022-22152-3

Shelar RS et al (2022) Sub-watershed prioritization of Koyna river basin in India using multi criteria analytical hierarchical process, remote sensing and GIS techniques. Phys Chem Earth 128:103219. https://doi.org/10.1016/j.pce.2022.103219

Senthamizhan M, Balamurugan P, Shunmugapriya K (2016) Morpho-metric analysis of Thadayampatti watershed, Madurai district, Tamilnadu, India. Int J Eng Develop Res 4(3):103–107

Singh WR, Barman S, Tirkey G (2021) Morphometric analysis and watershed prioritization in relation to soil erosion in Dudhnai Watershed. Appl Water Sci 11(9):1–12

Smith KG (1950) Standards for grading texture of erosional topography. Am J Sci 248(9):655–668. https://doi.org/10.2475/ajs.248.9.655

Sreedevi PD, Owais S, Khan HH, Ahmed S (2009) Morphometric analysis of a watershed of South India using SRTM data and GIS. J Geol Soc India 73(4):543–552. https://doi.org/10.1007/s12594-009-0038-4

Strahler A (1964) Quantitative geomorphology of drainage basins and channel networks. In: Chow V (ed) Handbook of applied hydrology. McGraw Hill, New York, pp 439–476

Tarboton DG, Bras RL, Rodriguez-Iturbe I (1991) On the extraction of channel networks from digital elevation data. Hydrol Process 5(1):81–100. https://doi.org/10.1002/hyp.3360050107

A GIS-Based DRASTIC Approach for Aquifer Vulnerability Assessment: Study Conducted in the Municipal Corporation Region of Ranchi, Jharkhand

Shivam Saw, Prasoon Kumar Singh, Rohit Patel, Vaibhav Deoli, and Deepak Kumar

Abstract The idea of Groundwater Vulnerability is based on the supposition that the physical environment may shield groundwater to some extent from impacts from the environment and people, particularly concerning contaminants that reach the subsurface environment. The present study develops the aquifer vulnerability map of the Ranchi Municipal Corporation area through the GIS-based DRASTIC method by calculating Drastic Value Index (DVI) values. The greater the DVI value, the greater the potential for aquifer contamination. The DVI values are categorized into five classes from low to high. The study area has a varied range of Drastic Value Index (DVI) values. It has been found that 47.08% of the total research region comes under the moderately low vulnerability class, followed by the Moderate vulnerability class with 29.49% of the total research region, and 18.57% of the total area comes under the high vulnerability class. Hence, the DVI map could be helpful in the environmental risk assessment aspects during town planning or during the proposal for new developmental or industrial activities. It could also be useful in the site selection procedure for activities that could directly impact Groundwater, like landfill sites.

S. Saw (✉) · P. K. Singh · R. Patel · V. Deoli
Department of Environmental Science and Engineering, Indian Institute of Technology (ISM), Dhanbad 826004, India
e-mail: shivamsaw@gmail.com

P. K. Singh
e-mail: pks0506@iitism.ac.in

R. Patel
e-mail: rohitpatelp7@gmail.com

V. Deoli
e-mail: deolivaibhavdeoli@gmail.com

D. Kumar
Department of Soil and Water Conservation Engineering, GB Pant University of Agriculture and Technology, Pantnagar 263145, India
e-mail: deepak.swce.cot.gbpuat@gmail.com

C. B. Pande et al. (eds.), *Surface and Groundwater Resources Development and Management in Semi-arid Region*, Springer Hydrogeology,
https://doi.org/10.1007/978-3-031-29394-8_3

Keywords Groundwater · DRASTIC · GIS · Vulnerability · DVI

1 Introduction

The most important form of naturally existing freshwater on earth i.e. groundwater, is in a very vulnerable position in terms of both its quantitative and qualitative evaluations. Aquifer conservation is now necessary since groundwater quality is continuously declining around the world (Mallik et al. 2021). Due to fast and substantial growth in population, careless planning, unplanned city modernization, various land use-land category trends, and improper drainage facilities, which would include sewage disposal from industry sectors, rural fields, and urban areas, the issue of water groundwater pollution has increased to alarming levels in the past few decades (Singh et al. 2017; Pande et al. 2017; Pande et al. 2023). Groundwater pollution remediation procedures and methods are frequently difficult and extremely costly. The most precise and efficient methods of groundwater preservation are those that safeguard it from various forms of pollution (Bera et al. 2022). Managing this precious resource is crucial for its protection, but it can be difficult to delineate and demarcate the amount of contamination on a broad scale (Bai et al. 2012). Thus, in order to set up and control numerous development planning needs, underground aquifer vulnerability examinations have become crucial from the viewpoint of groundwater preservation measures (Ghosh et al. 2015). The capability for contaminants to move from the soil profile to the groundwater's vadose zone is estimated via vulnerability mapping of the aquifer against contamination (Connell and Daele 2003). Researchers have attempted to include geologic formations in groundwater vulnerability evaluations due to the significant impact they may have on the vulnerability of severely fractured zones (Saranya and Saravanan 2021). The popular models used are DRASTIC approach (Aller 1985), AVI approach (Stempvoort et al. 1993), GALDIT approach (Mitra 2011), GOD approach (Foster et al. 2002), and IRISH approach (Daly and Drew 1999).

DRASTIC approach is among the most prominent for determining how vulnerable groundwater is to various possible pollutants (Evans and Myers 1990; Anshumala et al. 2021). The results obtained using the DRASTIC model are reliable and can be used for even complex areas such as mining areas (McLay et al. 2001). The DRASTIC model has transformed its indexes to suit aquifers all around the globe (Baalousha 2011; Yu et al. 2022). Groundwater depth, rainfall recharge, aquifer class, soil characteristics, topography, the impact of the vadose zone, and the hydraulic conductivity of the aquifer are among the geographical datasets incorporated in this approach (Navulur and Engel 1998). A few disadvantages of the DRASTIC technique include the possibility for dispute in the rankings and weights applied as well as the exclusion of several crucial factors indicating contamination loading (Saranya and Saravanan 2021; Nair et al. 2022).

The aim and objective of the present study is (i) to carryout the aquifer vulnerability assessment of study area using DRASTIC approach and (ii) to prepare a thematic map of the study area to identify the high-risk zone using ArcGIS 10.5, software. The study provides immense utility to the groundwater board, Industries, and researchers for identifying the vulnerable zone of the study area.

2 Study Area

In the present research Aquifer Vulnerability evaluation is being conducted for the Ranchi Municipal Corporation Area. The Municipal corporation was established in 1979. Presently, the Municipal Corporation consists of 52 wards. The municipal corporation area extent is 174.837 sq km. The area is located between 23°14′ to 23°26′ North latitude and 85°15′ to 85°24′ East longitude. Conferring to the 2011 Census, the population of India is 10,73, 427. Ranchi has a subtropical Climate. The summer temperature varies from 42 to 20 °C, while in the winter it varies from 25 to 0 °C. The coldest months are December as well as January, and certain urban areas experience freezing temperatures. The annual rainfall in Ranchi District is about 1430 mm (56.34 in.). The study area boundary, the Ranchi Municipal Corporation boundary, was digitized in ArcGIS at a scale of 1:40,000 using the municipal corporation ward map as a base reference. The Municipal corporation ward map was obtained from the Ranchi municipal corporation official website as a shapefile. The location map of present research is illustrated in Fig. 1.

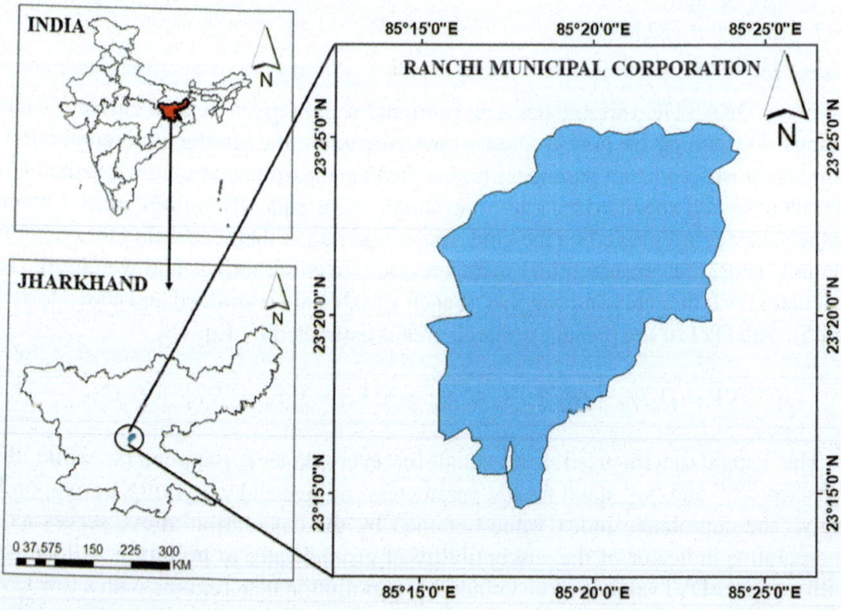

Fig. 1 Location Map of present work

Between 2013–14 and 2019–20, the Indian government launched the drinkable water and sanitation initiative for low-income states, which include Jharkhand State, with assistance from World Bank. It is heavily reported that this area has good precipitation, but due to the hardy rock area, the water retention capacity is low; hence net recharge of water is poor. Also, water quality is deteriorating due to unplanned urbanization and the migration of people from rural areas.

3 Material and Method

3.1 Methodology

Due to the fact that it takes numerous important hydrogeological elements into account when computing results, the DRASTIC approach is a numerical rating technique for assessing the possibility of groundwater pollution on diverse levels (local, regional, and global). The word DRASTIC is made up of the first letters of 7 model parameters which are as under.

- D—Depth to water table
- R—Net recharge
- A—Aquifer media
- S—Soil media
- T—Topography
- I—Impact of vadose zone
- C—Hydraulic conductivity

Every DRASTIC variable has a proportional weight given to it, so each class has ratings. Depending on how it affects how vulnerable the aquifer is to contamination, the most significant parameter is specified the maximum weight, while the least significant one is specified the minimum weight. Consequently, ratings are also given to the subcategories based on the kind, scope, and rate of their capability for contamination. Thus, the corresponding parameters are given set ratings and weighting. To calculate DVI, the rates and weights of each variable are combined and added (Aller 1985). The DVI of the present research area is estimated by Eq. 1:

$$\text{DVI} = D_r D_w + R_r R_w + A_r A_w + S_r S_w + T_r T_w + I_r I_w + C_r C_w \quad (1)$$

The capital letters used here stand for every layer's parameters, while the subscripts "r" and "w" stand for the parameters' ratings and weighting, correspondingly. The cumulative index value obtained by the calculation above serves as a comparative indicator of the susceptibility of groundwater to pollution. A location with a greater DVI value is more vulnerable to pollution than regions with a low DVI value. The methodology involved in DRASTIC approach of aquifer vulnerability assessement is illustrated in Fig. 2.

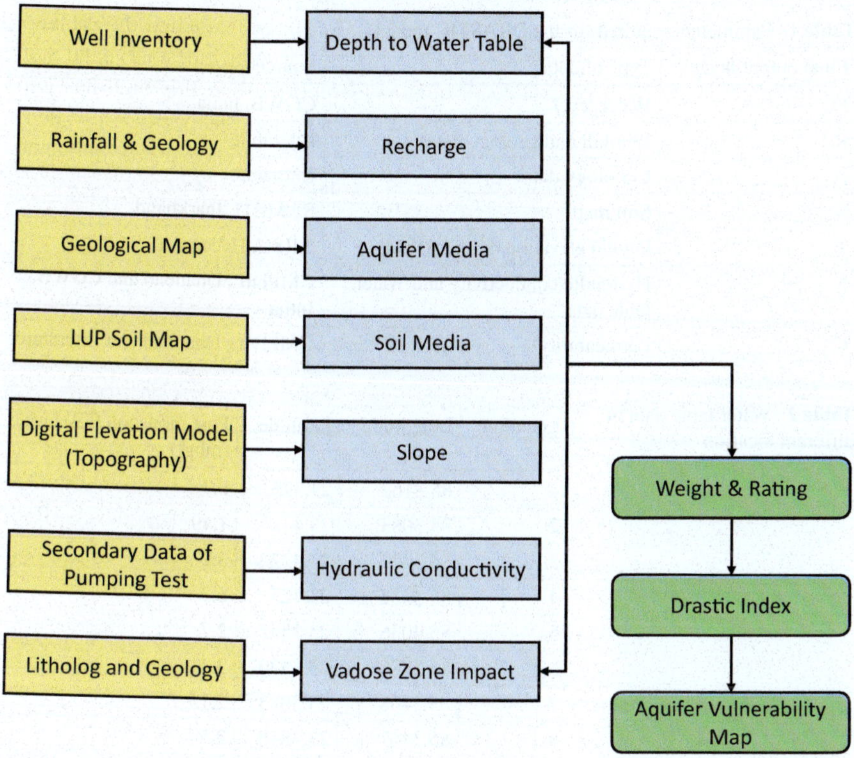

Fig. 2 Flow chart of methodology undertaken in the present study

3.2 Data Collection

First of all, to create thematic maps of the DRASTIC parameters, various kinds of data were necessary. The data was collected from published reports and websites of Government departments, Education & Research institute, and others. The data collected in various formats were processed using GIS software, QGIS, and ArcGIS (Pande et al. 2023). The required parameters used in DRASTIC models are illustrated in Table 1.

3.3 Depth to Water Table

The data for post-monsoon (2021) groundwater depth of 25 observation wells within the research area within a distance of 2 km from the research area's boundary was obtained from the CGWB website, as shown in Table 2. The latitudes and longitude values were also obtained; hence it was incorporated to plot the Depth to Water Table map.

Table 1 Parameters required for the DRASTIC model

Final output layer	Type of data	Source
D	Water level	CGWB, India
R	Rainfall and geology data	Literature
A	Lithology data	Literature
S	Soil map	SAMETI, Jharkhand
T	Digital elevation model (DEM)	NRSC, Hyderabad
I	Hydraulic conductivity and Water table data	CMPDIL, Dhanbad and CGWB, India
C	Permeability	CMPDIL, Dhanbad and Literature

Table 2 Water table data of different location

Location	Longitude	Latitude	Post monsoon GW level (mbgl)
1	85.3167	23.375	6.7
2	85.3083	23.4	1.49
3	85.3022	23.4222	3.2
4	85.3322	23.425	6.3
5	85.4036	23.3542	2.7
6	85.3786	23.3539	3.4
7	85.3408	23.3683	2.15
8	85.3467	23.3858	2.5
9	85.3481	23.3881	3.19
10	85.3908	23.4047	5.2
11	85.4086	23.4203	4.15
12	85.3194	23.343	1.8
13	85.3105	23.3136	1.4
14	85.3063	23.278	2.05
15	85.2777	23.3013	1.9
16	85.2972	23.2958	2.9
17	85.2805	23.3819	2.25
18	85.30833	23.3361	1.6
19	85.32	23.241	4.3
20	85.295	23.3777	1.85
21	85.31	23.3561	1.9
22	85.3686	23.4047	2.2
23	85.3688	23.3916	1.05
24	85.315	23.3069	3.45
25	85.3425	23.3138	2.1

3.4 Digital Elevation Model

The Digital Elevation Model (DEM) for the Ranchi region was obtained from the BHUVAN website, from the 'CartoDEM Version-2 R1' product. From the DEM raster, the Topography (Slope) Raster was created.

3.5 Hydrogeological Map

From the "Groundwater Information booklet" for Ranchi District, Published by CGWB, the Hydrogeological map of present research area was taken. The various aquifer types were plotted on this map. This map was used for creating Aquifer Map for the research area.

3.6 Soil Type

The National Bureau of Soil Survey and Land Use Planning (ICAR) and Department of Agriculture jointly released a Study on Soil Properties for the State of Jharkhand, in which the information on soil characteristics was included.

3.7 Rainfall Data

Rainfall data for the present research was collected from the Indian Metrological Department Website, and a land use distribution map has been prepared, as shown in Fig. 3a.

3.8 Land Use Data

The land Use Map of the Ranchi for the year 2017 was obtained from a study report on Land Use change analysis, and a rainfall distribution map has been prepared, as shown in Fig. 3b.

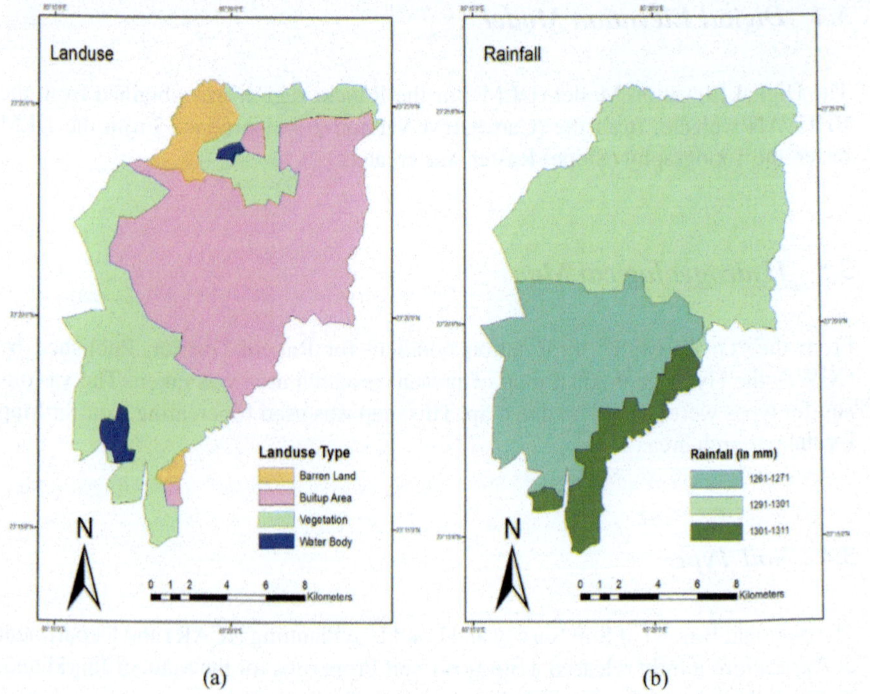

Fig. 3 Spatial distribution map of **a** land use pattern, **b** rainfall intensity

3.9 Weights and Ratings

The weights of the various DRASTIC constraints are given in Table 3. The Ratings for the different ranges of each of these parameters are given in Table 4. These weights and ratings were employed for calculating the DRASTIC value.

Table 3 Weight given to seven DRASTIC constraints (Aller 1985; Karan et al. 2018)

Parameter	Weight
Depth to groundwater	5
Net recharge	4
Aquifer type	3
Soil type	2
Topography or slope	1
Impact of vadose zone	5
Hydraulic conductivity	3

Table 4 Rating assigned to each parameters of DRASTIC approach

Range	Rating
Depth to groundwater	
0–1.52	10
1.52–4.57	9
4.57–9.14	7
9.14–15.24	5
15.24–22.86	3
22.86–30.48	2
30.48+	1
Net recharge	
0–5.08	1
5.08–10.16	3
10.16–17.78	6
17.78–25.4	8
25.4+	9
Aquifer type	
Metamorphic/igneous	3
Glacial till	5
Soil type	
Clay loam	3
Silty loam	4
Slope (topography)	
0–2	10
2–6	9
6–12	5
12–18	3
18+	1
Impact of vadose zone	
Metamorphic/igneous	4
Silt/clay	3
Hydraulic conductivity	
0.04–4.07	1
4.07–12.22	2
12.22–28.52	4
28.52–40.74	6
40.74–81.49	8
81.49+	10

4 Result and Discussion

4.1 Preparation of Thematic Map of DRASTIC Parameters

4.2 Depth to Water Table (D)

The groundwater depth data table of the 25 observation wells was exported to ArcGIS software as a.csv file. With the wells' latitude and longitude, the wells, the locations of the wells were converted into a point layer using the 'Display XY data' tool. The Groundwater depth values were automatically added as an attribute to the point layer. In ArcGIS, Spatial Interpolation was done using the point layer having groundwater depth data for creating the Groundwater depth. IDW technique of spatial interpolation was used. The map for Groundwater Depth was clipped to the research area boundary and its range of values were reclassified to the range according to which ratings were given. The thematic map was created in the Layout view, as illustrated in Fig. 4.

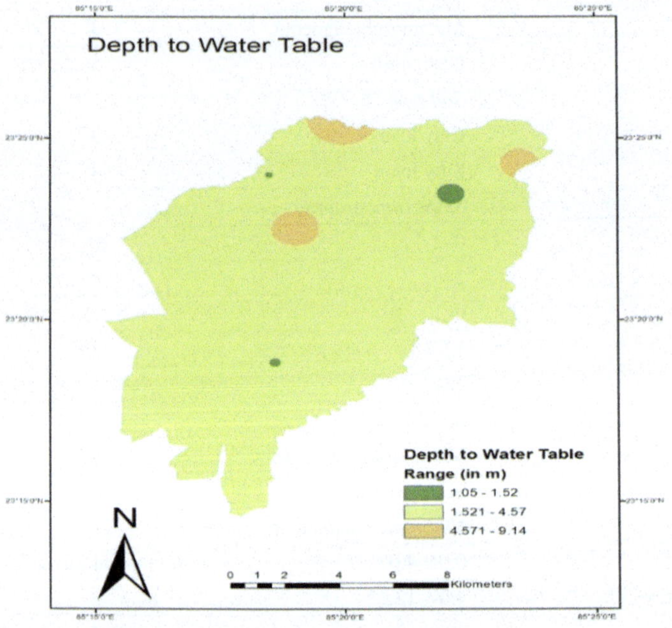

Fig. 4 Depth to water table (D) distribution map

Table 5 Runoff coefficient value of different land use category

Land use category	Runoff coefficient
Barren land	0.6
Built up area	0.9
Vegetation	0.5
Water body	0.01

4.3 Net Recharge (R)

It is calculated with the help of the Eq. 2 given as under.

$$\begin{aligned}\text{Net Recharge} = & [\text{Precipitation(Rainfall)}] - [\text{Precipitation(Rainfall)} * 0.05] \\ & - [\text{Precipitation(Rainfall)} * (\text{Runoff Coefficients})]\end{aligned} \quad (2)$$

The Indian Meteorological Division's rainfall data was used to create the thematic map of precipitation. It was estimated that 5% of the precipitation would be lost through evapotranspiration (this value is gathered from a report of IMD, Ranchi). Land use map was also taken, which had four classes of land use: Barren land, Built up area, Vegetation, and Water body. The runoff coefficients were allocated to the various land use classes ranging from 0 to 1, as shown in Table 5. The runoff coefficient values were selected from the Land use and Rainfall maps as rasters were used to derive the Net Recharge raster, according to the formula, using the Raster Calculator tool. Furthermore, the Net Recharge spatial distribution map is derived as shown in Fig. 5

4.4 Aquifer Type (A)

The hydrogeological map depicting the Aquifer types from the Groundwater booklet was Georeferenced using the 'Georeferenced' tool in QGIS. It was then clipped to the study area. Using this map as a base the Aquifer Map was digitized at a scale of 1:40,000. The Aquifer media distribution thematic map is illustrated in Fig. 6. In Ranchi municipal region, only two types of Aquifer Media were observed: glacial till, metamorphic/igneous.

4.5 Soil Type (S)

Birsa Agriculture University (BAU), Ranchi, provided the various soil map. For the purpose of creating the thematic map of soil media, it was georeferenced and then digitalized at a scale of 1:40,000 in ArcGIS software. The research region

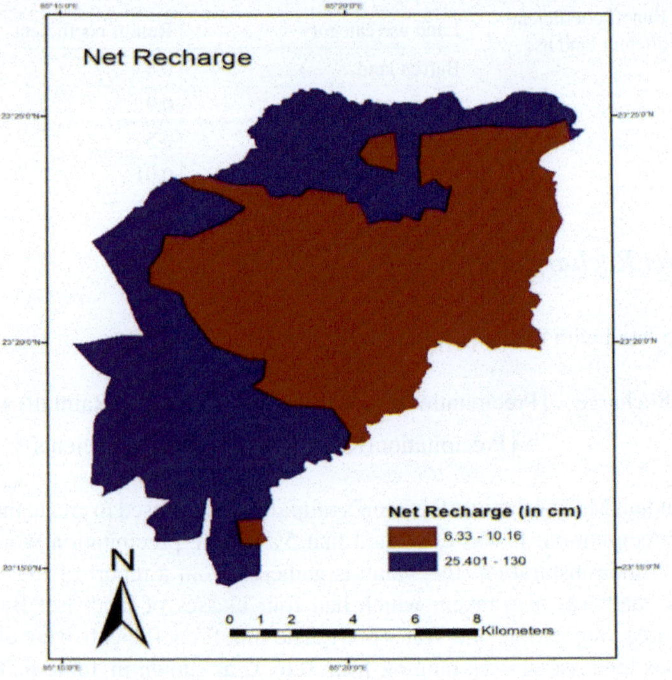

Fig. 5 Net recharge (R) distribution map

comprises of Clay loam and Silty loam soil. The two types of soil were categorized, and associated ratings were given to each category. Figure 7 displays the map created for soil media.

4.6 *Topography (T)*

The Digital Elevation Model (DEM) raster image was employed to develop the Topography (slope) map. Thematic map for slope was obtained using the DEM raster using the 'Slope' tool in ArcGIS software, as shown in Fig. 8. The Slope values obtained were in degrees, which was also reclassified into the range by which ratings were given.

4.7 *Impact of Vadose Zone (I)*

Since there was no evidence on the vadose zone in the study location, the estimated ratings for the vadose zone were derived employing the information on the soil media.

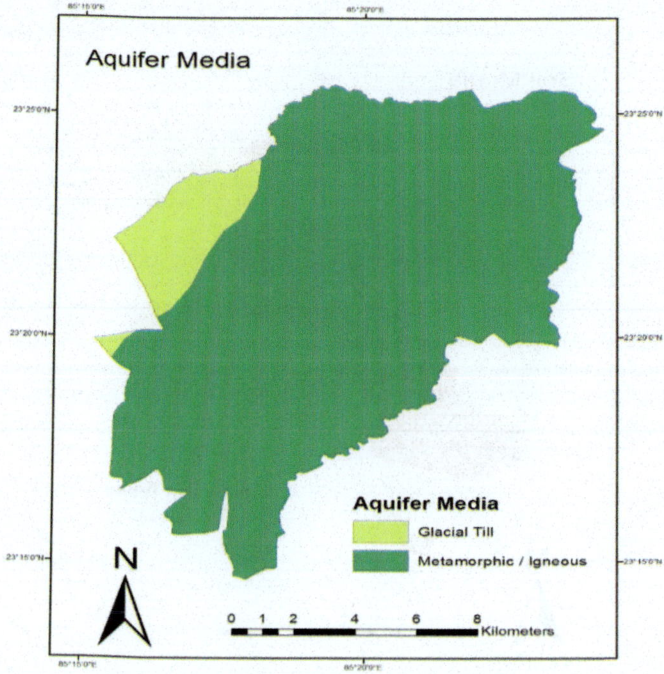

Fig. 6 Aquifer media (A) distribution map

Ratings for the vadose zone media were defined in order to convert the map to raster data (using soil media data). Figure 9 depicts the thematic map of the vadose zone's effects.

4.8 Hydraulic Conductivity (C)

Since there was no data on hydraulic conductivity in the research region, the estimated ratings for hydraulic conductivity were derived using data on the aquifer medium. In compliance with the required ratings, it was transformed into raster data. According to Table 4, ratings for hydraulic conductivity were provided (instead of utilizing aquifer media data here). Figure 10 displays the hydraulic conductivity map.

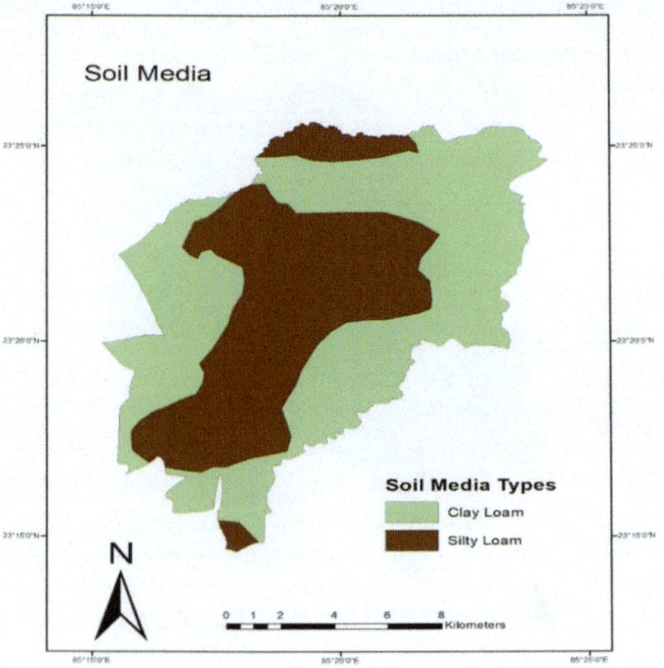

Fig. 7 Soil media (S) distribution map

4.9 *Aquifer Vulnerability Index Map Using DRASTIC Parameters*

The thematic map for DVI, as revealed in Fig. 11, is created by overlapping different thematic maps of the seven constraints. According to the DRASTIC value, Groundwater Vulnerability is alienated into five category. They are Low, Moderately low, Moderate, Moderately High, and High class. The total research region is 174.837 km^2. The area extent in each class of vulnerability and the percentage of it concerning the total study area is given in Table 6.

5 Conclusion

To create the map of Ranchi District's aquifer vulnerability, the GIS-based DRASTIC approach was equipped. The estimation of the DRASTIC index specified that the southern and some northern areas of the research region are highly vulnerable, and the Eastern part of research area is less vulnerable in comparison. This will illustrate the aquifer's intrinsic susceptibility for contamination. The final map

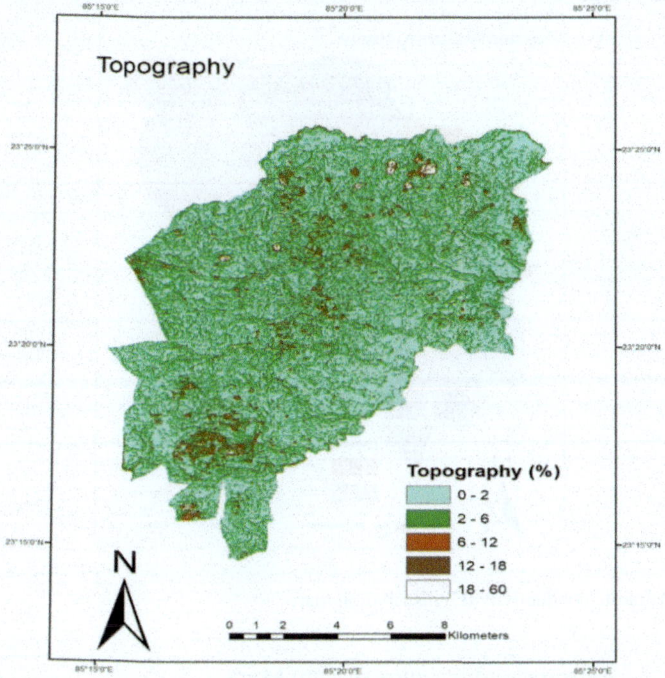

Fig. 8 Topography (slope) (T) distribution map

Fig. 9 Vadose zone (I) distribution map

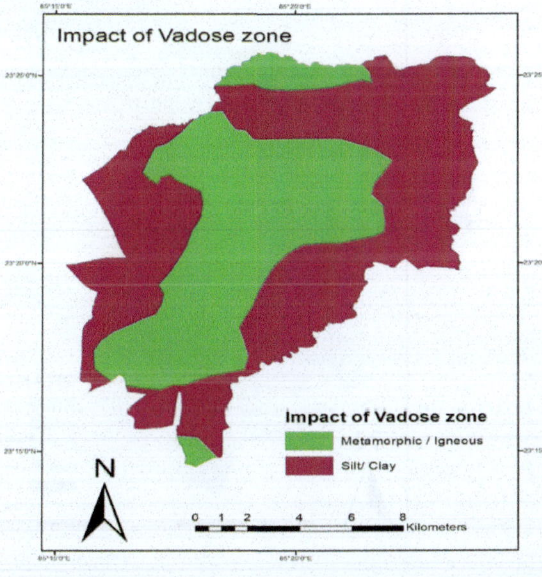

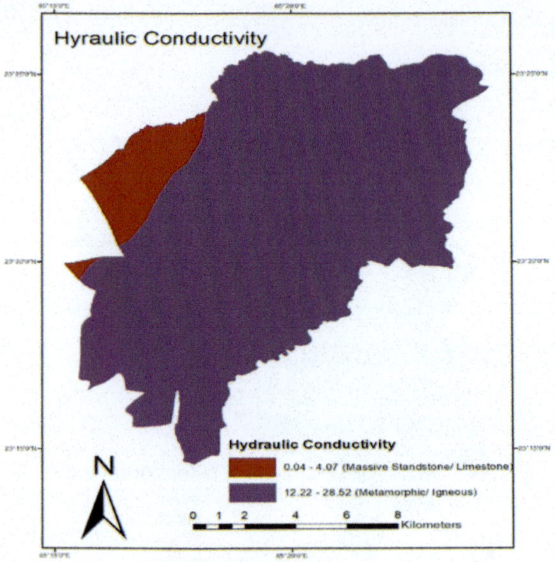

Fig. 10 Hydraulic conductivity (C) distribution map

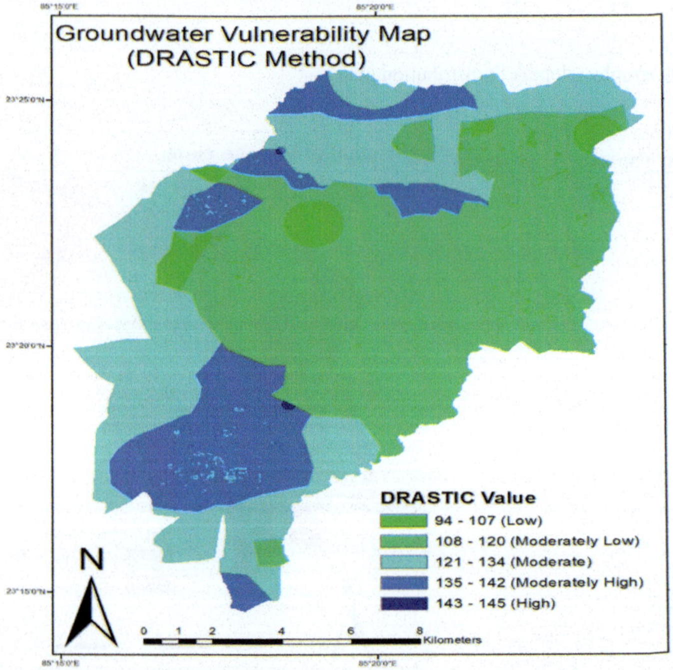

Fig. 11 DRASTIC DVI map of the research area

Table 6 Area under the vulnerability class as a percentage (%)

Vulnerability class	Area under the vulnerability class in km^2	Area under the vulnerability class as a percentage (%)
Low	7.134	4.08
Moderately low	83.569	47.08
Moderate	51.558	29.49
Moderately high	32.468	18.57
High	0.108	0.06

displays the whole spectrum of vulnerability indexes. The ability of the hydrogeologic situation to quickly transfer pollutants from surfaces to subsurface increases with increasing vulnerability index. On the other hand, low indices showed that the natural environment is doing a better job of protecting groundwater from pollution leaching.

References

Aller L (1985) DRASTIC: a standardized system for evaluating ground water pollution potential using hydrogeologic settings. In: Kerr RS (ed) Environmental research laboratory, office of research and development. US Environmental Protection Agency

Anshumala K, Shukla JP, Patel SS, Singh A (2021) Assessment of groundwater vulnerability zone in mandideep industrial area using DRASTIC model. J Geol Soc India 97(9):1080–1086

Baalousha HM (2011). Mapping groundwater contamination risk using GIS and groundwater modelling. A case study from the Gaza Strip, Palestine. Arab J Geosci 4(3):483–494

Bai L, Wang Y, Meng F (2012) Application of DRASTIC and extension theory in the groundwater vulnerability evaluation. Water Environ J 26(3):381–391

Bera A, Mukhopadhyay BP, Das S (2022) Groundwater vulnerability and contamination risk mapping of semi-arid Totko river basin, India using GIS-based DRASTIC model and AHP techniques. Chemosphere 307:135831

Connell LD, Van den Daele G (2003) A quantitative approach to aquifer vulnerability mapping. J Hydrol 276(1–4):71–88

Daly D, Drew D (1999) Irish methodologies for karst aquifer protection. In: Hydrogeology and engineering geology of sinkholes and karst, pp 267–272

Evans BM, Myers WL (1990) A GIS-based approach to evaluating regional groundwater pollution potential with DRASTIC. J Soil Water Conserv 45(2):242–245

Foster S, Hirata R, Gomes D, D'Elia M, Paris M (2002) Groundwater quality protection: a guide for water utilities, municipal authorities, and environment agencies. World Bank, Washington, DC

Ghosh A, Tiwari AK, Das S (2015) A GIS based DRASTIC model for assessing groundwater vulnerability of Katri Watershed, Dhanbad, India. Modeling Earth Syst Environ 1(3):1–14

Karan SK, Samadder SR, Singh V (2018) Groundwater vulnerability assessment in degraded coal mining areas using the AHP–modified DRASTIC model. Land Degrad Dev 29(8):2351–2365

Mallik S, Bhowmik T, Mishra U, Paul N (2021) Local scale groundwater vulnerability assessment with an improved DRASTIC model. Nat Resour Res 30(3):2145–2160

McLay CDA, Dragten R, Sparling G, Selvarajah N (2001) Predicting groundwater nitrate concentrations in a region of mixed agricultural land use: a comparison of three approaches. Environ Pollut 115(2):191–204

Mitra D (2011) Coastal hazard and Risk analysis in the gulf of Cambay, India: using remote sensing data and GIS technique. VDM Publishing

Nair AM, Prasad KR, Srinivas R (2022) Groundwater vulnerability assessment of an urban coastal phreatic aquifer in India using GIS-based DRASTIC model. Groundw Sustain Dev 19:100810

Navulur KCS, Engel BA (1998) Groundwater vulnerability assessment to non-point source nitrate pollution on a regional scale using GIS. Trans ASAE 41(6):1671

Pande CB (2022) Land use/land cover and change detection mapping in Rahuri watershed area (MS), India using the google earth engine and machine learning approach. Geocarto Int. https://doi.org/10.1080/10106049.2022.2086622

Pande CB, Moharir KN, Varade A (2023) Water conservation structure as an unconventional method for improving sustainable use of irrigation water for soybean crop under rainfed climate condition. In: Pande CB, Moharir KN, Singh SK, Pham QB, Elbeltagi A (eds) Climate Change Impacts on Natural Resources, Ecosystems and Agricultural Systems, pp 629–641. Springer Climate. Springer, Cham. https://doi.org/10.1007/978-3-031-19059-9_28

Pande CB, Patode RS, Moharir KN (2017) Morphometric analysis using remote sensing and GIS techniques (a case study of Devdari Watershed, Patur Tq., Akola District, Maharashtra). Trends Biosci 10(1): 219–223. ISSN 0974-8431.

Saranya T, Saravanan S (2021) Evolution of a hybrid approach for groundwater vulnerability assessment using hierarchical fuzzy-DRASTIC models in the Cuddalore Region, India. Environ Earth Sci 80(5):1–25

Singh R, Syed TH, Kumar S, Kumar M, Venkatesh AS (2017) Hydrogeochemical assessment of surface and groundwater resources of Korba coalfield, Central India: environmental implications. Arab J Geosci 10(14):1–16

Stempvoort DV, Ewert L, Wassenaar L (1993) Aquifer vulnerability index: a GIS-compatible method for groundwater vulnerability mapping. Can Water Resourc J 18(1):25–37

Yu H, Wu Q, Zeng Y, Zheng L, Xu L, Liu S, Wang D (2022) Integrated variable weight model and improved DRASTIC model for groundwater vulnerability assessment in a shallow porous aquifer. J Hydrol 608:127538

Evaluation of Infiltration Models in an Agricultural Catchment Using Guelph Permeameter in Mysore District

Y. Harshith and Abhishek A. Pathak

Abstract Infiltration process has a significant impact on many hydrological aspects of agricultural watersheds, including runoff generation, soil erosion, irrigation planning and management, contaminant transport, vadose zone hydrology and ground water management. Modelling of infiltration process in catchment scale is extremely complex, because it is influenced by numerous factors such as rainfall, soil physical properties, vegetative cover, various tillage practises, etc. Studies that emphasise the importance of infiltration modelling in agricultural catchments using guelph permeameter are minimal. The main goal of this research is to estimate and compare infiltration models that are used to measure infiltration rates in agricultural catchment located in Baradanapura village of Mysore district, Karnataka state, India. For this investigation, field infiltration experiments were conducted using a Guelph permeameter at a depth of 15 cm in 37 locations throughout the catchment of area 216 ha. Two infiltration models, Philip's and Kostiakov, were chosen for this study to test their dependability in catchment scale with measured values of infiltration rates. Parameters of two infiltration models were estimated using linear regression analysis. Statistical performance measures such as Coefficient of determination (R^2), Root mean square error (RMSE), Mean absolute error (MAE), and Nash Sutcliffe efficiency (NSE). Similarly, visual comparison methods such as box plots, Taylor's diagrams and scatter plots are used to evaluate model performance. Results indicates that Philip's model performed well for silty clay and sandy loam soils and were in good agreement with observed data. From this study it can be inferred that dominance of Philip's model will highlight the effect of soil texture on Sorptivity(s) and transmissivity factor(k) of Philip's equation. Efficiency and reliability of Philip's model in predicting infiltration rates for this study area and that will be helpful in irrigation system planning and management.

Keywords Philip's and Kostiakov models · Guelph permeameter · Visual comparison methods · Agricultural catchment

Y. Harshith (✉) · A. A. Pathak
Department of Civil Engineering, The National Institute of Engineering, Mysore, Karnataka 570008, India
e-mail: harshithharshi40@gmail.com

C. B. Pande et al. (eds.), *Surface and Groundwater Resources Development and Management in Semi-arid Region*, Springer Hydrogeology,
https://doi.org/10.1007/978-3-031-29394-8_4

1 Introduction

Stipulation for food, water, and land has increased due to the world's fast expanding population. Furthermore, the restricted amount of water accessible for agriculture worldwide, particularly in developing countries, is a result of climate change's major impact on the equilibrium between demand and supply for water resources (Pande et al. 2023). While irrigated agriculture is thought to generate 40% of the world's food, it produces more than 70% of food grains in India. Numerous studies have concentrated on increasing agricultural yield per unit of water use, but a small gain in water-use efficiency could result in substantial cost savings. In order to alleviate the current issue, surface irrigation systems must be designed properly and efficiently. Evaluation and modelling of soil infiltration characteristics are pivotal for quantifying catchment water storage capacity and also the final steady infiltration rates for varying soil texture classes throughout the catchment for planning and betterment of irrigation system. Infiltrometers and permeameters are two of the many field techniques used to measure infiltration rates in soils (Pande et al. 2022).

There are several well-known infiltrometers, including the single ring, double ring, and disc infiltrometers, as well as permeameters like the Guelph and tension permeameters etc. However, due to the high degree of soil variability, precise field measurement of infiltration is challenging, restricting their application in data scarce situations, and is a costly process in practise, particularly on large scale. As a result, modelling of infiltration behaviour gained much attention, which has given rise to variety of diverse analytical models. In past years many researchers conducted infiltration-based studies worldwide. These studies are mainly based on field experiments, laboratory experiments, modelling of infiltration process in catchment scale (Dahak et al. 2022), field scale (Jha et al. 2019; Mahapatra et al. 2020; Singh et al. 2018; Machiwal et al. 2006 etc.) different field conditions such as tillage and no tillage practices (de Almeida et al. 2018), different soil textures (Sajjadi et al. 2016; Amami et al. 2021; Thomas et al. 2020, etc.).

Three infiltration models applicability (Horton, Philips, and Kostiakov) in a catchment scale is assessed in Northern Algeria (Dahak et al. 2022). It demonstrates the relevance of initial and final infiltration measurements, as well as soil moisture, in determining the superiority of Horton model for estimating infiltration rates in Algerian catchment. It also exemplifies that performance criteria of Kostiakov and Philips, which showed similar variations. Some researchers analysed the effect of soil texture on infiltration rate. For example, Thomas et al. (2020) used DRI to analyse the impact of soil texture on soil infiltration characteristics for two distinct types of soil, namely silty loam and sandy loam. Performance metrics reveal that, in comparison to Green ampt, Horton, and Kostiakov models, Philip's model has a better relationship with observed data. The performance of cumulative infiltration models among four fine grained soil classes i.e., loam, clay loam, silty clay and silty loam soils was evaluated by Mirzaee et al. (2014). Revised Kostiakov model (RMK) performed better for loam, clay loam and silty clay. For silty loam soils Modified Kostiakov model fits better than RMK model. Mahapatra et al. (2020) conducted grid-based infiltration experiments using DRI to assess the effect of land use/land cover and soil texture on soil infiltration parameters. It was found that the properties of infiltration are not solely dependent on soil texture, bulk density, and patterns of land use and land cover.

Other elements that affect it include the existence of macropores and complex flow dynamics in vadose zones. (de Almeida et al. 2018) assessed effectiveness of infiltration models and effects of land-use/land cover changes, cultivation practises (conventional and no-tillage), and cultivation practises (conventional and no-tillage) on soil permeability. Compared to soil tillage techniques, soil infiltration characteristics are more susceptible to different land use patterns. In addition, the Horton model outperformed the Kostiakov-Lewis and Philip models. By conducting 72 double-ring infiltration experiments across several sites, Jha et al. (2019) aimed to study the infiltration features of lateritic soil–water zones. Five infiltration models were ranked as the Swartzendruber model, Brutsaert model, Kostiakov-Lewis model, Kostiakov model, and Philip Two-Term model, respectively, after being examined for their ability to model infiltration behaviour in lateritic vadose zones. Duan et al. (2012) on the Texas Tech University campus to assess efficacy of five conventional infiltration models in three turf soil types. A theoretically based model, Philip, was explored and compared, as of four empirical models, Kostiakov, Mezencev, NRCS, and Horton. The Mezencev model and Horton model outperformed other three infiltration models.

As per aforementioned literature review, only a limited number of studies have investigated how soil texture affects infiltration model performance and the characteristics of soil infiltration on catchment scale. The primary goal of this research is to measure infiltration rates in an agricultural watershed using a Guelph permeameter. There are few research works that highlight the use of Guelph permeameter data for modelling purposes. As a result, two renowned infiltration models i.e., Philip's and Kostiakov were chosen for this study. These models are compared using statistical and visual comparison approaches. On the basis of a larger number of data sets obtained in the field for diverse soil types, a dominating model suitable for selected study area is chosen.

2 Materials and Methods

2.1 Study Area

Study area selected for this research is an agricultural watershed near Baradanapura a region in Mysore district, South Karnataka Fig. 1 shows the location of study area map. The catchment located in Arkavathi River basin which is tributary of Cavery River. The catchment lies Between 12°13′2.72″N latitude and 76°32′59.51″E longitude and covers an area of 216 ha (2.16 km^2). Study area falls under semi-arid climate, usually hot during summers (temperature vary from 24 to 29 °C) and cold during winters (temperature vary from 18 to 22 °C). Average annual Rainfall over the last 10 years of 852 mm. More than 70% of the annual average rainfall occurs during

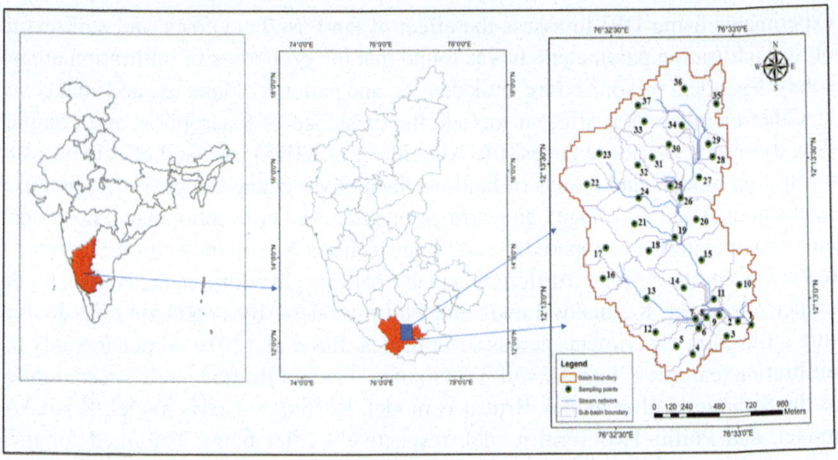

Fig. 1 Location map of study area

June to October. The evaporation ranges from 6.5 to 7.5 mm/day. The catchment has a wide range of soil textures as well as geological formations that are heterogeneous. The catchment shows topographic variation with elevations ranging from 650 and 728 m. Agriculture has always been the most important socio-economic activity in Baradanapura Catchment. Agricultural land makes up roughly 90% of overall catchment area, with 70% of it being irrigated cropland, 15% non-irrigated cropland, and 5% forest land. Remaining 10% includes barren land. Ground water supplies all irrigated croplands. Horticulture, dry farming, communal gardening, and polyculture are the most prevalent agricultural activities in the study area.

2.2 *Field Investigations*

The catchment was divided into 37 sub-catchments based on drainage pattern. Within the boundaries of sub-catchments, test locations are identified, covering the majority of soil types in the area. Up to 30 cm of root zone was considered for infiltration experimentation. In each sub catchment, two core samples were taken from upper vadose zone layer to evaluate soil texture in laboratory. Sieve analysis is used to determine the grain size of soil particles, whereas hydrometer analysis is utilised to determine percentage of silt and clay. Maps are developed that highlight the spatial heterogeneity of soil textures throughout the watershed (Fig. 2).

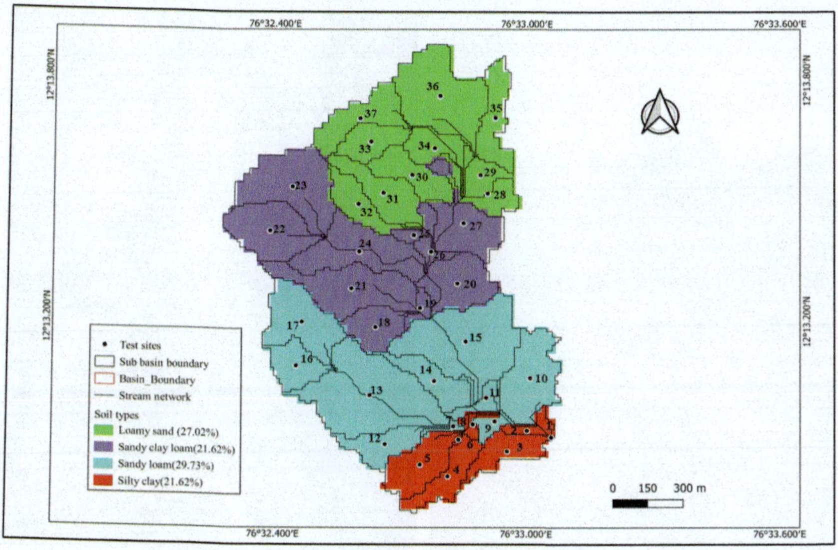

Fig. 2 Soil map showing variability of soils over the catchment

2.3 Determination of Field Infiltration Rates

Infiltration experiments were conducted on 37 sampling points using Guelph permeameter (Fig. 1) inside each sub-catchment. Infiltration tests were conducted in March and April of 2022. Reynolds and Elrick designed and developed Guelph Permeameter (1985). Marriotte Principle is used in Guelph Permeameter, which is an in-hole Constant-Head Permeameter (Fig. 3). The approach entails monitoring steady-state rate of water recharge into unsaturated soil from a cylindrical well hole with a constant water depth (head). Before using a Guelph permeameter, it is crucial to prepare the well of requisite depth on ground with an auger and cleaning equipment. The tripod should be centred over the cleaned well hole, as shown in Fig. 3, and water-filled Guelph permeameter should be lowered gradually into well hole. Verify again that inner and outer reservoirs are connected using reservoir knob. air inlet tip is then steadily increased by grasping upper air tube, resulting in an initial well head height (H_1) of 5 cm. At regular intervals, measure the flow rate from reservoir. Keep an eye on the flow rate until it stabilises. Readings can be taken when there is no change in flow rate over three consecutive time intervals. Now, if necessary, refill inner reservoir with water and set the second well head height (10 cm) by slowly raising air inlet tip. Make a reading by following steps as described above.

Fig. 3 Guelph permeameter and field experimentation

3 Infiltration Models and Parameters

3.1 *Philip's Model*

Philip (1957) model is given by:

$$F(t) = St^{-0.5} + Kt \tag{1}$$

where F(t) is cumulative infiltration capacity at time t, S is sorptivity, and k is constant proportional to Hydraulic conductivity.

Philip's two term model is given by equation:

$$f(t) = \frac{1}{2}St^{-0.5} + K \tag{2}$$

where, f(t) is infiltration rate at time t.

3.2 *Kostiakov Model*

Kostiakov (1932) model is given by:

$$F(t) = at^{b} \tag{3}$$

where F(t) indicates cumulative infiltration capacity at time t. a and b are constants, which represents soil infiltration characteristics.

4 Models Performance Evaluation Parameters

4.1 *Coefficient of Determination (R^2)*

The correlation coefficient is a measure of linear regression between anticipated values and model. R^2 is calculated as.

$$R^2 = \left\{ \frac{N\sum XY - (\sum X)(\sum Y)}{\sqrt{N(\sum X^2) - (\sum X^2)}\sqrt{Z(\sum Y^2) - (\sum Y^2)}} \right\} \quad (4)$$

where, X, Y represents predicted and observed infiltration rates and N represents number of observations.

4.2 *Root Mean Square Error (RMSE)*

By comparing predicted value to observed value, this approach calculates prediction error. Root Mean Squared Error (RMSE) is determined by:

$$RMSE = \sqrt{\frac{1}{N}(\sum_{i=1}^{N} (X_i - Y_j)^2)} \quad (5)$$

where X_i and Y_j are Predicted and observed values and N is number of measued values.

4.3 *Mean Absolute Error (MAE)*

The absolute error depicts magnitude of difference between predicted and observed values. Absolute error is estimated as.

$$MAE = \frac{1}{N}\left(\sum_{i=1}^{N} |X_i - Y_j|\right) \quad (6)$$

where X_i and Y_j are Predicted and observed values and N is number of measured values.

4.4 Nash Sutcliffe Efficiency (NSE)

A normalised statistic called the Nash–Sutcliffe efficiency (NSE) measures how much residual variation there is in relation to the variance of the measured data.

$$NSE = 1 - \left[\frac{\sum_{i=1}^{N}(X_i - Y_j)^2}{\sum_{i=1}^{N}(X_i - \overline{X}_1)^2}\right] \tag{7}$$

where, X_i and Y_j are Predicted and observed infiltration rate values, $\overline{X}$ represents mean of observed values and N is number of observations.

5 Results and Discussions

5.1 Determination of Soil Physical Properties

The infiltration tests was carried out in an agricultural watershed near Mysore district, located in southern portion of Karnataka using a Guelph permeameter to account for infiltration rate spatial variability. The catchment under consideration for this study is 216 ha in area. The experiments are conducted in 37 sub-catchments (P1–P37), which represent the most common soil types found across the catchment. The soil samples collected were analysed to determine the percentages of sand, silt, and clay. Soil samples collected throughout the catchment are evaluated in laboratory and classified according to USDA (United States Department of Agriculture) soil textural classification (Table 1). It reveals that study area has four major soil types i.e., silty clay (21.62%), sandy loam (29.73%), sandy clay loam (21.62%), and loamy sand (27.02%). Spatial variation of soil textures throughout the catchment are depicted in Fig. 2.

The Statistical description of results reveal that the study area has four major soil types (Fig. 2) i.e., silty clay (21.62%), sandy loam (29.73%), sandy clay loam (21.62%), and loamy sand (27.02%). In silty clay, the mean values of sand, silt, and clay content were 25.4%, 45.5%, and 29.1% respectively. Sand and silt show a lesser coefficient of variation (CV) of 3.6% and 7.1% compared to clay which shows a much higher CV of 12.6%. In Sandy Loam texture mean values of sand, silt, and clay content were 61.5%, 24.6%, and 13.9% respectively. Sand depicts a lesser CV of 2% compared to silt (7.8%) and Clay (13.8%). For Sandy Clay Loam mean values of sand, silt, and clay content were 56.6%, 20%, and 23.4% respectively. Silt and

Table 1 Statistical description of soil physical properties

Soil texture	Parameters	Range	Mean	SD	CV (%)
Silty Clay (P1–P8)	Sand (%)	24.0–26.0	25.4	0.9	3.6
	Silt (%)	39.0–48.0	45.5	3.3	7.1
	Clay (%)	26.0–32.0	29.1	3.7	12.6
Sandy Loam (P9–P19)	Sand (%)	59.0–63.0	61.5	1.2	2.0
	Silt (%)	22.0–27.0	24.6	1.9	7.8
	Clay (%)	12.0–17.0	13.9	1.9	13.8
Sandy Clay Loam (P20–P27)	Sand (%)	54.0–58.0	56.6	1.5	2.7
	Silt (%)	17.0–24.0	20.0	2.5	12.5
	Clay (%)	21.0–28.0	23.4	2.4	10.5
Loamy Sand (P28–P37)	Sand (%)	66.0–69.0	67.9	1.2	1.7
	Silt (%)	20.0–27.0	24.6	2.5	10.2
	Clay (%)	6.0–11.0	7.6	1.8	24.0

Clay show much higher CV of 12.5% and 10.5% compared to sand (2.7%). The mean values of sand, silt, and clay content in Loamy sand were 67.9%, 24.6%, and 7.6% respectively. Clay shows a higher CV of 24% compared to silt (10.2%) and sand (1.7%). It reveals that Clay and silt have a high degree of variability in different soil textures compared to sand in this agricultural watershed.

6 Determination of Model Parameters

The parameters of Philip's and Kostiakov models are estimated using linear regression analysis and average values of model paramteres for distinct soils are summarized Table 2. In comparison to silty clay soil, which has the lowest average value of 7.70 cm/h at sites (P1–P8), the final infiltration rate in loamy sand soil was higher, with an average value of 80.73 cm/h at sites (P28–P37). Minimum, maximum, and average parameter values of selected models for each soil were computed to gain an understanding of the model behaviour in the case of parameters variations (Table 2). The average values of S in the Philips model range from 2.61 to 8.88 cm, and k from 6.13 to 71.71 cm^{-h} for diverse soils. The Kostiakov model parameter a range from 14.40 to 87.50, and b ranges from 0.82 to 0.91, which is consistent with theory of infiltration, which states that value must be positive and smaller than one (Ogbe et al. 2011). The observations of field experimental tests were analysed and individual infiltration curves for Observed and modelled infiltration rates for four representative sites per individual soil type have been developed Fig. 4a–d.

Table 2 Parameters of Philip's and Kostiakov models

Test site	Soil type	Statistics	Philip's model		Kostiakov model	
			S (cm)	K (cm/hr)	A	b
P1–P8	Silty clay	Minimum	1.23	0.75	5.47	0.72
		Maximum	5.76	14.60	34.24	0.91
		Average	2.61	6.13	14.40	0.82
P9–P19	Sandy loam	Minimum	2.62	25.21	34.11	0.84
		Maximum	6.93	45.03	58.62	0.96
		Average	4.42	37.36	45.61	0.90
P20–P27	Sandy clay loam	Minimum	2.33	9.58	22.63	0.77
		Maximum	7.62	27.18	34.45	0.89
		Average	5.22	18.57	29.56	0.83
P28–P37	Loamy sand	Minimum	4.86	52.27	64.35	0.84
		Maximum	14.11	107.31	113.24	0.99
		Average	8.88	71.71	87.50	0.91

7 Statistical Performance Evaluation of Infiltration Models

Statistical performance indices methods such as Maximum absolute error (MAE), Root mean square error (RMSE), coefficient of determination (R^2), and Nash Sutcliffe Efficiency (NSE) were used to evaluate infiltration models. Best fit model was chosen based on higher values of the coefficient of determination (R^2), lowest possible Root Mean Square Error, and higher possible Mean Absolute Error Criteria. Computed average values of MAE, RMSE, R^2 and NSE for four different types of soils are in Table 3. For silty clay (P1–P8), average values of MAE were 2.24, 2.68 and RMSE were 2.90, 3.60 and R^2 were 0.80, 0.70 and those of NSE were 0.74, 0.57 for Philip's, Kostiakov respectively. From the above results it can be inferred that Philip's model show least values of MAE, RMSE and higher R^2 and NSE demonstrating that this model accurately predicted infiltration rates for the silty clay in this catchment. For sandy loam (P9–P19), the average values of MAE were 2.58, 3.29 and RMSE were 3.15, 4.11 and R^2 were 0.84, 0.73 and those of NSE were 0.80, 0.63 for Philip's, Kostiakov models respectively. As per the above statistical results, Philip's model had the lowest MAE, RMSE, and higher R^2 and NSE values, showing that infiltration rates for silty clay in this study area were accurately characterised by this model.

For sandy clay loam (P20–P27), the average values of MAE were 2.88,3.53 and RMSE were 3.55, 4.19 and R^2 were 0.86, 0.80 and those of NSE were 0.83, 0.74 for, Philip's, Kostiakov respectively. The above findings show that Philip's model accurately characterised the infiltration rates for sandy loam in this study area because it had lowest MAE, RMSE, and higher R^2 and NSE values. It was determined that Kostiakov model shows much similar variations compared to Philip's infiltration model. For loamy sand (P28–P37), average values of MAE were 4.31, 6.71 and

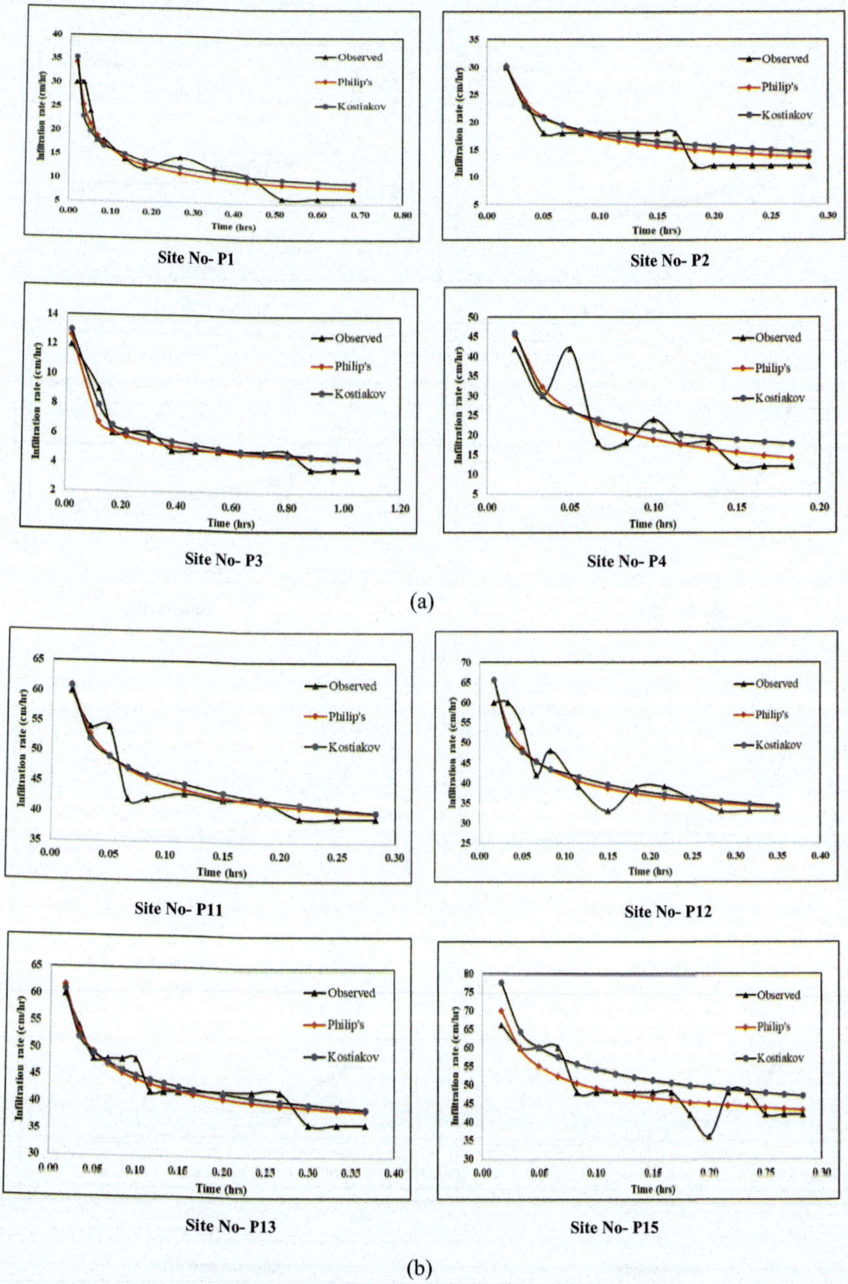

Fig. 4 **a** Comparison of observed and modeled infiltration rates for silty clay soil. **b** Comparison of observed and modelled infiltration rates for sandy loam soil. **c** Comparison of observed and modelled infiltration rates for sandy clay loam soil. **d** Comparison of observed and modelled infiltration rates for loamy sand soil

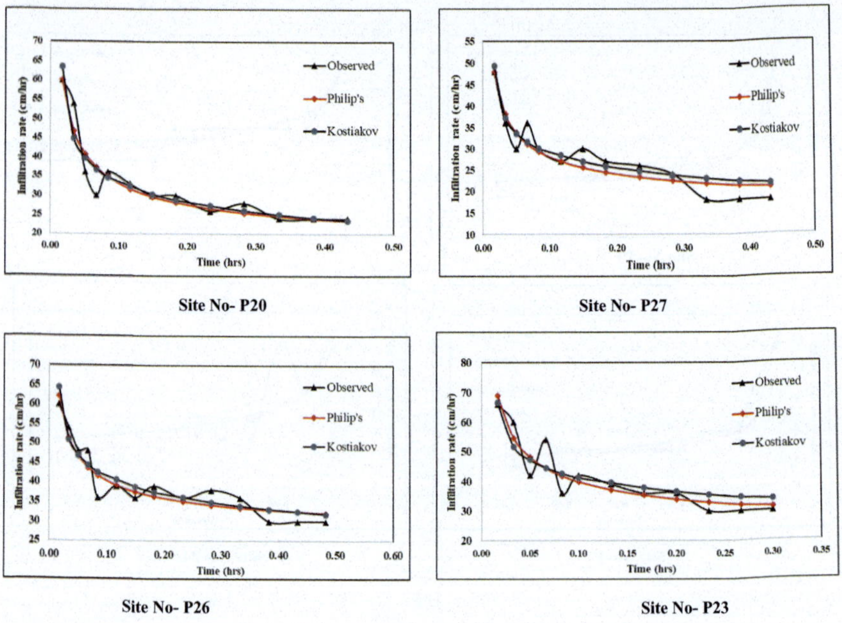

(c)

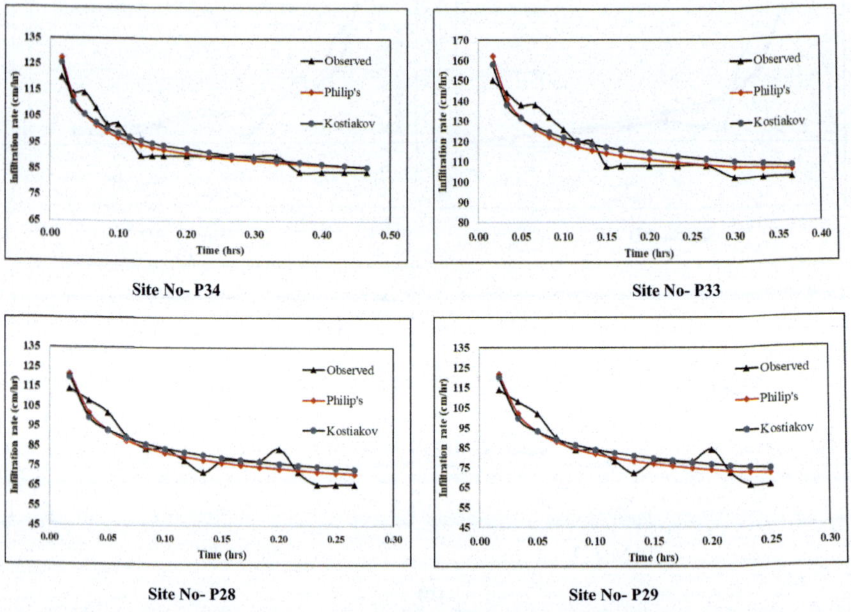

(d)

Fig. 4 (continued)

Table 3 Statistical performance evaluation of infiltration models

Soil type	Statistics	Philip's model				Kostiakov model			
		MAE	RMSE	R-squared	NSE	MAE	RMSE	R-squared	NSE
Silty clay	Minimum	0.53	0.76	0.72	0.60	1.72	1.99	0.63	0.32
	Maximum	4.04	5.58	0.90	0.88	5.02	6.40	0.90	0.87
	Average	2.24	2.90	0.80	0.74	2.92	3.62	0.76	0.61
Sandy loam	Minimum	1.73	2.01	0.77	0.70	1.72	1.99	0.38	0.35
	Maximum	3.20	3.77	0.90	0.89	10.41	14.27	0.90	0.87
	Average	2.58	3.15	0.84	0.80	4.65	5.93	0.74	0.64
Sandy clay loam	Minimum	2.17	2.87	0.75	0.67	2.36	2.87	0.74	0.65
	Maximum	3.60	4.37	0.91	0.90	4.00	5.02	0.89	0.89
	Average	2.88	3.55	0.86	0.83	3.03	3.78	0.84	0.79
Loamy sand	Minimum	1.80	2.11	0.73	0.63	1.77	2.29	0.69	0.31
	Maximum	8.09	9.51	0.94	0.93	8.19	9.40	0.92	0.92
	Average	4.31	5.08	0.85	0.82	4.52	5.36	0.84	0.72

RMSE were 5.08, 7.01 and R^2 were 0.85, 0.72 and those of NSE 0.82, 0.59 for Philip's and Kostiakov models respectively. It reveals that Philip's model, which had lowest MAE and RMSE values and highest R^2 and NSE values, accurately described the infiltration rates for loamy sand in this research region. Kostiakov model was observed to be the second-best infiltration model for loamy sand.

Box plots are adopted to evaluate the variation of R^2 (Fig. 5a, b). In silty clay and sandy loam (Fig. 5a), the R^2 Percentile of Philip's model are higher compared to Kostiakov model, and similarly, in Sandy clay loam and loamy sand soils (Fig. 5b), Philip's model shows higher R^2 percentiles and Kostiakov shows some similar variations in comparison to Philip's model. Box plots for variation of RMSE (Fig. 5c) for silty clay and sandy loam show lower values in Philip's compared to Kostiakov model. For sandy clay loam and loamy sand soils (Fig. 5d), Philip's model shows lower values. MAE values for silty clay and sandy loam (Fig. 5e) Philip's and model show lower values compared to Kostiakov model. For sandy clay loam and loamy sand soils (Fig. 5f), Philip's and Kostiakov models show lower values. NSE values for silty clay and sandy loam (Fig. 5g) were higher for Philip's compared to Kostiakov model. For sandy clay loam and loamy sand soils (Fig. 5h), Philip's model shows much higher values.

To investigate scatter around the line of perfect agreement. The Scatter plots between observed and estimated (predicted) values of infiltration rate was drawn in the resulting graph (Fig. 6a–d) showing that majority of values predicted by Philip's model for silty clay and sandy loam soil lies close to line of agreement. For Sandy clay loam and Loamy sand values of Philip's and Kostiakov model values are lies close to line of perfect agreement. Visual comparison methods of model's performance have some disadvantages that those with high performance are easily identified, but it is difficult to recognize and rank models based on performance. Hence, some

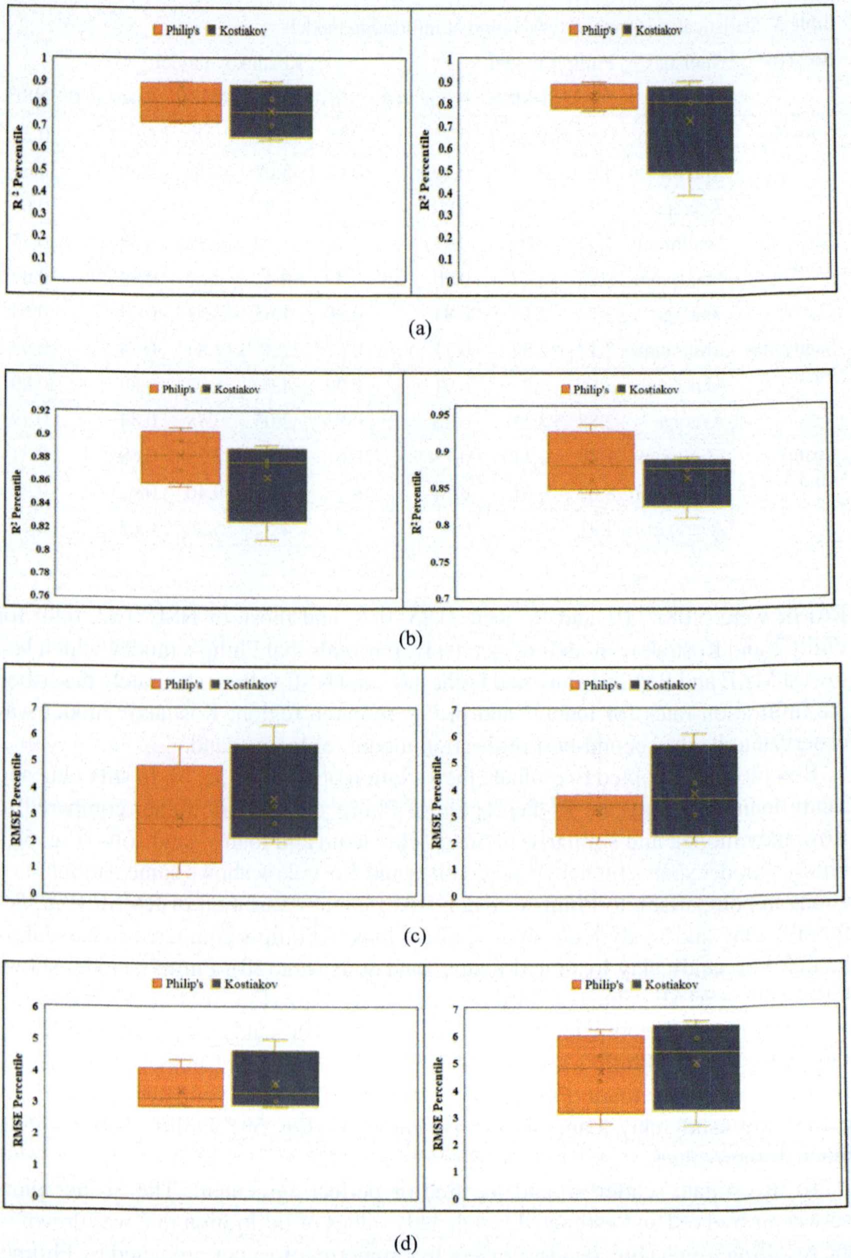

Fig. 5 **a** Box plots for R^2 percentile for Silty clay and Sandy loam soils. **b** Box plots for R^2 percentile for Sandy clay loam and Loamy sand soils. **c** Box plots for RMSE percentile for Silty clay loam and Sandy loam soils. **d** Box plots for RMSE percentile for Sandy clay loam and Loamy sand soils. **e** Box plots for MAE percentile for Silty clay loam and Sandy loam soils. **f** Box plots for MAE percentile for Sandy clay loam and Loamy sand soils. **g** Box plots for NSE percentile for Silty clay loam and Sandy loam soils. **h** Box plots for NSE percentile for Sandy clay loam and Loamy sand soils

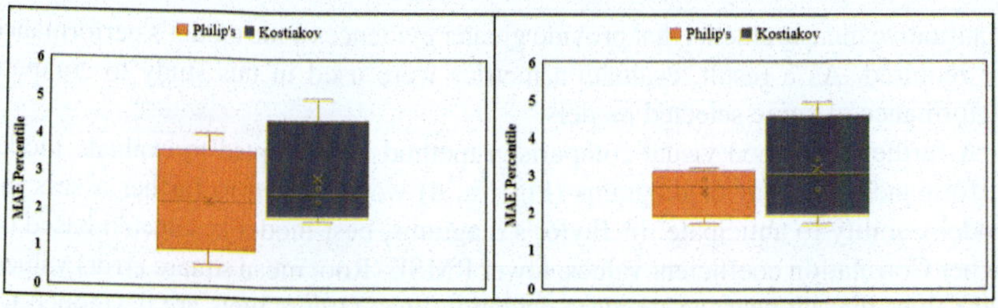

(e)

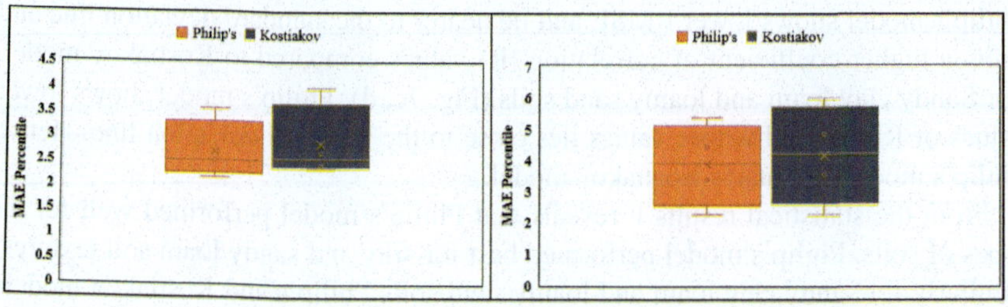

(f)

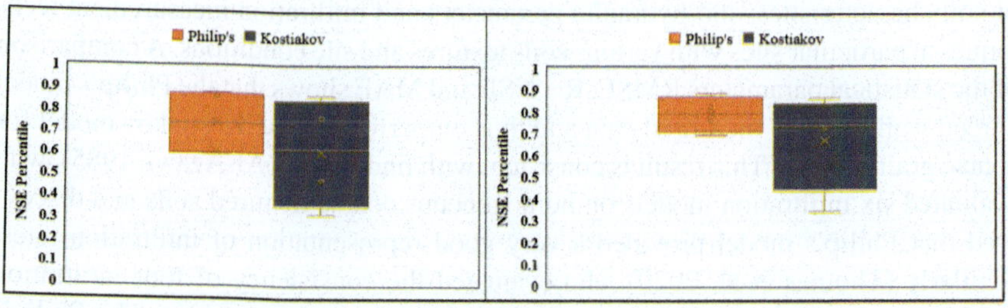

(g)

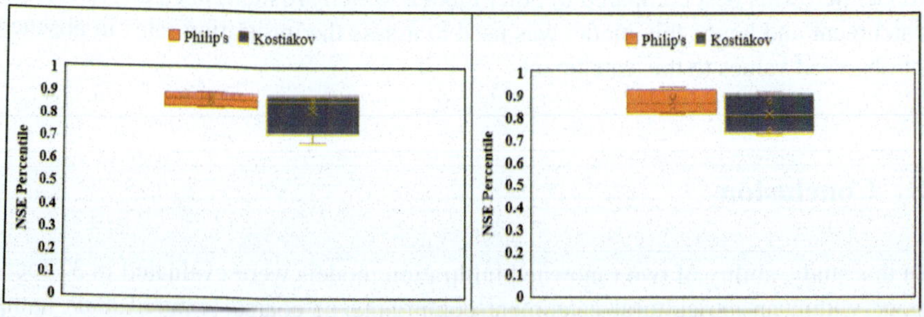

(h)

Fig. 5 (continued)

quantitative measurements that provide greater evidence of the model's performance are required. As a result, evaluation metrics were used in this study to compare performance of some selected models.

A further renowned visual comparison method was adopted to evaluate model performance. The Taylor diagrams (Fig. 7a, d) were created to further assess the model's ability to anticipate. In Taylor's diagrams, best model is selected based on higher Correlation coefficient values, lower RMSE (Root mean square error) values, and values that lie near the standard deviation line. Taylors plots are developed for four representative sites (Fig. 7a, b) in silty clay and sandy loam textures showing that Philip's model shows lower RMSE and lie nearer to the standard deviation line and show a higher coefficient of correlation (R) values compared to Kostiakov model. For Sandy clay loam and loamy sand soils (Fig. 7c, d), Philip's model shows lower values of RMSE and whose values lies close to the standard deviation line. Hence Philip's model dominates Kostiakov model.

From the statistical results it reveals that Philip's model performed well for all types of soils. Philip's model performed best for silty and sandy loam soil textures. Similarly for sandy clay loam and loamy sand soils Philip's and Kostiakov models show similar variations. The modelling outcomes will also highlight significance of soil characteristics, soil hydraulic parameters and infiltration measurement techniques at particular sites with various soils textures and site conditions. A comparison of the statistical parameters RMSE, R^2, NSE and MAE shows that the Philip's model agreed well with the measured data and thus outperformed the Kostiakov model for coarse grained soils. This result is consistent with findings of (Al-Azawi, 1985), who evaluated six infiltration models on homogeneous coarse textured soils and discovered that Philip's model provided a very good representation of infiltration rates. Similarly (Thomas et al. 2020) who evaluated the consistency of four infiltration models in coarse grained soils out of which Philip's performed best. The Philip's model performs best compared to other models which are mentioned above for this catchment and hence this model was used to assess the infiltration rates in absence of observed values to this catchment.

8 Conclusion

In this study, ability of two renowned infiltration models were evaluated in 37 locations within an agricultural catchment comprising of coarse textured soils using Guelph Permeameter. The results of statistical performance indices and visual comparison methods indicated that Philip's models are predominant for silty clay and sandy loam soil. Philip's and Kostiakov models show better performance in sandy clay loam and loamy sand. Dominance of Philip's model in this study area specifies the effect of soil texture on Sorptivity(S) and transmissivity factor(K) of Philip's equation. Soil physical properties such as soil texture, porosity, plays a crucial role in affecting parameters of Kostiakov models. Thus, even if the measured infiltration data sets are not available, Philip's model for prediction of infiltration rates is used

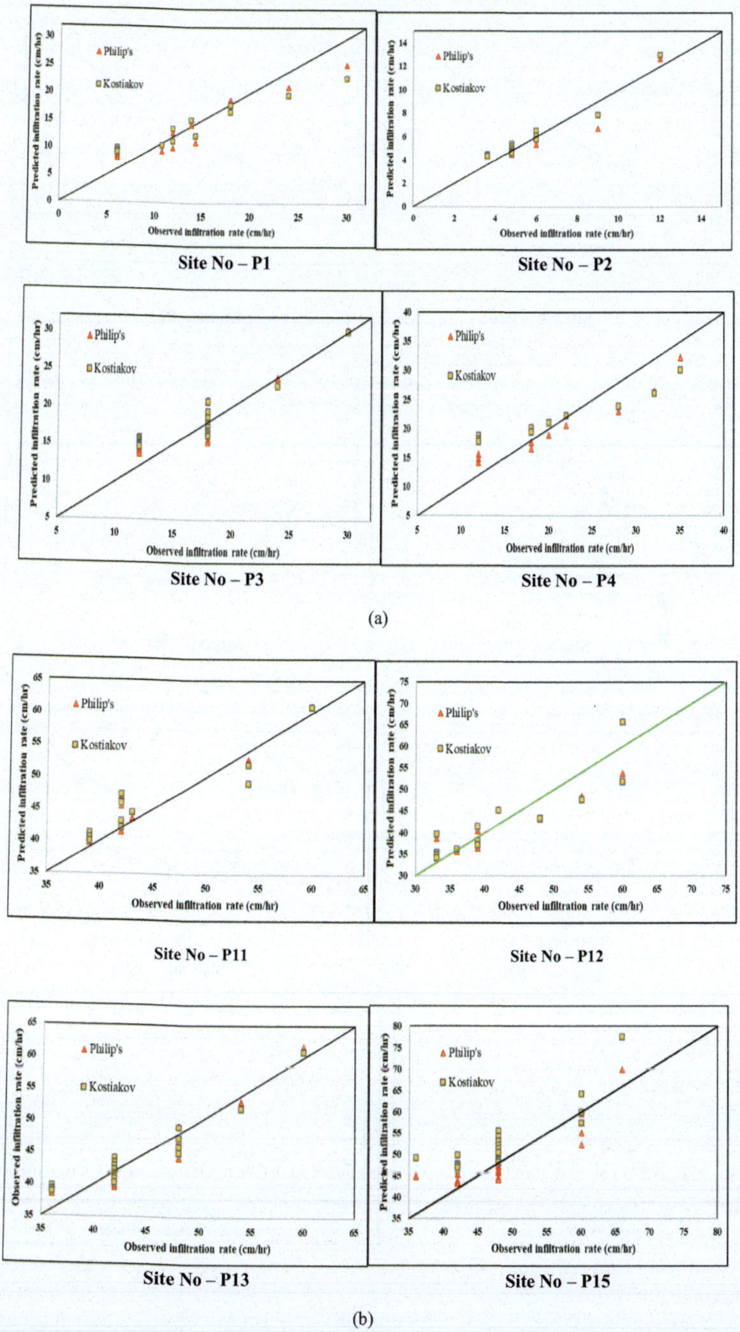

Fig. 6 **a** Observed vs predicted infiltration rates by various models in Silty clay soil. **b** Observed vs predicted infiltration rates by various models in Sandy loam soil. **c** Observed vs predicted infiltration rates by various models in Sandy clay loam soil. **d** Observed vs predicted infiltration rates by various models in Loamy sand soil

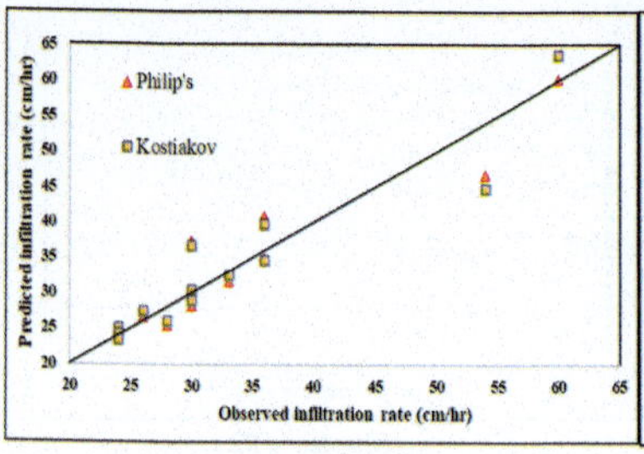

Site No – P20

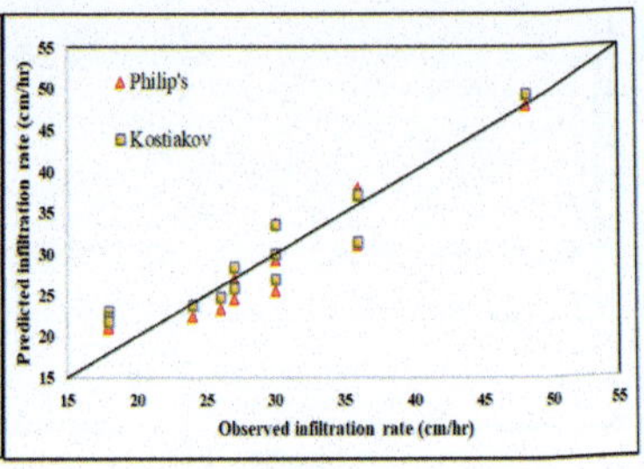

Site No – P23

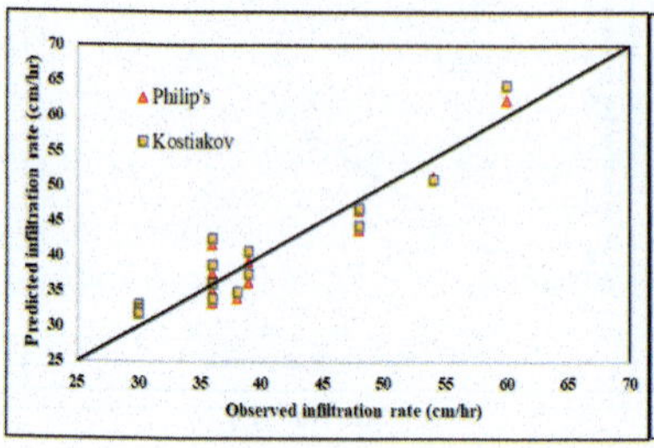

Site No – P26

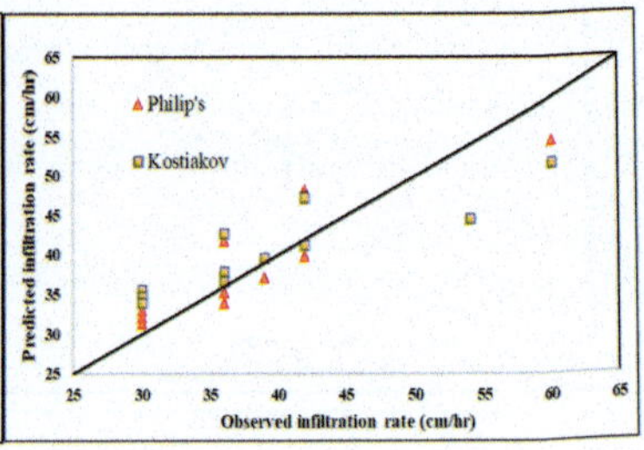

Site No – P27

(c)

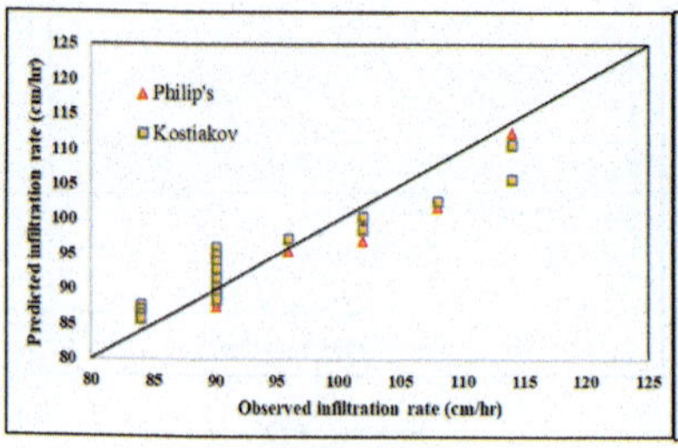

Site No – P28

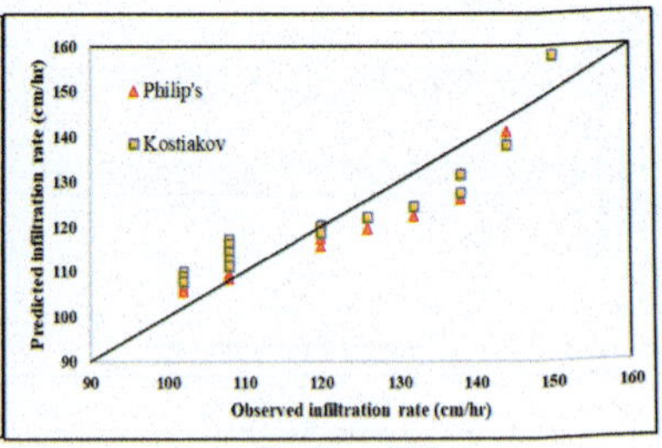

Site No – P29

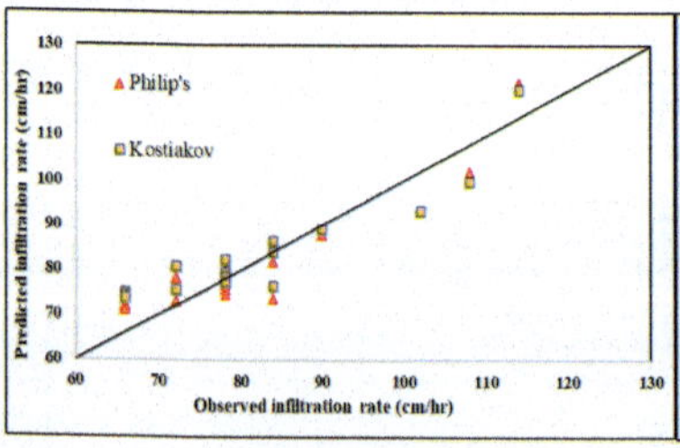

Site No – P33

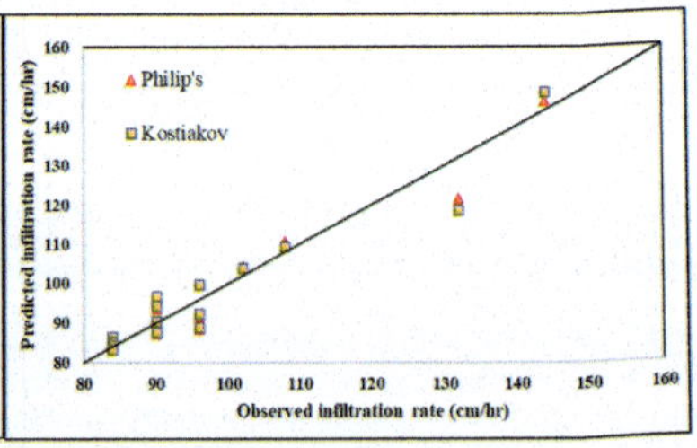

Site No – P34

(d)

Fig. 6 (continued)

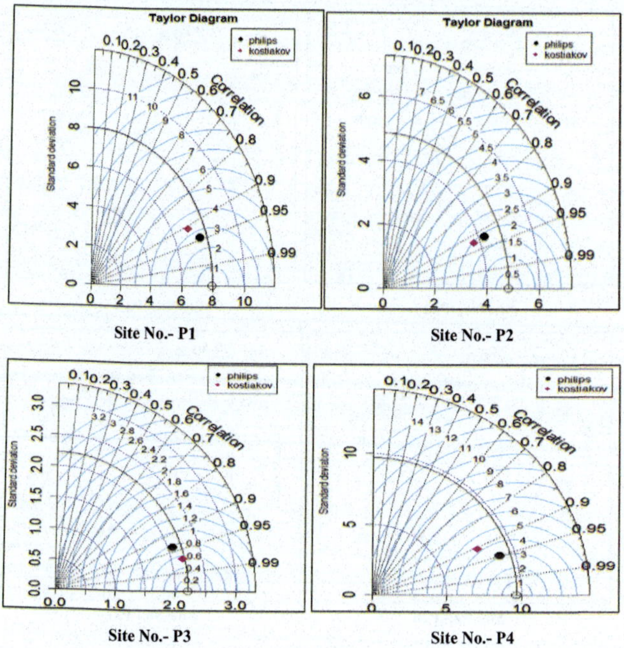

(a)

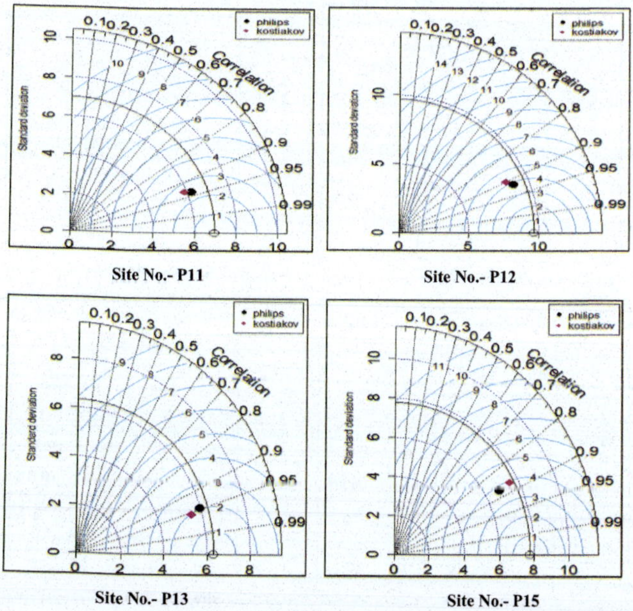

(b)

Fig. 7 **a** Taylors diagrams for Comparison of models in silty clay soil. **b** Taylors diagrams for comparison of models in sandy loam soil. **c** Taylors diagrams for Comparison of models in sandy clay loam soil. **d** Taylors diagrams for Comparison of models in Loamy sand soil

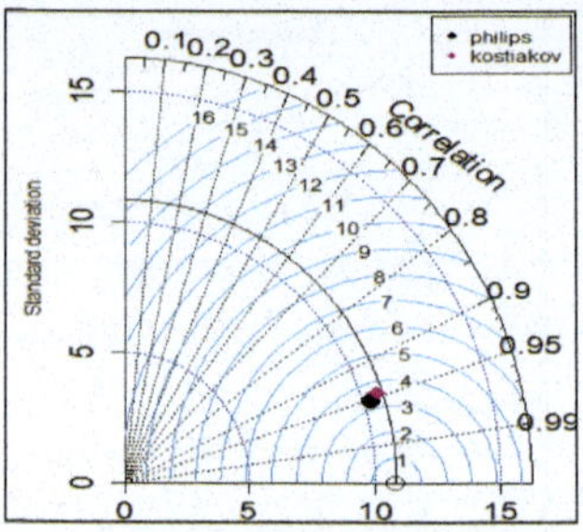

Site No.- P20

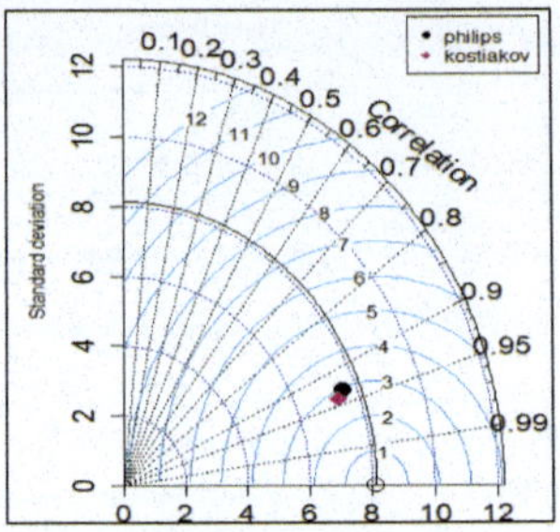

Site No.- P23

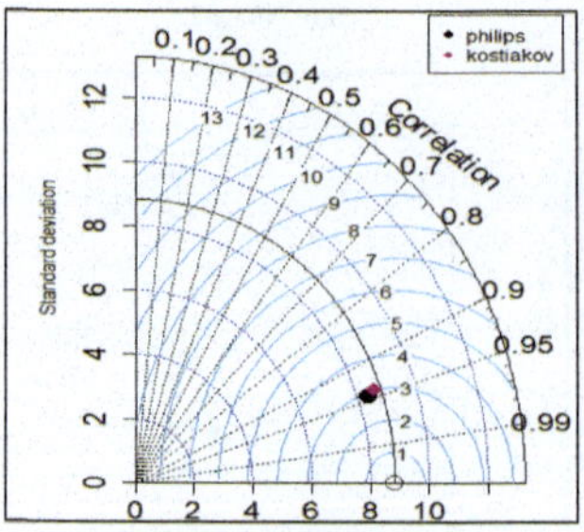

Site No.- P26

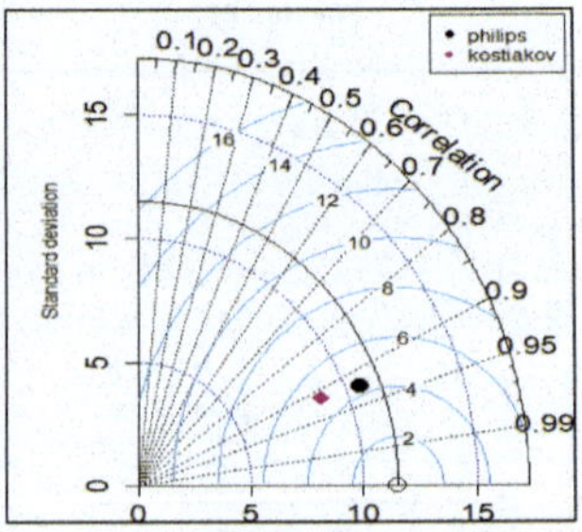

Site No.- P27

(c)

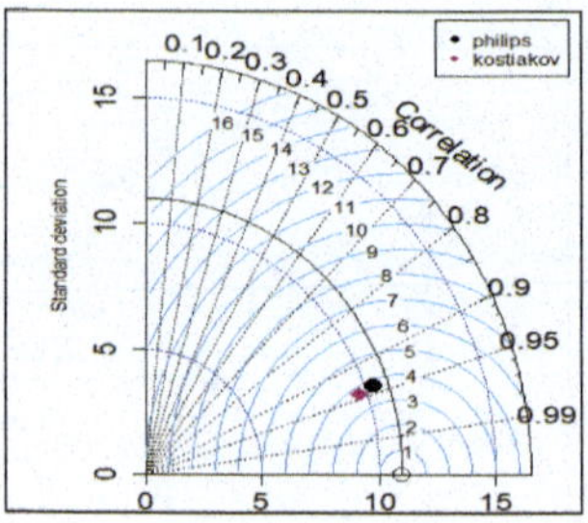

Site No – P28

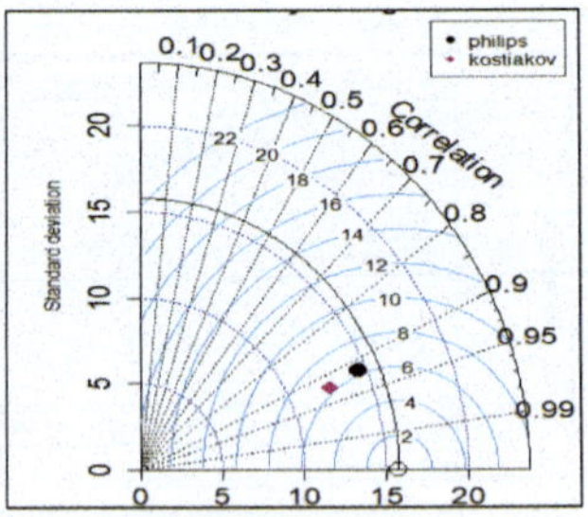

Site No – P29

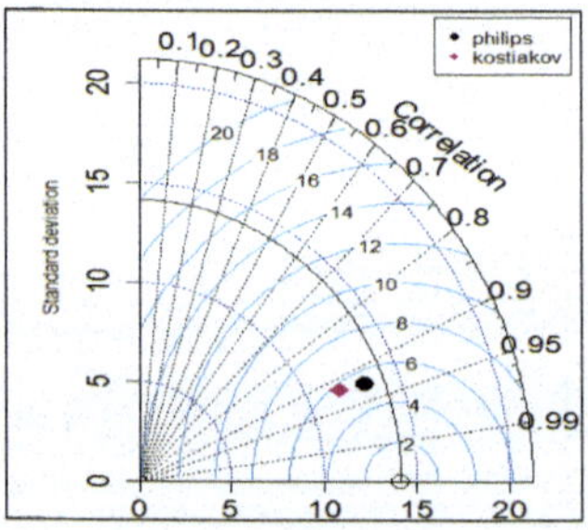

Site No – P33

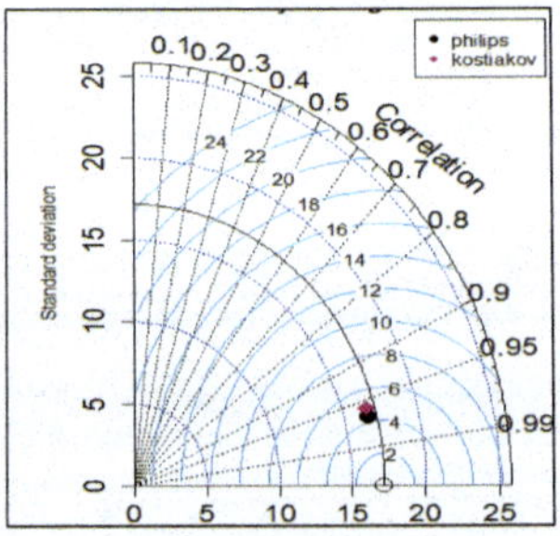

Site No – P34

(d)

Fig. 7 (continued)

in this study area. Further investigations in the direction of finding different soil physical properties which will affect the cumulative infiltration models parameters, applicability of three parameter models in agricultural catchments comprising of both fine textured and coarse textured soils are highly recommended.

References

Al-Azawi SA (1985) Experimental evaluation of infiltration models. J Hydrol (New Zealand), 24(2):77–88

Amami R, Ibrahimi K, Sher F, Milham P, Ghazouani H, Chehaibi S, Hussain Z and Iqbal HM (2021) Impacts of different tillage practices on soil water infiltration for sustainable agriculture. Sustainability, 13(6), p.3155

Dahak A, Boutaghane H, Merabtene T (2022) Parameter estimation and assessment of infiltration models for Madjez Ressoul catchment, Algeria. Water 14(8):1185. https://doi.org/10.3390/w14081185

de Almeida WS, Panachuki E, de Oliveira PTS, da Silva Menezes R, Sobrinho TA, de Carvalho DF (2018) Effect of soil tillage and vegetal cover on soil water infiltration. Soil Tillage Res 175:130–138

Duan R, Fedler CB, Borrelli J (2012) Comparison of methods to estimate saturated hydraulic conductivity in Texas soils with grass. J Irrig Drainage Eng 138(4)

Jha MK, Mahapatra S, Mohan C, Pohshna C (2019) Infiltration characteristics of lateritic vadose zones: field experiments and modelling. Soil Tillage Res 187:219–234

Kostiakov AN (1932) On the dynamics of the coefficient of water-percolation in soils and on the necessity of studying it from a dynamic point of view for purposes of amelioration. In: Trans. 6th cong. international soil science, Russian part A, pp 17–21

Machiwal D, Jha MK, Mal BC (2006) Modelling infiltration and quantifying spatial soil variability in a wasteland of Kharagpur, India. Biosyst Eng 95(4):569–582

Mahapatra S, Jha MK, Biswal S, Senapati D (2020) Assessing variability of infiltration characteristics and reliability of infiltration models in a tropical sub-humid region of India. Sci Rep 10(1):1–18

Mirzaee S, Zolfaghari AA, GorjiM, DyckM, Ghorbani Dashtaki S (2014) Evaluation of infiltration models with different numbers of fitting parameters in different soil texture classes. Arch Agron Soil Sci 60(5):681–693

Ogbe VB, Jayeoba OJ, Ode SO (2011) Comparison of four soil infiltration models on a sandy soil in Lafia, Southern Guinea Savanna Zone of Nigeria. Prod Agric Technol 7(2):116–126

Pande CB, Kadam SA, Jayaraman R, Gorantiwar S, Shinde M (2022) Prediction of soil chemical properties using multispectral satellite images and wavelet transforms methods. J Saudi Soc Agric Sci 21(1):21–28

Pande CB, Moharir KN (2023) Application of hyperspectral remote sensing role in precision farming and sustainable agriculture under climate change: A Review. In: Pande CB, Moharir KN, Singh SK, Pham QB, Elbeltagi A (eds) Climate change impacts on natural resources, ecosystems and agricultural systems. Springer Climate. Springer, Cham. https://doi.org/10.1007/978-3-031-19059-9_21

Philip JR (1957) The theory of infiltration: 1. the infiltration equation and its solution. Soil Sci 83(5):345–358

Sajjadi SAH, Mirzaei M, Nasab AF, Ghezelje A, Tadayonfar G, Sarkardeh H (2016) Effect of soil physical properties on infiltration rate. Geomech Eng 10(6):727–736

Shukla MK, Lal R, Unkefer P (2003) Experimental evaluation of infiltration models for different land use and soil management systems. Soil Sci 168(3):178–191

Singh B, Sihag P, Singh K (2018) Comparison of infiltration models in NIT Kurukshetra campus. Appl Water Sci 8(2):1–8

Thomas AD, Ofosu AE, Emmanuel A, De-Graft AJ, Ayine AG, Asare A, Alexander A (2020) Comparison and estimation of four infiltration models. Open J Soil Sci 10(2):45–57

Investigation of Trends and Variability Associated with the SPI and SPEI as a Drought Prediction Tools in Gujarat Regions, India

Paras Hirapara, Manthan Brahmbhatt, and M. K. Tiwari

Abstract Drought is a well-known yet incredibly difficult to understand hydro-meteorological natural hazard that occurs around the globe as a result of major climate change occurrences. For the central Gujarat region, we examined the drought periodicities during the previous 30 years for this study. The patterns of drought conditions are a sign of climatic and environmental change, and recognizing these trends is crucial for the sustainable management of water resources. Application of the MK test to the first SPI series revealed that the post-monsoon SPI series had a negligible upward trend. The MK test on the original SPEI series indicated several time series with large declining trends prior to the monsoon, whereas every post-monsoon SPEI series displayed an insignificant growing trend. The findings demonstrate that (1) due to the various time series, the SPI and SPEI's identification of the characteristics of drought were quite distinct in space at various timescales, (2) The SPI and SPEI differed most at the shortest time scale, and (3) The drought represented by the two indicators may be consistent over long periods of time. (4) The SPEI may be more suitable than the SPI for drought monitoring in the study region when compared to typical drought occurrence. It should be emphasized that future research will need to examine whether the SPI and SPEI's adaptability varies among regions and historical periods.

Keywords Drought · Standardized Precipitation Index (SPI) · Standardized Precipitation Evapotranspiration Index (SPEI) · MK test · Discrete wavelet transform (DWT)

P. Hirapara (✉)
Junagadh Agricultural University, Junagadh, India
e-mail: parashirapara972@gmail.com

M. Brahmbhatt · M. K. Tiwari
Anand Agricultural University, Anand, India

C. B. Pande et al. (eds.), *Surface and Groundwater Resources Development and Management in Semi-arid Region*, Springer Hydrogeology,
https://doi.org/10.1007/978-3-031-29394-8_5

1 Introduction

Climate change could have a significant and enduring impact on all major natural disasters known to mankind (Krishnamurthy et al. 2014). A perception of climate change can be obtained by examining the long-term trend and variability associated to hydro-climatic variables such as rainfall, temperature, and drought (Rashed et al. 2015; Yerdelen et al. 2021). In general, it is anticipated that drought intensity and frequency would rise due to global warming, decreased precipitation, and similarly increased thermometer readings (Das et al. 2020; Paras et al. 2022). Understanding the main physical processes affecting droughts can help improve current drought monitoring, forecasting, and management systems as well as reduce the vulnerability to and effects of drought. An investigation of changes in drought magnitude, severity, and duration can provide insight into the governing physical and atmospheric phenomena (Joshi et al. 2016; Luhaim et al. 2021; Menna et al. 2022). Drought, one of the most significant hydrometeorological conditions, is typically characterised as a prolonged shortage of freshwater on Earth (Yang et al. 2018; Menna et al. 2022).

Droughts can be classified as meteorological, agricultural, hydrological, or economical in nature. Each has distinctive qualities all of its own. The most severe of them is the meteorological drought, which is most clearly caused by a decrease in precipitation. The effects of the other three forms of drought on individuals and society are more severe. One may contend that the meteorological drought is what causes the other three types of drought. Due to the complexity and severity of drought, defining and evaluating drought features is particularly challenging (Asadi et al. 2015). To evaluate and monitor drought occurrences, numerous drought indices have been developed in recent years. Among these are the composite meteorological drought index, standardised precipitation evapotranspiration index, Palmer drought index, and standardised precipitation index (CI). Two of them, the SPI and SPEI, both have the characteristics of several timeframes, which can represent various types of droughts and more accurately illustrate the variations in drought aspects (Salehnia & Ahn 2022). These indexes are therefore widely used across the globe. Although the concepts behind the SPI and SPEI are similar, the parameters used in each computation are very different (Pande et al. 2023a, b). The SPEI is based on the cumulative difference between precipitation (P) and potential evapotranspiration (PET), which may properly reflect changes in the surface water balance. The SPI solely considers precipitation, which is easy to compute and has strong geographical and temporal flexibility.

However, the increase in evaporation brought on by warming is not trivial for an accurate assessment of drought with global warming (Bera et al. 2021). As a result, the SPEI is substantially better than the SPI in tracking drought, but its application in arid areas may be limited (Ojha et al. 2021). Furthermore, the SPI is still widely used worldwide. Therefore, there is still room for debate regarding the distinctions between the SPI and the SPEI in terms of tracking droughts as well as their geographical applicability in light of overall climate change.

India is one of the world's most susceptible and drought-prone nations (Mishra & Singh 2010). In India, the south-west monsoon, which lasts from June to September, is responsible for 70–90% of the nation's average annual precipitation (Kumar et al. 2013). Failure of the monsoon season could result in a lack of water accessible across the nation, which could cause droughts (Fung et al. 2019). According to the Indian Meteorological Department, a drought year occurs when the seasonal rainfall anomaly averaged throughout the nation as a whole is less than 10% of its long-term norm (IMD).

The standardised precipitation index (SPI), which was created at Colorado State University to measure precipitation deficiencies (Mckee et al. 1993; Asadi et al., 2015), has gained widespread acceptance as a tool for studying droughts (Mundetia & Sharma 2015; Qaisrani et al. 2021). The World Meteorological Department uses the SPI as the index for analyzing agricultural and hydrological drought because of its simplicity, stability, and adaptability (Mundetia & Sharma 2015). Despite SPI's widespread acceptance and use, it ignores other elements that can cause drought, such as temperature, evapotranspiration volume, wind speed, and soil water-holding capacity (Vicente-serrano et al. 2010). Consequently, the SPEI will be used to counteract the benefits of the SPI (Ojha et al. 2021). SPEI is particularly useful for analyzing, observing, and researching how drought is affected by climate change (Ghasemi et al. 2021).

Because its performance is unaffected by the assumptions of uniform data distribution and the requirement for skewed data distribution, the Mann–Kendall test is determined to be durable (Onoz & Bayazit, 2003; Afshar et al., 2022). The MK test has a noteworthy drawback in that it cannot handle data with serial correlation, which is typically present in hydro-climatic data (Salehnia & Ahn 2022). When examining trends in hydroclimatic time series, it's crucial to consider how these changes fluctuate across various time periods in addition to whether the trend direction is increasing or decreasing (Onoz & Bayazit 2003; Yerdelen, et al. 2021). A common method for identifying oscillatory signals is the Fourier Transform (FT), which uses sine- and cosine-based functions (Tefera et al. 2019).

A relatively new method for processing time and frequency domain signals for time series analysis is the wavelet transformation (WT) (Araghi et al. 2015; Mundetia & Sharma 2015). WT is employed to break down time series data into several sub-time series data with various periodicities using various scales and amplitudes (Labat 2005; Wang & Lu 2009). Time series data naturally have physical properties including trends, periodicity, discontinuities, and change points. Wavelet analysis has been shown to be an advanced technique for capturing these properties (Partal 2010; Nikhil raj & Azeez 2012; Araghi et al. 2015). Having tremendous possibilities of SPI and SPEI indices for drought assessment, these Due to the climate change and its variability, which could be accounted to the undulating topography of the research area, and variation in drought ratings from SPI and SPEI is expected (Bera et al. 2021). As a result, this study looked at the level of agreement between the SPI and SPEI ratings as instruments for assessing the drought in the Gujarat region.

This chapter has been focus on four main sections, which includes the introduction as 1st Sec. In Sect. 2 we present the materials and methods used in this study, while in Sect. 3 we present detail results of our study. In Sect. 4 we present the main conclusions and outcomes of the research objective.

2 Materials and Methods

2.1 Data and Methods

The main region of analysis for this study is Middle Gujarat, but for most of the analyses employed through the paper we split the Middle Gujarat domain into three separate micro-regions such as Dahod, Chhota Udaipur, Vadodara. Monthly rainfall data of 17 Grids under study area of 30 years (1986–2015) is used for calculate the seasonal and annual SPI drought index (Table 1 and Fig. 1). Monthly precipitation and mean temperature are used for calculate seasonal and annual SPEI drought index for 42 Grids. The data quality was rigorously monitored, and the average values of the data from the adjacent meteorological stations were used to fill in any missing values. We acquired the gridded dataset (Precipitation & Temperature) from the Climate Research & Services, Pune (IMD, Pune Lab) (https://www.imdpune.gov.in/). For detection of seasonal SPI and SPEI, four seasons are considered as Winter (December to February), Pre-monsoon (March to May), Monsoon (June to August) and Post-monsoon (September to November).

2.2 Description of Drought Indices

2.2.1 Standardized Precipitation Index (SPI)

The SPI is a popularly used drought index that is based only on rainfall measurements and may be summed up as the likelihood of precipitation being seen at a particular time range. Regardless of climate and land use, this probabilistic measure is independent and can be applied in various regions. Applying the following equation, the data were initially fitted to the gamma probability distribution function and transformed into a normal distribution:

If the amount of precipitation fell during a specific time period was x, the G distribution of the probability density function was as follows:

$$g(x) = \frac{1}{\beta^{\alpha}\tau(\alpha)}x^{\alpha-1}e^{\frac{-x}{\beta}},\ for x > 0 \quad (1)$$

where, x is the monthly accumulated precipitation amount, and α and β are the shape and scale parameters of gamma distribution, respectively.

Table 1 Study area grids location details

Grid code		Lat (°N)	Long (°E)	District
For SPI	For SPEI	From—To	From—To	
A01	AE01	22.00–22.25	72.75–73.00	Vadodara
A02	AE02	22.00–22.25	73.00–73.25	Vadodara
A03	AE03	22.00–22.25	73.25–73.50	Vadodara
A04	AE04	22.00–22.25	73.50–73.75	Chhota Udepur
A05	AE05	22.00–22.25	73.75–74.00	Chhota Udepur
A06	AE06	22.00–22.25	74.00–74.25	Chhota Udepur
A07	AE07	22.25–22.50	73.00–73.25	Vadodara
A08	AE08	22.25–22.50	73.25–73.50	Vadodara
A09	AE09	22.25–22.50	73.75–74.00	Chhota Udepur
A10	AE10	22.25–22.50	74.00–74.25	Chhota Udepur
A11	AE11	22.50–22.75	73.25–73.50	Vadodara
A12	AE12	22.50–22.75	73.75–74.00	Dahod
A13	AE13	22.50–22.75	74.00–74.25	Dahod
A14	AE14	22.75–23.00	73.75–74.00	Dahod
A15	AE15	22.75–23.00	74.00–74.25	Dahod
A16	AE16	23.00–23.25	74.00–74.25	Dahod
A17	AE17	23.25–23.50	74.00–74.25	Dahod

The Z or SPI values is more easily obtained computationally using an approximation that converts cumulative probability to the standard normal random variable Z;

$$\boldsymbol{Z} = \boldsymbol{SPI} = -\left(\boldsymbol{t} - \frac{\boldsymbol{C}_0 + \boldsymbol{C}_1\boldsymbol{t} + \boldsymbol{C}_2\boldsymbol{t}^2}{1 + \boldsymbol{d}_1\boldsymbol{t} + \boldsymbol{d}_2\boldsymbol{t}^2 + \boldsymbol{d}_3\boldsymbol{t}^3}\right)\text{for } 0 < \boldsymbol{H}(\boldsymbol{x}) \leq 0.5 \quad (2)$$

$$\boldsymbol{Z} = \boldsymbol{SPI} = +\left(\boldsymbol{t} - \frac{\boldsymbol{C}_0 + \boldsymbol{C}_1\boldsymbol{t} + \boldsymbol{C}_2\boldsymbol{t}^2}{1 + \boldsymbol{d}_1\boldsymbol{t} + \boldsymbol{d}_2\boldsymbol{t}^2 + \boldsymbol{d}_3\boldsymbol{t}^3}\right)\text{for } 0.5 < \boldsymbol{H}(\boldsymbol{x}) < 1 \quad (3)$$

where, $\boldsymbol{t} = \sqrt{\boldsymbol{ln}\left(\frac{1}{(\boldsymbol{H}(\boldsymbol{x}))^2}\right)}$ and $\boldsymbol{t} = \sqrt{\boldsymbol{ln}\left(\frac{1}{(1-\boldsymbol{H}(\boldsymbol{x}))^2}\right)}$ are for $0 < \boldsymbol{H}(\boldsymbol{x}) \leq 0.5$ and $0.5 < \boldsymbol{H}(\boldsymbol{x}) < 1$, , respectively. (The constants are $\boldsymbol{C}_0 - 2.515517$, $\boldsymbol{C}_1 = 0.802853$, $\boldsymbol{C}_2 = 0.010328$, $\boldsymbol{d}_1 = 1.432788$, $\boldsymbol{d}_2 = 0.189269$, and $\boldsymbol{d}_3 = 0.001308$).

Values from SPI indicate both dry and wet situations. Extreme wet (> 2), severely wet (1.5 to 1.99), moderately wet (1.00 to 1.49), slightly wet (0 to 0.99), mildly dry (0 to -0.990), moderately dry (−1.00 to -1.49), severely dry (−1.5 to −1.99), and extreme dry (–2.00) are the eight classifications that the SPI drought portion is randomly divided. When SPI approaches 0.0, a drought event begins, and it ends when SPI turns positive (World Meteorological Organization, 2012).

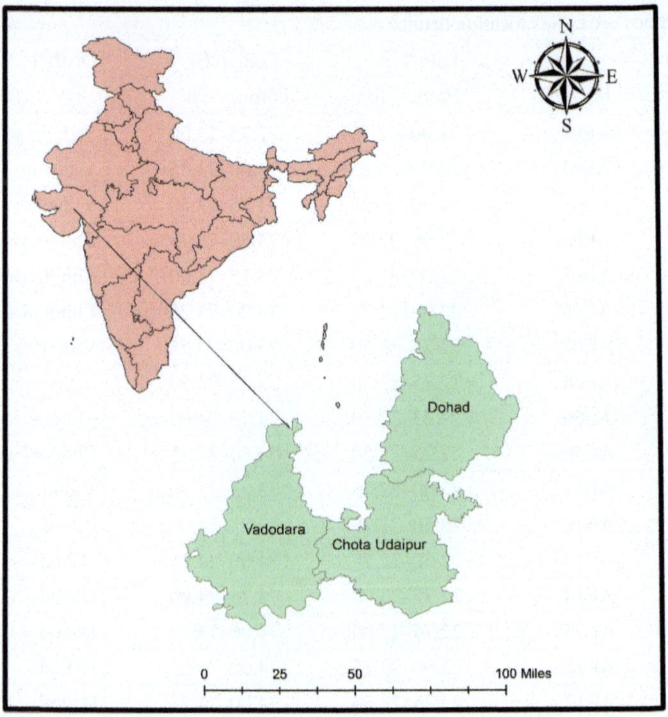

Fig. 1 A Geographical location map of the study location

2.2.2 Standardized Precipitation Evapotranspiration Index

Vicente Serrano and associates suggested the SPEI in 2010 as an improvement of the SPI. Based on the water balance principle, the SPEI assesses the dry and wet conditions of the region using the difference between precipitation (P) and potential evapotranspiration (PET) (Tirivarombo et al. 2018). The SPEI has a significant advantage over other commonly used drought indices that take the impact of PET on drought severity into account because its multi-scalar properties make it possible to identify various drought types and their effects in the context of global warming.

Using a discrepancy between precipitation and evapotranspiration from the average condition, the SPEI is used to detect drought in a region (Vicente-serrano et al. 2010). The Penmen Monteith (PM) equation is used to calculate PET (Allen et al., 1998) as follows;

$$\boldsymbol{ETo} = \frac{0.408\Delta(\boldsymbol{Rn} - \boldsymbol{G}) + \boldsymbol{\gamma}\left[\frac{900}{T+273}\right]\boldsymbol{u}_2(\boldsymbol{e}_s - \boldsymbol{e}_a)}{\Delta + \boldsymbol{\gamma}(1 + 0.34\boldsymbol{u}_2)} \quad (4)$$

where, ETo represents evapotranspiration (mm/day); Δ = saturated vapor pressure slope (kPa/°C); G = heat flux density of soil (MJ/m^2 /day); Rn = net radiation

(MJ/m^2 per day); T = mean temperature (°C); u_2 = average daily speed of wind (m/s); es − ea = deficit of vapor pressure; γ = psychrometric constant (kPa/°C).

After a standard normal distribution process, the $\boldsymbol{SPEI}$ can be obtained by below equation:

$$\boldsymbol{SPEI} = \boldsymbol{W} - \frac{\boldsymbol{C}_0 + \boldsymbol{C}_1\boldsymbol{W} + \boldsymbol{C}_2\boldsymbol{W}}{1 + \boldsymbol{d}_1\boldsymbol{W} + \boldsymbol{d}_2\boldsymbol{W} + \boldsymbol{d}_3\boldsymbol{w}'}$$

$$\boldsymbol{W} = \sqrt{-2\ln(\boldsymbol{P})} \tag{5}$$

When, $\boldsymbol{P \le 0.5, P = 1 - F(x); P > 0.5, P = 1 - P}$. ($\boldsymbol{C}_0$ = 2.515517, $\boldsymbol{C}_1$ = 0.802853, $\boldsymbol{C}_2$ = 0.010328, $\boldsymbol{d}_1$ = 1.432788, $\boldsymbol{d}_2$ = 0.189269, and $\boldsymbol{d}_3$= 0.001308 (Vicente-serrano et al. 2010).

2.3 *Mann–Kendall (MK) Trend Tests*

The Mann Kendall (MK) statistical test is used to detect trends in hydrological time series data that have been recorded (precipitation, runoff, water quality, temperature). Analysis of time series data trends has proven to be a useful technique for efficient management, planning, and design of water resources (Mallick et al. 2021). For each study location, the MK test and its variance were identified in order to derive the test's standard normal value (Z). The essential two-tailed Z-value (area under the normal curve) corresponding to the significant level of α /2 (this study used = 5%) then was compared to the absolute value of this Z. (Ali et al. 2019). The Z values for α of 5% in a two-tailed test are ± 1.96. If the Z-value obtained from the MK calculation is found outside the -1.96 and + 1.96 boundaries, then this indicates that the trends detected are significant.

The calculation of the MK test statistic, which is also known as the Kendall's tau, is as follows (Yue et al. 2002):

$$\boldsymbol{S}_t = \sum\nolimits_{c=1}^{n-1}\sum\nolimits_{d=c+1}^{n} \boldsymbol{sign}(\boldsymbol{X}_d - \boldsymbol{X}_c) \tag{7}$$

where, $\boldsymbol{X}_c$ and $\boldsymbol{X}_d$ are data points, and $\boldsymbol{n}$ is the number of the dataset.

$\boldsymbol{sign}(\boldsymbol{t})$ is defined as:

$$\boldsymbol{sign}(\boldsymbol{X}_d - \boldsymbol{X}_c) = \begin{cases} +1, \boldsymbol{X}_d > \boldsymbol{X}_c \\ 0, \boldsymbol{X}_d = \boldsymbol{X}_c \\ -1, \boldsymbol{X}_d < \boldsymbol{X}_c \end{cases} \tag{8}$$

The Mann–Kendall statistic $\boldsymbol{Z}$ is given as:

$$Var(S_t) = (n(n-1))(2n+5) - \sum\nolimits_{c=1}^{n} t_c(c)(c-1)(2c+5)/18$$

$$\boldsymbol{Z} = \begin{cases} (\boldsymbol{S_t} - 1)/\sqrt{\boldsymbol{Var}(\boldsymbol{S_t})}, \boldsymbol{S_t} > 0 \\ 0, \boldsymbol{S_t} < 0 \\ (\boldsymbol{S_t} + 1)/\sqrt{\boldsymbol{Var}(\boldsymbol{S_t})}, \boldsymbol{S_t} = 0 \end{cases} \quad (9)$$

where, $\boldsymbol{t_c}$ indicates the cumulative sum of $\boldsymbol{t}$, which is the size of ties or duplicates of the extent $\boldsymbol{c}$.

2.4 *Magnitude of Trend Using Sen's Slope Estimator*

In order to determine the magnitude and variability of the trend in time series of SPI and SPEI Sen's slope estimator was used (Ali et al. 2019) with the following set of equations:

The slope of the dataset is obtained from:

$$m_i = (x_j - x_k)/(j - k)$$

$$(i = 1, 2, \ldots, N) \quad (10)$$

where, $\boldsymbol{m_i}$, $\boldsymbol{N}$, $\boldsymbol{x_j}$, and $\boldsymbol{x_k}$ denote the slope, the number of data points and $\boldsymbol{j}$ and $\boldsymbol{k}$ $(\boldsymbol{j} > \boldsymbol{k})$ represent the time points, respectively.

The median of these $\boldsymbol{N}$ value of $\boldsymbol{m_i}$ is termed as Sen's estimator of slopes and is calculated by the following formulae:

$$\beta = \begin{cases} m_{(N+1)/2}, & when\ N\ is\ odd \\ 1/2(m_{N/2} + m_{(N+1)/2}, & when\ N\ is\ even \end{cases} \quad (11)$$

A positive value means an increasing or upward trend, and a negative value means a decreasing or downward trend (Jain et al. 2013; Agarwal et al. 2021).

3 Results and Discussion

3.1 *Temporal Evolution of the SPI and SPEI Analysis*

3.1.1 Pre Monsoon

Since the turn of the century, the frequency and severity of droughts in the middle Gujarat region have increased, with the tendency becoming more obvious the further back in time one goes. In this study, we examined the differences between the pre-monsoon (May) and post-monsoon seasons using 3-month SPI & SPEI values

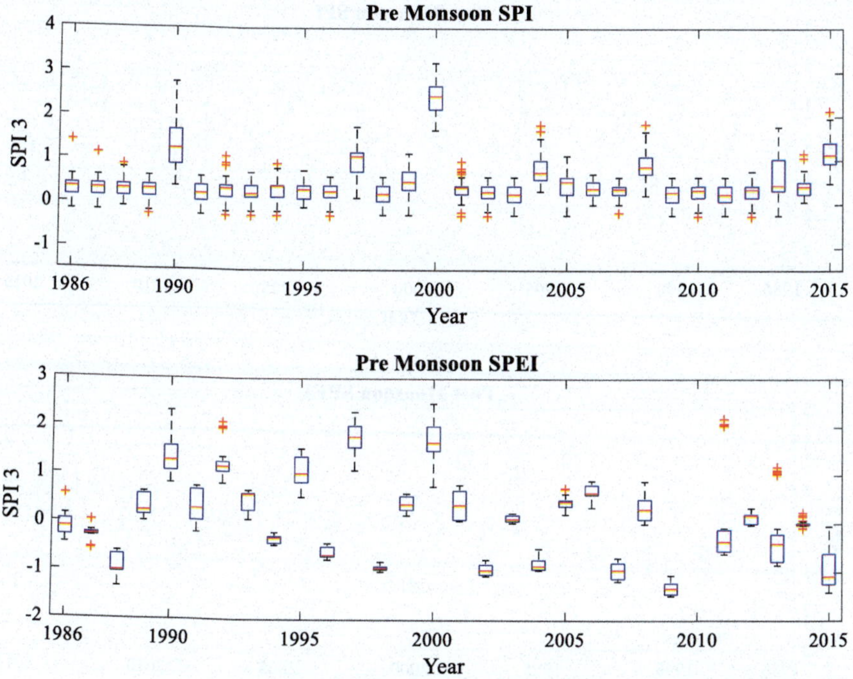

Fig. 2 The time series of the pre monsoon SPI and SPEI in the study region during 1986–2015

(November). The 12-month SPI values (SPI-12) and 12-month SPEI values (SPEI-12) for December are used to detect annual SPI and SPEI values (Yang et al. 2018; He et al. 2015). The temporal distribution of SPI & SPEI for pre monsoon are clearly indicated in Fig. 2.

The SPI based analysis indicates that pre monsoon indices from 3.22 to -0.25 (Extremely wet to Mildly dry state), while SPEI vary between 2.49 to – 4.39 which means extremely wet to extremely dry state. At 3-month time scale, strongest wet year identify by SPI & SPEI were 2000. Unlikely for the dentification of extremely driest year found by SPEI is 2010. Temporal distribution of SPI & SPEI are clearly indicated in Fig. 2 and it reviled that the maximum drought occurrence at 3-month time scale is highest in SPEI than SPI. Less fluctuation of extreme wet to extreme dry observed in SPI compared to SPEI, which means it give linear response-based rainfall data. While SPEI give quite good response by combination of rainfall and evapotranspiration data.

3.1.2 Post Monsoon

Figure 3 shows that post monsoon SPI & SPEI drought apical severity vary from 2.51 to –3.08 and 2.42 to –2.40 (Extremely wet to Extremely dry) respectively. In SPI

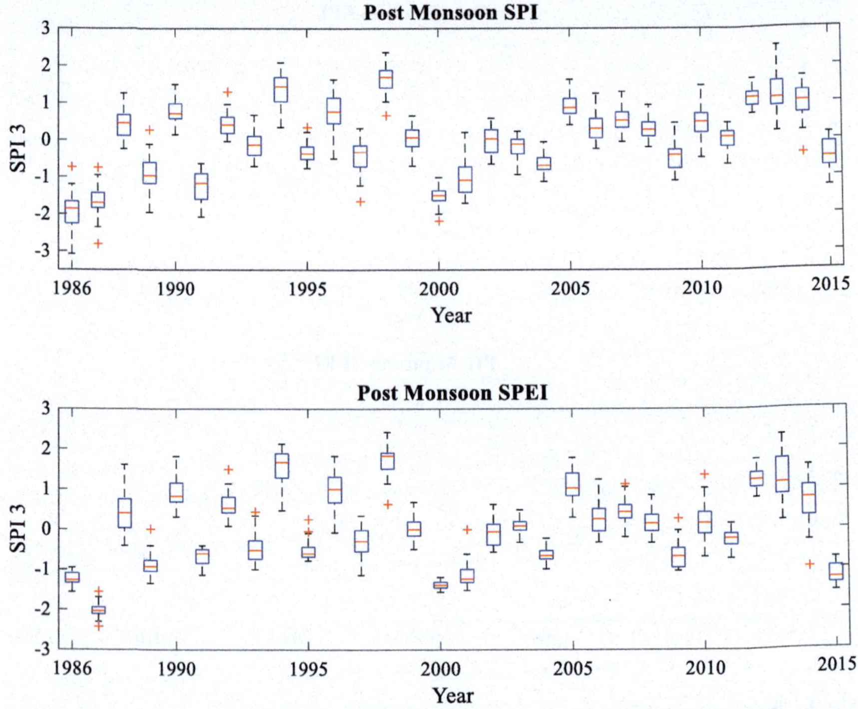

Fig. 3 The time series of the post monsoon SPI and SPEI in the study region during 1986–2015

identify year 2013 (Gride A05) is an extremely wet year while 1998 (AE10) year is based on SPEI. 3-month time scale SPI found 1986 is extremely dry drought severity while in SPEI 1987 year is extremely dry drought severity and intensity respectively.

Additionally, SPEI performed well in capturing the drought in 3-month time scale. Figure 2 shows that the SPEI identify higher draught year in all year compared to SPI. Regardless of their difference, however SPI and SPEI able to identify major recorded drought year in the study area including 1986, 1987, 2000, & 2015 in most cases. The SPI and SPEI were essentially consistent in representing drought periods, but the SPI's reflection of the severity of the drought was higher than the SPEI's.

3.2 Serial Correlation and Seasonality Factors

Here we used lag-1 autocorrelation for assessment of seasonality patterns, and it is applied to each and every SPI and SPEI data sets. Figures 4, 5 illustrate the lag-1 correlation graph for pre monsoon and post monsoon SPI. The frequency of statistical significance correlation at 0.4 and –0.4 levels is marked with red and grey color respectively (Figs. 4, 5) and investigated time scale of the analysis indicated

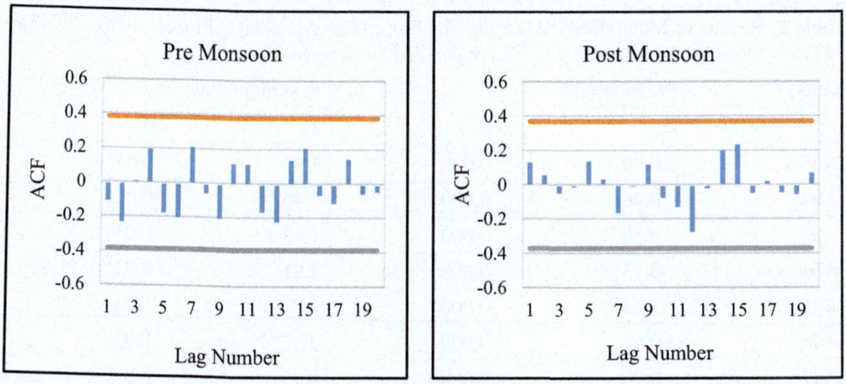

Fig. 4 Lag-1 correlograms for pre & post monsoon SPI

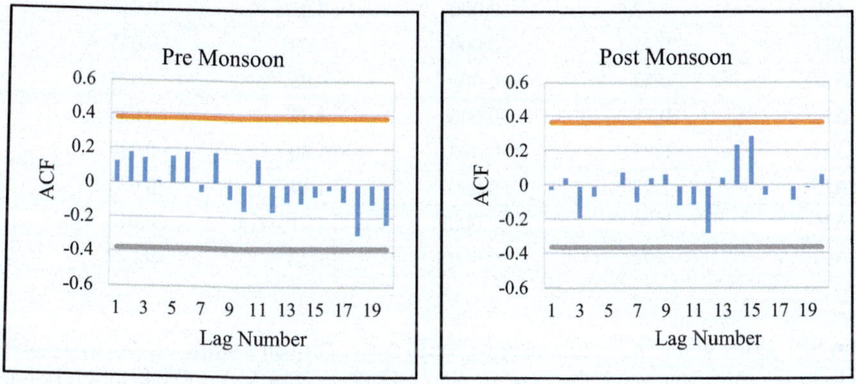

Fig. 5 Lag-1 correlograms for pre & post monsoon SPEI

an acceptable degree of agreement between SPI & SPEI. Correlograms for whole dataset was also prepared and we observed that no any significant seasonality pattern follows by SPI pre and post monsoon time series data set.

3.3 *Seasonal and Annual Trends in SPI*

MK test is employed to analyze the trend for SPI and SPEI drought indices of all stations over the period of 1986 to 2015. This is done with 3-month cumulative time series data and 5% significance level. Results of the Mann Kendall trend test and Sen's slope for 3-month time scale seasons of SPI & SPEI are exhibited in Tables 2, 3. Pre monsoon SPI series significantly increases in its trend in most of the grids, only A05, A06 indicates negative trend which means trend of these two grids are not at its significant level. In pre monsoon Sen's slope estimator indicates no change

Table 2 Results of Mann–Kendall (Z) and Sen Slope (β) for seasonal SPI from 1986 to 2015 for 17 Grids. (Significant values at α = 5% are denoted by an asterisk)

Grid	Pre monsoon		Post monsoon	
	Z	β	Z	β
A01	0.90	0.000	1.94	0.045
A02	0.46	0.000	0.46	0.011
A03	0.00	0.000	1.52	0.036
A04	0.13	0.000	1.21	0.031
A05	−1.21	0.000	1.87	0.045
A06	−0.61	0.000	2.48*	0.049
A07	0.98	0.000	0.95	0.028
A08	0.63	0.000	1.93	0.046
A09	1.87	0.000	1.70	0.036
A10	1.22	0.000	2.03*	0.045
A11	0.44	0.000	2.09*	0.045
A12	0.52	0.000	0.95	0.024
A13	0.75	0.000	1.37	0.031
A14	0.83	0.000	1.30	0.038
A15	2.22*	0.000	1.14	0.022
A16	0.86	0.000	1.71	0.043
A17	1.99*	0.000	1.82	0.029

on SPI series while in post monsoon all SPI series for sen's slope are positive trend. Highest significant tread observed in A06 grid (Z = 2.48 & β = 0.049). Sen's slope indicates weighted median of the difference in all values. Here is a simplest possibility illustration of zero is shorter upward trending segments.

Seasonal SPEI trend series period of 1986 to 2015 for 17 different grids into the study area (Table 3). Based on results, it found that the pre monsoon were significant decreasing trend in the study areas. On other hand post monsoon analysis indicates insignificantly increasing trend of SPEI over the study locations. Sen's slope indicates difference between major and minor value is significant in both pre monsoon and post monsoon. SPEI capture tall the major drought years between 1986 to 2015 at time scale, while SPI fail to capture extreme drought events (Tables 2, 3). SPI value appears little lower than the SPEI values. This type of dissimilarity observed in our study and in other studies also. Wang et al. 2014 reported difference between SPI and SPEI based drought assessment. SPI predict extreme event based only precipitation data, while SPEI uses precipitation and temperature variability to predict climate extreme events (Mckee et al. 1993; Guttmann 1999; Vicente, et al. 2010). However, difference between SPI and SPEI values doesn't mean that they give complete opposite results. In lower temperature variation regions, SPI can give equal results as strong as SPEI

Table 3 Results of Mann–Kendall (Z) and Sen Slope (β) for seasonal SPEI from 1986 to 2015 for 17 Grids. (Significant values at $\alpha = 5\%$ are denoted by an asterisk)

Grid	Pre monsoon		Post monsoon	
	Z	β	Z	β
AE01	−2.39*	−0.043	1.23	0.028
AE02	−2.34*	−0.044	0.23	0.006
AE03	−2.44*	−0.042	0.68	0.017
AE04	−2.48*	−0.042	0.79	0.023
AE05	−0.05	−0.003	1.15	0.034
AE06	0.12	0.005	1.93	0.041
AE07	−2.41*	−0.042	0.14	0.007
AE08	−2.39*	−0.041	1.18	0.037
AE09	0.00	0.000	1.53	0.031
AE10	0.00	0.001	1.12	0.019
AE11	−2.37*	−0.041	1.28	0.032
AE12	0.05	0.003	1.84	0.032
AE13	0.00	0.004	0.70	0.014
AE14	−0.95	−0.025	1.09	0.030
AE15	−1.07	−0.026	0.91	0.024
AE16	−0.98	−0.027	0.98	0.018
AE17	−0.95	−0.019	1.03	0.019

does. This indicates inability of SPI to count the effect of global warming in drought monitoring.

Sen's slope results agreed with (Dogan 2018), which concluded that the good relationship can be expected if scattering of points is diminished, and lie relatively close to line which represent mean. Here, existed good level of agreement for most of scattering points lie between set range of upper and lower limit.

3.4 Linear Relationship Between SPI and SPEI for Pre and Post Monsoon Reasons

Using Pearson's correlation coefficient, it was found that SPI and SPEI have a linear relationship for different seasons. The results showed that the pre- and post-monsoon seasons did not yield significant results for the SPI and SPEI. Pre and post monsoon SPI & SPEI correlation is displayed using R^2, which has values of 0.1928 & 0.1298, respectively. Scattered plot revealed that the relationship between SPI and SEPI and time was linearly decrementing (Fig. 6). This is partially due to the fact that the outcomes of the two distinct approaches do not agree, which could lead to a

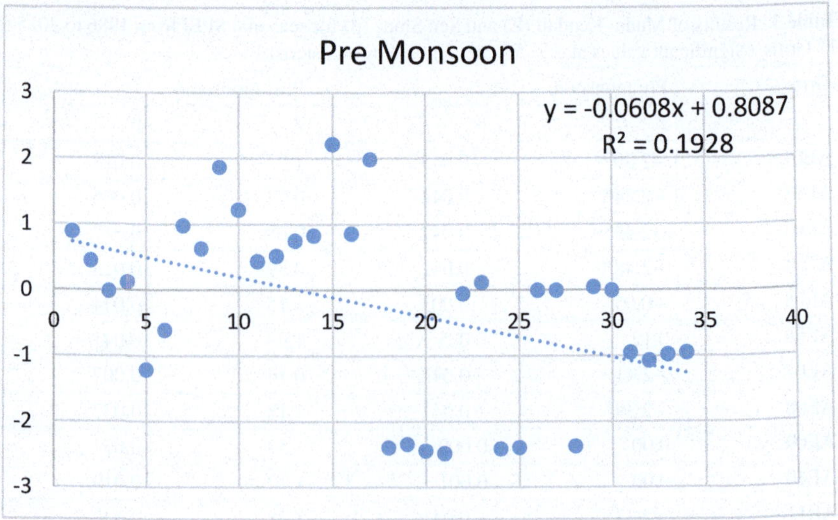

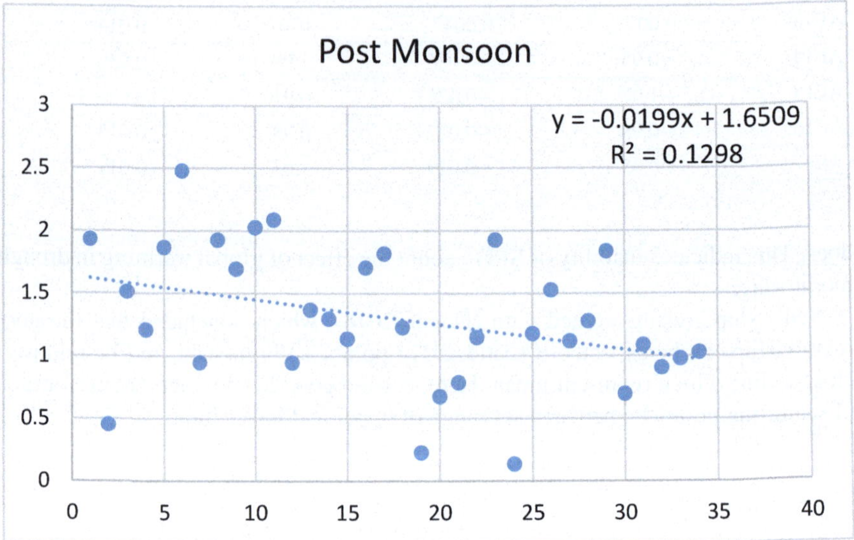

Fig. 6 Scatter plot showing the linear relationship of SPI & SPEI for all grids in pre monsoon & post monsoon season

variety of conclusions that are either good or negative. This was supported by Dogan NO 2018, which cautioned against using correlation to gauge the comparability of different methodologies. The results clearly demonstrate that SPI and SPEI have a linear connection in this case, however there is not necessarily a degree of agreement. Furthermore, the independent Sen's slop test clearly demonstrates that the SPI and SPEI in this study do not differ significantly from one another.

4 Conclusion

SPI and SPEI indicators were utilized in the current study to analyze the spatial and temporal characteristics of climate extreme occurrences that occurred in Gujarat, India, before during and after the monsoon season. Drought is a devastating and slow-moving event that is influenced by both human and natural activity, in contrast to other climatic calamities. The number of severe occurrences in the study area was trended using both Sen's slope and the MK test. We can therefore infer from the results of this study that SPEI outperformed SPI in the Vadodara regions of Gujarat. However, the research for additional study locations also showed that SPEI can be used in place of SPI at time scales. SPI cannot be calculated in the absence of temperature data and a suitable analytical instrument; hence it is reasonable to assume that SPI can be used to measure drought in the research region at all-time scales.

This study's focus was on meteorological conditions based historical drought indicators. In order to make well-informed decisions for sustainable basin planning and management and to maximize the operational guidelines of existing water resources, trends in drought indices would be projected to future periods based on the anticipated outcomes from global climate models. Future studies that consider hydrologic, agricultural, and socioeconomic droughts will produce more insightful findings.

References

Afshar MH, Bulut B, Duzenli E, Amjad M, Yilmaz MT (2022) Global spatiotemporal consistency between meteorological and soil moisture drought indices. Agric for Meteorol 316:108848

Agarwal S, Suchithra AS, Singh SP (2021) Analysis and interpretation of rainfall trend using Mann-Kendall's and Sen's slope method. Indian J. Ecol 48:453–457

Ali R, Kuriqi A, Abubaker S, Kisi O (2019) Long-term trends and seasonality detection of the observed flow in Yangtze River using Mann-Kendall and Sen's innovative trend method. Water 11(9):1855

Araghi A, Baygi MM, Adamowski J, Malard J, Nalley D, Hasheminia SM (2015) Using wavelet transforms to estimate surface temperature trends and dominant periodicities in Iran based on gridded reanalysis data. Atmos Res 155:52–72

Asadi Zarch MA, Sivakumar B, Sharma A (2015) Droughts in a warming climate: a global assessment of standardized precipitation index (SPI) and reconnaissance drought index (RDI). J Hydrol 526:183–195

Bera B, Shit PK, Sengupta N, Saha S, Bhattacharjee S (2021) Trends and variability of drought in the extended part of Chhota Nagpur plateau (Singbhum Protocontinent), India applying SPI and SPEI indices. Environ Challs 5:100310

Das S, Ghosh A, Hazra S, Ghosh T, de Campos RS, Samanta S (2020) Linking IPCC AR4 & AR5 frameworks for assessing vulnerability and risk to climate change in the Indian Bengal Delta. Prog Disaster Sci 7:100110

Dogan NO (2018) Bland-Altman analysis: a paradigm to understand correlation and agreement. Turk J Emerg Med 18(4):139–141

Fung KF, Huang YF, Koo CH (2019) Coupling fuzzy–SVR and boosting–SVR models with wavelet decomposition for meteorological drought prediction. Environ Earth Sci 78(24):1–18

Ghasemi P, Karbasi M, Nouri AZ, Tabrizi MS, Azamathulla HM (2021) Application of Gaussian process regression to forecast multi-step ahead SPEI drought index. Alex Eng J 60(6):5375–5392
Guttmann NB (1999) Accepting the standardized precipitation index: a calculation algorithm 1. J Am Water Resour Assoc 35(2):311–322
He Y, Ye J, Yang X (2015) Analysis of the spatio-temporal patterns of dry and wet conditions in the Huai River Basin using the standardized precipitation index. Atmos Res 166:120–128
Jain SK, Kumar V, Saharia M (2013) Analysis of rainfall and temperature trends in northeast India. Int J Climatol 33(4):968–978
Joshi S, Li Y, Kalwani RM, Gold JI (2016) Relationships between pupil diameter and neuronal activity in the locus coeruleus, colliculi, and cingulate cortex. Neuron 89(1):221–234
Krishnamurthy PK, Lewis K, Choularton RJ (2014) A methodological framework for rapidly assessing the impacts of climate risk on national-level food security through a vulnerability index. Glob Environ Chang 25:121–132
Kumar KN, Rajeevan M, Pai DS, Srivastava AK, Preethi B (2013) On the observed variability of monsoon droughts over India. Weather Clim Extrem 1:42–50
Labat D (2005) Recent advances in wavelet analyses: Part 1. A review of concepts. J Hydrol, 314 (1–4), 275–288.
Luhaim ZB, Tan ML, Tangang F, Zulkafli Z, Chun KP, Yusop Z, Yaseen ZM (2021) Drought variability and characteristics in the muda river basin of Malaysia from 1985 to 2019. Atmosphere 12(9):1210
Mallick J, Talukdar S, Alsubih M, Salam R, Ahmed M, Kahla NB, Shamimuzzaman M (2021) Analysing the trend of rainfall in Asir region of Saudi Arabia using the family of Mann-Kendall tests, innovative trend analysis, and detrended fluctuation analysis. Theoret Appl Climatol 143(1):823–841
Mckee TBT, Doesken NJNJ, Kleist J (1993) The relationship of drought frequency and duration to time scales. Eighth conference on applied climatology, 17(22). Am Meteorol Soc, Boston, pp 179–183
Menna BY, Mesfin HS, Gebrekidan AG, Siyum ZG, Tegene MT (2022) Meteorological drought analysis using copula theory for the case of upper Tekeze river basin, Northern Ethiopia. Theor Appl Climatol, 1–18
Mishra AK, Singh VP (2010) A review of drought concepts. J Hydrol 391(1–2):202–216
Mundetia N, Sharma D (2015) Analysis of rainfall and drought in Rajasthan State. India. Glob NEST J 17(1):12–21
Nikhil Raj PP, Azeez PA (2012) Trend analysis of rainfall in Bharathapuzha river basin, Kerala. India. Int J Climatol 32(4):533–539
Ojha SS, Singh V, Roshni T (2021) Comparison of meteorological drought using SPI and SPEI. Civ Eng J 7(12):2130–2149
Onoz B, Bayazit M (2003) The power of statistical tests for trend detection. Turk J Eng Environ Sci 27(4):247–251
Pande CB, Costache R, Sammen SS et al (2023a) Combination of data-driven models and best subset regression for predicting the standardized precipitation index (SPI) at the Upper Godavari Basin in India. Theor Appl Climatol 152:535–558. https://doi.org/10.1007/s00704-023-04426-z
Pande CB, Kushwaha NL, Orimoloye IR et al (2023b) Comparative assessment of improved SVM method under different Kernel functions for predicting multi-scale drought index. Water Resour Manage 37:1367–1399. https://doi.org/10.1007/s11269-023-03440-0
Paras H, Sanjay P, Vaibhav R (2022) The significance impact assessment of morphological parameters on watershed: a review. Agric Rev 43(1):110–115
Partal T (2010) Wavelet transform-based analysis of periodicities and trends of Sakarya basin (Turkey) streamflow data. River Res Appl 26(6):695–711
Qaisrani ZN, Nuthammachot N, Techato K (2021) Drought monitoring based on standardized precipitation index and standardized precipitation evapotranspiration index in the arid zone of Balochistan province. Pakistan. Arab J Geosci 14(1):1–13

Rashed M, Idris Y, Shaban M (2015) Integrative approach of GIS and remote sensing to represent the hydrogeological and hydro chemical conditions of Wadi Qena Egypt. In the 2nd International Conference On Water Resources And Arid Environment, Saudi Arabia. pp 26–29
Salehnia N, Ahn J (2022) Modelling and reconstructing tree ring growth index with climate variables through artificial intelligence and statistical methods. Ecol Ind 134:108496
Tefera AS, Ayoade JO, Bello NJ (2019) Comparative analyses of SPI and SPEI as drought assessment tools in Tigray Region. Northern Ethiopia. SN Applied Sciences 1(10):1–14
Tirivarombo S, Osupile D, Eliasson P (2018) Drought monitoring and analysis: Standardized Precipitation Evapotranspiration Index (SPEI) and Standardized Precipitation Index (SPI). Phys Chem Earth Parts 106:1–10
Vicente-Serrano SM, Beguería S, López-Moreno JI (2010) A multiscalar drought index sensitive to global warming: the standardized precipitation evapotranspiration index. J Clim 23(7):1696–1718
Wang N, Lu C (2009) Two-dimensional continuous wavelet analysis and its application to meteorological data. J Atmos Oceanic Tech 27(4):652–666
Wang W, Zhu Y, Xu R, Liu J (2014) Drought severity change in China during 1961–2012 indicated by SPI and SPEI. Nat Hazards 75(3):2437–2451
World Meteorological Organization (2012) Standardized Precipitation Index User Guide. Svoboda, M., Hayes, M., & Wood, D, Geneva
Yang P, Xia J, Zhan C, Zhang Y, Hu S (2018) Discrete wavelet transform-based investigation into the variability of standardized precipitation index in Northwest China during 1960–2014. Theoret Appl Climatol 132(1–2):167–180
Yerdelen C, Abdelkader M, Eris E (2021) Assessment of drought in SPI series using continuous wavelet analysis for Gediz Basin. Turkey. Atmospheric Research 260:105687
Yue S, Pilon P, Phinney B, Cavadias G (2002) The influence of autocorrelation on the ability to detect trend in hydrological series. Hydrol Process 16(9):1807–1829

Rashid M, Iqbal Y, Kousar M (2019) Integrated approach of [illegible]
[illegible]
[illegible]
[illegible]
Wang R, Liu S (2020) [illegible]
Wang W, Zhao X, [illegible] (2021) [illegible]
World Meteorological Organization (2020) [illegible]
Zhou X, Zhang [illegible] (2019) [illegible]
Zou C, [illegible]
[illegible]

Fluoride Mobilization and Provenance Identification in Semi-arid Conditions: A Hydrochemical and Isotopic Approach

Abhinav Patel, Abhinesh Kumar Singh, Rajesh Singh, Nijesh Puthiyottil, and Shive Prakash Rai

Abstract Health ailments due to fluoride rich groundwater is a major threat to millions of people around the globe. Fluorosis is more common in the arid and semi-arid regions of the world (USA, China, Argentina, India, etc.). This chapter emphasizes the significance of various factors deciding the fate of fluoride release in the groundwater. Fluoride release is mainly attributable to geogenic processes such as fluoride-bearing minerals in the aquifer matrix, prevailing climatic conditions, pH conditions, ion-exchange reactions, residence time etc. A significant positive correlation of F^- with parameters like pH, Na^+, HCO_3^-, SO_4^{2-}, and Cl^- describes the influence of evaporation and the role of chemical weathering on fluoride-bearing rocks. Further, saturation to over saturation of calcite while undersaturation of fluorite mineral under alkaline condition is indicative of silicate weathering and ion exchange reactions facilitating F^- release in the semi-arid conditions. Anthropogenic inputs include use of phosphate fertilizers, brick kiln, aluminum smelting, cement industries etc. which are capable of enriching nearby groundwater system with fluoride. Stable isotopes ^{18}O, ^{2}H, and ^{34}S have been utilized in the semi-arid regions to decipher the provenance and mechanism of F^- release in the groundwater. Evaporation plays a crucial part in the enrichment of F^- in arid and semi-arid zones globally. It is evident from various studies that slope of the Local Meteoric Water Line (LMWL) in the semi-arid regions is manifested by a lower slope than the Global Meteoric Water Line (GMWL), reflecting the importance of evaporative enrichment in F^- release in the semi-arid conditions.

Keywords Fluoride · Semi-arid · Geospatial · Hydrochemical · Stable isotopes

---

A. Patel · A. K. Singh · N. Puthiyottil · S. P. Rai (✉)
Department of Geology, Institute of Science, Banaras Hindu University, Varanasi 221005, Uttar Pradesh, India
e-mail: sprai1966roorkee@gmail.com

R. Singh
Environmental Hydrology Division, National Institute of Hydrology, Roorkee 247667, Uttarakhand, India

C. B. Pande et al. (eds.), *Surface and Groundwater Resources Development and Management in Semi-arid Region*, Springer Hydrogeology,
https://doi.org/10.1007/978-3-031-29394-8_6

1 Introduction

Fluorine is the 13th most prevalent element in the Earth's crust and is widely distributed in the environment. It is the most electronegative and lightest halogen family member, and a non-metallic, pale yellow-green gas with a strong odor (Hem 1959). It naturally occurs as fluoride ion in water, soil and aquifer sediments. It has no other oxidation states which are found in the aqueous systems. Due to its strong reactivity, it commonly occurs as compounds. Fluoride ion (F^-) has the same charge and radii as of hydroxyl ion (OH^-) and as a result, these ions exchange each other in many rock forming minerals. Due to the toxicity, fluoride pollution in groundwater has drawn a lot of attention in recent decades (Mukherjee et al. 2018). The F^- is mainly utilized by the higher forms of life and it is very essential for the development of teeth and bones of humans. However, both deficiency and excess concentration of F^- in the potable water may lead to several complications as dental and skeletal fluorosis which has been widely reported across the globe (Vithanage and Bhattacharya 2015). Fluoride concentration in potable water below 0.6 mg/L and above the 1.5 mg/L have negative impacts on human health when consumed for a considerable period of time. Groundwater having Fluoride concentration above the 1.5 mg/l shows considerable toxicological effects on human wellbeing, which has been reported globally (Fawell et al. 2006; Liu et al. 2014). Most common effects are on the bones, teeth, skeletal muscles, and nervous system. High F^- in pregnant women not only leads to fluorosis but also causes Anemia, which reduces absorption of iron, affecting both mother and the newborn (Bello 2020). Globally, approximately 260 million people face the problem of high fluoride levels (>1.5 mg/L). Elevated F^- in groundwater is mainly attributable to geogenic in origin through dissolution and ion exchange of F^- bearing minerals such as fluorite, apatite, fluorapatite, sphene, microlite, pyrochlore, topaz, tourmaline, spodumene, cryolite, etc. found in many granitic and metamorphic rocks. Volcanic, geothermal activities, and mixing of hot springs rich in F^- also contribute high fluoride to groundwater system in several regions (Kundu et al. 2001; Vithanage and Bhattacharya 2015). Anthropogenic sources include both point and nonpoint sources such as phosphate mining and fertilizers, cement industries, ceramic, aluminum smelting, brick kiln, burning of coal, glass and tile industries etc. which contribute fluoride to the natural water system (Datta et al. 1996; Brindha and Elango 2011). Other sources are atmospheric air and precipitation, though insignificant. Therefore, broadly fluoride concentration in groundwater is a function of (i) paleoclimate which facilitates dry deposition in arid to semi-arid climate, (ii) geology of the region, and (iii) anthropogenic activities. Brunt et al. (2004) has reported the likelihood of elevated concentration of F^- globally. According to Brunt, American continents possess low to moderate risk based on F^- enriched groundwater. Likewise, Europe and Oceania possess low risk of F^- concentration in groundwater. While, Africa and Asia possess medium to high threat of F^- in groundwater. All these geographic domains not only reflect the significance of bedrock geology but also conveys the importance of climatic variability which decides the fate of fluoride release in the groundwater. Continents, Asia and Africa and their arid/semi-arid regions with limited occurrence of groundwater are often characterized by enriched groundwater in fluoride which possess great health risk to the inhabitants. Worldwide elevated enriched groundwater with Fluoride concentration and related problems has been reported around the countries like Kenya, Norway, North Jordan, Ethiopia, Mexico, Malawi, Iran, Brazil, Japan, Sri Lanka, Turkey, Argentina, India, Pakistan, China, USA, Ghana, Italy, Canada (Nair et al. 1984; Saether et al. 1995; Rukah and Alsokhny 2004; Tekle-Haimanot et al. 2006; Valenzuela-Vasquez et al. 2006; Msonda et al. 2007; Asghari Moghaddam and Fijani 2008; Viero et al. 2009; Abdelgawad et al. 2009; Chandrajith et al. 2012; Yeşilnacar et al. 2016; Pilar Alvarez and Carol 2019; Ali et al. 2019; Younas et al. 2019; Li et al. 2020; McMahon et al. 2020; Zango et al. 2021; Fuoco et al. 2021; Bondu et al. 2022). In India groundwater rich in fluoride has been reported from the states of Odisha, Gujarat, Kerala, Andhra Pradesh, Uttar Pradesh, Karnataka, Tamil Nadu, Rajasthan, Haryana, and Punjab (Kundu et al. 2001; Gupta et al. 2005; Shaji et al. 2007; Rao 2009; Sajil Kumar et al. 2014; Choubisa 2018; Yadav et al. 2019; Ahada and Suthar 2019, Pant et al. 2021, Nijesh et al. 2021). Fluoride release and mobilization in groundwater has been rigorously studied through various tools like hydrogeological conditions of a given aquifer which allow us to know the groundwater occurrence, flow direction and hydrogeological parameters such hydraulic conductivity, porosity and permeability persisting in a given geographical domain (Ansari et al. 2022). Likewise, groundwater chemistry is used to understand the groundwater flow path, its source and chemical reactions due to rock-water interactions including chemical weathering such as silicate and carbonate, ion exchange, reverse ion exchange, evaporation etc. (Jampani et al. 2018; Ansari et al. 2022). To get insight into the hydrogeochemical evolution and extent of groundwater maturity, use of bivariate plots, trilinear plots, hydrochemical modelling, and various statistical analysis has been employed worldwide (Ansari et al. 2022; Pant et al. 2021). Stable isotopes ($\delta^{18}O$ and $\delta^{2}H$) inherit unique isotopic signature in the various hydrological domains and therefore are used to decipher recharge sources, and zones. Further, stable isotopes ($\delta^{18}O$ and $\delta^{2}H$) are also utilized to understand provenance, release mechanism and conditions in F^- rich groundwater (Clark and Fritz 2013; Younas et al. 2019; Pant et al. 2021). Thus, an integrated approach for the assessment of mobilization factors and source of fluoride release, and mitigation measures in the semi-arid regions would suffice proper water resource management in the semi-arid regions. For the present chapter, a detailed literature review has been done by compilation of various manuscripts related to Fluoride contamination in the natural water system with special emphasis on arid/semi-arid regions of India. The present chapter deals with the following objectives (i) to understand the geospatial variability of F^- rich groundwater in India, (ii) to identify the provenance and physicochemical controls on F^- release in the groundwater, and (iii) application of stable isotopes in identification of F^- releasing processes. For more insight into the hydrogeochemical and isotopic analysis various plots were replicated (based on the published plots) in the Grapher software version 17, while the spatial maps were prepared in the ArcGIS version 10.5.

2 Study Area

India is a large country encompassing an area of 3.28 million km^2, out of which 12% of the geographical area is characterized by arid zones. Out of total 32 million ha which is hot and dry regions covers the part of Rajasthan, Gujarat, Karnataka, Punjab, Andhra Pradesh, and some part of Haryana. Lahul Spiti region of Himachal Pradesh, Ladakh, and Jammu and Kashmir, are characterized by cold arid zones of India. The largest hot arid region of the country (20 million ha area) is covered by western part of Rajasthan is often affected with high F^- in groundwater. The arid and semi-arid zones are manifested by low annual precipitation and very high evapotranspiration, and the Aridisol type of major soil. India is among the major consumer of groundwater globally (Saha et al. 2018) and is marked by intensive agricultural activities which was promoted since the mid-twentieth century after the Green Revolution, launched by the Government of India to secure food demand of the country. As a result, a sharp rise in groundwater irrigation-based agricultural activities increased after 1970 has resulted many groundwater depleted regions in the Country. Groundwater is mainly exploited for irrigation purposes apart from domestic and agriculture usage.

3 Factors Responsible for Fluoride Release

Following section deals with the major contributing factors and processes related to the enrichment of Fluoride in groundwater.

3.1 Natural Causes

3.1.1 Geology and Hydrogeology

The geological and hydrogeological setup is integral factor in deciding liberation of F^- in the natural water system. Chowdhury et al. (2019) reported 5 major fluoride belts of the world and identified major rocks associated with these belts and tectonic settings. The geographical regions dominated by rocks such as granites, gneiss, shale, volcanic igneous rocks, hydrothermal fluids rich in F^-, meta-rhyolites, calc-alkaline, alkaline granites, pegmatite veins, hot spring, volcanic craters, and volcanic ash all contribute a significant amount of F^- to the groundwater. The main fluoride minerals which constitute the rock are fluorite (CaF_2), fluorapatite ($Ca_5(PO_4)_3(OH, F)$, amphiboles, micas, cryolite (Na_3AlF_6), and topaz ($Al_2(SiO_4)$ F_2). The concentration of F^- in the various lithologies follows this order: Potash feldspar rich granitoid > hornblende-biotite granitoid > basalts > hornblende-biotite tonalite > hornblende biotite granodiorite > biotite granitoid (Sunkari et al. 2021). The structural control of enriched fluoride groundwater can be validated through the examples of the East African Rift valley and West African mobile belt where the elevated concentration of F^- lies along these belts (Chowdhury et al. 2019). Fracture and topography control the groundwater hydrogeology in various tectonic settings and hence the groundwater chemistry. It is generally observed that groundwater is low in F^- concentration even if the geology has fluoride-bearing minerals in the highlands (unless marine sediments in Highland areas are found) and gradually increases in the flow direction as a result of more chemical weathering of alumina silicate rocks and dissolution of fluoride bearing aquifer matrix, and increasing residence time in the flow direction favoring more rock water interaction (Li et al. 2020). Enriched groundwater with F^- in the proximity of various tectonic settings has been observed around the globe (Chowdhury et al. 2019). A geographical area affected by a fault has a chance of exposing deeper rocks with high F^- content onto the surface and their weathering by various agents increases F^- concentration in the nearby natural water system. The various hotspots of F^- are often associated with some kind of tectonic setting. The aborted rifts, intra-continental hotspots, and Andean Magmatic belts have been reported with the highest F^- content. A classic example is East African rift valley and enriched groundwater with F^- content along this belt. The co-occurrence of volcanic belts such as in the Andes and Japan and groundwater rich in F^- is associated with pyroclasts ejected from the volcanic activities which are often rich in Na^+ and F^- while low in Ca^{2+} and Mg^{2+}.

3.1.2 Climate

Climate is another very important factor apart from the geology of a region which decides the extent of release and mobility of F^- in the groundwater. Various workers around the globe have reported elevated value of F^- in natural water system as a function of climate and geology (Chowdhury et al. 2019; Mandal et al. 2021). The arid and semi-arid regions of the tropical belt have shown more potential for groundwater F^- enrichment than other parts of the world. Countries like India, Pakistan, Sri Lanka, China, Ethiopia, Kenya, Tanzania, Syria, Jordan, Egypt, Sudan, Somali, etc. are part of the tropical region (Chandrajith et al. 2012; Younas et al. 2019; Chowdhury et al. 2019; Mandal et al. 2021). The arid and semi-arid belts of these countries have reported enriched groundwater with Fluoride. The typical soil type in the arid and semi-arid regions has been identified as Aridisol. However, recent insight to better comprehend the climatic control on F^- rich groundwater has reported other climatic zones with enriched fluoride groundwater such as Temperate Humid zones of Japan and Indonesia which is manifested by Andisols type of soil. High F^- groundwater in this belt is attributable to numerous active and dormant volcanoes in this belt, thus known as the volcano fluoride belt of the world. The active subduction along Japan, volcanic rocks, volcanic ash, and hot springs contributes high F^- to the groundwater (Chowdhury et al. 2019). However, the majority of F^- hotspots lie in the tropical belt of the world and are manifested by arid/semi-arid climate and Aridisol soil type. The release of F^- in the groundwater is influenced by the regional and local geology of arid to semi-arid regions, as well as the groundwater flow pattern (Su et al. 2013; Saxena and Saxena 2014). The spatial variability map Fig. 1, shows that most of the enriched groundwater with fluoride is attributable to arid/semi-arid regions of the country which favors the mobilization and release of fluoride from the F^- bearing aquifer material to the groundwater. States of Rajasthan, Gujarat, Punjab and some parts of South India are characterized by arid and semi-arid zones with high evaporation rates and lower annual precipitation have groundwater rich in Fluoride concentration. The majority of the population depends on the groundwater enriched with F^- in these regions and thus affected by Fluorosis. The absence and low concentration of fluoride in the other regions of the country is either because of the absence of fluoride-bearing minerals in the aquifer matrix, or freshwater recharge and minimal evaporation rates (Handa 1975).

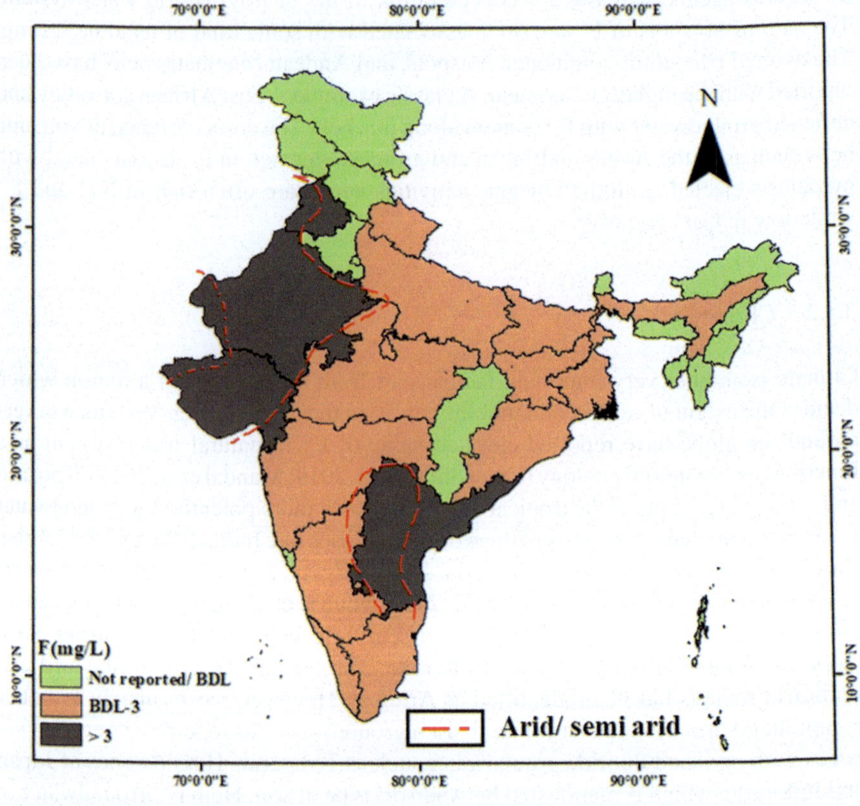

Fig. 1 India map (modified after CGWB 2018; Ali et al. 2019) highlighting the states affected by Fluoride, and the semi-arid/arid regions of the country (figure not to scale)

3.1.3 Adsorption/Desorption

Soil sorption capacity is affected by physicochemical properties like activity of hydrogen ion (pH) and salinity of the soil, as well as the type of sorbent present (Cronin et al. 2000). Processes like adsorption and desorption play a vital role in elevating F^- concentration in the groundwater even if the total fluorine concentration is low in the natural water system. Oxides and oxyhydroxides of metals (Al_2O_3 and Fe $(OH)_2$) in the aquifer matrix act as an adsorbing and desorbing surface for similar charged ions such as F^- and OH^-. Under acidic to neutral conditions F^- replaces OH^- onto the metal oxides and hydroxide surfaces and F^- adsorption capacity decreases from acidic to alkaline conditions and as well from wet to arid climatic conditions (Li et al. 2015). The leaching process is comparatively higher in alkaline soil and in both arid and semiarid zones worldwide. Because most of the fluoride in the soil is insoluble, it is merely available for plant intake. As a result, maximum fluorine compounds are adsorbed by aquifer matrix and oxyhydroxide in an alkaline condition, with only a small percentage dissolving in the soil (Hong et al. 2016). The competitive adsorption by other anions, such as bicarbonate, also causes fluoride desorption from the surfaces of minerals and organic materials in groundwater.

3.2 *Anthropogenic Inputs*

Fluoride-rich groundwater is generally attributable to geogenic causes as discussed in the previous sections. However, there are few notable human-induced sources of F^- in the groundwater which has been studied worldwide. Phosphate mining and the use of phosphate fertilizers in agricultural practices and industries are capable of increasing F^- concentration in the nearby groundwater. Phosphate fertilizer, NPK, is very commonly used and irrigation return flow has shown considerable enrichment in F^- concentration, especially in shallow groundwater. In this context, a positive correlation of F^- with K^+, Cl^-, $SO_4{}^{2-}$, and $NO_3{}^-$ is indicative of the same source of release through chemical fertilizers enriching the F^-concentration in shallow groundwater around the agricultural fields (Mukherjee and Singh 2022). Other noteworthy human-induced F^- contamination includes the combustion of coal, aluminum smelting, brick kiln, ceramic firing, cement industries, glass, and tile (Vithanage and Bhattacharya 2015). A bivariate plot of F^- versus $NO_3{}^-$ (Fig. 2) distinguishes the samples affected by geogenic and anthropogenic factors, and it becomes evident that most of the F^- enriched samples are geogenic derived and as evident with low $NO_3{}^-$ concentration in the groundwater.

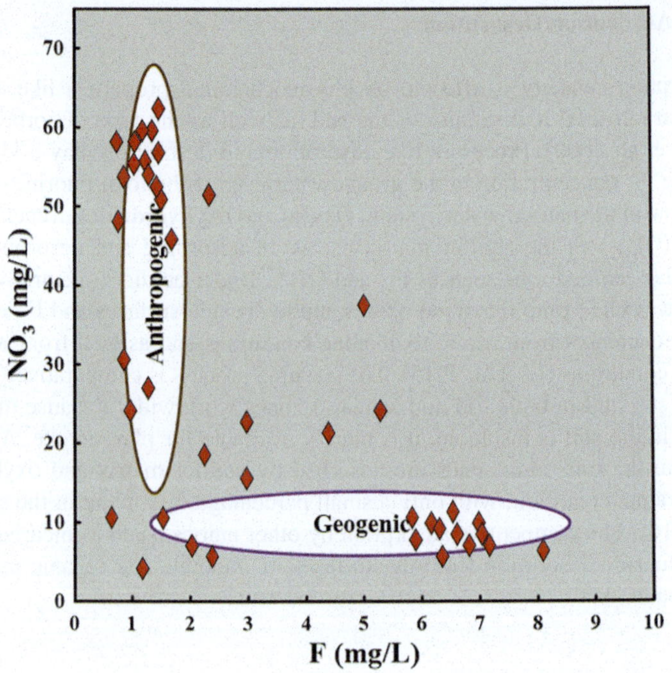

Fig. 2 F versus NO_3, distinguishing samples from geogenic and anthropogenic sources

4 Fluoride Release and Mobilization: A Hydrochemical Approach

The following sub-sections deals with the hydrogeochemical approaches to decipher the governing mechanism in the fluoride rich groundwater using hydrochemical tool.

4.1 pH/Alkalinity/Temperature

The hydrogen ion activity of water is a vital factor that decides the mobilization of different ions in the groundwater. Numerous research indicates that pH controls the release and mobility of F^- in the groundwater system. Under neutral and acidic conditions F^- is preferentially adsorbed onto a clay mineral surface while the alkaline environ is highly favorable for fluoride dissolution and gets desorbed from the clay into the groundwater and thus elevating the F^- concentration (Raj and Shaji 2017). The release of fluoride, can be understood with through following points.

Under the alkaline condition, groundwater is rich in HCO_3^- and Na^+ generally lowers the concentration of Ca^{2+} due to the precipitation as $CaCO_3$ and thereby increasing the relative concentration of F^-. In an alkaline environment, OH^- can exchange F^- adsorbed on surfaces of soil colloids and clay minerals. The distribution of F^- in groundwater is thus partly explained by the relationship of pH and fluoride. The presence of high pH condition affects the mobility of F^- and the material surfaces which help in desorption/adsorption process of anions. An excessive amount of OH^- in groundwater can promote precipitation of Ca^{2+}, Fe^{3+}, and Al^{3+}, which prevents F^- from complexing with cations and leads to a high release of F^- in groundwater. The elevated temperature of hot springs laden with high F^- such as in South America is a key factor influencing the nearby groundwater chemistry (Chowdhury et al. 2019). The groundwater which is in contact with calcite and fluorite remains in thermodynamic equilibrium which can be calculated with the help of Eq. (1)

$$CaF_{2(s)} \leftrightarrow Ca^{2+} + 2F^- \quad (1)$$

$$CaCO_{3(s)} + H^+ + 2F^- \leftrightarrow CaF_{2(s)} + HCO_3^- \quad (2)$$

$$K_{cal-fluor} = \frac{a_{HCO_3^-}}{a_{H^+} \times (a_{F^-})^2} \quad (3)$$

Since the K value is constant for a constant pH value, Eq. 3 predicts that any change in the concentration of HCO_3^- will change the concentration of fluoride. This supports a positive correlation between these two ions.

4.2 Rock Water Interaction

The chemistry of groundwater relies on the water bearing formation and its residence time in aquifer. Groundwater acquires its chemistry by the dissolution of mineral matter present in the aquifer. The rain water dissolves the CO_2 and forms carbonic acid. During the infiltration, carbonic acid reacts with subsurface rock material and dissolve the chemical constituents. The silicate rocks containing fluoride bearing minerals are very prone to dissociate/dissolution and release fluoride under the bicarbonated groundwater. Other than fluorite and fluor-apatite, micas after dissolution may release F^- in groundwater under alkaline water condition Eqs. (4 and 5)

$$KAl_3Si_3O_{10}(OH, F)_2 + CO_2 + 2.5H_2O \rightarrow 1.5Al_2Si_2O_5(OH)_4 + K^+ + 2F^- + HCO_3^- \quad (4)$$

$$NaMg_3AlSi_3O_{10}(OH, F)_2 + 7CO_2 + 7.5H_2O \rightarrow 0.5Al_2Si_2O_5(OH)_4 + Na^+ + 3Mg^+ 2F^- + 2H_4SiO_4 + 7HCO_3^- \quad (5)$$

The base and cation exchange process are common in the sedimentary aquifers of the world. For example, Ca^{2+} and Mg^{2+} in the aquifer matrix can exchange with Na^+ and K^+ present in groundwater and vice-versa is a common process. However, in reverse ion exchange, Na^+ and K^+ present in the groundwater exchange with Ca^{2+} and Mg^{2+} of the aquifer material. Due to the ion exchange process, the concentration of Na^+ and K^+ levels would be higher in comparison to Ca^{2+} and Mg^{2+} and would result

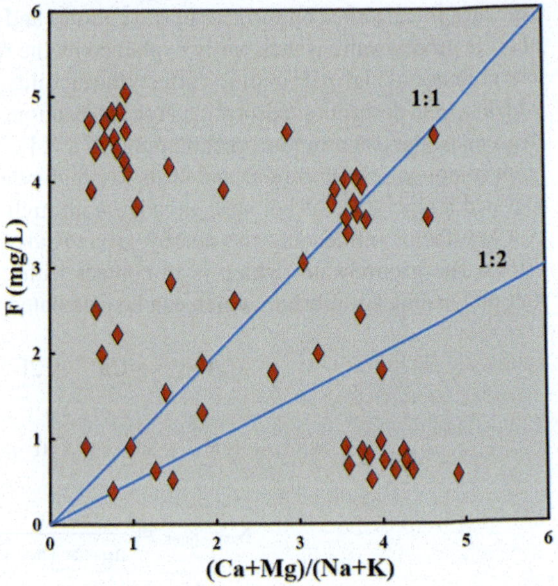

Fig. 3 F versus (Ca + Mg)/(Na + K)

in higher Na/Ca molar ratio values and result in water-type Na-HCO_3. The results of geochemical simulation reveal that under alkaline conditions, Na-HCO_3 water type is associated with increased F^- and low Ca^{2+} ion concentration through base ion exchange and calcite precipitation (Jha et al. 2013). The bivariate plot (Fig. 3) of F versus (Ca + Mg)/(Na + K) shows F^- enriched groundwater is generally manifested by low (Ca + Mg)/(Na + K) which is either due to the exchange of (Ca + Mg) from the groundwater with (Na + K) of aquifer matrix lowering the concentration of Ca^{2+} or precipitation of calcium carbonate and additionally dissolution of F^- bearing mineral under the alkaline condition as stated by Eqs. 1 and 2 (Sect. 4.1).

4.3 *Ion Effects*

In the upper section it is now proven that under natural water systems F^- is in positive correlation with HCO_3^-, and Na^+ while negatively correlated with Ca^{2+}. It is due to precipitation of calcium ion as carbonate releasing more F^- into the groundwater. Therefore, dissolution evaporites such as gypsum add up Ca^{2+} to the groundwater and will enrich groundwater in F^- through ion exchange. It is due to similar ion effect which causes enrichment of groundwater with F^-. The precipitation of calcite might also result from dissolution of dolomite, gypsum and from cation exchange reactions. Within the groundwater system, calcite precipitation removes Ca^{2+} from the water, raising the HCO_3^- concentration through further dolomite dissolution. This also causes continual rise in the Saturation Index (SI) of fluorite in groundwater.

This condition indicates faster dissolving rate of fluorite, and simultaneous rise in fluoride concentration. Thus, the ion effect in groundwater is also responsible for the higher fluoride concentration.

4.4 *Evaporation and Salt Effects*

Elevated salt concentration in groundwater is common in geographical areas having arid and semi-arid climatic conditions such as Rajasthan, Gujarat, and in parts of Southern states of India (Fig. 1). The excessive evapotranspiration and low precipitation aided with dry conditions increases the relative abundance of ions such as Na^+, SO_4^{2-} and Cl^- concentration in groundwater. The hydrogeochemical studies show a positive correlation of F^- with Na^+, HCO_3^-, pH, SO_4^{2-} and Cl^- in arid and semi-arid regions (Mandal et al. 2021). It explains the influence of evaporation along with the silicate weathering. The positive linear relationship of F^- with Cl^- is indicative of the evaporation process which enriches F^- concentration in the groundwater particularly in arid and semi-arid zones (Fig. 4a, b). The Cl^- versus F^- and F^- versus F^-/Cl^- bivariate plots trend (Fig. 4a, b) shows the reason of high fluoride concentration generally in the shallow groundwater systems because of evaporation or dissolution of Fluoride bearing minerals in the aquifer matrix (Luo et al. 2018).

The source of salts may be either evaporites or saline lake water intrusion (Gao et al. 2007). Salt effects are the dissolution of evaporites with fluoride bearing minerals (such as gypsum, halite, bloedite, mirabilite etc.). It also reduces the fluoride ion activity through ion complexation in groundwater. Thus, the availability of F^- along with increasing concentration of Cl^-, SO_4^{2-}, Na^+, and TDS suggests salt effect processes. Thus, these processes can be responsible for occurrence of fluoride in the groundwater in arid and semi-arid regions.

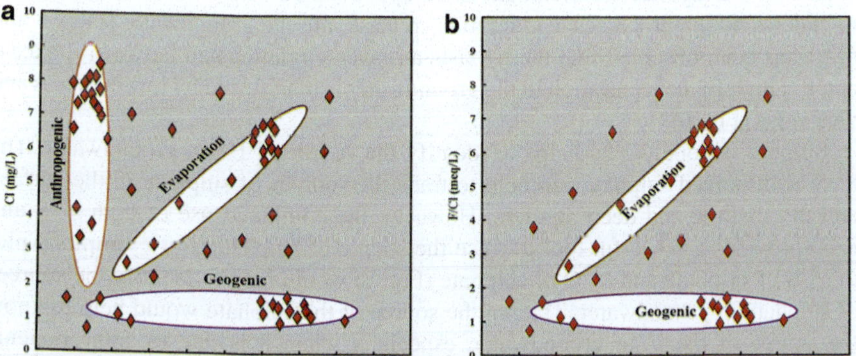

Fig. 4 **a** F versus Cl and **b** F/Cl versus F

5 Application of Stable Isotopes in Provenance and Process Identification

The application of Stable isotopes δ^2H and $\delta^{18}O$ has proven successful worldwide in many aquifer systems in provenance identification as recharge zones and the identification and quantification of dominant end members (Precipitation, River, Glacier, etc.) responsible for the recharge. The retention of the peculiar signature of the source inherited by the Stable isotopes makes them a special tool. Stable isotopic data aided with hydrochemical and hydrogeological data, particularly in arid and semi-arid zones have been utilized worldwide to identify the governing processes responsible for high F^- groundwater. The arid and semi-arid zones where evapotranspiration plays a noteworthy role in F^- enrichment and therefore the effect of evapotranspiration cannot be neglected in an arid and semi-arid area. The stable isotopic composition of such groundwater and precipitation is helpful in unravelling the controlling factors. The hydrochemical data analysis reveals that the rock water interaction under alkaline conditions is a governing factor for the mobilization of fluoride in a given aquifer. The hydrogen and oxygen isotope ratio ($\delta^{18}O$ and δ^2H) provides clues about the evaporation and other related processes during the recharge. The groundwater sample falling along the evaporation line in the plot of $\delta^{18}O$ versus δ^2H reveals that the groundwater is isotopically enriched due to evapotranspiration and/or water–rock interactions (Fig. 5). Usually, the $\delta^{18}O$ and δ^2H signatures of groundwater reflect the history of hydrological processes of groundwater from recharge to reaching the aquifer. A right deviation of water samples in $\delta^{18}O$ and δ^2H plot from the GMWL and an increase in F^- concentration confirms evaporative enrichment processes (Fig. 4). Evaporation intensifies the concentrations of all ionic species to some extent. In arid and semi-arid conditions oversaturation of $CaCO_3$ in groundwater occurs under areas of high evaporation, which endorses the breakdown of F^- containing minerals and thus releases F^- into the groundwater (Li et al. 2015). Evaporative enrichment of isotopes coupled with d-excess can also help to understand the processes leading to F^- release into the groundwater (Figs. 5 and 6). It has been observed that there exists an inverse relationship between d-excess and F^-, high F^- concentration in the groundwater is manifested by low d-excess and vice versa (Fig. 6).

Sulphur isotope ($\delta^{34}S$) helps to identify the source of F^- in groundwater. The $\delta^{34}S$ of dissolved sulphate can help to trace the sources of sulphate in the surface and the shallow and deep aquifers. However, the combined use of both $\delta^{34}S$ and $\delta^{18}O$ provides a powerful tool to trace the source of sulphate. For example, a plot of $\delta^{34}S$ of sulphate and $\delta^{18}O$ of sulphate (Fig. 7) would help to identify the sources of sulphate in groundwater. Tracing the source of the sulphate would help to know the source of the F^-in groundwater, especially when it is due to anthropogenic activities. The various processes acting together are responsible for the variations of the isotopic composition of $SO_4{}^{2-}$ in aquifers, which help in tracing the sources. The negative correlation between $SO_4{}^2/Cl^-$ and $\delta^{34}S$ show the presence of bacterial reduction processes, particularly in high F^-conditions (Fig. 8). Groundwater samples

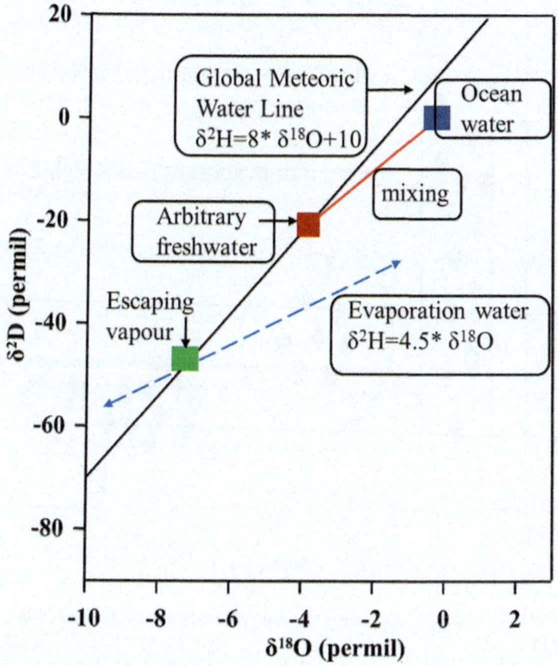

Fig. 5 The plot between $\delta^{18}O$ versus $\delta^{2}H$ shows the relation between evaporated and non-evaporated water (modified from Mook 2000)

of the phreatic aquifer condition have low F^{-}concentration without any evidence of bacterial sulphate reduction (Marimon et al. 2007).

6 Mitigation Measures

6.1 *In-Situ*

Groundwater is the most abundant natural resource on the earth and clean water is a basic need of every human. Every 1 out of 3 individuals lacks access to clean water (WHO 2019). The potability of groundwater depends on the dissolved ions and many times on the organic constituents as well. Any excess of these constituents and their prolonged intake through drinking water may introduce many serious health ailments. Therefore, groundwater with an excess of F^{-} must be treated before it can be consumed. Like any treatment, it involves both in-situ (treatment on site) and ex-situ (laboratory) treatment. Finding an alternate source of drinking water works in some cases as well. In-situ measures involve treating groundwater/affected aquifers on-site rather than collecting groundwater samples and treating them separately in the

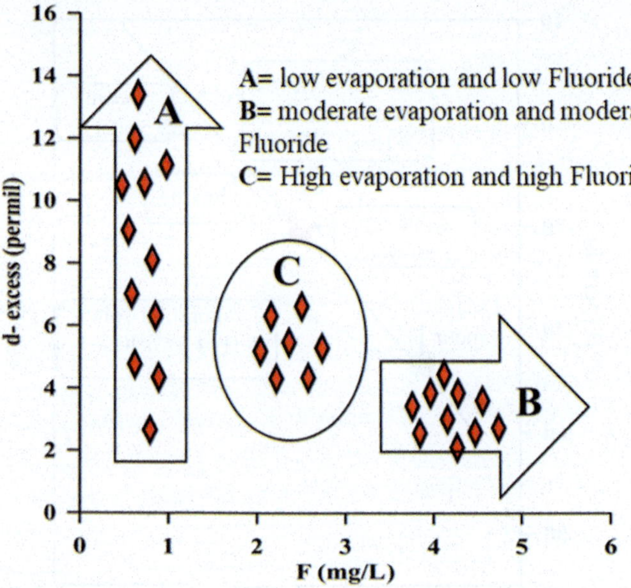

Fig. 6 Plot of d-excess versus F^- indicates the role of evaporation on groundwater (modified from Mandal et al. 2021)

Fig. 7 $\delta^{34}S$ and $\delta^{18}O$ of sulphate used sourcing of sulphate of various origins dissolved in groundwater (modified from Mook 2000)

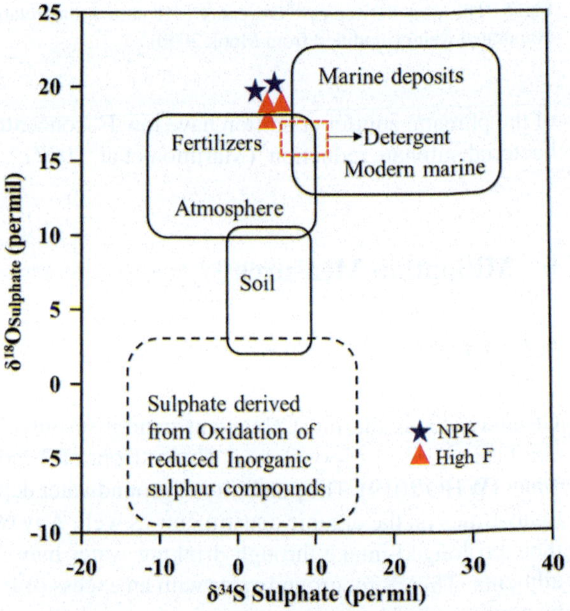

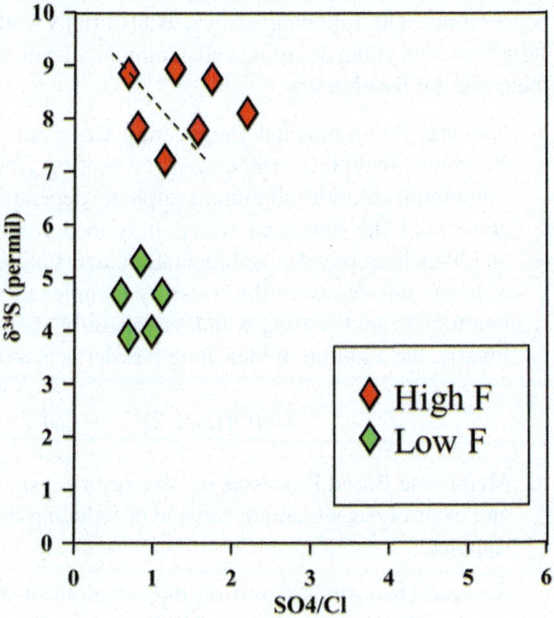

Fig. 8 $\delta^{34}S$ sulphate versus normalized sulphate concentrations show a reverse correlation with high fluoride (modified from Marimon et al. 2007)

laboratory. It is advantageous in terms of little maintenance and cost-effectiveness with minimum waste generation and provides a long-term solution. The various onsite reduction options for F^- can be achieved through:

(a) Managed aquifer Recharge (MAR) has worked very well around the globe by dilution of groundwater rich in F^-. However, the challenge lies in those areas where an alternate source is difficult to locate with lower levels of F^- for the dilution of the affected aquifer.
(b) Rainwater harvesting works in those areas where annual rainfall is enough for direct vertical recharge with minimal surface runoff.
c) The construction of check dams in many parts of India (Anantapur district) has significantly reduced the fluoride concentration in the groundwater.
(d) Mixing high F^- groundwater with groundwater lower in F^- concentration.
(e) Using various adsorbing materials such as H_3PO_4, apatite, etc. can significantly lower the concentration of F^-, however, adsorbing and desorbing is a pH-dependent process, and maintaining an optimum level of pH condition in natural water systems is often encountered with challenges.
(f) Minerals like natrolite, magnesite, goethite, serpentine, apophyllite, bentonite, etc. have significantly reduced the F^- concentration.

6.1.1 Ex-Situ

Ex-Situ Remedial Measures used for treating groundwater elevated in Fluoride concentration generally involve the introduction of an adsorbing material or chemical constituents. The following section deals briefly with techniques that are successfully deployed at the domestic and community level. Various methods and removing materials are listed below.

a. National Environmental Engineering Research Institute, India has developed Nalgonda Technique which involves treating groundwater with aluminum salts (aluminum chloride, aluminum sulphate) depending on the chloride and sulphate content of the untreated water. It is then followed by the addition of lime and bleaching powder, sedimentation, and finally filtration. The aluminum salt removes fluoride from the water by complexation, while lime helps in flocculation and rapid settling which is generally about 1/20th of the aluminum salt. Finally, the addition of bleaching powder acts as a disinfectant.

$$Ca(OH)_2 + 2F^- \rightarrow CaF_2 + 2(OH)^- \qquad (6)$$

b. Membrane Based Processes involve reverse osmosis (RO), Nanofiltration (NF), and electrolysis which are capable of reducing the concentration of F^- in water samples.

   Reverse Osmosis is based on the principle of movement of solvent from the region of high concentration of solvent to a region of low concentration of solvent through a semipermeable membrane by applying high pressure which is capable of removing the dissolved solids and thus removing F^-.
c. The Ion Exchange Method uses a strongly basic anion exchange resin with the quaternary ammonium functional groups to help with fluoride removal from the water. The F^- ion at all the sites in the resin is replaced by OH^- group thus reducing the concentration of F^-.
d. Adsorbing Media has proved to be useful in the reduction of F-concentration. Materials such as activated carbon, activated sawdust, activated alumina, carbon-activated fly ash, coffee husk, calcite, etc. are capable of adsorbing Fluoride (Jagtap et al. 2012).

$$\text{Matrix} - NR^{3+}Cl^- + F^- \rightarrow \text{Matrix} - NR^{3+}F^- + Cl^- \qquad (7)$$

7 Conclusion

The spatial variability map suggests most of the fluoride rich groundwater is found in the arid/semi-arid regions of Rajasthan, Gujarat, and some parts of south India where evaporation plays a vital role in enrichment of F^- in the shallow groundwater system. Geologically, Fluoride is concentrated in rocks like granites, gneiss, shale, and volcanic igneous rocks and minerals like fluorite, fluoroapatite, micas, amphiboles, cryolite, etc. The arid and semi-arid regions of the tropical belt have shown more potential for groundwater F^- enrichment than other parts of the world and typical soil associated with these regions is aridisol. However, F^- rich groundwater has also been reported from other climatic zones such as the Temperate Humid zones of Japan and Indonesia which is manifested by Andisols type of soil. Under neutral to acidic environ, fluoride is preferentially adsorbed onto the clay mineral surfaces, while the alkaline environment is highly conducive for fluoride dissolution. Moreover, it is often observed that enriched groundwater with F^- is associated with low Ca^{2+} and high Na^+ and HCO_3^- under alkaline conditions. Ion exchange reactions confirm that enriched groundwater is generally manifested by low (Ca + Mg)/(Na + K) which is due to the exchange of (Ca + Mg) from the groundwater with (Na + K) of aquifer matrix lowering the concentration of Ca^{2+} and increasing F^- concentration in the groundwater. A positive correlation of F^- with Na^+, HCO_3^-, pH, SO_4^{2-} and Cl^- in arid and semi-arid regions shows the influence of evaporation along with the silicate weathering. The linear positive relationship of F^- with Cl^- is indicative of the evaporation process which enriches F^- concentration in the groundwater. The hydrogen and oxygen isotope ratio ($\delta^{18}O$ and δ^2H) reflects the significance of evaporation in the enrichment of F^- in arid and semi-arid regions. The Local Meteoric Water Line (LMWL) in the semi-arid regions is manifested by a lower slope than the Global Meteoric Water Line (GMWL), also the inverse relationship of d-excess with F^- confirms the importance of evaporative enrichment. On the other hand, the negative correlation between $\delta^{34}S$ and SO_4^2/Cl^- shows the presence of bacterial reduction processes, particularly in high F^-conditions. Anthropogenic sources include the use of phosphate fertilizers, brick kilns, glass and cement industries, etc. Many in situ and ex situ mitigations are available whose success depends on the various geological and hydrogeological factors. Use of an integrated approach would thus suffice proper water management in the arid and semi-arid zones of the world.

Acknowledgements The authors would like to express their gratitude to the Head, Department of Geology, BHU, Varanasi and Institute of Eminence (IoE), BHU for their constant support.

References

Abdelgawad AM, Watanabe K, Takeuchi S, Mizuno T (2009) The origin of fluoride-rich groundwater in Mizunami area, Japan—mineralogy and geochemistry implications. Eng Geol 108(1–2):76–85

Ahada CP, Suthar S (2019) Assessment of human health risk associated with high groundwater fluoride intake in southern districts of Punjab, India. Exposure Health 11(4):267–275

Ali S, Fakhri Y, Golbini M, Thakur SK, Alinejad A, Parseh I, Bhattacharya P (2019) Concentration of fluoride in groundwater of India: a systematic review, meta-analysis and risk assessment. Groundwater Sustain Dev 9:100224

Ansari MA, Noble J, Deodhar A, Kumar US (2022) Isotope hydrogeochemical models for assessing the hydrological processes in a part of the largest continental flood basalts province of India. Geosci Front 13(2):101336

Asghari Moghaddam A, Fijani E (2008) Distribution of fluoride in groundwater of Maku area, northwest of Iran. Environ Geol 56(2):281–287

Behera UK, France J (2016) Integrated farming systems and the livelihood security of small and marginal farmers in India and other developing countries. Adv Agron 138:235–282

Bondu R, Humez P, Mayer B, Chaste E, Naumenko-Dèzes MO, Cloutier V, …, Kloppmann W (2022) Estimating natural background concentrations for dissolved constituents in groundwater: a methodological review and case studies for geogenic fluoride. J Geochem Explor 233:106906

Brindha K, Elango L (2011) Fluoride in groundwater: causes, implications and mitigation measures. Fluoride Propert Appl Environ Manage 1:111–136

Brunt R, Vasak L, Griffioen J (2004) Fluoride in groundwater: probability of occurrence of excessive. Netherlands Institute of Applied Geoscience TNO—National Geological Survey, Utrecht, The Netherlands

Chandrajith R, Padmasiri JP, Dissanayake CB, Prematilaka KM (2012) Spatial distribution of fluoride in groundwater of Sri Lanka. J Nat Sci Found Sri Lanka 40(4)

Choubisa SL (2018) Fluoride distribution in drinking groundwater in Rajasthan, India. Curr Sci 1851–1857

Chowdhury A, Adak MK, Mukherjee A, Dhak P, Khatun J, Dhak D (2019) A critical review on geochemical and geological aspects of fluoride belts, fluorosis and natural materials and other sources for alternatives to fluoride exposure. J Hydrol 574:333–359

Clark ID, Fritz P (2013) Environmental isotopes in hydrogeology. CRC Press

Cronin SJ, Manoharan V, Hedley MJ, Loganathan P (2000) Fluoride: a review of its fate, bioavailability, and risks of fluorosis in grazed-pasture systems in New Zealand. N Z J Agric Res 43(3):295–321

Datta PS, Deb DL, Tyagi SK (1996) Stable isotope (18O) investigations on the processes controlling fluoride contamination of groundwater. J Contam Hydrol 24(1):85–96

Del Bello L (2020) Fluorosis: an ongoing challenge for India. Lancet Planetary Health 4(3):e94–e95

del Pilar Alvarez M, Carol E (2019) Geochemical occurrence of arsenic, vanadium and fluoride in groundwater of Patagonia, Argentina: sources and mobilization processes. J S Am Earth Sci 89:1–9

Fawell J, Bailey K, Chilton J, Dahi E, Magara Y (2006) Fluoride in drinking-water. IWA Publishing

Fuoco I, Apollaro C, Criscuoli A, De Rosa R, Velizarov S, Figoli A (2021) Fluoride polluted groundwaters in Calabria region (Southern Italy): natural source and remediation. Water 13(12):1626

Gao X, Wang Y, Li Y, Guo Q (2007) Enrichment of fluoride in groundwater under the impact of saline water intrusion at the Salt Lake area of Yuncheng basin, northern China. Environ Geol 53(4):795–803

Gupta SK, Deshpande RD, Agarwal M, Raval BR (2005) Origin of high fluoride in groundwater in the North Gujarat-Cambay region, India. Hydrogeol J 13(4):596–605

Handa BK (1975) Geochemistry and genesis of fluoride-containing ground waters in India. Groundwater 13(3):275–281

Hem JD (1959) Study and interpretation of the chemical characteristics of natural water

Hong BD, Joo RN, Lee KS, Lee DS, Rhie JH, Min SW, Chung DY (2016) Fluoride in soil and plant. Korean J Agric Sci 43(4):522–536

Jagtap S, Yenkie MK, Labhsetwar N, Rayalu S (2012) Fluoride in drinking water and defluoridation of water. Chem Rev 112(4):2454–2466

Jampani M, Huelsmann S, Liedl R, Sonkamble S, Ahmed S, Amerasinghe P (2018) Spatio-temporal distribution and chemical characterization of groundwater quality of a wastewater irrigated system: a case study. Sci Total Environ 636:1089–1098

Jha SK, Singh RK, Damodaran T, Mishra VK, Sharma DK, Rai D (2013) Fluoride in groundwater: toxicological exposure and remedies. J Toxicol Environ Health, Part B 16(1):52–66

Kundu N, Panigrahi M, Tripathy S, Munshi S, Powell M, Hart B (2001) Geochemical appraisal of fluoride contamination of groundwater in the Nayagarh District of Orissa, India. Environ Geol 41(3):451–460

Li C, Gao X, Wang Y (2015) Hydrogeochemistry of high-fluoride groundwater at Yuncheng Basin, northern China. Sci Total Environ 508:155–165

Li P, He X, Li Y, Xiang G (2019) Occurrence and health implication of fluoride in groundwater of loess aquifer in the Chinese loess plateau: a case study of Tongchuan, Northwest China. Exposure Health 11(2):95–107

Liu H, Gao Y, Sun L, Li M, Li B, Sun D (2014) Assessment of relationship on excess fluoride intake from drinking water and carotid atherosclerosis development in adults in fluoride endemic areas, China. Int J Hyg Environ Health 217(2–3):413–420

Li J, Wang Y, Zhu C, Xue X, Qian K, Xie X, Wang Y (2020) Hydrogeochemical processes controlling the mobilization and enrichment of fluoride in groundwater of the North China Plain. Sci Total Environ 730:138877

Luo W, Gao X, Zhang X (2018) Geochemical processes controlling the groundwater chemistry and fluoride contamination in the Yuncheng Basin, China—an area with complex hydrogeochemical conditions. PLoS ONE 13(7):e0199082

Mandal R, Das A, Sudheer AK, Kumar S, Verma S, Gaddam M, Deshpande RD (2021) Sources, controls, and probabilistic health risk assessment of fluoride contamination in groundwater from a semi-arid region in Gujarat, Western India: an isotope–hydrogeochemical perspective. Environ Geochem Health 1–17

Marimon MP, Knöller K, Roisenberg A (2007, June) Anomalous fluoride concentration in groundwater—is it natural or pollution? A stable isotope approach. Isotopes Environ Health Stud 43(2):165–75. https://doi.org/10.1080/10256010701360132. PMID: 17558753

McMahon PB, Brown CJ, Johnson TD, Belitz K, Lindsey BD (2020) Fluoride occurrence in United States groundwater. Sci Total Environ 732:139217

Mook WG (2000) Environmental isotopes in the hydrological cycle: principles and applications

Mook W, Rozanski K (2000) Environmental isotopes in the hydrological cycle. IAEA Publish 39

Msonda KWM, Masamba WRL, Fabiano E (2007) A study of fluoride groundwater occurrence in Nathenje, Lilongwe, Malawi. Physics and Chemistry of the Earth, Parts A/B/C 32(15–18):1178–1184

Mukherjee I, Singh UK (2018) Groundwater fluoride contamination, probable release, and containment mechanisms: a review on Indian context. Environ Geochem Health 40(6):2259–2301

Mukherjee I, Singh UK (2022) Exploring a variance decomposition approach integrated with the Monte Carlo method to evaluate groundwater fluoride exposure on the residents of a typical fluorosis endemic semi-arid tract of India. Environ Res 203:111697

Nair KR, Manji F, Gitonga JN (1984) The occurrence and distribution of fluoride in groundwaters of Kenya. East Afr Med J 61(7):503–512

Nijesh P, Akpataku KV, Patel A, Rai P, Rai SP (2021) Spatial variability of hydrochemical characteristics and appraisal of water quality in stressed phreatic aquifer of Upper Ganga Plain, Uttar Pradesh, India. Environ Earth Sci 80(5):1–15

Pant N, Rai SP, Singh R, Kumar S, Saini RK, Purushothaman P, Pratap K (2021) Impact of geology and anthropogenic activities over the water quality with emphasis on fluoride in water scarce Lalitpur district of Bundelkhand region, India. Chemosphere 279:130496

Rai SP (2021) Technical guidelines on groundwater pollution investigation (including Deep Groundwater) Using Isotope Techniques; IAEA Technical Cooperation Expert Mission (unpublished)

Raj D, Shaji E (2017) Fluoride contamination in groundwater resources of Alleppey, southern India. Geosci Front 8(1):117–124

Raju NJ, Dey S, Das K (2009) Fluoride contamination in groundwaters of Sonbhadra district, Uttar Pradesh, India. Curr Sci 979–985

Rao NS (2009) Fluoride in groundwater, Varaha River Basin, Visakhapatnam District, Andhra Pradesh, India. Environ Monitor Assess 152(1):47–60

Rukah YA, Alsokhny K (2004) Geochemical assessment of groundwater contamination with special emphasis on fluoride concentration, North Jordan. Geochemistry 64(2):171–181

Saether OM, Reimann C, Hilmo BO, Taushani E (1995) Chemical composition of hard-and soft-rock groundwaters from central Norway with special consideration of fluoride and Norwegian drinking water limits. Environ Geol 26(3):147–156

Saha D, Marwaha S, Mukherjee A (2018) Groundwater resources and sustainable management issues in India. In: Clean and sustainable groundwater in India. Springer, Singapore, pp 1–11
Sajil Kumar PJ (2017) Geostatistical modeling of fluoride enrichment and nitrate contamination in the groundwater of Lower Bhavani Basin in Tamil Nadu, India. Model Earth Syst Environ 3(1):1–10
Sajil Kumar PJ, Jegathambal P, James EJ (2014) Factors influencing the high fluoride concentration in groundwater of Vellore District, South India. Environ Earth Sci 72(7):2437–2446
Saxena U, Saxena S (2014) Ground water quality evaluation with special reference to Fluoride and Nitrate contamination in Bassi Tehsil of district Jaipur, Rajasthan, India. Int J Environ Sci 5(1):67
Shaji E, Viju J, Thambi DS (2007) High fluoride in groundwater of Palghat District, Kerala. Curr Sci 240–245
Sharma BS, Agrawal J, Gupta AK (2011) Emerging challenge: fluoride contamination in groundwater in Agra District, Uttar Pradesh. Asian J Exp Biol Sci 2(1):131–134
Su C, Wang Y, Xie X, Li J (2013) Aqueous geochemistry of high-fluoride groundwater in Datong Basin, Northern China. J Geochem Explor 135:79–92
Tekle-Haimanot R, Melaku Z, Kloos H, Reimann C, Fantaye W, Zerihun L, Bjorvatn K (2006) The geographic distribution of fluoride in surface and groundwater in Ethiopia with an emphasis on the Rift Valley. Sci Total Environ 367(1):182–190
Valenzuela-Vasquez L, Ramirez-Hernandez J, Reyes-Lopez J, Sol-Uribe A, Lazaro-Mancilla O (2006) The origin of fluoride in groundwater supply to Hermosillo City, Sonora, Mexico. Environ Geol 51(1):17–27
Viero AP, Roisenberg C, Roisenberg A, Vigo A (2009) The origin of fluoride in the granitic aquifer of Porto Alegre, Southern Brazil. Environ Geol 56(8):1707–1719
Vithanage M, Bhattacharya P (2015) Fluoride in the environment: sources, distribution and defluoridation. Environ Chem Lett 13(2):131–147
Yadav S, Bansal SK, Yadav S, Kumar S (2019) Fluoride distribution in underground water of district Mahendergarh, Haryana, India. Appl Water Sci 9(3):1–11
Yeşilnacar Mİ, Demir Yetiş A, Dülgergil ÇT, Kumral M, Atasoy AD, Rastgeldi Doğan T, Aydoğdu M (2016) Geomedical assessment of an area having high-fluoride groundwater in southeastern Turkey. Environ Earth Sci 75(2):1–14
Younas A, Mushtaq N, Khattak JA, Javed T, Rehman HU, Farooqi A (2019) High levels of fluoride contamination in groundwater of the semi-arid alluvial aquifers, Pakistan: evaluating the recharge sources and geochemical identification via stable isotopes and other major elemental data. Environ Sci Pollut Res 26(35):35728–35741
Zango MS, Pelig-Ba KB, Anim-Gyampo M, Gibrilla A, Sunkari ED (2021) Hydrogeochemical and isotopic controls on the source of fluoride in groundwater within the Vea catchment, northeastern Ghana. Groundw Sustain Dev 12:100526
1 in 3 people globally do not have access to safe drinking water—UNICEF, WHO 2019. https://www.who.int/news/item/18-06-2019-1-in-3-people-globally-do-not-have-access-to-safe-drinking-water-unicef-who

Simulation–optimization Models for Aquifer Parameter Estimation

Sharad Patel and **T. I. Eldho**

Abstract An accurate estimation of aquifer parameters is important for effective groundwater management and future scenario prediction. These parameters are mostly obtained through different time-consuming and cumbersome field pumping tests. The inverse problem is a recently developed widely accepted mathematical approach to obtain the representative optimal aquifer parameters, particularly in large heterogeneous aquifer systems. For the inverse problem solution, the simulation–optimization (SO) model approach has been effectively used. The efficiency of these SO models depends mainly on two factors like, the accuracy of the simulation model and the ability of the optimization algorithm to explore the solution space. In this study, we selected the combination of two simulation models (i.e., FEM and Meshfree method) and four optimization algorithms (i.e., Particle Swarm Optimization (PSO), Differential Evolution (DE), a hybrid version of DE and PSO (DE-PSO) and Co-variance Matrix Adaptation Evolution Strategy (CMA-ES)) which resulted into the development of total eight number of SO models. These models are successfully applied to a synthetic confined aquifer problem. The obtained results showed the better performance of the Mfree-CMA-ES compared to its other counterparts like: FEM-DE, Mfree-DE, FEM-PSO, Mfree-PSO, FEM-CMA-ES and Mfree-DE-PSO in terms of convergence and a higher degree of unanimity with the known values of transmissivity and hydraulic conductivity.

Keywords Aquifer · Transmissivity · Metaheuristic optimization · Finite element method · Meshfree method · Objective function

S. Patel (✉) · T. I. Eldho
Department of Civil Engineering, Indian Institute of Technology Bombay, Mumbai, Maharashtra 400076, India
e-mail: sharadptl8@gmail.com

T. I. Eldho
e-mail: eldho@civil.iitb.ac.in

S. Patel
Department of Earth and Environmental Sciences, Indian Institute of Science Education and Research, Bhopal, Madhya Pradesh, India

C. B. Pande et al. (eds.), *Surface and Groundwater Resources Development and Management in Semi-arid Region*, Springer Hydrogeology,
https://doi.org/10.1007/978-3-031-29394-8_7

1 Introduction

Groundwater is a significant natural resource in India, accounting for 45% of urban water supply, 85% of rural water supply, and 62% of irrigation demand (The Comptroller and Auditor General of India 2021). However, as a result of unchecked pumping, groundwater levels have recently plummeted in several parts of the world. The same trend can be seen in India, where a World Bank report predicted that in 15–20 years, 60% of Indian aquifers would be in unsafe conditions due to unchecked groundwater over-exploitation (The World Bank 2009). The best possible use of groundwater should thus be a priority for ensuring its sustainability, and strict adherence to groundwater management policies is required to achieve this goal.

Groundwater management policies are chosen in accordance with an anticipated future scenario for groundwater. The temporal variations in groundwater head are estimated by simulating groundwater governing flow equations with various numerical techniques such as the finite difference method (FDM), meshfree method (Mfree; Patel and Rastogi 2017) and the finite element method (FEM). These groundwater simulation models rely heavily on the accuracy of estimated aquifer parameters like hydraulic conductivity, transmissivity, and storage coefficient. As a result, accurate estimation of aquifer parameters is critical, which indirectly aids in the formulation of groundwater management policies (Thangarajan 2007). In-situ tests or graphical matching-based pumping tests are commonly used to estimate these aquifer parameters. The former tests are limited to homogeneous and isotropic aquifer domains and are based on governing equations with closed-form solutions (Theis, 1935). Aside from that, these pumping-based methods necessitate nearly 24–72 h of nonstop pumping in order to collect the data required for graphical matching, which is an inefficient and time-consuming solution (Michael 2009). As a result, researchers frequently employ a purely mathematical process known as inverse groundwater modelling. Simulation-optimization (SO) is a commonly used approach to solve these inverse problems. The SO approach assigns distributed parameters to a mathematical model with known boundary conditions in such a way that the error between observed and simulated state variables is minimised (Lakshmi Prasad and Rastogi 2001). This entire process is an optimization process in which aquifer parameters are decision variables, least square difference is the objective function, and possible parameter limits are the constraints.

According to Mahinthakumar and Sayeed (2005), the optimization methods used in the SO approach are broadly classified as derivative-based and non-derivative-based optimization. In the former, a derivative of the objective function improves the initial guess of parameters until the required objective function value is obtained. Previous research on inverse groundwater modelling demonstrated that the objective functions for parameter estimation problems are discrete, have multiple optima, and are non-convex. Because these objective function-related peculiarities cannot be addressed by derivative-based local optima methods and have a higher likelihood of becoming stuck in local minima, population-based stochastic search methods were introduced to inverse problems. These methods are non-derivative based optimization methods that do not require an initial guess of the parameter to be estimated. Global stochastic population-based metaheuristic optimizations have gradually replaced traditional numerical optimization methods in the SO approach due to their superior ability to handle discrete problems. By solving the inverse groundwater problem, these characteristics of metaheuristics-based optimization are successfully explored. Ant colony optimization (ACO; Abbaspour et al. 2001), Particle swarm optimization (PSO; Ch and Mathur 2012), differential evolution (DE; Rastogi et al. 2014), and, cat swarm optimization (CSO; Thomas et al. 2018), among others, are examples of this class of optimization that have been successfully applied to estimate aquifer parameters. However, these optimization methods have their own limitations. For example, DE explores the space with a higher multiplicity, making it more susceptible to unstable convergence (Wu et al. 2011); PSO typically becomes stuck to the previous best value (pbest), and eventually all remaining particles begin to follow it, resulting in a suboptimal solution (Jiang et al., 2010). Above all, the accuracy of the previously discussed heuristic-based global search methods is highly dependent on their manually adjusted control parameters. The control parameters of various popular and traditional metaheuristic algorithms are problem specific, and their tuned values are obtained after numerous model runs, which is the main reason for the higher model run cost. The Covariance Matrix Adaptation Evolutionary Strategy (CMA-ES) developed by Hansen (2006) is a quasi-parameter free global stochastic optimization algorithm in which population size is the only parameter that must be tuned. As a result, it may be a viable option to replace the existing optimization model with CMA-ES optimization in the parameter estimation problem.

In this paper, we proposed a novel approach to estimate aquifer parameters by combining the multiquadric-based Mfree approach with CMA-ES optimization. It is anticipated that this coupling will enhance the estimation of aquifer parameters, particularly in regional aquifer systems. Here, the Mfree is able to produce accurate head values (Patel et al. 2022), and CMA-ES optimization achieves objective function convergence faster with fewer generations, so this ultimate combination as a SO model yields accurate aquifer parameter values.

2 Materials and Methods

2.1 *Mfree Based Groundwater Simulation Model*

The groundwater flow governing equation for confined aquifer for the transient condition, including variabilities like anisotropy, non-homogeneity, areal recharge including pumping or draft or both is represented as Willis and Yeh (1987):

$$\frac{\partial}{\partial x}\left(T_x\frac{\partial H}{\partial x}\right)+\frac{\partial}{\partial y}\left(T_y\frac{\partial H}{\partial y}\right)=S\left(\frac{\partial H}{\partial t}\right)+Q_w\left(x-x_p,y-y_p\right)-R \quad (1)$$

With an initial condition as:

$$H(x, y, 0) = H_0(x, y) \ \ x, y \in \Omega \tag{2}$$

The constant groundwater head (Dirichlet boundary) and boundary flux (Neumann boundary) are described as following:

$$H(x, y, t) = H_1(x, y, t) \ x, y \in \partial\Omega_1 \tag{3}$$

$$\left.\begin{array}{l} T\left(\frac{\partial H}{\partial n}\right) \\ or\left[T_x\left(\frac{\partial H}{\partial x}\right)\right] + \left[T_y\left(\frac{\partial H}{\partial y}\right)\right] \end{array}\right\} = q_2(x, y, t) \ \ x, y \in \partial\Omega_2 \tag{4}$$

where $H(x, y, t)$ is the groundwater head (m); T, transmissivity (m^2/day); T_x and T_y are transmissivity values along principal axes (m^2/day); S, storativity (dimensionless); Q_w, source (−) or sink (+) term (m/day); (x_p, y_p), ,coordinate for the well location (m); δ, Dirac delta function with the property that if $x = x_p$ and $y = y_p$ then $\delta = 1$ else $\delta = 0$; R, areal- recharge (m/day); t, time (day); H_0, initial known groundwater head distribution (m); H_1, known groundwater head values at the boundary (m); q_2, known boundary flux (m^3/day/m); (l_x, l_y), direction cosine of the outward normal at certain node on Neumann boundary (dimensionless); Ω, the computational domain; $\partial\Omega$, the boundary $\partial\Omega_1 \cup \partial\Omega_2 = \partial\Omega$ of computational domain; and $\left(\frac{\partial}{\partial n}\right)$, the normal derivative.

Using the global-collocation based Mfree method (Patel and Rastogi 2017) the governing groundwater flow Eq. (1) is approximated by scattered data interpolation, which is explained as follows:

$$\left[\frac{S_j}{\ddot{A}t}\left(\sum_{j=1}^{N}\phi_j(x_i, y_i)\right) - T_{x_j}\left(\sum_{j=1}^{N}\frac{\partial^2\phi_j(x_i, y_i)}{\partial x^2}\right) - T_{y_j}\left(\sum_{j=1}^{N}\frac{\partial^2\phi_j(x_i, y_i)}{\partial y^2}\right)\right]$$
$$\times\{h_j\}^{t+1} = \left[\frac{S_j}{\Delta t}\sum_{j=1}^{N}\phi_j(x_i, y_i).\{h_j\}^t\right]$$
$$-Q_w(x_i - x_p, y_i - y_p) + R_j \text{ where } i = 1, 2 \ldots N_I \tag{5}$$

where ϕ_j is a matrix of basis or shape function. Multiquadric is used as a shape function and can be express as (Hardy 1971) $\phi_j(x, y) = \sqrt{(x - x_j)^2 + (y - y_j)^2 + C_s^2}$; $\{(x_j, y_j)\}_{j=1}^{N}$ are coordinates of N collocation nodes in Ω. C_s is a free parameter referred as shape parameter (Cheng et al. 2003) given as $C_s = \alpha_s d_s$, where α_s is support size for radial basis function (dimensionless) and d_s is the nodal spacing. Nodal spacing (Liu and Gu 2005) for two- dimensional case is computed as: $d_s = \frac{\sqrt{A}}{(\sqrt{N}-1)}$ (where A, is an area of the whole computational domain and N

is the total number of nodes distributed over the domain). The optimum value of α_s is known by performing numerous simulation on different benchmark problems. Liu and Gu (2005) suggested that the value between 2–3 is giving good results for a variety of problems.

2.2 *Inverse Groundwater Modelling: As an Optimization Problem Using SO Approach*

The inverse model generates initial natural guesses of upper and lower bounds of aquifer parameters based on random numbers. These values are used as input for the simulation model, which calculates the aquifer state variables. To calculate the objective function, the observed values are compared to the calculated values at the observation well location. If the termination criteria is met, the initial guess will be the optimum aquifer parameters; if not, the optimization model will modify previous input parameter values until the required termination criteria is met, and the corresponding modified input parameter will be the optimum aquifer parameters. Minimizing the fitting error between observed and simulated aquifer state variables at specific monitoring well locations yields the representative optimal parameter values. Because the fitting error-based objective function is nonlinear (NL) and non-continuous, it cannot be expressed explicitly in terms of decision variables (i.e. aquifer parameters). This study's objective function is the sum of squared differences (SSD), which can be expressed as:

$$Min\, E(P) = \beta_{l,t} \sum\nolimits_{l=1}^{L} \sum\nolimits_{t=t_0}^{t_t} \left[H_{l,t}^{obs} - H_{l,t}^{sim}(P) \right]^2 \tag{6}$$

Subjected to:

$$P_i^{lb} \leq P_i \leq P_i^{ub} \tag{7}$$

where E(P) represents objective function to be minimized; $H_{l,t}^{sim}$ is calculated groundwater head at observation well l at time t with parameter (P) as input [L]; $H_{l,t}^{obs}$ is observed groundwater head at observation well l at time t [L]; P_i is aquifer parameter at zone i; L is total number of observation wells; t_0 and t_t are beginning and ending time of observations [day]; l_b and u_b are the superscripts representing the lower and upper bounds on the parameters and $\beta_{l,t} \in [0, 1]$ is the weighing coefficient whose value is chosen according to confidence on measured groundwater head at a certain observation well location.

3 Optimization Models

A number of population-based global metaheuristic optimizations (i.e., DE, PSO and CMA-ES) have been successfully applied to the varieties of groundwater problems such as aquifer parameter estimation, groundwater management problems, pollutant source detection, and many others. Since DE and PSO are commonly used in this class of problems, they are not discussed in depth; however, CMA-ES optimization, which is relatively new, has been extensively discussed.

3.1 Working of CMA-ES Optimization

CMA-ES belongs to the family of evolutionary algorithm, like GA and mimics the characteristics of Darwin's evolution theory. To generate new candidate solutions, a typical CMA-ES employs initialization, evaluation, and mutation (selection with recombination) operators. Like other stochastic search methods in the evolutionary strategy (ES), the possible solution is known as individuals. Mutation is a main step to generate the new individuals by adding the random vector from multivariate random distribution to parent vector. In CMA-ES, the possible solution moves with in the fitness landscape by rotating and scaling of the covariance matrix. This whole procedure is controlled by different strategy parameters which also evolve with each generation (Bayer and Finkel 2007) and hence there is no need to pre-calibrate them as they update themselves by utilizing the internal mechanism of CMA-ES. Eventually this iterative updation of covariance matrix leads the individual towards the convergence at an optimum value. Here it is noteworthy that CMA-ES uses the information of number of previous generations (called as evolution path) instead of only the last one (Bayer and Finkel 2004).

4 Initialization

In CMA-ES, new search points within the fitness landscape is produced by multivariate normal distribution. The equation to produce the sampling point is represented as (Hansen 2006):

$$p_k^{g+1} \sim \mathrm{N}\left[m^g, \left(\sigma^g\right)^2, C^g\right] \, where \, k = 1, 2 \ldots \lambda \tag{8}$$

The equation further simplified as:

$$p_k^{g+1} \sim m^g + \sigma^g \mathrm{N}[0, C^g] \sim m^g + \sigma^g B^g D^g \mathrm{N}[0, I] \tag{9}$$

where λ represent population size or total number of search points generated by multivariate distribution; g is generation number; $p_k^{g+1} \in \mathbb{R}^n$ is kth member of population from g + 1 generation; $m^g \in \mathbb{R}^n$ is mean value of p^g; $\sigma^g \in \mathbb{R}_+$ is overall standard deviation or step size at generation g; $C^g \in \mathbb{R}^{n \times n}$ is covariance matrix at generation g; $B^g \in \mathbb{R}^n$ represents eigenvectors of C^g and $D^g \in \mathbb{R}^{n \times n}$ is diagonal matrix of eigenvalues of C^g.

4.1 *Selection and Recombination for Calculation of Mean Vector*

After generation, all λ vectors (population) are evaluated based on the problem specific objective function. Subsequently, μ (≤λ) numbers of best parental vectors are selected from total λ vectors. This selection may be random or based on evaluated fitness value of each vector. Later these selected μ vectors recombined together and its weighted (based on fitness) mean vector is calculated using weighted recombination (represented by μ_w, λ-CMA-ES). This operation keeps the mean vector nearer to better individuals. This entire procedure is mathematically expressed as (Hansen 2006):

$$m^{g+1} = \frac{\sum_{i=1}^{\mu} w_i p_i^{g+1}}{\sum_{i=1}^{\mu} w_i} \tag{10}$$

where $w_{i=1:\mu}\left(=ln(\mu+1) - ln\left(\sum_{i=1}^{\mu} i\right)\right)$ represents weight coefficient for recombination of μ selected vectors; μ (≤ λ) is size of parent population and calculated as $\frac{\lambda}{2}$ where $\lambda = 4 + 3logD$ (here D is number of parameter or dimension of the problem); p_i^{g+1} is ith individual out of $p_1^{g+1} \ldots. p_\lambda^{g+1}$; $i : \lambda$ denotes the index of ith ranked individual with $E\left(p_{1:\lambda}^{g+1}\right) \prec E\left(p_{2:\lambda}^{g+1}\right) \prec \ldots\ldots E\left(p_{\lambda:\lambda}^{g+1}\right)$ where E is objective function to minimize.

Equation (10) implements recombination by taking weighted sum of μ individuals and selection by choosing μ ≤ λ and assigning different weights w_i.

4.2 *Adapting the Covariance Matrix*

According to Eq. (9), covariance matrix and step size are the other terms which are need to be estimated. Initially, the covariance matrix is estimated from single population and one generation. This matrix further needs to be modified because it is considered as unreliable due to its small population size by adaptation procedure. Further inclusion of successive step size also enhances the estimation of covariance matrix.

4.3 Estimating Covariance Matrix

According to the assumption by Hansen (2006), population contains enough information to estimate a covariance matrix. The covariance matrix can be estimated using generated sample population, p_1^{g+1}, p_2^{g+1} ....p_λ^{g+1} as:

$$C_{emp}^{g+1} = \frac{1}{\lambda - 1}\sum_{i=1}^{\lambda}\left(p_i^{g+1} - \frac{1}{\ddot{e}}\sum_{j=1}^{\lambda} p_j^{g+1}\right)\left(p_i^{g+1} - \frac{1}{\ddot{e}}\sum_{j=1}^{\lambda} p_j^{g+1}\right)^T \quad (11)$$

Here C^{g+1} is unbiased estimator of covariance matrix by assuming p_i^{g+1} is randomly distributed using normal distribution. Since it is based on the population of single generation we will try to further modify it to include the effect of previous generation as:

$$C_\lambda^{g+1} = \frac{1}{\lambda}\sum_{i=1}^{\lambda}\left(p_i^{g+1} - m^g\right)\left(p_i^{g+1} - m^g\right)^T \quad (12)$$

The above equation represents an unbiased maximum likelihood estimator of covariance matrix. Here we can see the difference between Eqs. (11) and (12) in terms of mean value. In first one the mean value is calculated from actually realized sample while in second it is true mean value of gth population distribution. Subsequently, Eq. (11) represents the deviation within the sampled points while in Eq. (12) it is within the sampled steps. Hence Eq. (12) is more prominent representation of adapted covariance matrix.

For further improvement in Eq. (12) weightage (similar as mean) can be given to better to more successful μ vectors and can be represented as:

$$C_\mu^{g+1} = \sum_{i=1}^{\mu} w_i\left(p_{i:\lambda}^{g+1} - m^g\right)\left(p_{i:\lambda}^{g+1} - m^g\right)^T \quad (13)$$

To maintain a reliable estimation of covariance matrix, variance effective selection mass $\mu_{eff}\left(=\left[\sum_{i=1}^{\mu} {w_i}^2\right]^{-1}\right)$ should be large enough get condition smaller than 10. To avoid this restriction the upcoming next step modification is essential.

4.4 Rank-μ Update

Large population helps to estimate reliable values of covariance matrix but it will also increase the number of generation to achieve convergence criteria. As current form of Eq. (13) is not capable to estimate C^{g+1} value, therefore as a remedy, information from previous generation is added. Mathematically, it can be presented as:

$$C^{g+1} = \frac{1}{g+1}\sum\nolimits_{i=0}^{g} \frac{1}{(\sigma^{i})^{2}} C_{\mu}^{i+1} \tag{14}$$

It is further modified by assigning different weight to different generation which is known as learning rate. Then C^{g+1} reads:

$$\begin{aligned} C^{g+1} &= (1 - c_{cov})C^{g} + c_{cov}\frac{1}{(\sigma^{g})^{2}}C^{g+1} \\ &= (1 - c_{cov})C^{g} + c_{cov}\sum\nolimits_{i=1}^{\mu} w_{i}\,OP\left(\frac{p_{i:\lambda}^{g+1} - m^{g}}{\sigma^{g}}\right) \end{aligned} \tag{15}$$

where $c(\in [0, 1])$ is learning rate for updating the covariance matrix. If c_{cov} is 1 then no information from previous generation will be incorporated and it is zero then learning take place; OP denotes the outer product of a vector by itself. Here it is noteworthy that covariance matrix is initiated as identity matrix (i.e. $C^{0} = \mathbf{I}$).

As sum of the outer product in Eq. (14) is of rank μ, therefore, this modification for covariance is called as rank-μ-update.

The value of c_{cov} is very crucial for rank-μ-updation. Small value of c_{cov} leads towards slow convergence, while large value leads to premature convergence. Hansen (2006) applied CMA-ES on different classical optimization problem and found that c_{cov} is only dependent on dimension of the problem and suggested an approximated value as μ_{eff}/D.

4.5 *Cummulation: Utilizing the Evolution Path*

In Eq. (15) a term of outer product doesn't use sign information, as $OP(x) = xx^{T} = OP(-x)$. Therefore, a concept of evolution path in which represents sequence of steps and the strategy over number of generation is introduced. It is expressed in terms of sum of consecutive steps and its summation is called as cummulation. For example, an evolution path of three steps can be constructed by the sum as:

$$\frac{m^{g+1} - m^{g}}{\sigma^{g}} + \frac{m^{g} - m^{g-1}}{\sigma^{g-1}} + \frac{m^{g-1} - m^{g-2}}{\sigma^{g-2}} \tag{16}$$

Similar as Eq. (15) using the exponential smoothing Eq. (16) can be utilized to write an expression for evolution path of generation (g + 1) for covariance matrix as (Hansen 2006):

$$p_{c}^{g+1} = (1 - c_{c})p_{c}^{g} + \sqrt{c_{c}(2 - c_{c})\mu_{eff}}\frac{m^{g+1} - m^{g}}{\acute{\sigma}^{g}} \tag{17}$$

where $p_c^g \in \mathbb{R}^n$ is evolution path at gth generation; $\sqrt{c_c(2-c_c)\mu_{eff}}$ is a normalize constant for p_c^g

If $c_c = 1$ and $\mu_{eff} = 1$, then $p_c^{g+1} = \frac{p_{1:\lambda}^{g+1} - m^g}{\sigma^g}$

$$p_c^{g+1} \sim \mathbb{N}(0, C) \tag{18}$$

If

$$p_c^{g+1} \sim \frac{p_{1:\lambda}^{g+1} - m^g}{\sigma^g} \sim \mathbb{N}(0, C) \text{ for all i=1,2...,}\mu \tag{19}$$

In Eq. (17)

$$(1-c_c)^2 + \sqrt{c_c(2-c_c)\mu_{eff}}^2 = 1$$
$$\sum_{i=1}^{\mu} w_i \mathbb{N}(0, \mathrm{C}) \sim \frac{1}{\sqrt{\mu_{\mathrm{eff}}}} \mathbb{N}(0, \mathrm{C}) \tag{20}$$

Now utilizing the concept of evolution of path in Eq. (15) the ultimate equation can be read as:

$$C^{g+1} = (1-c_{cov})C^g + c_{cov} p_c^{g+1} p_c^{g+1^T} \tag{21}$$

Empirical value of learning rate for rank-1-update of C, (c_{cov}) and time cummulation of C (c_c) are respectively $\frac{2}{D^2}$ and $\frac{4}{D}$ provides optimal value of covariance matrix.

4.6 Combining Rank-μ Update and Commulation

Now by combining Eqs. (15) and (16) the ultimate equation for covariance matrix is:

$$C^{g+1} = (1-c_{\mathrm{cov}})C^g + \frac{c_{\mathrm{cov}}}{\mu_{\mathrm{cov}}} \underbrace{p_c^{g+1} p_c^{g+1^T}}_{rank-1\,update} + c_{\mathrm{cov}}\left(1 - \frac{1}{\mu_{\mathrm{cov}}}\right)$$
$$\times \underbrace{\sum_{i=1}^{\mu} W_i \left(\frac{p_{i:\lambda}^{g+1} - m^g}{\sigma^g}\right)\left(\frac{p_{i:\lambda}^{g+1} - m^g}{\sigma^g}\right)^T}_{rank-\mu\,update} \tag{22}$$

Where $\mu_{\mathrm{cov}} \geq 1$ and $\mu_{\mathrm{cov}} = \mu_{eff}$.

The rank-one-update part of Eq. (22) uses the information of correlation between generations using evolution path while rank-μ-update uses the information within the population to reach optimal value of covariance matrix.

4.7 Step Size Control

Similar to covariance matrix, step size also utilizes an evolution path (sum of successive steps). To define the length of step size as 'long' or 'short', the length of evolution path is compared with its expected length under selection.

The conjugate evolution path for step size is defined as (based on exponentially smoothed sum):

$$p_\sigma^{g+1} = (1 - c_\sigma)p_c^g + \sqrt{c_\sigma(2 - c_\sigma)\mu_{eff}}\, C^{g^{\frac{-1}{2}}} \frac{m^{g+1} - m^g}{\sigma^g} \tag{23}$$

The updation of step size based on the comparison of $\| p_\sigma^{g+1} \|$ with its expected length $E\|\mathbb{N}(0, I)\|$, and can be represented mathematically as:

$$\sigma^{g+1} = \sigma^g exp\left(\frac{c_\sigma}{d_\sigma}\left(\frac{\| p_\sigma^{g+1} \|}{E\|\mathbb{N}(0, I)\|} - 1\right)\right) \tag{24}$$

where σ^{g+1} is global step size; $E\|\mathbb{N}(0, I)\|$ is expectation of Euclidean norm of a $\mathbb{N}(0, I)$ distributed random vector $\left(=\sqrt{D}\left(1 - \frac{1}{4D} + \frac{1}{21D^2}\right)\right)$; d_σ is damping parameter $\left(=\frac{4}{D}\right)$ and C_σ is backward time horizon of evolution path $\left(=1 + \sqrt{\frac{\mu_{eff}}{D}}\right)$.

5 Results and Discussion

5.1 Problem Description: Synthetic Confined Rectangular Problem

A confined hypothetic problem (6 km × 6 km) similar to Carrera and Neuman (1986) is selected in this study as shown in Fig. 1. Here, the assumed distance between two consecutive nodes is 1000 malong X and Y-directions. This rectangular confined region has an area of 36 sq. km. which is bounded by two impervious, a constant head and one inflow boundaries. The northern part of the aquifer is getting areal recharge at the rate of 0.15×10^{-3} m/d (AR-1) and 0.25×10^{-3} m/d (AR-2) through two distinct aquitard-layers. This aquifer is assumed to have three zones of known transmissivity values varying within the range of 5 to 150 m^2/d. An inflow rate of

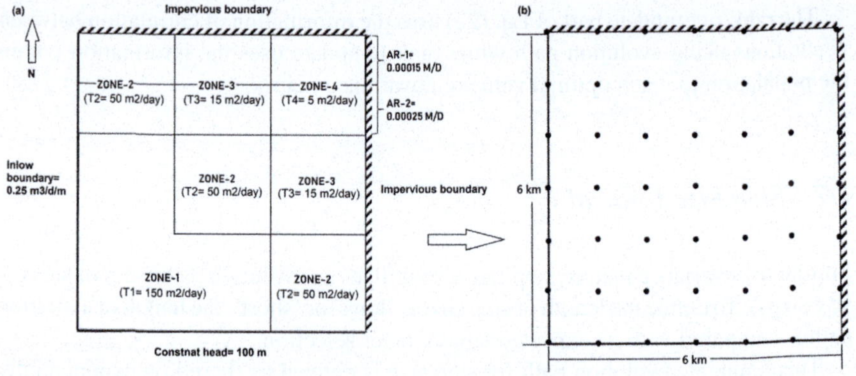

Fig. 1 Synthetic confined aquifer domain showing zonation pattern (indexed in red), boundary conditions and two areal recharge regions

0.25 m3/d/m across the western boundary is also considered. A uniform storage coefficient value for entire aquifer region is assumed as 0.001.

5.2 *Model Input*

For testing of all the developed SO models, the known transmissivity values of selected rectangular confined synthetic problem are considered to be unknown. The objective here is to determine the transmissivity values using known data i.e. storativity, boundary conditions and zonation pattern by minimizing the error between observed and simulated head values at certain monitoring well locations. The inputs in terms of predefined upper and lower limits of unknown aquifer parameters are kept between 1 to 150 m^2/day.

FEM and Mfree simulators are used to estimate the head values by discretizing entire domain using uniformly distributed 49 nodes and 72 triangular elements. In case of FEM simulator, the coordinates of distributed nodes and elemental area are used to form various element-based coefficient matrices which are further assembled to form a global coefficient of matrix. On the other hand, the estimated value of average nodal distance (d_s) and shape parameter value (α_s) are utilized to calculate the elements of shape parameter which eventually forms a coefficient matrix in Mfee simulator Eq. (5). In this synthetic problem, the estimated value of d_s is 1000 m. Since Mfree model is successfully applied on different synthetic problems with α_s as 3, the same is adopted for present case also. Total 49 nodes as shown in Fig. 1 are used. Flux vector contains the known values like inflow flux and constant groundwater head values. Using both the simulators (i.e. FEM and Mfree) model runs are performed for 25 days as total simulation period with 1 day as time-step size.

Table 1 The range and calibrated values of control parameters used in the DE, PSO and DE-PSO based SO models for synthetic aquifer problem

Control parameter	Suggested range	Calibrated value	Optimization method	
Population size (N)		50		
Mutation weighing factor (F)	0.3–0.5	0.4	DE	DE-PSO
Crossover rate (Cr)	0.8–1	0.8		
Inertia weight (ω)	0.8–0.3	Linearly varying from 0.8 to 0.2	PSO	
Acceleration constants ($C_1 = C_2$)	1.5–2	1.8		

5.3 *Parameter Setting for Developed SO Models*

In the developed SO models, the population evolution guides the ultimate algorithm towards the optima. This navigation is controlled by certain problem dependent parameters allied to that specific optimization models. These problem specific parameters are needed to be tuned or estimate empirically which are discussed in the upcoming sub-sections.

5.4 *DE, PSO and DE-PSO Based Model Setting*

The heuristic algorithms are dependent on various weighting factors for their best performance, which are commonly known as control parameters. These control parameters are fine-tuned to extract the best performance prior to their application. In the whole study, the possible range of DE based control parameters is investigated based on the literature of Storn and Price (1997) and Price et al. (2005). Similarly for PSO parameters, the range proposed by Eberhart and Kennedy (1995) and Kennedy and Eberhart (2010) are explored for their optimum values. For DE-PSO, the appropriate tuned control parameter values of both the individual heuristic are used directly. The optimum values of these control parameters for DE, PSO and DE-PSO are presented in Table 1.

5.5 *CMA-ES Based Model Setting*

The main strength of CMA-ES optimization lies on the capacity of self-adaption with each generation (Bayer et al. 2009). Unlike pre-calibrated control parameters of prior discussed metaheuristics, the strategy parameters of CMA-ES are calculated by certain empirical formulae. These parameters are obtained by researchers after

Table 2 The empirical equation and estimated values of control parameters used in the CMA-ES based SO models for synthetic aquifer problem

Control parameter	Equation	Estimated value
Population size (λ)	$\lambda = 4 + 3\log D$ where D =Problem dimension	8
Parent population (μ)	$\mu = \frac{\lambda}{2}$	4
Time const. for cummulation of C (Cc)	$C_c = \frac{4}{D+4}$	0.5
Damping parameter for cummulation of σ (dσ)	d_σ	1

numerous past experiments on different classical benchmark problems. Some of these like λ, μ and c_c are function of dimension of the problem. Remaining parameters like c_{cov}, c_μ and c_σ adapt their values based on rank- μ-update with cummulation, and vary with the progress of each generation. The calculated value of CMA-ES strategy parameters for rectangular synthetic confined problem is presented in Table 2.

5.6 *Comparative Performance of Results Obtained Through the Developed SO Models*

Using the prior-tuned values of optimization specific parameters, all the developed SO models i.e. FEM-DE, FEM-PSO, FEM-DE-PSO, FEM-CMA-ES, Mfree-DE, Mfree-PSO, Mfree-DE-PSO and Mfree-CMA-ES are applied to selected rectangular synthetic problem. The obtained results are presented in terms of a convergence graph as shown in Fig. 2. It is visible that Mfree-CMA-ES produces best convergence with lowest value of objective function as compared to others. It takes nearly 153 generations to get steady global convergence of all 4 transmissivity values. The second-best performer is Mfree-DE-PSO which tries to explore the solution space intensely by switching between DE and PSO phases hence at initial stage some oscillation is observed in the fitness function (Fig. 2). It can be seen in convergence graph that individual versions of selected optimizations i.e. DE and PSO lagged behind their hybrid version due to lack of multiplicity after certain generations. Use of MQ based Mfree simulator with different optimizations also strengthened the solution-space exploration capacity of a SO model due to its higher accuracy compared to conventional FEM simulator. It also compels the specific SO model to explore the solution space faster with less number of generations.

Apart from functional evaluation, computational time is also an important criterion to judge the performance of a SO model. The time required to perform one iteration of all the developed models is presented in Table 3. It concludes that Mfree-CMA-ES is a better performing algorithm due self-adaptive internal mechanism. On the contrary, DE-PSO model consumes slightly higher time as population generated on

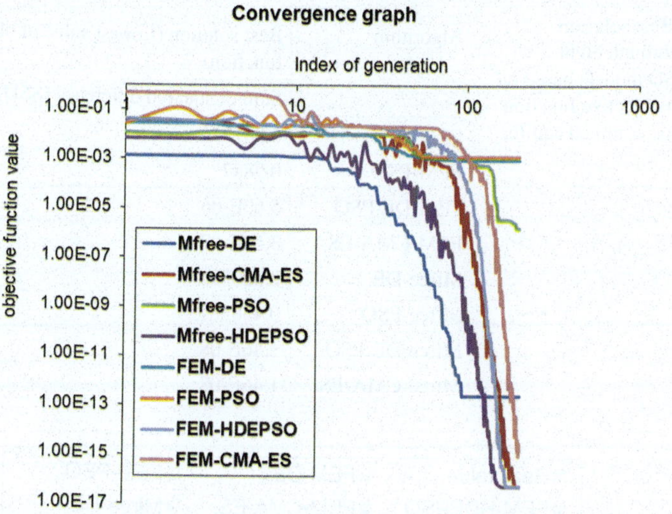

Fig. 2 The variation of objective function with iteration using eight different SO models for synthetic confined aquifer problem

each iteration is passed to DE and PSO phase in series manner for objective function evaluation.

In above described discussion, for all the analysis the SSD (Eq. 6) is used as an objective function. The results obtained are presented in Table 4 and reaffirms the superiority of Mfree-CMA-ES over other seven models.

As all these SO models are random number based stochastic search methods therefore each model run is performed 10 times and its mean value is taken as representative zonal transmissivity value as shown in Fig. 3 and Table 5. It was showed greater agreement with real value in all the developed eight models.

Table 3 Time needed to complete one iteration of the synthetic confined aquifer problem utilising eight developed SO models

SI. no	SO model	Time required for one generation (sec.)
1	FEM-DE	0.27
2	FEM-PSO	0.28
3	FEM-DE-PSO	2.96
4	FEM-CMA-ES	0.06
5	Mfree-DE	0.19
6	Mfree-PSO	0.18
7	Mfree-DE-PSO	2.75
8	Mfree-CMA-ES	0.05

Table 4 Best solution obtained through eight different SO models based on SSD as an objective function on synthetic confined aquifer problem

Algorithm	Best solution (Lowest value of objective function)
	Sum of squared difference (SSD) (m)
FEM-DE	0.001356
FEM-PSO	0.9887
FEM-DE-PSO	5.60E-06
FEM-CMA-ES	3.50E-07
Mfree-DE	0.000825
Mfree-PSO	0.458387
Mfree-DE-PSO	1.46E-08
Mfree-CMA-ES	1.46E-08

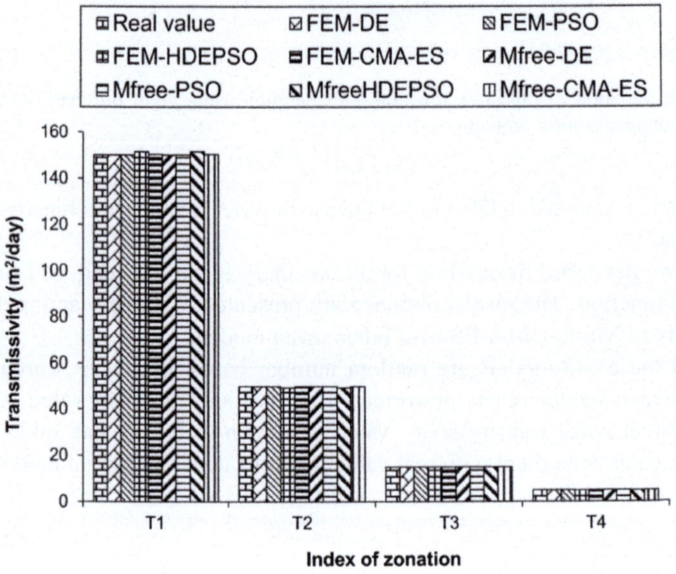

Fig. 3 Average values of each parameter after 10 times model run by 8 different methods and their comparison with the known value

It is clearly evident from present study that all the developed models are able to estimate the aquifer parameter values. This selected problem is relatively small in dimension where parameter estimation using different SO models is fairly accurate and easy to implement.

Table 5 Average values of each parameter after 10 times model run by 8 different methods

SO models	T_1	T_2	T_3	T_4
FEM-DE	150.00	50.00	15.00	5.00
FEM-PSO	150.00	50.00	15.00	5.00
FEM-DE-PSO	151.49	49.29	14.93	4.94
FEM-CMA-ES	150.29	49.98	15.00	4.98
Mfree-DE	150.00	50.00	15.00	5.00
Mfree-PSO	150.00	50.00	15.00	5.00
Mfree-DE-PSO	151.49	49.29	14.93	4.94
Mfree-CMA-ES	150.29	49.98	15.00	4.98
Real value	150.00	50.00	15.00	5.00

6 Conclusions

In this study eight different SO models are tested on a synthetic confined aquifer problem with known solution and found efficient and robust. Following are the conclusions that can be drawn from the present study:

1. DE, PSO and DEPSO based SO models require tuning of control parameters before its application to the problems while CMA-ES based models are free from such a limitation. Therefore, the CMA-ES is more efficient and robust algorithm and highly suitable to field problems.
2. Eight different combinations of SO models are applied to a synthetic confined aquifer problem. The obtain results proved that the developed CMA-ES based models are able to estimate the aquifer parameter values with the lowest value of objective function.
3. In terms of objective function evaluation, the accuracy-wise general pattern is CMA-ES > DE-PSO > DE > PSO and for time consumption criteria the general sequence is CMA-ES < DE < PSO < DE < DE-PSO.

References

Abbaspour KC, Schulin R, van Genuchten MT (2001) Estimating unsaturated soil hydraulic parameters using ant colony optimization. Adv Water Res 24:827–841. https://doi.org/10.1016/S0309-1708(01)00018-5

Bayer P, Finkel M (2004) Evolutionary algorithms for the optimization of advective control of contaminated aquifer zones. Water Resour Res 40(6). http://doi.wiley.com/https://doi.org/10.1029/2003WR002675

Bayer P, Finkel M (2007) Optimization of concentration control by evolution strategies: formulation, application, and assessment of remedial solutions. Water Resour Res 43(2): n/a-n/a. http://doi.wiley.com/https://doi.org/10.1029/2005WR004753

Bayer P, Duran E, Baumann R, Finkel M (2009) Optimized groundwater drawdown in a subsiding urban mining area. J Hydrol 365(1–2):95–104. http://linkinghub.elsevier.com/retrieve/pii/S0022169408005799

Carrera J, Neuman SP (1986) Estimation of aquifer parameters under transient and steady state conditions: 2. uniqueness, stability, and solution algorithms. Water Resour Res 22(2):211–227

Ch S, Mathur S (2012) Particle swarm optimization trained neural network for aquifer parameter estimation. KSCE J Civ Eng 16:298–307. https://doi.org/10.1007/s12205-012-1452-5

Cheng AHD, Golberg MA, Kansa EJ, Zammito G (2003) Exponential convergence and H-c multiquadric collocation method for partial differential equations. Numer Methods Part Differ Equ 19(5):571–594

Eberhart R, Kennedy J (1995) A new optimizer using particle swarm theory. In: MHS'95. Proceedings of the Sixth International Symposium on Micro Machine and Human Science IEEE; 39–43. https://doi.org/10.1109/MHS.1995.494215

Hansen N (2006) The CMA evolution strategy: a comparing review. In: Towards a new evolutionary computation. Springer-Verlag, Berlin/Heidelberg, pp 75–102. http://arxiv.org/abs/1604.00772

Hardy RL (1971) Multiquadric Equations of Topography and Other Irregular Surfaces. J Geophys Res 76(8):1905–1915. http://doi.wiley.com/https://doi.org/10.1029/JB076i008p01905.

Jiang Y, Liu C, Huang C, Wu X (2010) Improved particle swarm algorithm for hydrological parameter optimization. Appl Math Comput 217(7):3207–3215. https://doi.org/10.1016/j.amc.2010.08.053

Kennedy J, Eberhart R (2010) Particle swarm optimization. In: Proceedings of ICNN'95 - International Conference on Neural Networks. IEEE, Piscataway, NJ, pp 1942–1948

Lakshmi Prasad K, Rastogi AK (2001) Estimating net aquifer recharge and zonal hydraulic conductivity values for Mahi Right Bank Canal project area, India by genetic algorithm. J Hydrol 243:149–161. https://doi.org/10.1016/S0022-1694(00)00364-4

Liu GR, Gu Y (2005) An introduction to meshfree methods and their programming. Springer-Verlag, Berlin/Heidelberg. http://link.springer.com/https://doi.org/10.1007/1-4020-3468-7.

Mahinthakumar G, Mohamed S (2005) Hybrid Genetic Algorithm—Local Search Methods for Solving Groundwater Source Identification Inverse Problems. J Water Res Planning Manag 131(1):45–57. https://doi.org/10.1061/(ASCE)0733-9496(2005)131:1(45)

Michael AM (2009) Irrigation: theory and practice. Vikas Publishing House Pvt Limited

Patel S, Rastogi AK (2017) Meshfree multiquadric solution for real field large heterogeneous aquifer system. Water Resour Manage 31(9):2869–2884

Patel S, Eldho TI, Rastogi AK, Rabinovich A (2022) Groundwater parameter estimation using multiquadric-based meshfree simulation with covariance matrix adaptation evolution strategy optimization for a regional aquifer system. Hydrogeol J 30(7):2205–2221. https://doi.org/10.1007/s10040-022-02544-y

Price K, Storn RM, Lampinen JA (2005) Differential Evolution. Springer-Verlag, Berlin/Heidelberg. https://doi.org/10.1007/3-540-31306-0

Rastogi AK, Cyriac R, Munuswami V (2014) PSO and DE application in groundwater hydrology. LAP Lambert, Chisinau, Moldova

Storn R, Price K (1997) Differential evolution – a simple and efficient heuristic for global optimization over continuous spaces. J Glob Optim 11(4):341–359. https://doi.org/10.1023/A:1008202821328

Thangarajan M (2007) 1 National geophysical research institute, Hyderabad, India. In: Thangarajan M (ed) Groundwater resource evaluation, augmentation, contamination, restoration, modeling and management. Springer, Hyderabad

The World Bank (2009) Deep wells and prudence: towards pragmatic action for addressing groundwater overexploitation in India. Washington

Theis CV (1935) The relation between the lowering of the Piezometric surface and the rate and duration of discharge of a well using ground-water storage. Trans Amer Geophys Union 16(2):519–524. https://doi.org/10.1029/TR016i002p00519

Thomas A, Majumdar P, Eldho TI, Rastogi AK (2018) Simulation optimization model for aquifer parameter estimation using coupled meshfree point collocation method and cat swarm optimization. Eng Anal Bound Elem 91:60–72. https://doi.org/10.1016/j.enganabound.2018.03.004

Willis R, Yeh WW-G (1987) Groundwater systems planning and management. Prentice Hall Inc., Old Tappan

Wu Y, Lee W, Chien C (2011) Modified the performance of differential evolution algorithm with dual evolution strategy. In: International Conference on Machine Learning and Computing IACSIT press: Singa, pp 57–63

Water Resources and Irrigation Management Using GIS and Remote Sensing Techniques: Case of Multan District (Pakistan)

Ali Raza, Aftab Khaliq, Yongguang Hu, Nadeem Zubair, Siham Acharki, Muhammad Zubair, Neyha Rubab Syed, Fiaz Ahmad, Sadia Iqbal, and Ahmed Elbeltagi

Abstract Pakistan experiences extreme water scarcity, which has an impact on the sustainability of agricultural output. Irrigated agriculture could benefit from effective water management employing geographic information system (GIS) and remote sensing (RS) approaches. This study has shown that quantifying the transfer of soil-vegetation and atmosphere could aid in understanding rainfall estimation, evapotranspiration, soil fertality analysis, water status, planning and management of surface and groundwater resources with the combined GIS and RS approaches. These methods proved highly successful in mapping the present state of water resource availability and anticipating future requirements for agricultural use. In addition,

A. Raza (✉) · Y. Hu (✉)
School of Agricultural Engineering, Jiangsu University, Zhenjiang 212013, P.R. China
e-mail: alir3201@gmail.com

Y. Hu
e-mail: deerhu@ujs.edu.cn

A. Khaliq · N. Zubair · F. Ahmad · S. Iqbal
Department of Agricultural Engineering, Bahauddin Zakariya University, 60800 Multan, Pakistan
e-mail: fiazahmad@bzu.edu.pk

S. Acharki
Department of Earth Sciences, Faculty of Sciences and Techniques of Tangier, Abdelmalek Essaadi University, Tetouan, Morocco

M. Zubair
School of Transportation, Southeast University, Nanjing 21009, China
e-mail: 233217035@seu.edu.cnn

N. R. Syed
School of Energy & Environment, Power Engineering & Engineering Thermophysics, Southeast University, Nanjing 21009, China

A. Elbeltagi
Agricultural Engineering Department, Faculty of Agriculture, Mansoura University, Mansoura 35516, Egypt
e-mail: ahmedelbeltagy81@mans.edu.eg

C. B. Pande et al. (eds.), *Surface and Groundwater Resources Development and Management in Semi-arid Region*, Springer Hydrogeology,
https://doi.org/10.1007/978-3-031-29394-8_8

researchers, field consultants, and lawmakers can benefit from the crucial and accurate information that remote sensing and GIS tools can provide regarding managing water resources. The land use land coner (LULC) shows that how the water bodies effected with the setelment area, vegetation cover and soil fertility. The GIS mappig for the waters status clearified the effect of rainfall, and evapotranspiration on the goundwater profile. The effect of the water resouces and irrigation management have all been proven with varying degrees of precision using RS. By identifying significant issues that can be resolved by RS and GIS applications in the real world, this research fills the gap that currently exists between academics and policymakers. GIS/RS technologies will be used to conduct analysis on the Multan district's, which is located in a hyper-arid environment. This study specifically explains how the practical implementation of remote sensing is of pivotal significance in water resources.

Keywords Water management · Remote sensing · Geographic information system · Hyper-arid region

1 Introduction

The irrigation system is the world's greatest user of fresh water. The production of 30–40% of the world's staple crops uses about 70% of all water (Bastiaanssen 1998). The main sources of consumable water are groundwater, rainfall, and surface water bodies like rivers, ponds, and lakes (Pande et al. 2023). However, there are too many competitors in the home, agricultural, infrastructure, and industrial sectors. As a result, water resources must be managed to satisfy future food demands with a restricted water supply. When water resources are sufficient and environmental pollution and degradation are not a concern, water managers may afford to be negligent in their management. Nevertheless, there won't be many areas in the twenty-first century where we have this luxury because of population expansion and the associated water demand for food, health, and the environment. Reliable information is required for management and planning, and accurate information on water resource utilization is currently limited (Bastiaanssen et al. 2000). When resources are limited, effective planning and decision-making at all levels are necessary. The key to making decisions in the modern global environment is gathering and assembling various kinds of data into a manner that can be used (Abdelhaleem et al. 2021). From the farm to the river basin, however, extensive managerial expertise and understanding are essential to enhance water productivity at all scales. Moreover, it can be feasible if the system can be measured quantitatively and qualitatively, which will justify the investments to enhance productivity and improve the irrigation system (Pande et al. 2021).

From the trivial, it is no easy task to provide dependable and precise measurements on a scale ranging from individual farmer's fields to vast river basins, including irrigated land covering millions of hectares. Nonetheless, frequent data on agricultural and hydrological land surface properties may be obtained from spatially remote sensing observations across massive swaths. During the past 20 years, remote sensing's ability to detect and track crop development and other relevant biophysical characteristics has significantly improved, while a number of problems still need to be fixed (Stewart et al. 1999; Rango and Shalaby 1998).

Although remote sensing technology is cutting-edge, its offshoots, such as Geographical Information System (GIS)/Land Information System (LIS), have grown significantly in significance and utility when it comes to remotely sensed data computer-aided analysis for resource management. The application sectors have accelerated as additional satellite platforms gather data on natural resources from the earth at 10 m spatial resolution. When defining the training regions for classification and updating databases for spatially and temporally dynamic phenomena evaluation, the integration of remotely sensed data with GIS can be helpful guidance (Walsh et al. 1990). Despite having a lot of potential, the application of GIS thematic overlays as a tool for remotely sensed data interpretation is not extensively used. Enhancement methods that improve the interpretability and thematic information extraction from images include ratioing, principal components analysis, spatial filtering, and contrast stretching. Multi-spectral classification facilitates the quantitative estimations of land cover types, land use patterns, and crop water consumption (Montesinos and Fernández 2012).

This research describes the combined use of satellite images and georeferenced overlays, using GIS. It also exhibits typical land and water usage applications of selected coastal, alluvial, and hard rock environments. Furthermore, this research presents potential remote sensing applications in irrigation and water resources management in Multan district, Pakistan. It provides scientists with background information on the developments in irrigation-related remote sensing.

2 Study Area

The Multan district in Pakistan was chosen as the research region for GIS and RS approaches, as illustrated in Fig. 1. The research region is located at latitude of 30.29°, longitutde of 71.47° and altitude of 123 m (Ahsen et al. 2020). The sidhnai canal is the primary irrigation water supply source for the Multan district, which has a large control area of 0.349 Mha (Khattak 2006).The Multan district spread over the four tehsil namely, multan city, multan sadar, shujabad and jalalpur pirwala. It is bounded on the west by the Chenab River. In summer (winter), the minimum and maximum temperatures are 26 and 50 °C (4.8 and 23.4 °C), respectively. The Multan experiences 25.6 °C on average each year. The Multan district is situated in desert terrain and receives about 200 mm of precipitation annually. Rabbi and Kharif are the two main growing seasons in Multan. The most significant crop during the Kharif season is cotton. It is sown in April to May, and it is harvested between October and December. The main crop during Rabbi season is wheat. From October to December, wheat is sown, and it is harvested from April to May (Hussain et al. 2020).

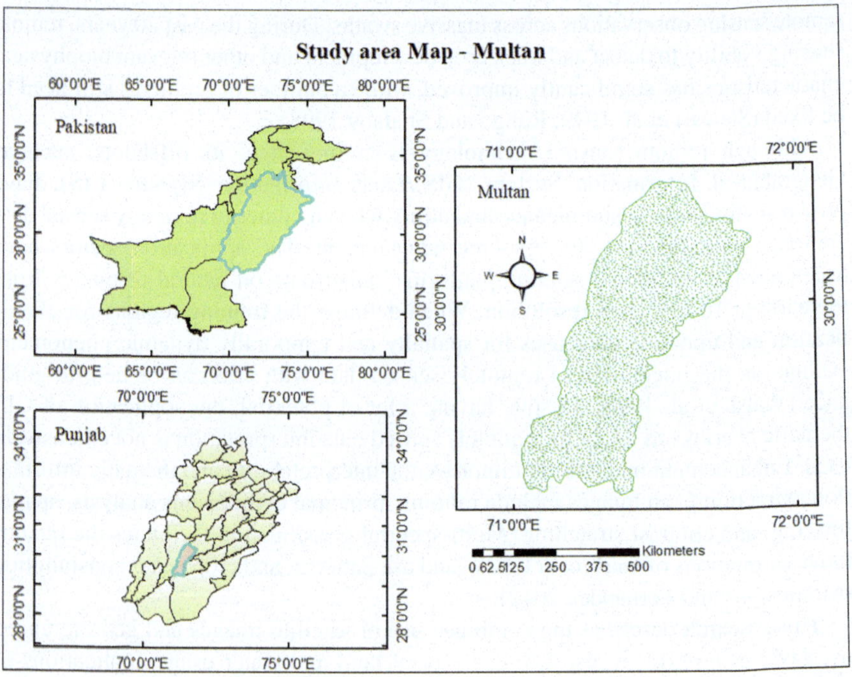

Fig. 1 GIS map for study area location

3 Materials and Methods

In June 2020, literature and reviews on different aspect of water resources and irrigation management using GIS and RS techniques were limited to global web searches using Google search engines. The literature study uncovered a few reports, research articles, and theses on the use of GIS in water and irrigation management that had been published or unpublished over the past two decades.

3.1 *Remote Sensing and Its Approaches*

Remote sensing is the collection of data (spectrum, topographical, or chronological) about real-world objects or locations without direct contact. As can be seen in Fig. 2, remote sensing utilises the electromagnetic spectrum to scan land, sea, and sky utilising electromagnetic waves (EMR) of different wavelength (visible, red, NIR, TIR, and microwave). Quantitative information on hydrological processes may be gleaned from the identification of the unique spectral features emitted by every item on Earth's surface at these wavelengths.

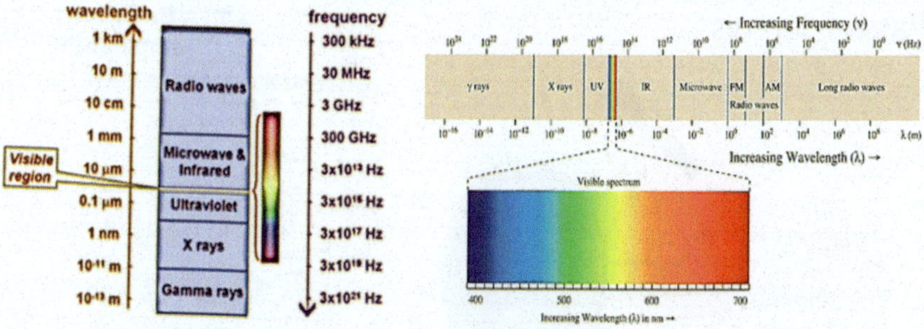

Fig. 2 Classification of EM spectrum frequency and wavelengths (Zhu et al. 2018)

Table 1 Polar orbiting satellite-based resolution categories characteristics (Jackson et al. 2010)

Resolution category	Pixel resolution (mm)	Swath width (mm)	Satellites
Very high	$0.5 * 10^3$–$2.5 * 10^3$	$5 * 10^3$–$40 * 10^3$	GeoEye, Ikonos, Worldview, Quick bird
High	$2.5 * 10^3$–$30 * 10^3$	$40 * 10^3$–$700 * 10^3$	RadarSat RapidEye, SPOT, LandSat, Aster, Chers, FormoSat, LISS
Moderate	$30 * 10^3$–$400 * 10^3$	$700 * 10^3$–$3000 * 10^3$	ASAR, MODIS, FY, AWIFS, MERIS
Low	$400 * 10^3$–$25{,}000 * 10^3$	$3000 * 10^3$	TRMM AMSRE,, MODIS, MERIS, FY, GRACE, ASAR, ASCAT

Several satellites circle the Earth, gathering information on climate and ecosystems. Table 1 shows the pixel sizes range from a few millimeters to kilometers, and the prediction accuracy ranges from three hours to many months. Projections of the 24 h rainfall at a dpi of 25 km since 1990 can help with this. Meteorological, terrestrial, and oceanic factors are monitored by the Advanced Microwave Scanning Radiometer-Earth Observing System (AMSR-E). Advanced Microwave Scanning Radiometer-Earth Observing System provides daily estimations of soil moisture at a 25 km pixel resolution (Jackson et al. 2010). AMSR-E and MODIS satellites at one-kilometergrids might be used to measure daily evapotranspiration. Additionally, MODIS, SPOT vegetation, land use, albedo, and biomass may all be estimated at a 1 km resolution.

3.1.1 Passive Remote Sensing Operation

Passive remote sensing (PRS) collects data by utilizing natural. Figure 3 shows how PRS sensors identify and monitor electromagnetic radiation, bounced or generated by entities that derive their power from the environment. Solar radiations are the

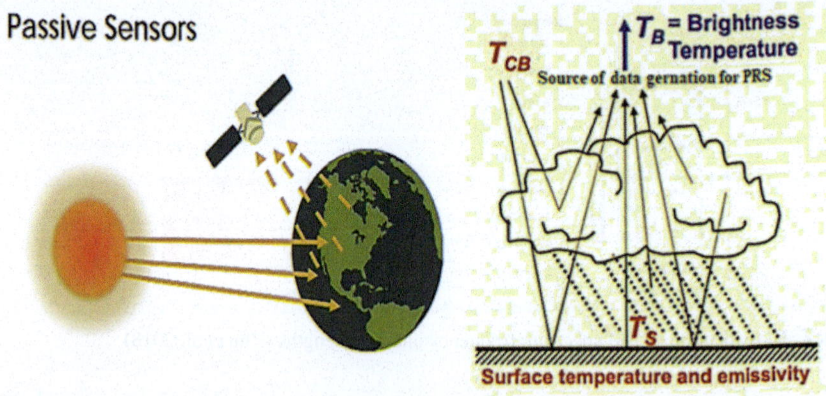

Fig. 3 Passive remote sensing operating system (Cheema and Bastiaanssen 2012)

main source for data collection for RS. The sun is the and largest energy source. Like thermal infrared wavelengths, optical wavelengths can be absorbed or reflected before being re-emitted. Passive sensors (PS), particularly those operating in the microwave range of the electromagnetic spectrum, can detect radiations emitted by Earth.

3.1.2 Active Remote Sensing

The sensors in active remote sensing are powered by independent sources. As seen in Fig. 4, radar is an example. They radiate in the direction of the item being examined while also detecting and recording radiances coming from it. Active sensors provide the advantage of taking measurements at any time of day or year. Active sensors are frequently employed at wavelengths where the sun's output is insufficient. In the case of these devices, there are significant energy requirements for greater illumination of the target. Active sensors include synthetic aperture radar (SAR), laser scanning earth observation (LASER), and European remote sensing satellites (ERS).

3.2 Geographic Information System (GIS)

The term "GIS" refers to a computerised system for managing and analysing geographical information (Fischer and Nijkamp 1992).Typically, GIS is defined as "an organised collection of dataset, applications, hardware, software, and trained personnel capable of acquiring, processing, maintaining, and analysing the geographically reference dataset and delivering output both in statistical and visual form," as seen in Fig. 5. In its broadest sense, a geographic information system (GIS) is a tool for

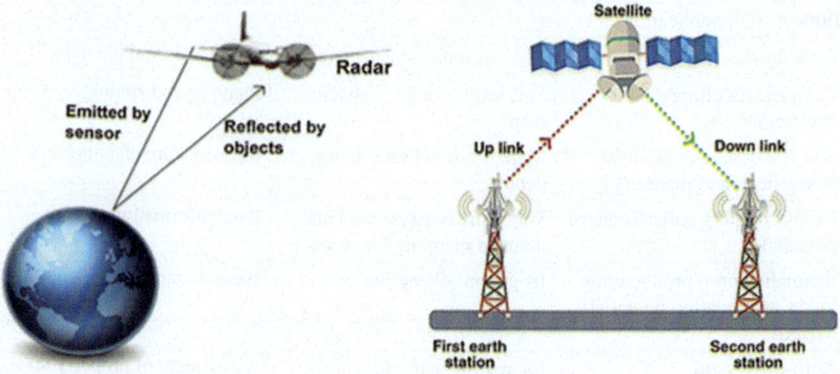

Fig. 4 Active remote sensing operation system (Islam et al. 2022)

conducting interactive searches, performing geographical analyses, and modifying existing information.

The main purposes of geographic information system are problem-solving and decision-making. Similar to other information systems, a geographic information system offers the following four capabilities for handling geospatial information:

- Input
- Data management
- Manipulation and analysis
- Output

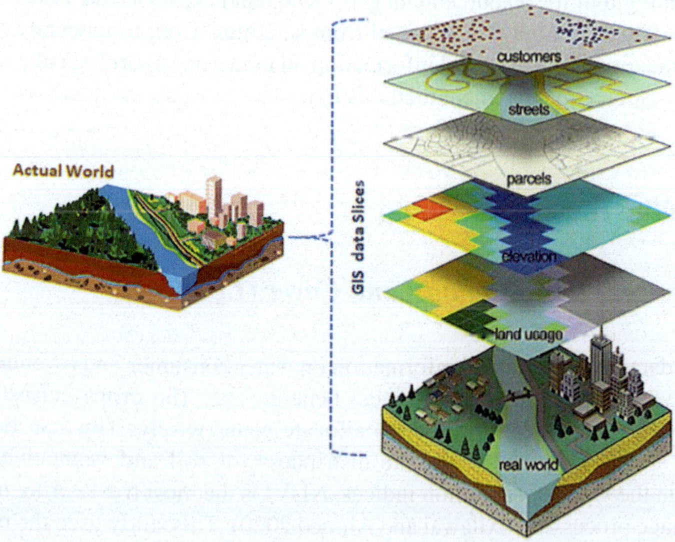

Fig. 5 GIS classification for data collection (Acharya and Lee 2019)

Table 2 GIS applications

Agriculture	Water resources management	Urban and rural planning
Crop growth climate constraints	Soil water-holding capacity map	Planning and zoning
Soil resources availability, assessment, and planning	Map of groundwater table depth	Infrastructure planning
Crops/cropping pattern pattern potential	Water stress assessment and demand mapping for crops	Land information system
Estimating crop productivity losses and locating potential hazards	Irrigation scheduling	Percale mapping
Agro-ecosystem characterization	Estimating water logging condition	Assessment of property tax based on current land usage

Additionally, a geographic information system is built for the purpose of gathering, storing, and analysing things and phenomena in which geography is a key aspect or analytical component. Geographic information systems (GIS) are distinguished by their capabilities for spatial searching and the superimposition of (map) layers. Using a GIS, it is possible to create a temporal and spatial map of crop/land by combining, like, a map of crop potential with a map of the ground/surface water condition. Table 2 represents the use of GIS in various sectors. Since complexity in real-world situations is great (for example, in agriculture, data on soil, land, crops, climatic condition, water, forest, cattle, fish stocks, and socioeconomic characteristics are necessary for making decisions), and since the physical computer capability to alter data is restricted and time-consuming, geographic information systems (GIS) are an ideal planning tool for resource managers (Knox and Weatherfield 1999).

Hydrological data may be obtained from satellites. High to moderate resolution satellite imagery offers essential information on numerous hydrological components for water resource management methods in terms of irrigation water concerns.

4 Satellite Image Processing

4.1 Spatial Land Use and Land Cover (LULC)

A LULC dataset must include information on water consumer and revenues in food, woods, hydroelectric power, ecological benefits, etc. The crops cultivated in the areas must be recognized in order to allocate water wisely. This can be done by selecting appropriate land uses. The distribution of soil and vegetative covers is reflected in the various vegetation indices. NDVI is the most trustworthy method for digital image processing (Allawai and Ahmed 2020). This study used the normalised difference vegetation index (NDVI) technique to analyse satellite datasets of Multan

Table 3 Explanation of remote sensing data used in spatial

Explanation	Remote sensing data of 2020
Satellite	Landsat 8
Sensor type	OLI_TIRS
Resolution	30 m
Cloud cover	5.3
Projection	UTM43N
Sun azimuth	109.3074943
Sun elevation	68.78002522
Spectral bands	B2 Blue: 0.45–0.51 B3 Green: 0.53–0.59 B4 Red: 0.64–0.67 B5 NIR: 0.85–0.88
Acquisition date	26 June

district in search of indicators of different land-use types. Classifying the 2020 Landsat data using NDVI yielded more accurate findings. Pixel-by-pixel values of the normalised difference vegetation index were calculated using the visible and near-infrared (NIR) spectrums of the sattelite images.

$$\text{NDVI} = \frac{\text{NIR} - \text{Red}}{\text{NIR} + \text{Red}} \tag{1}$$

where RED represents visible red reflectance (600–700 nm) and NIR represents near infrared reflectance (750–1300 nm). Table 3 lists the visible-to-infrared spectrum and picture properties. The NDVI value varied from –1 to 1. Values closer to 1 were associated with more intense vegetated areas, whereas values closer to 0 were associated with less or no greenery. Water was represented by negative values (Zaidi et al. 2017).

4.2 *Image Classification*

In this investigation, NDVI measurements were used for unsupervised classification. Table 4 lists the designated classifications, which include waterbody, settlements, barren land, crop land, spare, and dense vegetations. It can be seen in Fig. 6 that the blue areas represent water, the red areas represent human settlement, the orange areas represent barren land, the light green areas represent cropland, the moderate green areas reflect sparse vegetation, and the dark green areas represent dense vegetation. ERDAS Imagine was used to activate the Iterative Self-Organizing Data Analysis Technique Algorithm (ISODATA) clustering algorithm to group pixels with comparable attributes without any sample classes (Nelson et al. 2020). Pixel-by-pixel identifiers based on the DN values of various topographical elements are used to divide the region into six distinct categories.

Table 4 Unsupervised classification-based classes

Classes	Description
Waterbodies	Freshwater lakes, ponds, rivers, and oceans
Settlements	A variety of human-made structures, such as towns, cities, villages, and residential and commercial roadways, are included in this category
Barren land	Areas of the Earth's surface that are bare soil or incapable of supporting plant life
Crop land	Crops and grasslands
Dense vegetation	A huge region that is covered with mature trees and other flora
Spare vegetation	Low-density tree cover that precludes using the area as a forest

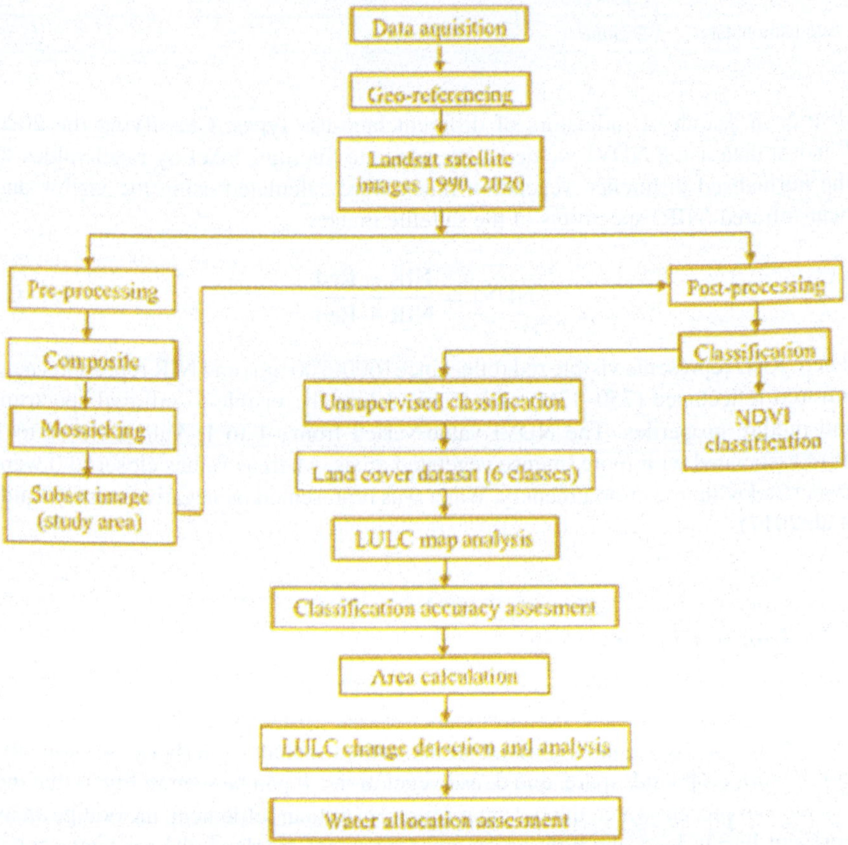

Fig. 6 Flowchart for the LUCL change detection and water allocation assisment (Dogru et al. 2020)

4.3 Water Allocation Assessment

Allocating water supplies has been identified as a top water management issue in response to rising demand, especially in the agricultural sector. In order to irrigate their crops, farmers need access to an adequate supply of water (Li et al. 2020). Most areas, including the Multan district, use the warabandi method to help farmers provide enough water for their crops. Due to rapid urbanisation and the rise of large businesses, the country's land usage and land cover have been shifting at an unprecedented rate. The high rate of change in land cover has resulted in an equal rate of change in water bodies, which in turn has led to water allocation issues (Saeidian et al. 2019).

The GIS assists in determinining area under various land-use classes and describes the classes along with their area in the Multan as shown in Fig. 7. In 1990, about 8.9% area of Multan District was under waterbody that remains 1.4% in 2020, 26.2% was under settlements that increase upto 51.5% in 2020, 2.6% was under barren land and increase upto 12.7% in 2020, 15.5% was under crop land and 20.1% in 2020, 13.4%area was covered by spare vegetation which decrease 11.7% in 2020, and 33.3% was under dense vegetation that highly effected with settlement area and remain 2.6% of the total area (Fig. 8).

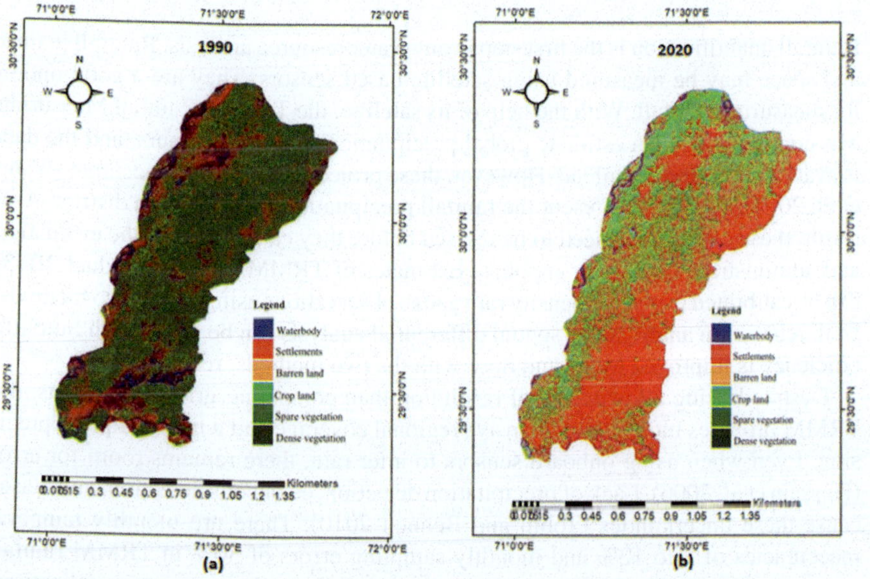

Fig. 7 LULC Map of Multan **a** for 1990 **b** for 2020 (Naeem et al. 2022)

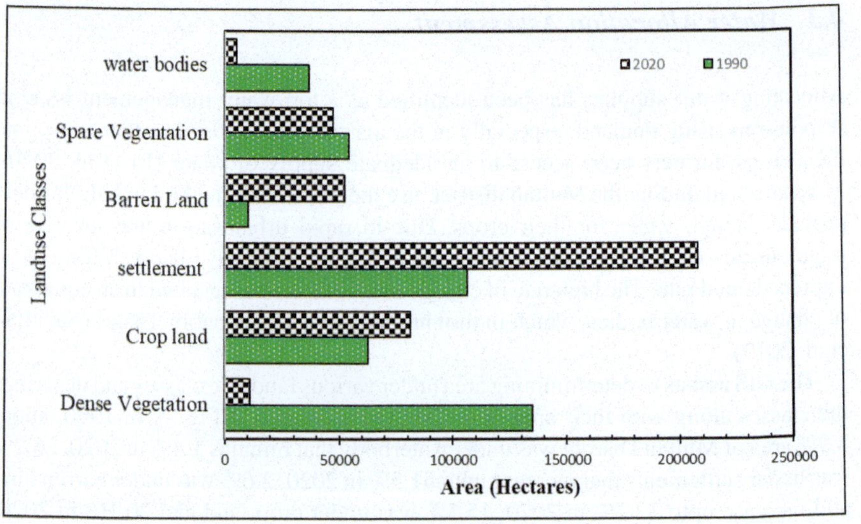

Fig. 8 Multan district land use area distribution (Naeem et al. 2022)

4.4 Spatio-Temporal Precipitation

Rainfall quantification is the first step in any water resource analysis. Rainfall in time and space may be measured using satellite-based sensors. They are a good option for measuring rainfall. With the help of its satellite, the Tropical Rainfall Measuring Mission (TRMM) can estimate global precipitation every three hours, and the data is available for free download. However, these projections are also inaccurate (Wang et al. 2005). Figure 9 represent the rainfall precipitation in the Multan district. As a result, these estimations need to be revised before they can be used in the evaluation and administration of water supplies. Estimates of TRMM rainfall (product 3B43) can be calibrated using low density rain gauge observations using one of two methods. Both regression analysis and spatial differential analysis can be used. Nash Sutcliffe efficiency is improved by 81 and 86% with the two methods, respectively.

Despite having a lower spatial resolution than competing gridded products, the TRMM provides more comprehensive regional coverage and a finer temporal precision. Even when using onboard sensors to infer rain, there remains room for error (Hossain et al. 2006). Lack of precipitation detection, erroneous detection, and biases cause these uncertainties (Tobin and Bennett 2010). There are monthly temporal inaccuracies of 8 to 12% and monthly sampling errors of 30% in TRMM rainfall estimates. If such flaws are not corrected, they might lead to incorrect applications (Gebremichael et al. 2010). To reduce such inaccuracies, TRMM satellite estimations require area-specific calibration.

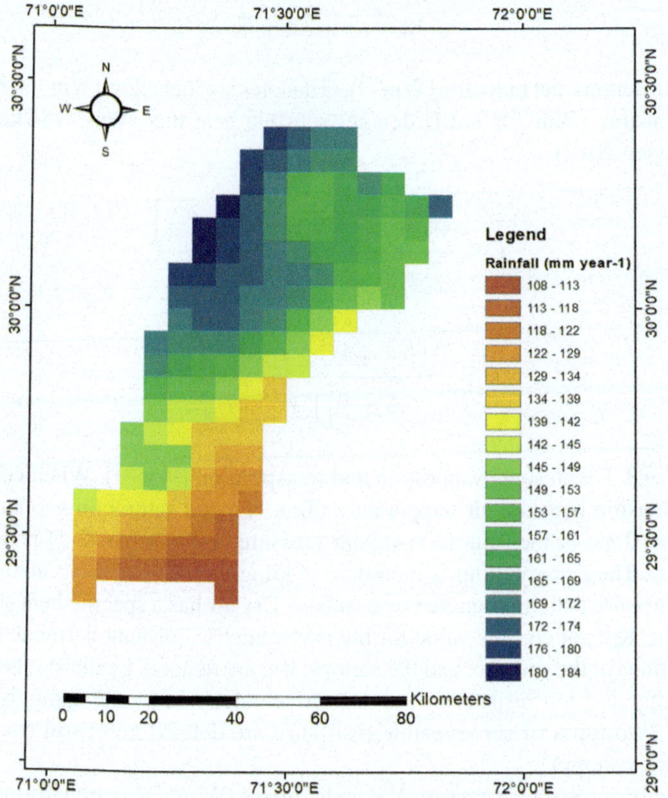

Fig. 9 Spatial distribution of calibrated TRMM rainfall for 2020

4.5 Evapotranspiration (ET_0) Spatial Distribution Map

Traditional point measurements, innovative modelling, and regionally distributed remote sensing estimations are only some of the methods used to calculate ET_0. Lysimeters, Bowen ratios, heat pulse velocity, eddy correlation, and surface renewal are often used to measure ET_0 in both plants and fields. The accuracy of these old tools, however, is less than 90% (Prasad and Mahadev 2006). Furthermore, considerable manpower, equipment costs, and coverage issues are seen as major barriers to implementing these strategies on a broad scale. Routine meteorological data cannot be used to calculate the real rate of evapotranspiration. With simple tools, rain can be easily detected, while ET_0 from terrestrial surfaces cannot (unless locations have energy balance equipment). Therefore, a novel Penman–Monteith approach (Calvache et al. 2015) has been evaluated, which gives geographic estimates of evapotranspiration using satellite observations as well as a surface energy balance. Surface energy balance can be represented as (Cheema and Bastiaanssen 2012)

$$R_n - G = \lambda E + H \tag{2}$$

where, Rn denotes net radiation (Wm^{-2}), G denotes soil heat flow (Wm^{-2}), E denotes latent heat flux (Wm^{-2}), and H denotes sensible heat flux (Wm^{-2}) (Cheema and Bastiaanssen 2012)

$$E = \frac{\Delta\left(R_{n,soil} - G\right) + \rho c_p\left[\frac{\Delta_e}{\gamma_{n,soil}}\right]}{\Delta + \gamma\left[1 + \frac{r_{soil}}{r_{a,soil}}\right]} \tag{3}$$

$$T = \frac{\Delta\left(R_{n,canopy}\right) + \rho c_p\left[\frac{\Delta_e}{\gamma_{n,canopy}}\right]}{\Delta + \gamma\left[1 + \frac{r_{canopy}}{r_{a,canopy}}\right]} \tag{4}$$

where E and T represent evaporation and transpiration (Wm^{-2}). When considering the relationship between air temperature (Tair, °C) and saturation vapour pressure (Psat), the slope of the saturation vapour pressure curve (mbar K^{-1}) is denoted as (es, mbar). The density of air is measured in kilogrammes per cubic metre, and the vapour pressure deficit is denoted by e (mbar). Dry air has a specific heat capacity of 104 J kg^{-1} K^{-1}, or cp. The value for the psychometric constant is (mbar K^{-1}). The net radiations at the soil, Rn, and the canopy, Rn, are denoted by the symbols Rn and Rn. Canopy and soil resistances are respectively denoted by the symbols rsoil and rcanopy. Two types of aerodynamic resistance are defined here: soil (ra, soil) and canopy (ra, canopy).

The units of resistance are s m^{-1}. E and T fluxes (W m^{-2}) are transformed to rates (mm d^{-1}) using a temperature-dependent LHV function.

There are other different ET_o algorithms described in the literature, but this one stands out because it provides reliable, weather-independent estimations of ET_o all year round. The predicted ET_o at 1 km dpi was closely correlated with lysimeter, Bowen ratio, and remote sensing observations (R2 of 0.70–0.76 at annual time scale; RMSE of 0.29- and 0.45-mm d^{-1}). It was shown that the ET_o fluxes at the pixel scale can be calculated on daily, 8-day, or monthly time periods.

Figure 10 shows that the Multan evapotranspiration rang from 1.2 to 10.1 mm/ year in 2020. Areas in the northwest and lower south have higher ET_o (8.62–10.1). The ET_o range 7.13–8.62 was found center of the northwest and below the center of east to west side however the lowest range (1.2–2.67) of the ET_o was shown in the northeast sides.

4.6 *Soil Fertility Status*

The ability of the soil to provide vital nutrients to plants is known as soil fertility. Currently, identifying of four factors (Organic matter, pH, Electric Conductivity

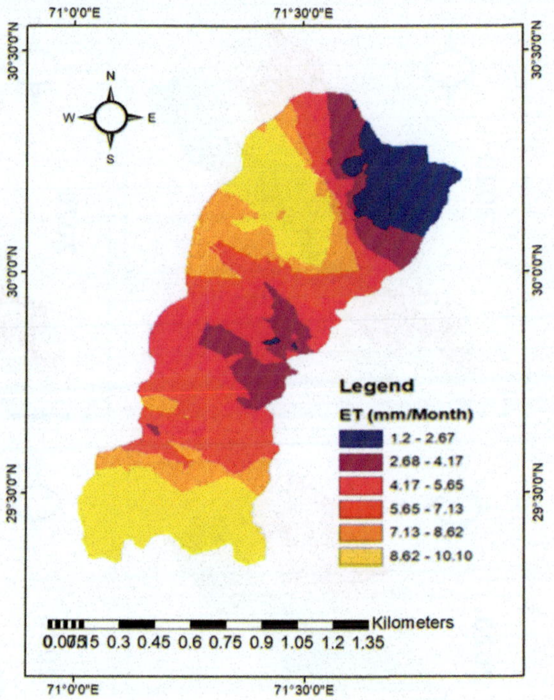

Fig. 10 Evapotranspiration (ET_O) in Multan

(EC), Phosphorus (P) that can be used to estimate soil fertility (Javed et al. 2022). We set the range in line with the soil fertility state by measuring the values of these parameters using remote sensing at various locations within the study region. From the map we can clearly see that soil fertility is changing over the different regions of Multan. Figure 11 represents that the Organic Matter map shows that the central area of Multan, North–West & North–East area of Multan Saddar and small area from central Jalapur Pir wala is low fertile, area of Multan Saddar around the Multan city and area around the central Jalapur pur Wala has medium fertile soil and the total area of Shujabad, Eastern area from center of Multan saddar and northern area of Jalalpur Pir wala has high fertile soil as shown in due to its high pH, Ec, P, ranges in these region.

4.7 *Groundwater Status*

Surface water supply for agricultural purposes is quite low, but water demand is extremely high. The groundwater is extracted by the farmer for agricultural growth. Water availability in Kharif 2020–21 remained at 65.1 million acre-feet (MAF), a slight decrease of 0.2% from 65.2 MAF in Kharif 2019–20. In comparison to Rabi

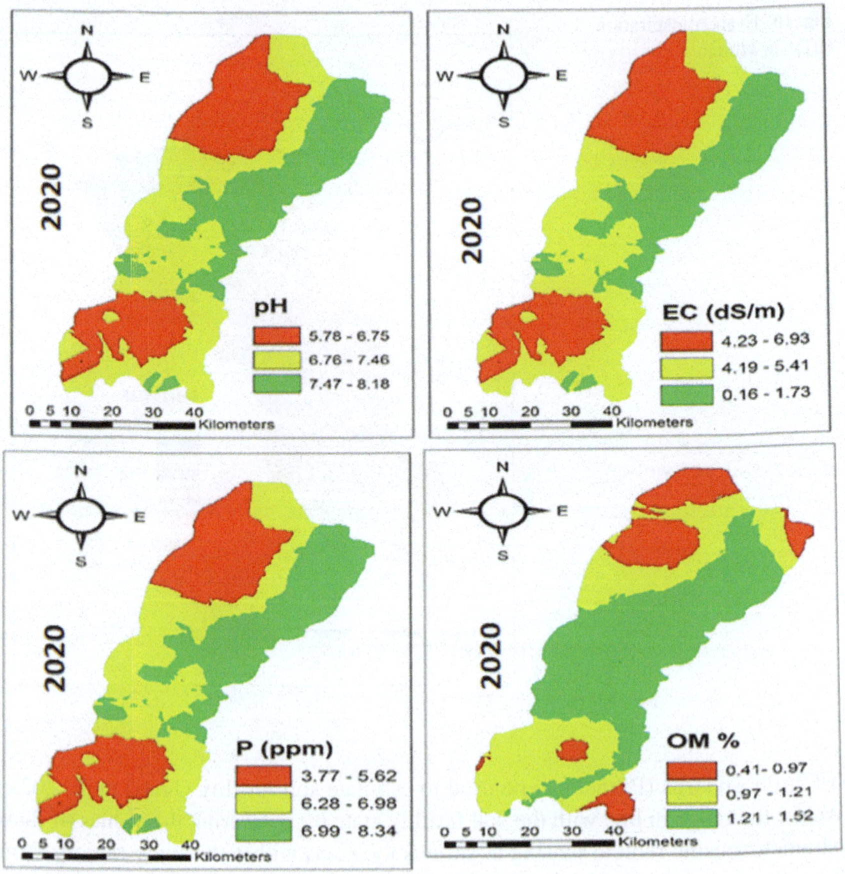

Fig. 11 Soil fertility status in Multan

Fig. 12 Groundwater status of Multan City (Imran et al. 2022)

2019–20, Rabi 2020–21 got 31.2 MAF, an increase of 6.9%. The total number of tubewells was less than 30,000 until the 1960s; today, there are more than 1 million (Watto and Mugera 2015). Over 80% of groundwater is extracted through small capacity private tubewells making it extremely difficult to establish control of the resource (Imran et al. 2022). Groundwater depletion is commonly defined as long-term water-level commonly defined continuous groundwater pumping. The study analyze the season-vise data of 16 years from 2004 to 2020. The outcomes are based on data from 101 observation wells with varying depths to the water table and more extraction of groundwater developed the depletion of groundwater. The GIS programme was used to process the pointed data, and the Inverse Distance Weighted method was employed to interpolate the missing data points (IDW). The interpolated data was then divided into groups based on well depth (Imran et al. 2022). The results depicted about 1.59 ft groundwater depletion rate per year presented in Fig. 11. Results depicted that the farmers in the upper area of the Multan region have taken more water from the surface water as well as groundwater whereas farmers can't easily uptake the surface and groundwater from the lower area with ultimately lower yield compared to upper area farmers (Fig. 12).

4.8 Accuracy Assessment

In remote sensing, accuracy is determined by whether or not the data collected from remote sensing accurately depicts what is really on the ground. It is basically comparison between user accuracy (actual field condition) and producer accuracy (values from remote sensing and software). ERDAS Imagine Software is one of the best tool which is very important for the accuracy assessment. It's useful because it streamlines the processes of radar processing, basic vector analysis, LIDAR analysis, and RS photogrammetry. Using a simple raster-based interface, ERDAS IMAGINE can extract data from imagery and compare it to the present. To determine the accuracy of producer with respect to user data, we select 53 points from the map which was develop from GIS and check the behavior of soil at these points that in which class these lie and what is frequency of these values individually. First, we check the producer values for year 2020, we found that 31 points out of total 53 were lie in high fertile class, 12 lie in medium and 10 lie in Low fertile class. Now, we select 53 points in all four tehsils (Multan Saddar, Multan City, Shujabad & Jalal Pur Pir wala) of district Multan and collect samples of soil from there along with their coordinates at that location and then analyze them in Soil and Water Testing Laboratory, Multan. After analysis, these user points were classified as 26 points lie in high fertile class, 13 points lie in medium fertile class and 14 points lie in low fertile class. When we compare both these points then we get correct points for each class where both (producer and user) match each other as high fertile class is same at 22 points, medium at 10 points and low fertile class matches at 8 points. With the help of these values, we get the User and Producer Accuracy and then finally determine the overall accuracy as 75.47% which means the 75% of our values determined with the help of RS and GIS are same as found in the actual field by soil sample analysis. Which is acceptable and show the accuracy that RS and GIS techniques are feasible to determine the soil fertility of any area without physical contact to that place.

5 Conclusion

Pakistan experiences extreme water scarcity, which has an impact on agricultural output sustainability. This study has shown that quantifying the transfer of soil-vegetation and atmosphere could aid in understanding with GIS, the RS approaches, how crop growth and water management are related. This is useful for crop categorization, rainfall estimation, soil moisture analysis, and planning and management of surface and groundwater resources. The results shows that settlement has increased by 25% because of development of Multan. The area under dense vegetation has been decreased by 30% because of changes in barren land and settlements. The total annual mean rainfall in the basin calculated was 187 mm yr^{-1} (or 213 km3 yr^{-1}). The lowest value was 108 mm yr^{-1} and the highest value was 184 mm yr^{-1}. The total ET of the Multan district was 140 mm yr^{-1}. It is also important to gain knowledge of net water producing (R > ETo) and water consuming areas (ET > R). The waters stuts shows that the due to less vegetation area and more ETo ground water pumping incerae which reduce the ground water table from 65 to 75 ft. This study specifically explained how the practical implementation of accurate and precise information provided by remote sensing is a pivotal significance in water resources.

Acknowledgements The authors are grateful to the reviewers for their comments and valuable suggestions.

Author Contributions A.R and Y.H conceptualized the book chapter. A.R. completed the original draft preparation. All the coauthors edited and reviewed the chapter. All authors have read and agreed to the published version of the book chapter.

Funding The research was supported by Key R&D program of Jiangsu Provincial Government (BE2021340) and the Priority Academic Program Development of Jiangsu Higher Education Institutions (PAPD-2018-87).

Ethics approval and consent to participate Not applicable.

Data Availability Statement Data available on reasonable request.

Conflicts of Interest The authors declare no conflict of interest.

Institutional Review Board Statement Not applicable.

Informed Consent Statement Not applicable.

References

Acharya TD, Lee DH (2019) Remote sensing and geospatial technologies for sustainable development: a review of applications. In: Sensors and materials, vol 31, Issue 11. M Y U Scientific Publishing Division, pp 3931–3945. https://doi.org/10.18494/SAM.2019.2706

Abdelhaleem FS, Basiouny M, Ashour E, Mahmoud A (2021) Application of remote sensing and geographic information systems in irrigation water management under water scarcity conditions in Fayoum, Eqypt. J Environ Manage 299:113683

Ahsen R, Khan ZM, Farid HU, Shakoor A, Ali I (2020) Estimation of cropped area and irrigation water requirement using Remote Sensing and GIS. JAPS: J Animal Plant Sci 30(4)

Allawai MF, Ahmed BA (2020) Using remote sensing and GIS in measuring vegetation cover change from satellite imagery in Mosul City, North of Iraq. In: IOP conference series: materials science and engineering, vol 757, no 1, IOP Publishing, p 012062

Bastiaanssen WG (1998) Remote sensing in water resources management: the state of the art. International Water Management Institute

Bastiaanssen WG, Molden DJ, Makin IW (2000) Remote sensing for irrigated agriculture: examples from research and possible applications. Agric Water Manag 46(2):137–155

Calvache Quesada ML, Sánchez Úbeda JP, Duque C, López Chicano M, Torre B (2015) Evaluation of analytical methods to study aquifer properties with pumping tests in coastal aquifers with numerical modelling (Motril-Salobreña Aquifer)

Cheema MJM, Bastiaanssen WGM (2012) Local calibration of remotely sensed rainfall from the TRMM satellite for different periods and spatial scales in the Indus Basin. Int J Remote Sens 33(8):2603–2627. https://doi.org/10.1080/01431161.2011.617397

Dogru AO, Goksel C, David RM, Tolunay D, Sözen S, Orhon D (2020) Detrimental environmental impact of large scale land use through deforestation and deterioration of carbon balance in Istanbul Northern Forest Area. Environ Earth Sci 79(11). https://doi.org/10.1007/s12665-020-08996-3

Fischer MM, Nijkamp P (1992) Geographic information systems and spatial analysis. Ann Reg Sci 26

Gebremichael M, Anagnostou EN, Bitew MM (2010) Critical steps for continuing advancement of satellite rainfall applications for surface hydrology in the Nile River basin 1. JAWRA J Am Water Resour Assoc 46(2):361–366

Hossain F, Anagnostou E, Bagtzoglou A (2006) On Latin Hypercube Sampling for efficient uncertainty estimation of satellite rainfall observations in flood prediction. Comput Geosci 32:776–792. https://doi.org/10.1016/j.cageo.2005.10.006

Hussain S, Mubeen M, Akram W, Ahmad A, Habib-ur-Rahman M, Ghaffar A, Nasim W (2020) Study of land cover/land use changes using RS and GIS: a case study of Multan district, Pakistan. Environ Monitor Assess 192(1):1–15

Imran M, Zaman M, Zahra SM, Misaal MA (2022) Impact of climate change on groundwater resources and adaptation strategies in arid zone mahl hydropower project view project Ph.D. project view project. www.cewre.edu.pk

Islam SU, Jan S, Waheed A, Mehmood G, Zareei M, Alanazi F (2022) Land-cover classification and its impact on Peshawar's land surface temperature using remote sensing. Comput Mater Continua 70(2):4123–4145. https://doi.org/10.32604/cmc.2022.019226

Jackson TJ, Cosh MH, Bindlish R, Starks PJ, Bosch DD, Seyfried M, Goodrich DC, Moran MS, Du J (2010) Validation of advanced microwave scanning radiometer soil moisture products. IEEE Trans Geosci Remote Sens 48(12):4256–4272. https://doi.org/10.1109/TGRS.2010.2051035

Javed A, Ali E, Binte Afzal K, Osman A, Riaz DS (2022) Soil fertility: factors affecting soil fertility, and biodiversity responsible for soil fertility. Int J Plant Anim Environ Sci 12(01). https://doi.org/10.26502/ijpaes.202129

Khattak EZ (2006) Pakistan: renewable energy development sector investment program-project I

Knox JW, Weatherfield EK (1999) The application of GIS to irrigation water resource management in England and Wales. Geog J 90–98

Li M, Xu Y, Fu Q, Singh VP, Liu D, Li T (2020) Efficient irrigation water allocation and its impact on agricultural sustainability and water scarcity under uncertainty. J Hydrol 586:124888

Montesinos S, Fernández L (2012) Introduction to ILWIS GIS tool. Erena M, López-Francos A, Montesinos S, Berthoumieu JP (eds) Otions Méditerranéennes, vol 67, pp 47–52

Naeem M, Farid HU, Madni MA, Ahsen R, Khan ZM, Dilshad A, Shahzad H (2022) Remotely sensed image interpretation for assessment of land use land cover changes and settlement impact on allocated irrigation water in Multan, Pakistan. Environ Monit Assess 194(2):98

Nelson SAC, Khorram S, Dorgan S (2020) Image processing and data analysis with ERDAS IMAGINE. Photogramm Eng Remote Sens 86(10):597–598. https://doi.org/10.14358/pers.86.10.597

Pande CB, Moharir KN, Panneerselvam B et al (2021) Delineation of groundwater potential zones for sustainable development and planning using analytical hierarchy process (AHP), and MIF techniques. Appl Water Sci 11:186. https://doi.org/10.1007/s13201-021-01522-1

Pande CB, Moharir KN, Varade A (2023) Water conservation structure as an unconventional method for improving sustainable use of irrigation water for soybean crop under rainfed climate condition. In: Pande CB, Moharir KN, Singh SK, Pham QB, Elbeltagi A (eds) Climate Change Impacts on Natural Resources, Ecosystems and Agricultural Systems. Springer Climate. Springer, Cham. https://doi.org/10.1007/978-3-031-19059-9_28

Prasad VH, Mahadev RH (2006) Estimating actual evapotranspiration using RS and GIS. In: Agriculture and hydrology applications of remote sensing, vol 6411, pp 116–126. SPIE

Rango A, Shalaby AI (1998) Operational applications of remote sensing in hydrology: success, prospects and problems. Hydrol Sci J 43(6):947–968

Saeidian B, Mesgari MS, Pradhan B, Alamri AM (2019) Irrigation water allocation at farm level based on temporal cultivation-related data using meta-heuristic optimisation algorithms. Water 11(12):2611

Stewart JB, Watts CJ, Rodriguez JC, De Bruin HAR, Van den Berg AR, Garatuza-Payan J (1999) Use of satellite data to estimate radiation and evaporation for northwest Mexico. Agric Water Manag 38(3):181–193

Tobin KJ, Bennett ME (2010) Adjusting satellite precipitation data to facilitate hydrologic modeling. J Hydrometeorol 11(4):966–978

Walsh SJ, Cooper JW, Von Essen IE, Gallager KR (1990) Image enhancement of Landsat Thematic Mapper data and GIS data integration for evaluation of resource characteristics. Photogramm Eng Remote Sens 56(8):1135–1141

Wang G, Gertner G, Anderson AB (2005) Sampling design and uncertainty based on spatial variability of spectral variables for mapping vegetation cover. Int J Remote Sens 26(15):3255–3274

Watto MA, Mugera AW (2015) Econometric estimation of groundwater irrigation efficiency of cotton cultivation farms in Pakistan. J Hydrol Reg Stud 4:193–211

Zaidi SM, Akbari A, Abu Samah A, Kong NS, Gisen A, Isabella J (2017) Landsat-5 time series analysis for land use/land cover change detection using NDVI and semi-supervised classification techniques. Pol J Environ Stud 26(6)

Zhu L, Suomalainen J, Liu J, Hyyppä J, Kaartinen H, Haggren H (2018) A review: remote sensing sensors. In: Multi-purposeful application of geospatial data. InTech. https://doi.org/10.5772/intechopen.71049

Coorelating Stream Guage Stations Using Multi Gene Genetic Programming and Random Forest

Preeti Kulkarni, Pradnya Dixit, and Shreenivas Londhe

Abstract Correlation stream gauge stations i.e. linking of discharge at upstream stations to find the discharge at the downstream station, is an important method which can be adopted. Corelating stations for discharge estimation plays a crucial role in the planning of hydrological applications, optimization of water resource allocations, pricing and water quality assessment, and agriculture and irrigation operations. Many data driven techniques have been seen to be utilized for this activity. The present study is an attempt to carry the baton forward with an aim of correlating the three stream gauging stations namely Ashti, Bhatpalli and Tekra which are situated in the Andhra Pradesh state at the Godavari River, India using Multi Gene Genetic Programming and Random Forest techniques. Previously measured streamflow values for the years of 1995–2013 at these three locations were used to develop the data driven models wherein stream flow at Tekra station is estimated using the stream flow values of the two upstream stations; Ashti and Bhatpalli. Monsoon monthly models and yearly models have been developed. All the models display better performance in estimating the stream flow at Tekra. The performance of developed models is judged by the traditional error measures along with the visual plots.

Keywords Coorelating · Discharge · Multi Gene Genetic Programming · Random Forest

---

P. Kulkarni (✉) · P. Dixit · S. Londhe
Vishwakarma Institute of Information Technology, Pune 411037, Maharashtra, India
e-mail: Preeti.kulkarni@viit.ac.in

P. Dixit
e-mail: pradnya.dixit@viit.ac.in

S. Londhe
e-mail: shreenivas.londhe@viit.ac.in

C. B. Pande et al. (eds.), *Surface and Groundwater Resources Development and Management in Semi-arid Region*, Springer Hydrogeology,
https://doi.org/10.1007/978-3-031-29394-8_9

1 Introduction

The stream flow measurements are used for a variety of water resources related studies and especially for developing stream flow prediction models. Unfortunately, the measurement records available for many stream gauging stations are either incomplete or too short. When there is no lateral flow between any two-stream flow measuring stations the traditional flood routing techniques can be implemented to streamflow at the downstream flow stations using the same at an upstream flow station. However, when one or more tributaries join the river in between two stations an unsteady and non-uniform flow analysis is necessary to predict the flow on the downstream side which though can be accomplished using the modern-day numerical methods still not very easy owing the exogenous data requirement. To counter this hurdle a technique of transfer of information from one or more nearby stream gauge stations to other can be implemented and is termed as correlating stream gauge stations. The bond of climate is the key for strong correlation between two stream gauge stations (William and Burns 1983). If the data for the upstream stations is available, the discharge at downstream station can be predicted using the recently available data driven techniques. With the invent of Machine Learning Techniques, Tree based techniques are paving their way in this area. These data-driven techniques learn from the input-output relationship without explicit knowledge of the physical processes or formulation of mathematical equations and try to give the reasonable solutions. To make up for their lack of ability to provide interpretation of the underlying mechanisms these techniques habitually require fewer data, demonstrate high accuracy in their performance, are computationally efficient, and can be used in real-time forecast (Mosavi et al. 2018; Adamowski 2008).

Researchers have attempted stream flow predictions using techniques like Artificial Neural Networks, Support vector Machines etc. Tongal and Booij (2018) forecasted daily streamflow in four rivers in the United 35 States with Support Vector Regression (SVR) and Artificial Neural Network (ANN) and Random Forest (RF) coupled with a baseflow separation method (Pande et al. 2022a; Pande et al. 2022b; Pande et al. 2023). In a study discharge was estimated at downstream station with dıscharge at upstream stations using Multigene Genetic Programming (MGGP) and concluded the work with satisfactory performance of the technique (Mengade et al. 2020). Mehr et al. (2017) suggested a new hybrid approach of Season Algorithm-Multigene Genetic Programming (SA-MGGP) to enhance timing accuracy of GP based rainfall-runoff models to enhance timing accuracy of the stand-alone GP-based models. Mehr et al. (2017), proposes a Pareto-optimal moving average multigene genetic programming (MA-MGGP) approach to develop a parsimonious model for single-station stream flow prediction. RF applications in hydrology include precipitation downscaling (Diez-Sierra et al. 2019) flood prediction (Muñoz et al. 2018), predicting flow characteristics at ungauged locations (Prieto 2019). Lanng et al. (2018) present long term Streamflow forecasting using Soil and Water Assessmet tool (SWAT) through the integration of random forests. In this study, a hybrid forecasting model is developed to improve accuracy of long-term stream flow forecasting by combining random (RF) forest and soil and water assessment tool (SWAT) (Pande et al. 2022c). The results show random forest gives accurate prediction. Mohamed et al. (2019) noticed that RF-regionalized models allowed slightly better prediction of streamflow in ungauged situations when compared with benchmark regionalization approaches. Ability of RF was assed to make 1-day lead time daily streamflow forecast at 86 watersheds in the Pacific Northwest and based on Kling-Gupta Efficiency scores (ranging from 0.62 to 0.99) and it is noticed that RF is able to produce useful forecasts across all the watersheds (Pham et al. 2021).

Thus, the earlier highlight the fact that Techniques like MGGP and RF re used in stream flow predictions and other hydrological applications. However, its application in the area of correlating gauging stations is seldom. Thus, the present work is an attempt to study the performance of techniques: MGGP and RF in discharge prediction at station which is at downstream with the discharge at upstream stations. In the current study three stations situated in Godavari River basin in India namely Ashti, Bhatpalle and Tekra are correlated using three Tree based techniques: Multi-Gene Genetic Programming (MGGP) and Random Forest (RF). All the developed models are trained and developed in two distinct ways in the current work. In the first case, separate monthly models are developed for monsoon months data whereas in second case, the models are developed of the entire data in the monsoon season. The models are assessed in testing by root mean square error, mean absolute error along with coefficient of determination. The present paper consists total of 6 sections in which study area and data is explained in Sect. 2, followed by brief explanation of techniques employed. Section 4 elaborated the methodology used for model development and immediately after those results are discussed in the Sect. 5. Conclusions are marked at the end in the Sect. 6 followed by the references.

2 Tree Based Techniques

Tree based approach is the basics of the three techniques utilized: Multi-Gene Genetic Programming (MGGP) and Random Forest (RF). The basic advantage of tree-based models is that they are easy to represent visually, making a complex predictive model much easier to interpret. Splitting and combining with a respective algorithm maximises the information gain.

2.1 Multi Gene Genetic Programming (MGGP)

Genetic programming was proposed by Koza (1992), being a generalization of the genetic algorithms. It is a search algorithm based on the principles of Darwin's Theory of Evolution which is often paraphrased as "survival of the fittest". MGGP is the most recent advancements of Genetic Programming (GP) that linearly combines low depth GP trees in order to improve fitness of classical GP (Searson 2007). Owing

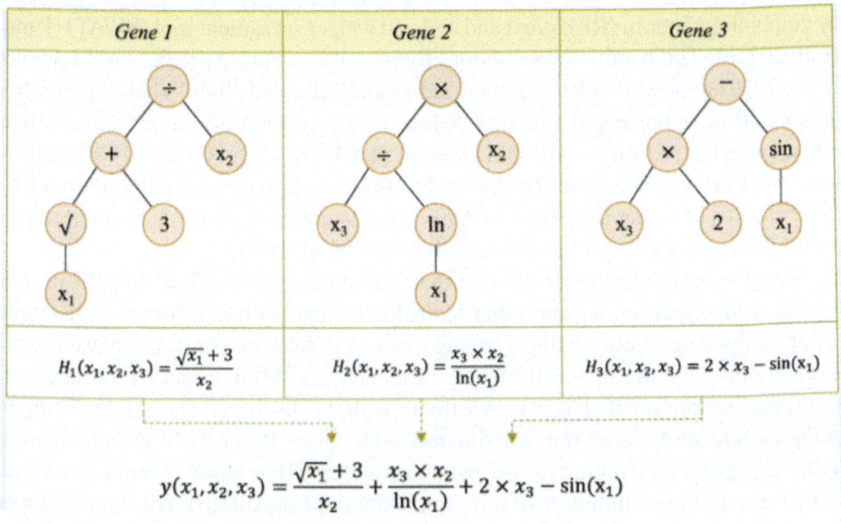

Fig. 1 Illustration of MGGP (Searson et al. 2007)

to the use of smaller trees, the MGGP is expected to provide simpler models than those of classical GP. In MGGP, predicted variables are computed by the weighted output of each gene in the multigene program plus a bias term. Multigene symbolic regression can be implemented using GPTIPS (is a free Explainable-AI machine learning platform and interactive modelling environment for MATLAB)toolbox for Matlab (Pandey 2015; Tzuc et al. 2019; Searson et al. 2007). The resulting pseudo-linear model can capture the non-linear behaviour. Forcing the transformations to be low order (by restricting the GP tree depth) allows the evolution of accurate, relatively compact mathematical models of predictor–response (input–output) data sets, even when there is a large number of input variables. Therefore, MGGP employs the power of classical linear regression method to capture nonlinear behaviours without need for pre-specified nonlinear structure. The highlighting characteristic of MGGP is that it provides a readily available equation which is can be used in excel for forecasting of output stream flow. Redears are referred to (Searson 2007) for further details. An illustration of MGGP is shown in Fig. 1 for 3 genes adapted from (Searson et al. 2007).

2.2 *Random Forest (RF)*

Random Forest (RF) proposed by Breiman in 2001, is an ensemble tree-based algorithm of semi-unsupervised and non-parametric type. It is from the decision tree family that comprises an ensemble of uncorrelated trees to yield prediction for classification and regression works. Since a single decision tree can produce high variance

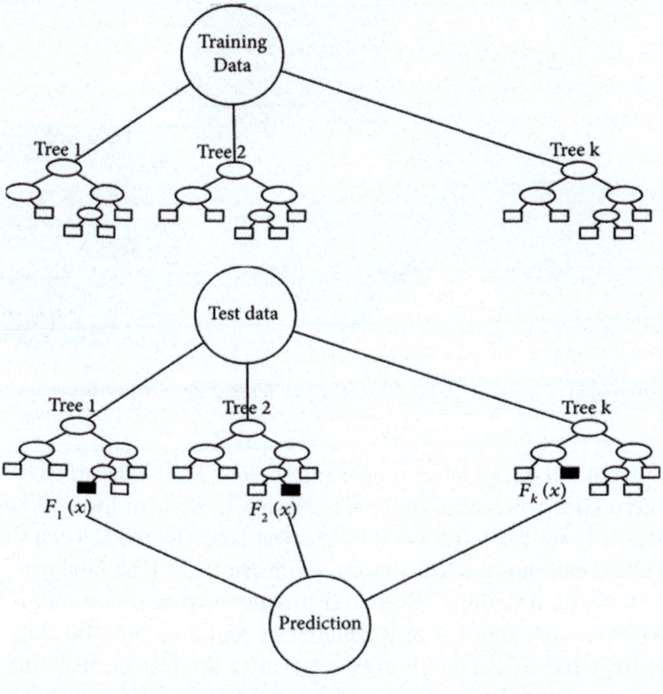

Fig. 2 Typical random forest tree (Granada et al. 2018)

and is prone to noise, RF addresses this limitation by generating multiple trees where each tree is built on a bootstrapped sample of the training data (James et al. 2013). Each time a binary split is made in a tree (also known as split node), a random subset of predictors (without replacement) from the full set of predictor variables is considered. One predictor from these candidates is used to make the split where the expected sum variances of the response variable in the two resulting nodes is minimized. The randomization process in generating the subset of the features prevents one or more particularly strong predictor from getting repeatedly chosen at each split, resulting in highly correlated trees (Granada et al. 2018). After all the trees are grown, each tree casts a vote on a label class for classification task or a prediction value for regression task. The output is the most popular class or the average of all regression values. In the recent days, the step-by-step of building a regression RF is explained in Pham et al. (2021) and readers are directed to refer these articles for more details. A typical Random Forest tree is shown in Fig. 2 adapted from (Granada et al. 2018).

3 Study Area and Data

The present study is carried out at the three stations of Godavari River namely Ashti, Bhatpalli and Tekra which are situated in Andhra Pradesh, India and hence a brief

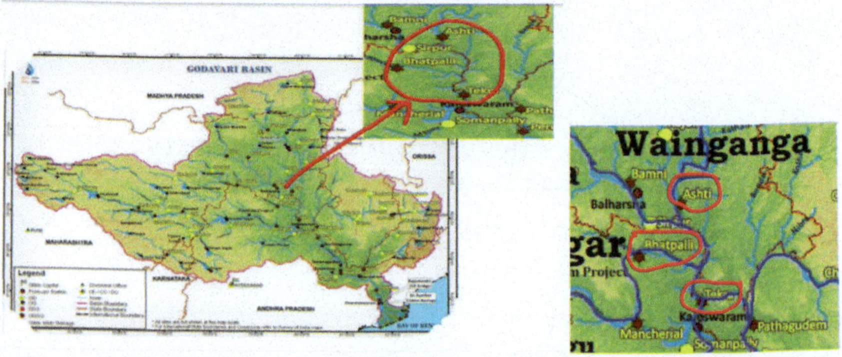

Fig. 3 Location Map of stations: Ashti, Bhatpalli and Tekra (http://indiawris.gov.in)

mention about the Godavari River is presented below. The Godavari River originates in the Western Ghats of central India near Nashik in Maharashtra, 80 km from the Arabian Sea. It flows for 1,465 km, first eastwards across the Deccan Plateau then turns southeast, entering the West Godavari district and East Godavari district of Andhra Pradesh, until it splits into two distributaries that widen into a large river delta at Dowleswaram Barrage in Rajahmundry and flow into the Bay of Bengal. The Godavari is India's second longest river after the Ganga. In terms of length, catchment area and discharge, the Godavari is the largest in peninsular India, and had been dubbed as the Dakshina Ganga (Ganges of the South). Tekra station is at the downstream side of Ashti and Bhatpalli and thus streamflow from the Ashti and Bhatpalli reaches at Tekra by natural gravity flow. The daily previously measured discharge (stream flow values in m^3/s) at these three locations is made available for the years of 1995–2013 from the India WRIS portal (http://indiawris.gov.in). Figure 3 and Table 1 presented below, showcases the location map of the study area and the data characteristic in detail respectively.

Examining the data from Table 1, one observes that the average discharge for monsoon months of July to October differ considerably. Though stations belong to the same river basin, there is a remarkable difference between measured values of stream flow. A comparatively high skewed data can be seen for station Bhatpalle indicating that the data distribution in the right tail is long relative to the left tail. Higher standard deviation at Ashti indicate variation in discharge values. For Bhatpalle the distribution shows an asymmetry as 12.07 in August and 18.30 in October. The data is deviated as shown in Table 1. Ashti has high standard deviation for all months which indicates that the data is spread out over a wider range. Bhatpalle shows a low standard deviation which means that the data spread is not in a wider range also it indicated that the values tend to be slightly close to the mean value of data set.

Table 1 Data characteristics of daily discharge values from 1995 to 2013 at the three locations

Station	Month	Max (m^3/s)	Min (m^3/s)	Mean	Skewness	Standard deviation
Ashti	July	24,276	11.40	203.65	3.85	2723.38
	August	20,613	59.88	1589.98	2.57	2981.89
	September	24,481	135.00	2715.63	3.7	3011.95
	October	5511	73.07	555.16	3.07	723.75
4 monthly Data	July–October	24,481	11.40	1827.68	3.604	2694.33
Bhatpalle	July	2307	1.00	70.58	7.06	198.54
	August	6652	1.32	121.66	12.07	365.31
	September	1567	0.94	88.19	5.94	156.73
	October	7862	2.94	61.12	18.3	361.12
4 monthly data	July–October	7862	1.00	90.92	16.09	295.19
Tekra	July	38,915	14.24	2509.66	4.28	4306.31
	August	32,665	126.92	4339.28	2.72	4324.25
	September	20,695	256.93	3291.41	2.30	3403.30
	October	17,223	92.30	1120.81	4.91	1512.73
4 monthly data	July–October	38,915	14.24	2811.42	3.48	3763.02

4 Methodology Adopted

Correlating stream gauging stations: Ashti, Bhatpalli and Tekra i.e. estimating the discharge at downstream station Tekra using the discharge data at upstream stations: Ashti and Bhatpalle; is the objective of the current work. Analysing the data used in the present work, two sets of models were developed. Set 1: Discharge prediction at Tekra (Output) with discharge at Ashti and Bhatpalle respectively (as input/s) for monsoon months i.e. separate models for July, August, September and October; Set 2 models with same input and output as in set 1, however the data now is for combined 4 monsoon months. Each set of models with its respective Input and output is shown in Table 2.

RF models were developed in Python through Jupyter notebook (a web-based interactive computing platform). The trees selected were between 100 and 500, and the bagging iterations and the tree size were selected, anticipating that these were those that yielded the best performance by a low Mean Square Error (MSE). Multi-Gene Genetic Programming was developed in Matlab 2017 using GPTIPS-2 (an open-source toolbox for MGGP). Readers are referred for features of GPTIPS to Searson et al. (2007, 2015). The following parameters were selected owing to the trials conducted for less complexity of olutions and experience of the authors:

Table 2 Methodology of models developed

Set No:	Input (I)–Output (O)	Model No. and respective month (Each model developed using MGGP and RF)
1	Ashti (I) and Tekra (O)	1 (July), 2 (August), 3 (September), 4 (October)
	Bhatpalle(I) and Tekra (O)	5 (July), 6 (August), 7 (September), 8 (October)
	Ashti and Bhatpalle (I) and Tekra (O)	9 (July), 10 (August), 11 (September), 12 (October)
2	Ashti (I) and Tekra (O)	13 (July–October)
	Bhatpalle (I) and Tekra (O)	14 (July–October)
	Ashti and Bhatpalle (I) and Tekra (O)	15 (July–October)

Number of generations: 200–500, Selection method: Tournament with size 13–15, Cross-over rate: 0.78–0.84, Mutation rate: 0.14–0.20, Termination criteria: 500 generation or fitness value less than 0.00 whichever is earlier, Maximum number of genes and tree depth: 4–5. All the models were developed using MGGP and RF techniques and 70% data was used for training the models whereas 30% data was used for the testing. Root Mean Squared Error (RMSE), Mean Absolute Error (MAE) and Correlation coefficient (r) along with the scatter plots and flow plots were used to judge the performance of the models in the testing phase. High prediction accuracy was expected for lower error statistics (RMSE, MAE), while the opposite was valid for the correlation coefficient. Further, the degree by which RMSE exceeded MAE was an indicator of the extent to which outliers (or variance in the differences between the modelled and observed values) existed in the data (Jain et al. 2008; Legates et al. 1991; Londhe 2008). Along with errors, the performance of the model in testing will be judged through visual technique as Scatter plot, Hydrograph and Box Plot.

Along with Scatter plot and Hydrograph, Box plot are effective visualisation method which provide statistical information like medians, ranges, and outliers. Box plot, will enable to see the data representation in their quartiles. The bottom and top of the box represent the first and third quartile, and the band inside the box represents the second quartile or median. The whisker in the box plot which extends to the most extreme data points; that is, the largest and smallest values which are not outliers, and the value should be no more than 1.5 * IQR (interquartile range) from the box. Any data outside the whisker is considered an outlier (https://www.simplypsychology.org/boxplots.html).

It is interesting to note that the techniques Random Forest and MGGP are data driven techniques and the crux of their performance lies in the data; specifically, the training data. Addition to the data given in table, the Fig. 4a, b shows the box plot of training data utilised for Model 3 and testing data utilised for Model 3. A higher variability in the data can be seen for Ashti in training phase with a greater number of outliers as compared to that in testing. The discharge at Tekra (which is the output) shows more variability in testing dataset with less outliers in testing phase, however

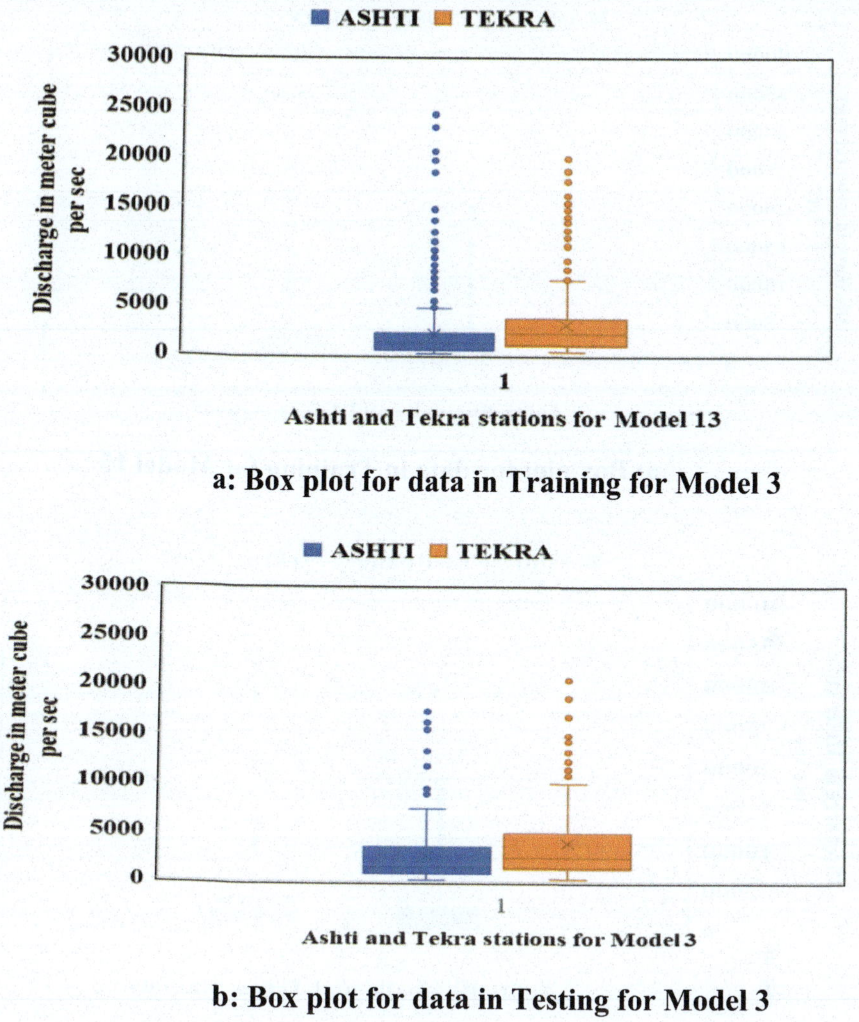

Fig. 4 **a** Box plot for data in Training for Model 3, **b** Box plot for data in testing for Model 3

with higher condensed data in between Q1 and Q3 in testing as compared to that in training. The standard deviation of Discharge value at Tekra in training is less (3175.81) as that of the same in testing (3822.75). The discharge data of Ashti and Tekra in Training also positive skewness, however less than that in the testing phase. All of this data characteristics affects the performance of all models and can provide hint/s towards performance of the model.

Similar characteristics of data can be seen in Data utilized for Model 15 (Refer Fig. 5a, b).

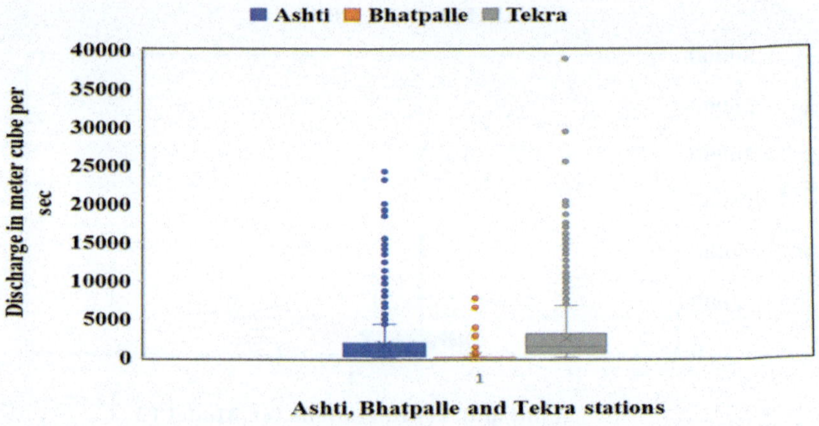

a: Box plot for data in Training for Model 15

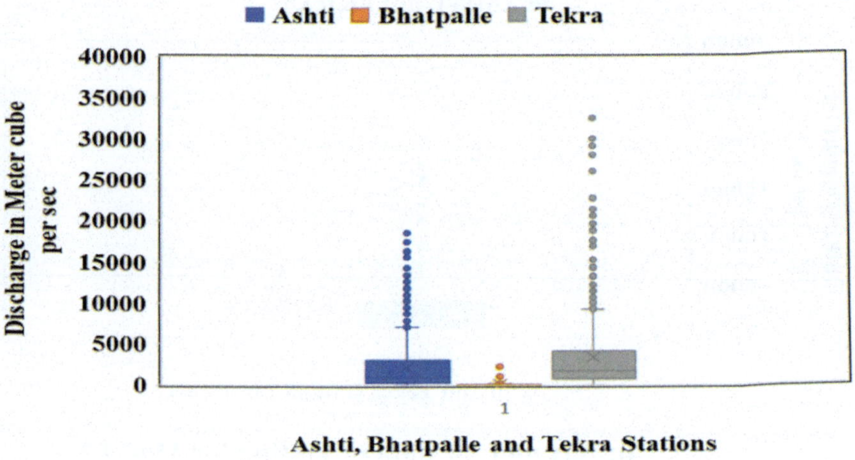

b: Box plot for data in Testing for Model 15

Fig. 5 **a** Box plot for data in Training for Model 15, **b** Box plot for data in Testing for Model 15

5 Results and discussion

The present work attempts estimating discharge at Tekra with inputs as discharge at Ashti and Bhatpalle respectively. 15 Models were developed in total each using MGGP and RF. The performance of each model is shown in Table 3.

Table 3 Performance of models developed using MGGP and RF

Model No	Input Data	Inputs (Station)	Output (Station)	RF (Correlation coefficient)	MGGP (Correlation coefficient)	RF (MAE) m^3/s	MGGP (MAE) m^3/s	RF (RMSE) m^3/s	MGGP (RMSE) m^3/s
Model 1	July	Ashti	Tekra	0.90	0.84	1220.45	1011.22	2444.91	1028.91
Model 2	August			0.81	0.86	1654.73	990.07	3173.40	1801.73
Model 3	September			0.93	0.91	910.58	1118.96	1429.87	1782.87
Model 4	October			0.90	0.56	306.84	183.81	606.43	303.15
Model 5	July	Bhatpalle	Tekra	0.49	0.35	2351.19	1339.29	4826.49	1999.05
Model 6	August			0.51	0.62	2462.44	1830.73	4586.76	2788.16
Model 7	September			0.51	0.53	2068.58	2258.56	3266.44	3550.29
Model 8	October			0.71	0.36	483.63	223.94	976.87	341.27
Model 9	July	Ashti, Bhatpalle	Tekra	0.92	0.84	1136.81	676.10	2191.74	1027.60
Model 10	August			0.80	0.92	1657.76	800.35	3222.86	1415.09
Model 11	September			0.93	0.93	839.10	899.56	1363.48	1568.59
Model 12	October			0.90	0.78	269.84	134.72	606.97	229.51
Model 13	July to October	Ashti	Tekra	0.83	0.80	1167.46	1517.09	2550.25	3269.86
Model 14		Bhatpalle	Tekra	0.58	0.50	1923.58	2500.44	3754.12	4753.14
Model 15		Ashti, Bhatpalle	Tekra	0.90	0.86	1238.78	1377.61	2461.28	2765.92

Table 3 shows that MGGP and RF have been performing good in estimation of discharge for both sets of models. However, RF shows an upper hand in performance as compared to MGGP which can be seen through higher correlation coefficient between actual and predicted values (r = 0.91–0.93), higher MAE and lower RMSE. MGGP models however display higher RMSE values. Model 1–4 and Model 9–12 perform well as compared to the models 5–8. Model 5–8 display lower performance as compared to other models in both the techniques. This can be attributed toward the fact that Bhatpalli station is at a distance from Tekra as seen in Fig. 3. Also as seen in Table 1 the skewness of discharge data at Bhataplli is more as compared to other stations suggesting its higher right tailed data. However, with discharge at Ashti and Bhatpalli as input parameters the performance of the models show a rising trend as seen in table. In monthly models, September models show a good performance for both the techniques. This can be attributed towards the standard deviation of the data at Ashti and Bhataplli and Tekra. The difference in the standard deviation between input and output for September models is less as compared to that of other models. October models are the next best models followed by others. Model 13–15 also display an acceptable performance with higher r (0.83 and 0.90) with station inputs as Ashti, Ashti and Bhatpalli. Results in model 14 are at lower side and is due to the location of station: Bhatpalle. Monthly models show better performance as compared to yearly models as the difference in the standard deviation between the stations for yearly data is more as compared to that of monthly data thus suggesting more dispersion of data. Higher skewness of yearly data suggesting asymmetric data at Bhatpalli as compared to the same at other stations also contribute towards lower performing models.

RF as a technique utilizes ensemble technique which combine the prediction of several models, which results in better performance compared to what could be obtained from a single model. This attribute of RF also contributes towards lower risk of overfitting of data. Standard Random Forest (SRF) is a powerful method for high-dimensional regression and classification. For Random Forest, SRF gives an accurate approximation of the conditional mean of a response variable, revealing that RF provides information about the full conditional distribution of the response variable, not only about the conditional mean One of the tree developed in RF for model 3 is shown in Fig. 6 along with the detail of one of leaf.

MGGP on other hand combines the weighted output of each gene to display a single model and bias (in form of a regression equation). The regression equation developed for Model 3 and Model 15 is shown in Eq. 1 and Eq. 2 respectively.

$$
\begin{aligned}
qt &= 67900.a^{0.125} + 19900.a^{0.25} - 55473.6a^{0.5} + 48000.(-a^{0.5} + a + 278)^{0.5} \\
&- \frac{1.524 * 10^{16}(a + 201.5)(a + 416.)}{1.37 * 10^{11}a^3 + 5.95 * 10^{13}} - 2222(\frac{-a^{0.5} + a^3 + a^2}{a + 329})^{0.5} \\
&+ 2222a + 6655(a - 15.2)^{0.5} - 711526
\end{aligned}
\tag{1}
$$

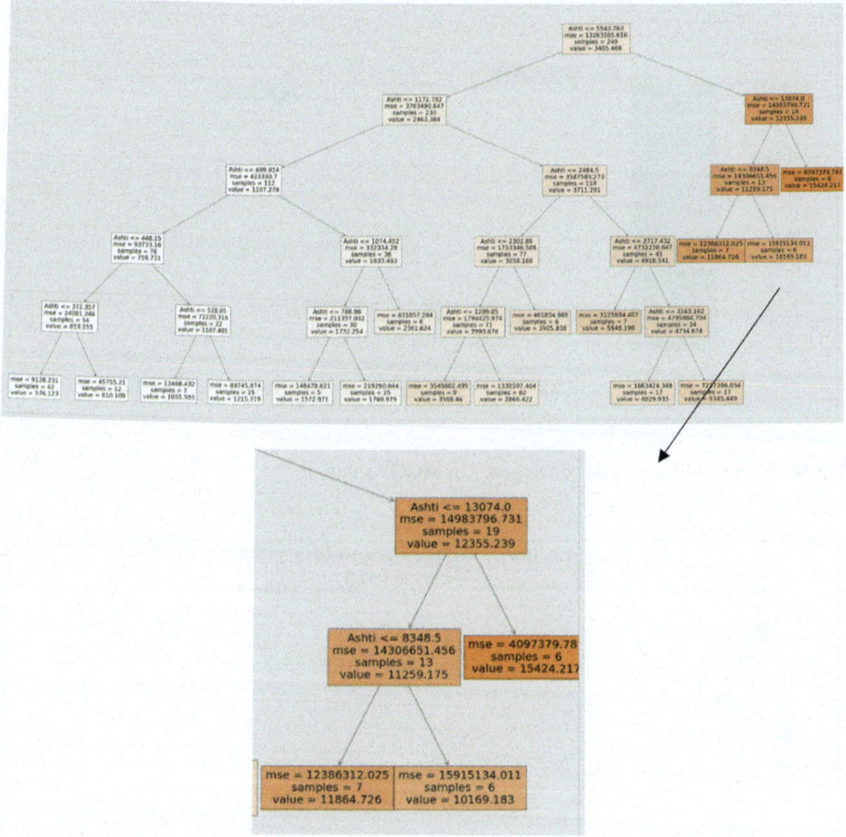

Fig. 6 One tree developed in RF for Model 3

$$qt = -0.000491b\left(-2a^{0.5} + \frac{b^3}{a^3} + \frac{3a}{b} + a\right) + \frac{(2.13742 * 10^{16}a^{0.5} + 2.36185 * 10^{16})(b^2)^{0.5} + 8.64 * 10^{15} * a^{0.5}b + 8.64 * 10^{15}a^2}{-8.84 * 10^{15}a^{1.5} + 1.41 * 10^{14}a^{0.5} + 4.41 * 10^{16}a^2 + a(8.82 * 10^{16}b - 3.53296 * 10^{16})(b + 4.87430417 * 10^{-15})(b + 1.92443)} + 2.04a \quad (2)$$

where qt = discharge at Tekra, a = Discharge at Ashti, b = Discharge at Bhatpalli.

The weight of respective genes offer knowledge about importance of that genes while predicting the output. The gene having highest weight has largest contribution towards the final prediction. It is seen that the weights of genes 1, 3 and bias term are higher than the other genes indicating that they have higher contribution to the prediction of stream flow (Fig. 7). The Pareto front report of MGGP model M-07 shown in Fig. 8 signals at the accuracy of each model against the complexity of evolved models. It is used to plot expressional complexity against the goodness of fit (R^2) for all the models. The Pareto front report enables the user to visualize the performance of solutions and select a solution that retains a balance between

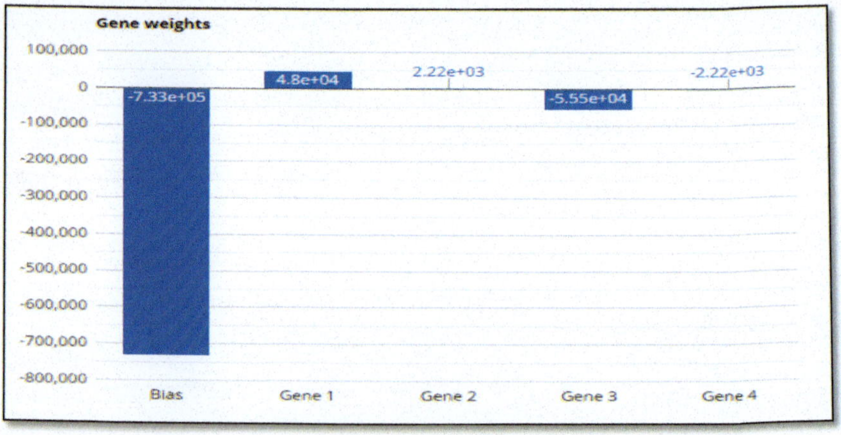

Fig. 7 Weights of Genes and convergence of MGGP solution

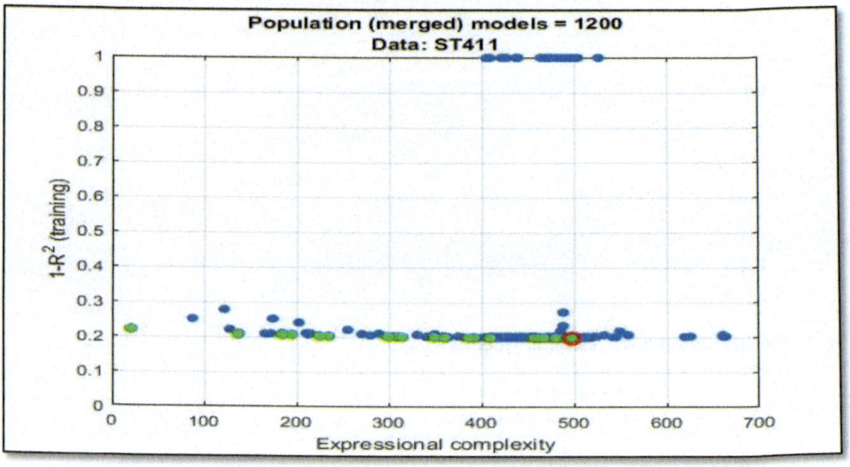

Fig. 8 Pareto front report of MGGP

complexity and accuracy. Green dots represent the Pareto front models in terms of model performance (1-R^2) and model complexity. Blue dots represent non-Pareto models. The red circled dot represents the best model in the population in terms of R^2 on training data set (Searson et al. 2007).

The performance of models discussed in Table 3 are also been visually interpreted through scatter plots and flow diagrams. Figure 9 shows the plot for model 3 and 15 which clearly indicates the close concur of discharge values estimated by MGGP and RF with the observed values. A slight difference in the estimation of the peak values in model 15 can be seen in Fig. 10.

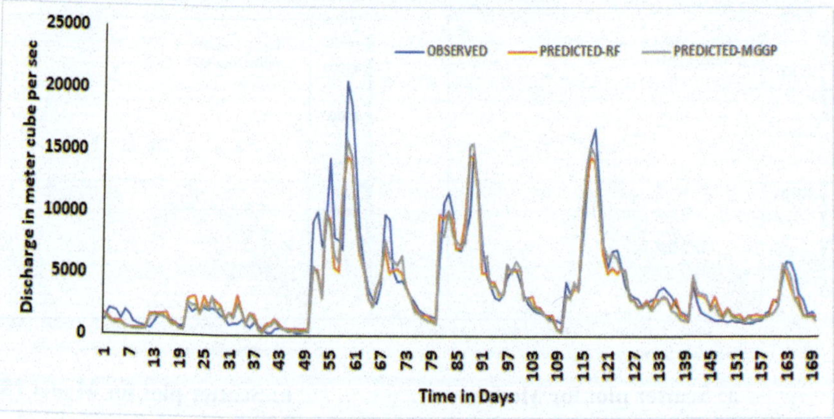

Fig. 9 Plot for observed and Estimated discharge using MGGP and RF for Model 3

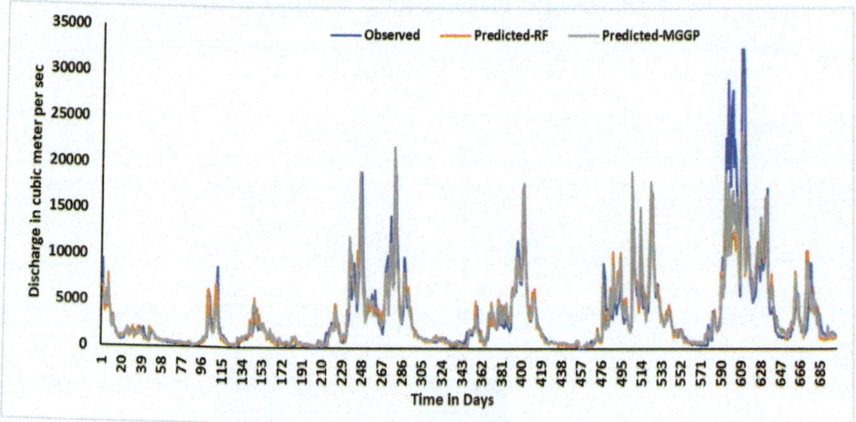

Fig. 10 Plot for observed and Estimated discharge using MGGP and RF for Model 15

Figure 11a shows the combined scatter plot for the observed and predicted values in Model 3 using MGGP and RF and Fig. 11b for model 15. Scatter plot for Model 3 shows an almost balanced scatter, however higher underprediction can be seen in scatter plot for Model 15, specially for the higher values. This also can be seen in figure above. This also explains the higher RMSE for Model 15 with 2461.28 for RF and 2765.92 with MGGP.

Figure 12 shows the box plot for Model 3 in Testing phase. The box length gives an indication of the sample variability and the line across the box shows where the sample is centred. A large variability in Observed discharge values can be seen followed by that of discharge values predicted by MGGP and then the discharge values predicted by RF. MGGP shows the highest value in its predicted data as compared to RF. The figure also shows observed values with higher outliers as

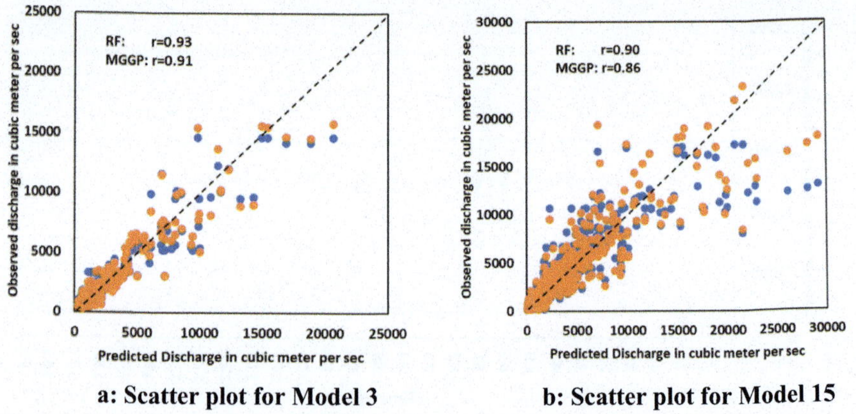

Fig. 11 **a** Scatter plot for Model 3, **b** Scatter plot for Model 15

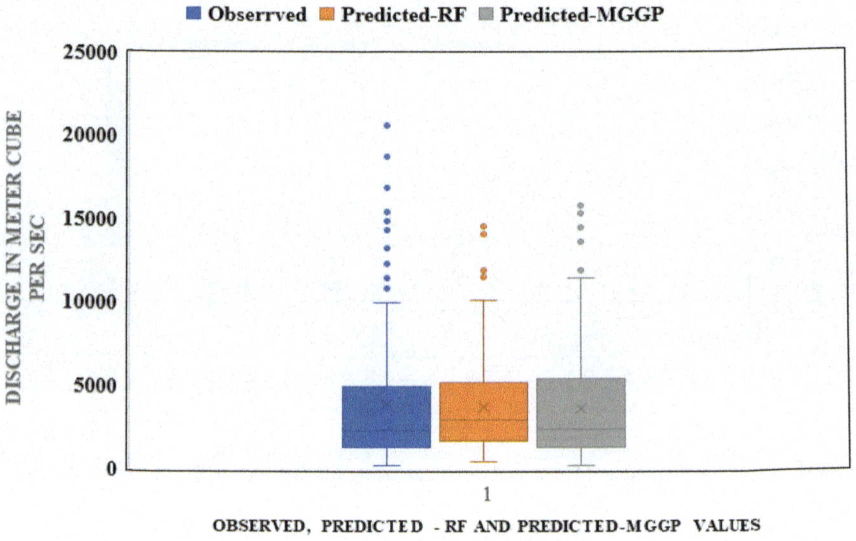

Fig. 12 Box plot for Observed and Predicted values for Model 3

compared to Predicted values by RF and MGGP respectively, which also hints towards the error between the observed and predicted values. The box lot also shows the median line towards the lower bound of the box plot and the upper whisker is longer than the lower one in observed and predicted values, indicating the right skewness of data. The box plot also shows more amount of data in the quartile 3 i.e. Q3 as compared to that in Q1. Thus the analysis of Observed and predicted values can help in providing insight towards the error analysis.

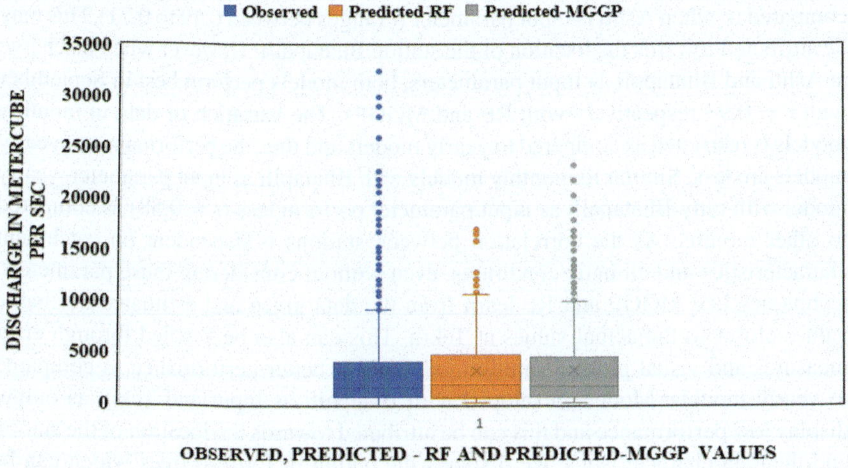

Fig. 13 Box plot for Observed and Predicted values for Model 15

Further analysis also shows that the highest discharge value in the testing dataset is 20694.84 and the predicted value for the same by RF is 14656.0308 and by MGGP is 15854.35, thus RF and MGGP shows underprediction for higher value, alias MGGP tries to predict closer to observed value. For the minimum observed value of 256.92, the predicted value by RF is 493.83 which is overprediction and by MGGP the value is 297.38 which is slight overprediction. The median value also shows an overprediction through RF.

A similar box plot of Observed and Predicted values for Model 15 is shown in Fig. 13. In the testing dataset, higher variability in predicted values can be seen for MGGP as compared to RF. Higher skewness of the data can be seen in Observed values as compared to that of RF and MGGP.

6 Conclusion

In the present research, the work of correlating stream gauge stations is carried out for three stations in Godavari River basin namely Ashti, Bhatpalle and Tekra. The models are formulated using Multigene Genetic Programming (MGGP) and Random Forest (RF) in two ways based on monthly and yearly arrangement of data. In case of Yearly correlating models, there are only four models one each from the respective combination while in case of Monthly correlating models; there are altogether 15 models with each technique. MGGP and RF display satisfactory results towards estimation of discharge at Tekra with RF showing best results.With input as discharge at Ashti and output as discharge at Tekra, September model displays a good result with RF as Techniue (r with RF = 0.93 and MGGP = 0.91) With Bhatpalli as input parameter and Tekra as output, both the moels perform poor as compared to when Ashti is input parameter (r ranges between 0.49 to 0.71).This may be attributed towards the location of the station Bhatapalli. However when discharge at Ashti and Bhatappli as Input parameters, both models perform best in September with r = 0.93 respectively with RF and MGGP). The variation of data in monthly models is restricted as compared to yearly models and thus the performance of yearly models are less. Similar to monthly models with Bhataplli as input parameter, yearly model with only Bhatapalli as input parameter perform less (r = 0.58) as compared to other months. As the correlation between stations is dependent on catchment characteristics and climatic conditions, even without considering these parameters, techniques like MGGP and RF learn from the data given and estimated discharge values closer to the actual values at Tekra. This can also be verified through error measures and visual plots. Monthly models show better performance as compared to yearly models. Models developed with Bhatpalli as input and Tekra as output display low performance and this can be attributed towards the location of the station and the data characteristics. RF displays the output in form of trees which can be easily interpreted and also MGGP shows the output in form of equation which can be readily used by the end users.

References

Adamowski JF (2008) Development of a short-term river flood forecasting method for snowmelt driven floods based on wavelet and cross-wavelet analysis. J Hydrol 353:247–266

Breiman L (2001) Random forests. Mach Learn 45:5–32

Diez-Sierra J, del Jesus M (2019) Subdaily rainfall estimation through daily rainfall downscaling using Random Forests in Spain. Water 11:1–25

Granada F, Saroli M, de Marinis G, Gargano R (2018) Machine learning models for spring discharge forecasting. Geofluids. Article ID: 832816

http://indiawris.gov.in

James G, Witten D, Hastie T, Tibshirani R (2013) An introduction to statistical learning, vol 103 XIV. Springer, New York, Heidelberg, Dordrecht, London

Jain A, KumarJha S, Misra S (2008) Modeling and analysis of concrete slump using Artificial Neural Networks. J Mater Civ Eng 20:628–633

Koza JR (1992) Genetic programming: on the programming of computers by means of natural selection. MIT Press, Cambridge, MA

Legates DR, McCabe GJ Jr (1991) Evaluating the use of "goodness of fit" measures in hydrological and hydro climatic model validation. Water Resour Res 35:233–241

Londhe SN (2008) Soft computing approach for real-time estimation of missing wave heights. Ocean Eng 35:1080–1089

Liu GR, Liu MB (2003) Smoothed Particle Hydrodynamics: a meshfree particle method. World Scientific Publishing, Singapore

Liang Z, Tang T, Li B, Liu T, Wang J, Hu Y (2018) Long-term streamflow forecasting using SWAT through the integration of the random forests precipitation generator: case study of Danjiangkou Reservoir. Hydrol Res 49:513–1523

Mosavi A, Ozturk P, Chau K (2018) Flood prediction using machine learning models: literature review. Water 10(11):1536,1–40.

Mengade A, Londhe SN, Dixit PR, Kulkarni P (2020) Correlating Stream Gauge stations using Multigene Genetic Programming. In: Proceedings of 25th international conference on

hydraulics, water resources and coastal engineering (HYDRO 2020) held online, hosted by NIT Rourkela, India during 26–28 March 2021
Mehr AD, Nourani V (2017) Season algorithm-multigene genetic programming: a new approach for rainfall-runoff modelling. Water Resour Manage 32:2665–2679
Mehr AD, Kahya E (2017) A Pareto-optimal moving average Multigene Genetic Programming model for daily stream flow prediction. J Hydrol 549:603–615
Muñoz P, Orellana-Alvear J, Willems P, Célleri R (2018) Flash-flood forecasting in an Andean Mountain catchment—development of a step-wise methodology based on the Random Forest Algorithm. Water 10(11):15–19
Mohamed S, Ludovic O, Ribstein P (2019) Random forest ability in regionalizing hourly hydrological model parameters. Water 11(10):1540, 1–22
MCLeod S (2019) What does a box plot tell you. https://www.simplypsychology.org/boxplots.html
Pham LT, Lifeng L, Andrew OF (2021) Evaluation of Random Forest for short-term daily streamflow forecast in rainfall and snowmelt driven watersheds. Hydrol Earth Syst Sci 25:2997–3015
Prieto C, Vine NL, Kavetski D, Garcia E, Medina R (2019) Flow prediction in ungauged catchments using probabilistic Random Forests regionalization and new statistical adequacy tests. Water Resour Res 55:4364–4392
Pandey DS, Pan I, Das S, Leahy JJ, Kwapinski W (2015) Multi-gene genetic programming based predictive models for municipal solid waste gasification in a fluidized bed gasifier. Biores Technol 179:524–533
Pande CB et al (2022a) Forecasting of SPI and meteorological drought based on the artificial neural network and M5P model tree. Land 11, no 11:2040. https://doi.org/10.3390/land11112040
Pande CB (2022b) Land use/land cover and change detection mapping in rahuri watershed area (ms), india using the google earth engine and machine learning approach. Geocarto Int 37(22):1–21
Pande CB, Kadam SA, Jayaraman R, Gorantiwar S, Shinde M (2022c) Prediction of soil chemical properties using multispectral satellite images and wavelet transforms methods. J Saudi Soc Agric Sci 21:21–28
Pande CB, Costache R, Sammen SS et al (2023) Combination of data-driven models and best subset regression for predicting the standardized precipitation index (SPI) at the Upper Godavari Basin in India. Theor Appl Climatol 152:535–558. https://doi.org/10.1007/s00704-023-04426-z
Searson DP, Willis MJ, Montague GA (2007) Co-evolution of non-linear PLS model components. J Chemom 21(12):592–603
Searson DP (2015) GPTIPS 2: an open-source software platform for symbolic data mining. In: Chapter 22 in handbook of genetic programming applications, New York
Tongal H, Booij MJ (2018) Simulation and forecasting of streamflows using machine learning models coupled with base flow separation. J Hydrol 564:266–282
Tzuc OM, Hernández-Pérez I, Macías-Melo EV, Bassam A, Xamán J, Cruz B (2019) Multi-gene genetic programming for predicting the heat gain of flat naturally ventilated roof using data from outdoor environmental monitoring. Measurement 138:106–117
William A, Burns A (1983) Mixed Station Extension of monthly stream flow records. J Hydraul Eng 109:10–11

Mapping and Trend Analysis in Water Spread Area of Upper and Lower Lakes of Bhopal, Using Remote Sensing Technique

Vaibhav Deoli, Deepak Kumar, Deep Shikha, Shivam Saw, and Rohit Patel

Abstract Surface water is essential for all forms of life. Identification and calculation of water bodies can be useful in various ways like drought mapping, flood mapping, drinking and irrigation water analysis. This study is done for revolving the affectivity of satellite data in water spread mapping and area estimation of Upper Lake and Lower Lake of Bhopal city, Madhya Pradesh. Water surface areas for 21 years from 2001 to 2021 are calculated. To calculate the water spread area of studied lakes, Landsat-7 imageries from 2001 to 2012 and Landsat-8 imageries from 2013 to 2021 were used. To calculate the area of lakes, 3 water indices namely Normalized Difference Water index (NDWI), Modified Normalized Difference Index (MNDWI) and Water Ratio Index (WRI) used. To check the accuracy of the water index physical GPS survey has been conducted. Based on that survey, MNDWI was the most accurate method in this study. Furthermore, to detect the trend and magnitude of the trend in the water spread area Mann–Kendall and Theil's Sen Slope Test has been used. Z-value for Upper Lake and Lower Lake was obtained as –3.23 and –3.097 respectively. Both Lake shows a decreasing trend in water surface area for study period which is significant at 10% as well as 5% level of significance. The result can be effective in water related problems like drought, flood or debris dams.

Keywords Landsat · MNDWI · Trend analysis · Lake · Remote sensing

1 Introduction

For the sustainability of life on earth, surface water is an important component for humans as well as other forms of life (Acharya et al. 2016). Lake is one of the

V. Deoli (✉) · D. Kumar
Department of Soil and Water Conservation Engineering, College of Technology, GB Pant University of Agriculture and Technology, Pantnagar 263145, India
e-mail: deolivaibhavdeoli@gmail.com

D. Shikha · S. Saw · R. Patel
Department of Environmental Science and Engineering, Indian Institute of Technology (Indian School of Mines), Dhanbad 826004, India

C. B. Pande et al. (eds.), *Surface and Groundwater Resources Development and Management in Semi-arid Region*, Springer Hydrogeology,
https://doi.org/10.1007/978-3-031-29394-8_10

essential sources of surface water. Lake water has an important role in the ecosystem which may include water storage, fishery production, agriculture production and flood mitigation (Deng et al. 2017). Due to fluctuations in lake water all of these ecosystems are severely affected. They are affected by climate change and human activities also on different time scales on the regional hydrological processes (Cao et al. 2021; Tao et al. 2015; Zhang et al 2011). Hence, the identification and mapping of lakes might be useful in various ways for surface water estimation. Remote sensing and GIS is an effective technique for mapping of lake water (Feng et al. 2018; Qin et al. 2021; Kaplan and Avdan 2017). Remote sensing provides a well-established and sufficient platform for monitoring and management of different surface water features without physical contact with them. Landsat series, Sentinel satellite data and MODIS are effectively used to estimate water resources.

To estimate the lake water surface area many researchers successfully used Landsat series imagery (Rover et al. 2012; Yang et al. 2015; Deoli and Kumar 2019; Kandekar et al. 2021). To analysis Landsat imageries, water indices have been used due to their simplicity, low cost and superior analysis process (Elsahabi et al. 2016; Du et al. 2016; Mohsen et al. 2018). Most commonly used water indices are the Normalized Difference Water Index (NDWI), Modified NDWI, Water Ratio Index (WRI), and General Water Index (GWI). Ozelkan (2019) analysed water body detection from Landsa-8 imagery using NDWI for Atikhisar Dam Lake situated in western Turkey. Yue et al (2017) studied dynamic change of Hongjiannao Lake. They used MNDWI to extract lakes from 1988 to 2014 and find that the lake water is decreasing in the studied period. Yang and Du (2017) used Landsat TM imagery to extract water bodies. To calculate water surface area, author used NDWI, GWI and MNDWI and concluded that MNDWI method has the highest accuracy. Dai et al. (2019) computed dynamic change in Bosten Lake in China. To evaluate the area, the authors used NDWI and MNDWI techniques. Thayammal et al. (2021) analysied water body segmentation from Landsat data and deep learning technique. Author used deep neural network and signet techniques. Deoli et al. (2022) studied water spread mapping of Nainital Lake situated in Nainital city of India for year 2001 to 2019. Authors used NDWI. MNDWI and WRI technique and ground truthing has been done using GPS survey of the lake. Based on the above review, this study has been carried out with following objectives: (i) to determine temporal change and detect lake water area in every year from 2001 to 2021 for Upper Lake and Lower Lake of Bhopal. (ii) to determine trend and magnitude of trend in lake water area for studied period.

2 Materials and Methods

2.1 *Study Area*

In current study, two lakes namely Upper Lake and Lower Lake have been studied to determine surface water change. Both lakes are in western end of the Bhopal city of Madhya Pradesh, India. Upper Lake has been situated at 23.250° N and 77.253° E and 509 m above mean sea level. Lower Lake is situated at 23.258° N and 77.405° E and 499 m above sea level (Fig. 1).

2.2 *Data Used*

The level 1 data of Landsat 7 and Landsat 8 sensors for every year for the month of April have been collected from the United States Geological Survey (USGS) website (Table 1). In case of no cloud-free Landsat image in the month of April, the image has been taken of nearby month. To carry out this study, Landsat 7 imageries from 2001 to 2012 and Landsat 8 imageries from 2012 to 2021 have been used. Specifications of Landsat-7 and Landsat-8 are given in Tables 2 and 3 respectively.

2.3 *Water Body Extraction*

To extract the water surface area of studied lakes from Landsat imageries, band rationing technique has been used. NDWI, MNDWI and WRI have been used for identification of water spread area of lakes. A model was developed in QGIS software for change detection. The equation for NDWI, MNDWI and WRI is given as below:

$$\text{NDWI} = (\text{Green} - \text{NIR})/(\text{Green} + \text{NIR}) \quad (1)$$

$$\text{MNDWI} = (\text{Green} - \text{SWIR\_1})/(\text{Green} + \text{SWIR\_1}) \quad (2)$$

$$\text{WRI} = (\text{Green} + \text{Red})/(\text{NIR} + \text{SWIR\_2}) \quad (3)$$

2.4 *Non-parametric Trend Analysis*

To find the significance of hydro-climatic data non parametric Mann–Kendall test gives valuable results and used by different researchers (Kumar and Kumar 2020;

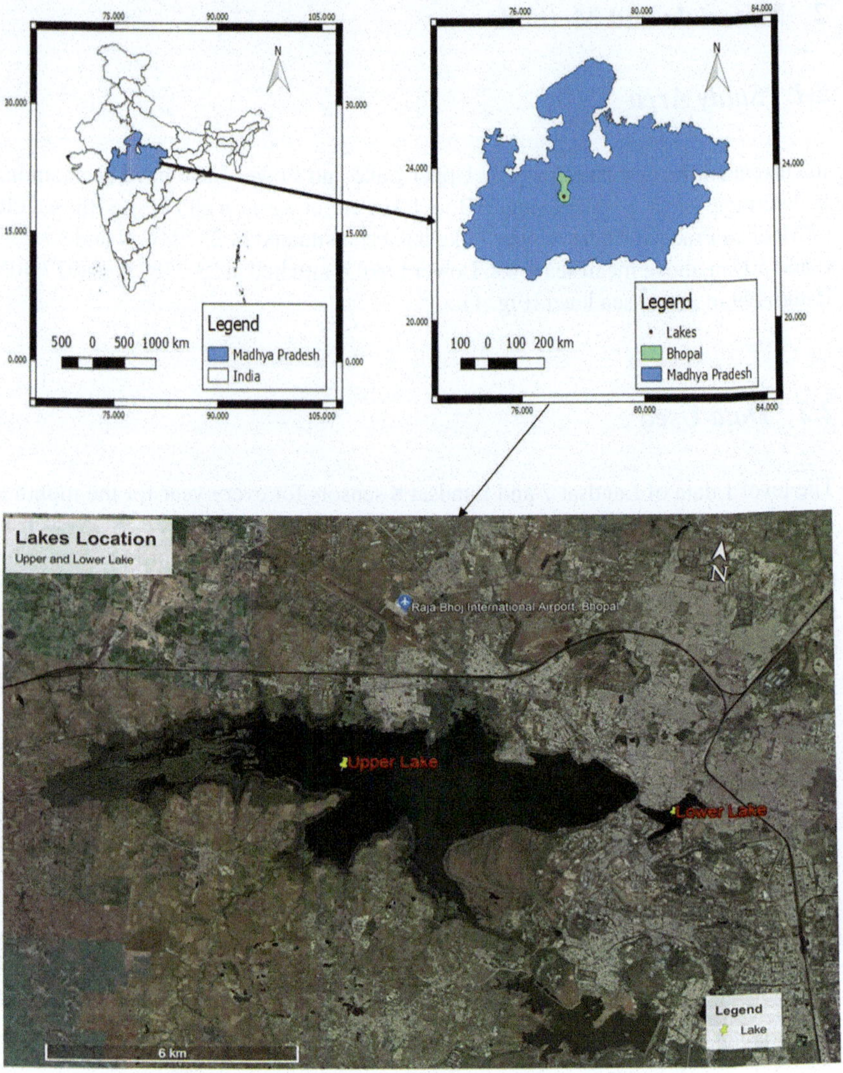

Fig. 1 Location map of studied lakes

Kuriqi et al. 2020; Rana et al. 2019). In this study Mann–Kendall test was used to detect trend in calculated lake surface area from 2001 to 2021. The magnitudes of trend were calculated using Theil's Sen Slope estimator test. The direction of change and the rate of increase or decrease in the lake surface water trend are revealed through the slope (Choudhury et al. 2012). The detailed mathematical expressions of the tests have been given in Rana et al. (2022), Deoli and Kumar (2020).

Table 1 Cloud free Landsat imageries used in study

Satellite	Year	Date of imagery
Landsat 7	2001	02 April, 2001
	2002	29 March, 2002
	2003	24 April, 2003
	2004	19 April, 2004
	2005	22 April, 2005
	2006	02 May, 2006
	2007	28 April, 2007
	2008	30 April, 2008
	2009	17 April, 2009
	2010	20 April, 2010
	2011	23 April, 2011
	2012	25 April, 2012
Landsat 8	2013	06 April, 2013
	2014	23 April, 2014
	2015	17 April, 2015
	2016	19 April, 2016
	2017	22 April, 2017
	2018	25 April, 2018
	2019	28 April, 2019
	2020	23 April, 2020
	2021	26 April, 2021

Table 2 Specification of Landsat-8

Bands	Wavelength (micrometres)	Resolution (meters)
Band 1—Ultra Blue	0.43–0.46	30
Band 2—Blue	0.46–0.51	30
Band 3—Green	0.53–0.59	30
Band 4—Red	0.64–0.67	30
Band 5—Near Infrared	0.85–0.88	30
Band 6—SWIR 1	1.57–1.65	30
Band 7—SWIR 2	2.11–2.29	30
Band 8—Panchromatic	0.50–0.68	15
Band 9—Cirrus	1.36–1.38	30
Band 10—Thermal Infrared 1	10.60–11.19	100
Band 11—Thermal Infrared 2	11.50–12.51	100

Table 3 Specification of Landsat-7

Bands	Wavelength (micrometres)	Resolution (meters)
Band 1—Blue	0.45–0.52	30
Band 2—Green	0.52–0.60	30
Band 3—Red	0.63–0.69	30
Band 4—NIR	0.77–0.90	30
Band 5—SWIR 1	1.55–1.75	30
Band 6—Thermal	10.40–12.50	60
Band 7—SWIR 2	2.09–2.35	30
Band 8—Panchromatic	0.52–0.90	15

3 Results

3.1 Comparison of Different Band Ratios for Correct Water Spread Area

In the present study, the water surface area of Upper Lake and Lower Lake has been determined every year by NDWI, MNDWI and WRI indices and shown in Tables 4, 5 and 6 respectively. From Table 4, it could be determined that the minimum and maximum area obtained by NDWI was 18.35 km^2 in the year 2020 and 31.87 km^2 in 2002 respectively for Upper whereas for Lower lake minimum and maximum water spread area was 0.72 and 1.01 km^2 in 2008 and 2011 respectively. The average water spread area for Upper Lake and Lower Lake is 26.5 and 0.8 km^2 for respectively.

Similarly, Table 5 depicted the maximum water surface area of Upper Lake as 31.50 km^2 in the year 2002 whereas the minimum water surface area was 19.35 km^2 in year 2020 by MNDWI. The average water spread area of Upper Lake in the studied period was obtained as 26.08 km^2. The maximum water surface area of Lower Lake was 0.98 km^2 in the year 2009 whereas the minimum water spread area was 0.65 km^2 in the year 2021. The average water spread area of Lower Lake is obtained as 0.77 km^2.

Further, the lake water spread areas calculated by WRI were tabulated in Table 6. The minimum and maximum water spread area calculated using WRI was 20.02 km^2 in 2015 and 36.48 km^2 in 2010 for upper lake whereas 0.63 km^2 in the year 2016 and 1.01 km^2 in 2011 for lower lake using the same method. This method's average water spread area for Upper Lake and Lower Lake was 27.04 and 0.81 km^2 respectively.

Table 4 Water spread area of Upper Lake and Lower Lake by NDWI for period 2001–2021

Year	Area (km^2)	
	Upper Lake	Lower Lake
2001	28.91	0.77
2002	31.87	0.89
2003	22.35	0.82
2004	30.24	0.91
2005	24.8	0.73
2006	31.25	0.82
2007	28.96	0.83
2008	30.99	0.72
2009	25.63	0.79
2010	26.3	0.88
2011	29.85	1.01
2012	22.22	0.88
2013	28.66	0.78
2014	28.67	0.74
2015	22.65	0.79
2016	22.06	0.58
2017	27.91	0.73
2018	22.65	0.78
2019	21.06	0.82
2020	18.35	0.78
2021	29.68	0.71

3.2 *Accuracy Analysis*

For the purpose of accuracy, a GPS survey of Upper and Lower Lakes was conducted on 25 April 2021, and the obtained areas are presented in Table 7. Landsat-8 image of 26 April, 2021, has been obtained from earth explorer which is the nearest to the day of the GPS survey and extracted to determine the area of studied lakes. Table 7 also shows the comparison between GPS surveyed area and the estimated water area by different indices. The overall errors in the estimated water spread area are also shown in the same table. It was also found that the accuracy of all used indices was more than 90%. However, the overall accuracy of MNDWI was higher. 2.49 and 2.94% deviation or errors were found using MNDWI for Upper Lake and Lower Lake respectively. In contrast, 8.59 and 6.07% deviation were found in the water spread area of Upper Lake using NDWI & WRI indices respectively and 4.41 and 8.82% deviations were determined in Lower Lake water spread area using NDWI and WRI respectively.

Table 5 Water spread area of Upper Lake and Lower Lake by MNDWI for period 2001–2021

Year	Area (km^2)	
	Upper Lake	Lower Lake
2001	27.21	0.78
2002	31.50	0.87
2003	25.70	0.77
2004	30.91	0.89
2005	23.78	0.78
2006	30.53	0.78
2007	28.01	0.85
2008	30.60	0.67
2009	25.67	0.86
2010	25.6	0.85
2011	28.83	0.98
2012	23.25	0.82
2013	28.24	0.69
2014	28.73	0.70
2015	21.91	0.75
2016	21.47	0.69
2017	27.08	0.67
2018	21.74	0.72
2019	20.06	0.74
2020	19.35	0.68
2021	28.01	0.66

Since this study that MNDWI estimated more accurate water spread area of studied lakes than other indices, MNDWI has been used to analyse further water spread area fluctuations. Figures 2 and 3 depicts the water spread area fluctuation for Upper Lake and Lower Lake respectively.

4 Trend Analysis

Trend in lake spread area has been determined using non-parametric test for 21 years (2001–2021) and tabulated in Table 8. To know the significance change in water spread area, trend analysis based on Mann-Kendal test has been calculated and the result indicate that the water spread area in both lakes are sharply decreasing. Z-value for Upper Lake and Lower Lake was obtained as –3.23 and –3.097 respectively. Both Lake shows decreasing trend in water surface area for studied period which is significant at 5% level of significance. Based on Theil's Sen Slope Estimator the

Table 6 Water spread area of Upper Lake and Lower Lake by WRI for period 2001–2021

Year	Area (km^2)	
	Upper Lake	Lower Lake
2001	28.92	0.81
2002	30.52	0.87
2003	25.63	0.82
2004	29.85	0.95
2005	24.68	0.82
2006	30.51	0.83
2007	29.19	0.89
2008	30.66	0.74
2009	24.68	0.98
2010	36.48	0.79
2011	28.83	1.01
2012	22.65	0.86
2013	30.94	0.66
2014	28.98	0.77
2015	20.02	0.78
2016	21.68	0.63
2017	28.0	0.68
2018	22.65	0.72
2019	23.66	0.86
2020	20.37	0.88
2021	28.99	0.74

Table 7 Water spread area using the GPS survey and water indices

Lake	Using GPS (km^2)	Using NDWI (km^2)	Using MNDWI (km^2)	Using WRI (km^2)
Upper Lake	27.33	29.68 (–8.59% deviation from GPS's area)	28.01 (–2.49% deviation from GPS's area)	28.99 (–6.07% deviation from GPS's area)
Lower Lake	0.68	0.71 (–4.41% deviation from GPS's area)	0.66 (2.94% deviation from GPS's area)	0.74 (–8.82% deviation from GPS's area)

magnitude of trend for Upper Lake and Lower Lake were determined as –0.32 and –0.0075 km^2/year respectively.

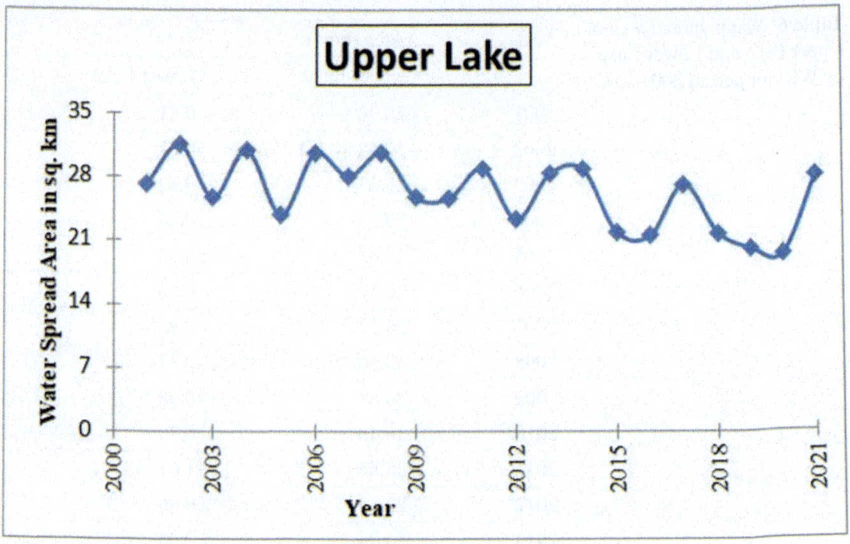

Fig. 2 Fluctuation in water spread for Upper Lake for period 2001–2021

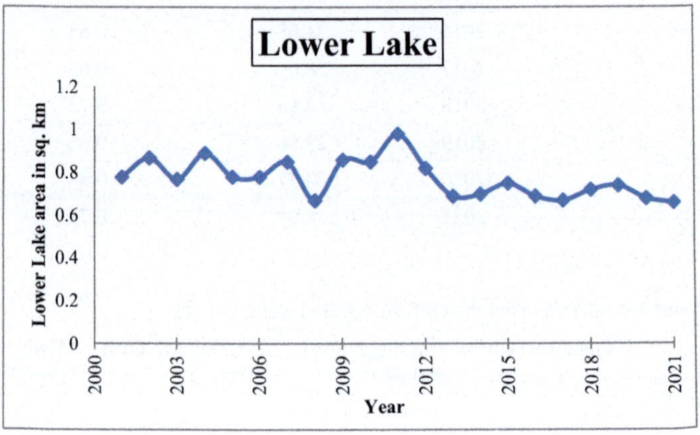

Fig. 3 Fluctuation in water spread for Lower Lake for period 2001–2021

Table 8 Trend detection for surface area change of lakes and magnitude of change at 5% significance level

Lake	Z-value (%)	Trend	Trend magnitude (km^2/year)
Upper Lake	–3.23	Yes (–)*	–0.32
Lower Lake	–3.09	Yes (–)*	–0.0075

* Significant at 5% level of significance

5 Discussion

In this study, the ratio indices, NDWI, MNSWI and WRI were computed. Furthermore, the best suitable band ratio technique was determined based on the GPS mapping of the lakes to predict the water extension in the lake between 2001 to 2018. The MNDWI was the accurate water index in our study. The obtained results agree with the finding of Gautam et al. (2015) who evaluated the surface water area in Bangalore using MNDWI, NDWI and WRI and found a high performance of MNDWI. Furthermore, the current study results also agree with Mukherjee and Samuel (2016) who investigated water bodies' extraction using MNDWI and WRI. The result of their study suggested that MNDWI is acceptable in flooded regions inside a metropolitan zone.

On the other hand, Deoli et al. (2021) observed that the WRI was better than other indices for high-altitude lakes and hence utilized for mapping and modeling spatiotemporal lake changes. Authors also find the trend in lakes water area for studied lakes based on Mann-Kendal and Sen's Slope test and found a strong downward tendency and their results align with this study. Rokni et al. (2014) concluded that NDWI was better than other indices and hence utilized for mapping changes in water bodies. They also present a new technique and assessed for changes in the multi-temporary NDWI components.

However, it should be noted that extracting water bodies using a multiband pixel-based approach has inherent difficulties because they need analysts' expertise to choose the precise training data set (Li et al. 2013). In order to gather information about water bodies and deliver it to investors, legislators, and decision-makers, it is essential to use a water index like the NDWI. In particular, the technique proposed in this study may be highly helpful for decision-makers and water managers to take effective steps, such as minimising pollution and overwater extraction from such significant and vulnerable water bodies.

6 Conclusions

In this study, the effectiveness of NDWI, MNDWI and WRI were applied and explored to identify and determine of water spread area using Landsat-7 and Landsat-8 imageries. Water surface area of Upper and Lower lake from 2001 to 2021 has been determined. The results evaluate that MNDWI is very close to the actual area of lakes based on GPS survey. The study successfully revolved that the water surface area in both of the lakes is sharply decreasing with time for study period based on trend analysis in lake water area. It has been observed that the lake water area was decreasing by 0.32 km^2/year for upper lake and 0.0075 km^2/year for lower lake. This type of study might be very useful in water spread mapping without physical contact with study area. The result can be effective in water-related problems like drought, flood or debris dams. A similar methodology may be obtained for agriculture land mapping, built-up area and glacier mapping using suitable indices. In future, we will try new indices with the machine-learning techniques for water surface mapping and monitoring.

References

Acharya TD, Lee DH, Yang IT, Lee JK (2016) Identification of water bodies in a Landsat 8 OLI image using a J48 decision tree. Sensors 16(7):1075. https://doi.org/10.3390/s16071075

Choudhury BU, Das A, Ngachan SV, Slong A, Bordoloi LJ, Chowdhury P (2012) Trend analysis of long term weather variables in mid altitude Meghalaya, North-East India. J Agricult Phys 12(1):12–22

Cao H, Han L, Liu Z, Li L (2021) Monitoring and driving force analysis of spatial and temporal change of water area of Hongjiannao Lake from 1973 to 2019. Eco Inform 61:101230. https://doi.org/10.1016/j.ecoinf.2021.101230

Dai X, Yang X, Wang M, Gao Y, Liu S, Zhang J (2019) The dynamic change of bosten lake area in response to climate in the past 30 years. Water 12(1):4

Deng Y, Jiang W, Tang Z, Li J, Lv J, Chen Z, Jia K (2017) Spatio-temporal change of lake water extent in Wuhan urban agglomeration based on Landsat images from 1987 to 2015. Remote Sens 9(3):270. https://doi.org/10.3390/rs9030270

Deoli V, Kumar D (2019) Remote sensing and GIS approach for spatiotemporal mapping of Ramganga reservoir. Int J Curr Microbiol Appl Sci 8(5):775–783

Deoli V, Kumar D (2020) Analysis of groundwater fluctuation using grace satellite data. Ind J Ecol 47:299–302

Deoli V, Kumar D, Kuriqi A (2022) Detection of water spread area changes in eutrophic lake using Landsat data. Sensors 22(18):6827. https://doi.org/10.3390/s22186827

Deoli V, Kumar D, Kumar M, Kuriqi A, Elbeltagi A (2021) Water spread mapping of multiple lakes using remote sensing and satellite data. Arab J Geosci 14(21):1–15. https://doi.org/10.1007/s12517-021-08597-9

Du Y, Zhang Y, Ling F, Wang Q, Li W, Li X (2016) Water bodies' mapping from Sentinel-2 imagery with modified normalized difference water index at 10-m spatial resolution produced by sharpening the SWIR band. Remote Sens 8(4):354. https://doi.org/10.3390/rs8040354

Elsahabi M, Negm A, El Tahan AHM (2016) Performances evaluation of surface water areas extraction techniques using Landsat ETM+ data: case study Aswan High Dam Lake (AHDL). Procedia Technol 22:1205–1212. https://doi.org/10.1016/j.protcy.2016.02.001

Feng W, Sui H, Huang W, Xu C, An K (2018) Water body extraction from very high-resolution remote sensing imagery using deep U-Net and a superpixel-based conditional random field model. IEEE Geosci Remote Sens Lett 16(4):618–622. https://doi.org/10.1109/LGRS.2018.2879492

Gautam VK, Gaurav PK, Murugan P, Annadurai MJAP (2015) Assessment of surface water Dynamicsin Bangalore using WRI, NDWI, MNDWI, supervised classification and KT transformation. Aquat Procedia 4:739–746. https://doi.org/10.1016/j.aqpro.2015.02.095

Jing Y, Zhang F, Wang X (2018) Monitoring dynamics and driving forces of lake changes in different seasons in Xinjiang using multi-source remote sensing. Eur J Remote Sens 51(1):150–165. https://doi.org/10.1080/22797254.2017.1413955

Kandekar VU, Pande CB, Rajesh J et al (2021) Surface water dynamics analysis based on sentinel imagery and Google Earth Engine Platform: a case study of Jayakwadi dam. Sustain Water Resour Manag 7:44. https://doi.org/10.1007/s40899-021-00527-7

Kaplan G, Avdan U (2017) Water extraction technique in mountainous areas from satellite images. J Appl Remote Sens 11(4):046002. https://doi.org/10.1117/1.JRS.11.046002

Kumar Y, Kumar A (2020) Spatiotemporal analysis of trend using nonparametric tests for rainfall and rainy days in Jodhpur and Kota zones of Rajasthan (India). Arab J Geosci 13(15):1–18. https://doi.org/10.1007/s12517-020-05687-y

Kuriqi A, Ali R, Pham QB, Montenegro Gambini J, Gupta V, Malik A, Dong X (2020) Seasonality shift and streamflow flow variability trends in central India. Acta Geophys 68(5):1461–1475. https://doi.org/10.1007/s11600-020-00475-4

Li H, Mao D, Li X, Wang Z, Wang C (2019) Monitoring 40-year lake area changes of the Qaidam Basin, Tibetan Plateau, using Landsat time series. Remote Sens 11(3):343. https://doi.org/10.3390/rs11030343

Li W, Du Z, Ling F, Zhou D, Wang H, Gui Y, Zhang X (2013) A comparison of land surface water mapping using the normalized difference water index from TM, ETM+ and ALI. Remote Sens 5(11):5530–5549. https://doi.org/10.3390/rs5115530

Mohsen A, Elshemy M, Zeidan BA (2018) Change detection for Lake Burullus, Egypt using remote sensing and GIS approaches. Environ Sci Pollut Res 25(31):30763–30771. https://doi.org/10.1007/s11356-016-8167-y

Mukherjee NR, Samuel C (2016) Assessment of the temporal variations of surface water bodies in and around Chennai using Landsat imagery. Ind J Sci Technol 9(18):1–7. https://doi.org/10.17485/ijst/2016/v9i18/92089

Özelkan E (2019) Comparison of remote sensing classification techniques for water body detection: a case studY in Atikhisar Dam Lake (Çanakkale). Cumhur Sci J 40(3):650–661

Qin P, Cai Y, Wang X (2021) Small waterbody extraction with improved U-Net using Zhuhai-1 hyperspectral remote sensing images. IEEE Geosci Remote Sens Lett 19:1–5. https://doi.org/10.1109/LGRS.2020.3047918

Rana S, Deoli V, Kashyap PS (2019) Temporal analysis of rainfall trend for Udaipur district of Rajasthan. Ind J Ecol 46(2):306–310

Rana S, Deoli V, Chavan SR (2022) Detection of abrupt change in trends of rainfall and rainy day's pattern of Uttarakhand. Arab J Geosci 15(7):1–17. https://doi.org/10.1007/s12517-022-09883-w

Rokni K, Ahmad A, Selamat A, Hazini S (2014) Water feature extraction and change detection using multitemporal Landsat imagery. Remote Sens 6(5):4173–4189. https://doi.org/10.3390/rs6054173

Rover J, Ji L, Wylie BK, Tieszen LL (2012) Establishing water body areal extent trends in interior Alaska from multi-temporal Landsat data. Remote Sens Lett 3(7):595–604. https://doi.org/10.1080/01431161.2011.643507

Tao S, Fang J, Zhao X, Zhao S, Shen H, Hu H, Guo Q (2015) Rapid loss of lakes on the Mongolian Plateau. Proc Natl Acad Sci 112(7):2281–2286. https://doi.org/10.1073/pnas.1411748112

Thayammal S, Jayaraghavi R, Priyadarsini S, Selvathi D (2022) Analysis of water body segmentation from Landsat imagery using deep neural network. Wirel Pers Commun 123(2):1265–1282. https://doi.org/10.1007/s11277-021-09178-5

Xu H (2006) Modification of normalised difference water index (NDWI) to enhance open water features in remotely sensed imagery. Int J Remote Sens 27(14):3025–3033. https://doi.org/10.1080/01431160600589179

Yang Y, Liu Y, Zhou M, Zhang S, Zhan W, Sun C, Duan Y (2015) Landsat 8 OLI image based terrestrial water extraction from heterogeneous backgrounds using a reflectance homogenization approach. Remote Sens Environ 171:14–32. https://doi.org/10.1016/j.rse.2015.10.005

Yang J, Du X (2017) An enhanced water index in extracting water bodies from Landsat TM imagery. Ann GIS 23(3):141–148. https://doi.org/10.1080/19475683.2017.1340339

Yue H, Liu Y, Wang H, Zhang W (2017) Analysis of dynamic change of Hongjiannao Lake based on MNDWI. In: IOP Conference series: earth and environmental science, vol 57, no 1, p 012005. IOP Publishing

Zhang G, Xie H, Kang S, Yi D, Ackley SF (2011) Monitoring lake level changes on the Tibetan Plateau using ICESat altimetry data (2003–2009). Remote Sens Environ 115(7):1733–1742. https://doi.org/10.1016/j.rse.2011.03.005

Water Resource Management for Alleviating Crop Diseases in Semi-Arid Regions

K. Kamesh Krishnamoorthy and K. Karthik Subramaniam

Abstract The semi-arid regions are characterized by long periods of sunshine, deficient rainfall and unpredictable drought-like situations. In addition, climate change accelerates the intensity of these problems by creating scanty rainfall leading to an excess dependency on groundwater. Many crop diseases arise in such regions due to improper management of water resources and therefore its management becomes a highly essential tactic. Water resource management by choosing the proper method of irrigation helps in reducing the occurrence and intensity of many plant diseases. Drip irrigation is the most sought in semi arid regions as this minimizes water usage and the spread of soil-borne and foliar plant diseases. In addition, chemigation as an alternative to the foliar spray of fungicides and bactericides results in a uniform application with better efficacy avoiding the leaching of chemicals which can be done using drip and trickle irrigation methods. The application of micronutrients through drip irrigation helps manage diseases directly by the antagonistic effect on the pathogen or indirectly by augmenting the defense mechanism of plants or by inciting antagonist organisms in the rhizosphere. Application of fungal and bacterial biocontrol agents to plants helps plants under semi-arid areas to tolerate drought stress by improving the photosynthetic ability and production of a broad range of enzymes and metabolites. Recycled waste water used for irrigation has to be evaluated for the presence of plant pathogens before use to avoid detrimental effects on plant health.

Keywords Semi-arid · Irrigation · Drip irrigation · Biocontrol agents

K. K. Krishnamoorthy (✉)
Department of Plant Pathology and Microbiology, Amrita Vishwa Vidyapeetham, Amrita School of Agricultural Sciences, Coimbatore, India
e-mail: kk_kamesh@cb.amrita.edu

K. K. Subramaniam
Centre for Soft Nanoscience (SoN), University of Munster, Münster, Germany

C. B. Pande et al. (eds.), *Surface and Groundwater Resources Development and Management in Semi-arid Region*, Springer Hydrogeology,
https://doi.org/10.1007/978-3-031-29394-8_11

1 Introduction

Water management using the proper choice of irrigation method is a highly crucial factor for efficient crop production. Proper irrigation assures good quality seeds and food materials. Around 30–40% of global food production comes from irrigated agriculture (Ondrasek 2014; Pande and Moharir 2023).

Groundwater resources form a key source of water for irrigating crops in semi-arid areas. As rainfall is an unpredictable factor, the majority of the water for irrigation is obtained from groundwater. Proper irrigation and drainage in such areas are highly imperative to prevent the predisposition of many plants to be attacked by pathogens. The major challenges in semi-arid regions include long periods of sunshine, and the absence of considerable rainfall. In addition to that evapotranspiration due to prolonged periods of hot temperature and dry winds accelerate the water loss from plants. Under semi-arid environments, plants are subjected to growth under higher temperatures as a result of which they display diminishing levels of carbohydrate and protein synthesis rendering them vulnerable to pests and plant pathogens (Obrepalska-Steplowska et al. 2015). Unpredictable rainfall patterns pose a threat to pulses in semi-arid regions which are easily predisposed to the attack of diseases like *Fusarium* wilt and Dry root rot (Pande and Sharma 2011). Some of the additional problems caused by semi-arid situations contributing to plant diseases include poor soil and water. A prolonged period of water scarcity may lead to desertification. This can be visualized in the semi-arid southern regions like Kolar and Chikballapur in Karnataka and Madanapalli and Anantapur in Andhra Pradesh which is prone to desertification as a result of consistent water scarcity. Such a kind of water-scarce ecosystem may lead to favoring the survival of certain plant pathogens (Ganeshamurthy et al. 2020). Drought-like situations enhance the severity of certain plant viruses (Clover et al. 1999) in addition to increasing the virulence of fungal plant pathogens like Armillaria (Lonsdale and Gibbs 1996).

2 Diseases of Crop Plants in the Semi-Arid Regions

The pathogens predominant in the semi-arid regions include *Fusarium*, *Sclerotium*, and *Macrophomina* (Sinha et al. 2019). Drought stress in semi-arid regions escalates the occurrence of certain pathogens such as *Aspergillus flavus* infecting groundnut (Wotton and Strange 1987) and *Fusarium solani* infecting beans (Agrios 1997). Dry root rot of chickpeas is caused by the pathogen *Rhizoctonia bataticola* and is severe under semi-arid regions in India. The conditions like low soil moisture and high temperature favor its development. In addition to that the soil pH also plays a role in the modulation of disease incidence (Bhatti and Kraft 1992; Sinha et al. 2019). Sinha et al. (2021) observed a positive correlation between the disease incidence of dry root rot of chickpeas with minimum and maximum atmospheric temperatures. The most common plant diseases observed in the semi arid regions include *Fusarium* wilt caused by *Fusarium* sp., Collar rot caused by *Sclerotium rolfsii*. The pathogens sustain such harsh semi-arid atmospheres by the formation of resting structures. Dry root rot caused by *Macrophomina phaseolina* affects crops in both semi-arid and temperate regions. The disease expression is enhanced during high day temperatures of 30 °C coupled with dry conditions. Moreover the prolonged persistence of the pathogen in the soil as resting and dormant sclerotial structures renders the control of the disease a challenging task (Nene et al. 1991). The soil-borne pathogen *Fusarium roseum* causing seedling blight thrives well in dry environment and creates severe symptoms in plants suffering from water stress (Agrios 1997). The occurrence of *Fusarium* wilt disease of chickpea is common in semi-arid regions characterized by warm and dry conditions. The very nature of the pathogen to sustain such harsh conditions lies in its capacity to form chlamydospores which help it remain dormant under a period of stress (Gopalakrishnan et al. 2005).

3 Irrigation and Plant Diseases

The method and choice of irrigation resources and irrigation method is deciding factor for the occurrence of many plant diseases. Irrigation water can spread pathogenic propagules through its physical action by dispersing pathogens or being a contaminated source of plant pathogens by itself. In the case of contaminated water, irrigation water has the likelihood of getting contaminated from various sources among which the most common are ponds, lakes, rivers, streams, and artificial storage units (Hong and Moorman 2005). A worldwide estimation of the irrigated agricultural fields receiving treated and untreated forms of irrigation water are found to be around 35.9 Mha among which most of them are located in regions where less than 75% of the wastewater is subjected to treatment before irrigating crops (Thebo et al. 2017). Surface runoff water passing through Such sources are most likely to get contaminated by plant debris or soil source present in the vicinity. Such contaminated water resources may ultimately end up in the irrigation channel. However, if the water source is still devoid of pathogens, it has the chance of getting contaminated by encountering various sources in its pathway of flow, such as crop debris, soil, and pathogen resting structures or propagules. Sometimes the unsterilized potting mixture serves as the best initial source of the pathogen (Hong and Moorman 2005). Irrigation has to be optimized as either excess or very low amount of irrigation can contribute to the occurrence of plant diseases. Surface irrigation systems mainly include flood irrigation, furrow irrigation, trickle irrigation, and seepage whereas overhead irrigation is mainly delivered by the use of the sprinkler method. Timely irrigation can serve as a means of protecting the crop from disease attack by evading the occurrence of plant pathogens whose survival is enhanced by drought-like conditions (Zentmeyer and Bald 1977). Furrow irrigation helps in carrying the pathogenic propagules and resting structures from one end of the field to another which helps in the rapid dispersal of plant diseases (Manda et al. 2021). When compared to irrigation water sources like pond and river water, groundwater is generally regarded as a safer source. This is mainly due to the susceptibility of pond and river water to contamination by various point sources like discharge pipes of sewage points and nonpoint sources like herbicide, fertilizer, sediments, and other kinds of contaminating materials which tend to end up on the ground in natural means or due to human activity (Ritter et al. 2002).

Surface water used for irrigation possess chance of being contaminated with the *Xanthomonas campestris* bacterial complex consisting of *X. campestris* pv. *begoniae* (Atmatjidon 1991), *X. campestris* pv. *pelargonii* (Wohanka 1995), and *X. axonopodis* complex which consists of *X. axonopodis* pv. *alli* (Gent et al. 2005), *X. axonopodis* pv. *begonia* (Hotink et al. 1991) and *X. axonopodis pv. phaseoli* (Steadman et al. 1975). The *Xanthomonas* bacteria are capable of causing diseases in a wide range of crops like rice, wheat, citrus, pepper and cabbage, spanning to the extent of more than 400 crops. Common disease symptoms incited by Xanthomonas include leaf spots, leaf streaks, blights and cankers. These bacteria can adapt to a wide range of conditions due to their high genome plasticity (Timilsina et al. 2019). Plant pathogenic bacteria such as *Pseudomonas syringae* complex (Monteil et al. 2014), *P. syringae* pv. *aptata* (Riffaud and Morris 2002), *Ralstonia solanacearum* complex (Caruso et al. 2005), *Pectobacterium* sp. And *Dickeya* sp. (Roozen 1990) ranged from a limit of 10^2 to 10^6 CFU L^{-1} in various water sources which included surface water, river water, subalpine catchment basin, river headwaters, river basins, and ponds. These densities were strongly affected by the geographical area, time of sampling, temperature, and chemistry of the water. The high levels of the pathogen were assumed to be favored by weed colonization in the water resources (Lamichhane and Bartoli 2015). An increased nitrogen fertilization makes the plants more susceptible to many diseases. In addition, the irrigation furrow serves as a suitable area for the dispersal of many soil-borne pathogens (Lopes et al. 2006). Sprinkler irrigation is cost-effective and more efficient when compared to furrow irrigation however it leads to wetting of the surface of the foliage. Prolonged wetting of crop canopy infested with foliar plant pathogens serves as a breeding area for sporulating fungal pathogens leading to their easy and increased dispersal. Such a wet condition further leads to an increased pathogen buildup when the crops are grown closely spaced resulting in a dense crop canopy. This kind of situations mostly arises due to the use of sprinkler systems for irrigation. (Rotem and Palti 1969). The incidence of gummy stem blight caused by *Didymella bryoniae* and anthracnose caused by *Colletotrichum gleosporoides* fsp. *cucurbitae* in watermelon were highly reduced when the method of irrigation was shifted from overhead to furrow. Such a change was attributed due to the reduction in the leaf wetness period (Lopes et al. 2006). An increased duration of leaf wetness creates a favourable microclimate for the propogation of fungi and bacteria on the leaf surface leading to their effective invasion and establishment within the plant host (Lee et al. 2016). Water supplied through overhead irrigation on the leaf surface also serves as a matrix for the bacterial population as their survival is highly dependent upon free moisture (Café-Filho et al. 2019). Moreover, droplet splash resulting from sprinkler irrigation helps in the dispersal of the soil-borne pathogen, *Ralstonia solanacearum* causing wilt disease in potatoes. However the contrary case is observed in powdery mildew pathogens whose conidial spores are damaged by the action of overhead irrigation thereby proving overhead irrigation to be detrimental to their survival (Café-Filho et al. 2019).

The zoospores of oomycete pathogens depend on free-flowing irrigation water for their movement and subsequent infection of other neighboring plants. In addition to transmitting plant pathogens, irrigation water also acts as a carrier of plant pathogens rendering many plant diseases to be water-borne. This is noticed in the case of diseases like club roots of cabbage caused by *Plasmodiophora brassicae*. An increase in the requirement for irrigation water as a result of the growing population has led to excessive use of groundwater resources leading to the depletion of groundwater. Moreover, climate change also has led to droughts and scarcity of rainfall which further demands increased usage of groundwater resources (Karthikeyan et al. 2020). Irrigation when given in excess, is also detrimental to the growth of crops in semi arid regions as they deprive the crop of sufficient respiration, increased salinity, and crop lodging (Wichelns and Qadir 2015). Use of poor-quality irrigation water for a prolonged period can alter soil pH values. A shift in the pH change favours the survival of certain plant pathogens (Hentati et al. 2014). The presence of physical forms of debris in irrigation water results in the clogging of emitters in drip irrigation systems. This kind of physical debris is often combined with bacterial slimes thereby posing a risk of transmission of bacterial diseases to the crops (Gilbert et al. 1981). The extremely mild concentration of plant viruses in water samples renders it cumbersome to detect their presence and the sample has to be concentrated in order to generate a detectable amount of viral inoculum (Haramoto et al. 2018; Hjelmo et al. 2017). Metagenomics and high throughput sequencing approach have helped in generating an idea of the constitution of viromes from various sources of water viz., wastewater (Aw et al. 2014), Sewage (Fernandez-Cassi et al. 2018), Reclaimed water (Rosario et al. 2019) and freshwater (Djikeng et al. 2009; Mohiuddin and Schelhorn 2015). The pepper mild mottle virus (PMMoV) holds the capacity to survive and pass along the digestive tract of humans. In addition to that, this pathogen has a global distribution and is present in various water resources, and is categorized as a water pollution indicator (Rosario et al. 2019; Kuroda et al. 2015; Kitajima et al. 2018).

4 Role of Biocontrol Microorganisms in Water Use Efficiency and Plant Disease Control

In order to manage highly resistant soil-borne fungal pathogens under semi-arid conditions, farmers have advocated the indiscriminate use of fungicides. Such an approach tends to be perilous to the farmer's health as well as the environment. Moreover, such a prolonged used of fungicidal-based approaches may lead to the development of fungicide-resistant strains of plant pathogens (Pavlou and Vakalounakis 2005). Under such situations, use of biological control agents for plant disease management offers a viable alternative.

Biological control refers to the use of microbial antagonists to suppress plant diseases. The microbes employed for biological control use various mechanisms like Antibiosis, lysis, competition, hyperparasitism and induction of host resistance to inhibit the development of plant pathogens. Inoculation of plants with such biocontrol microorganisms helps in strengthening the plant leading to drought tolerance and better water retention by the plants. Several benefricial microorganisms exist side by side in the rhizospheric region of the plant. Such organisms establish useful relationships with the plant and affect the growth, nutrient uptake and water use efficiency through a variety of mechanisms. The strains of *Azospirrilum brasilense, Aeromonas punctata, Bacillus megaterium, Pseudomonas fluroscens and Serratia marscescens* are capable of directly modulating crop physiology by the production of useful growth hormones while others lead to an upsurge of plant minerals and nitrogen in the soil as a way of accelerating crop growth under water-deficient conditions (Yadav 2010; Wang et al. 2012; Kaushal and Wani 2015; Barnawal et al. 2017). Plant growth-promoting rhizobacteria secrete multivarious phytohormones, which include abscisic acid (ABA), ethylene, gibberellins, auxins, cytokinins, and salicylic acid. Such hormones help in sustaining the water status of the plant under moisture deficit conditions and are necessary for plant growth and disease avoidance (Prasad et al. 2019). Some famous examples of such rhizobacteria include *Acetobacter*, *Bacillus*, *Herbaspirillum* and *Rhizobium* sp. Which produce gibberellins (Yang et al. 2009). Indole-3-acetic acid is produced by certain species of *Micrococcus*, *Pseudomonas* and *Staphylococcus*. The *Azotobacter chroococum* bacteria produce cytokinins that makes plant capable of maintaining proper moisture levels under extreme conditions of drought stress. In addition to the action against plant pathogens, biological control plays a key role in the management of water resources in the semi-arid regions due to the specific capabilities of biocontrol agents like imparting drought tolerance. The plant growth-promoting rhizobacteria strain *Paenibacillus poylmyxa* conferred drought tolerance to *Arabidopisis thaliana* by stimulation of transcription of a drought responsive gene termed ERD-15 and ABA-responsive gene RAB 18 (Timmusk and Wagner 1999). Moreover, biocontrol strains containing 1-aminocyclopropane-1carboxylate (ACC) deaminase generate induced systemic tolerance to drought-acquired stress conditions in several plants by the mechanism of ACC deaminase which tends to lower plant ethylene levels (Mayak et al. 2004). The use of *Trichoderma* strains for the management of soil-borne diseases in the arid regions of Saudi Arabia proved to be beneficial as the *Trichoderma* strains chosen for the purpose are native to the harsh soil conditions and tend to perform better than exotic strains of *Trichoderma* collected from elsewhere (Aleandri et al. 2015; El-Komy et al. 2020; Yu et al. 2021). Combined application of a mixture of native *Trichoderma* strains was found to be effective in managing *Fusarium* root and stem rot disease in cucumber in semi-arid regions of Saudi Arabia (El-Komy et al. 2022). Chickpeas and pigeonpea constitue the major pulse crops grown under semi-arid conditions in India. The major biocontrol agent used to manage pulse diseases in the semi-arid regions include *Streptomyces* spp.

Among them *Streptomyces tsusimanensis, S. caviscabies, S. setonii, S. africanus* were found quite effective in the management of *Fusarium* wilt of chickpea. Among these strains KAI-32 and KAI-90 was also equally effective in inhibiting the soil borne pathogen *Macrophomina phaseolina* causing charcoal rot of sorghum. In addition to their role in preventing plant pathogens the Streptomyces agents also exhibited dual role of promoting crop growth and yield. *Pseudomonas* protect plant health by a wide array of mechanisms which include nitrogen fixation, phosphate solubilization and plant hormone production (Penrose and Glick 2003; Mirza et al. 2006). The use of Rhizobacteria as a biocontrol agent finds great application among which Exopolysaccharide (EPS) production is well documented (Rashid et al. 2016) The exopolysaccharides in turn help in regulating the hydraulic conductivity and supplements the soil water holding capacity thereby protecting plants during drought stress. The EPS also protects the beneficial soil microbes from undergoing desiccation under conditions of extreme water stress in the soil (Colica et al. 2014).

In addition to aiding in drought tolerance, certain archaea and bacteria discovered in semi-arid regions possess additional features which help in strengthening the plant to tolerate harsh semi-arid conditions and improve water use capacity. Among them nitrogen fixation and phosphorus solubilization properties are exhibited by *Bradyrhizobium* spp. And *Serratia marcescens* found in the Northern part of Egypt (Badawi et al. 2011), nitrogen fixation alone by *Pelomonas saccharophila, Methylosinus trichosporium, Sinorhizobium* sp, *Methylococcus capsulatus* and *Azorhizobium caulenodans* found in The Thar desert of India (Chowdhury et al. 2009), nitrogen fixation, phosphate and zinc solubilization, production of hydrogen cyanide, siderophore, indole acetic acid and salicyclic acid production by *Bacillus, Enterobacter* and *Pseudomonas* from Saudi Arabia (El-Sayed et al. 2014). The mechanism of EPS in mitigating drought stree is through the improvement of soil texture and structure. Maize plants growing in the semi-arid regions of Pakistan showed an improvement in plant growth despite being under drought stress. This was mainly attributed to treatment of maize seeds with bacteria such as *Pseudomonas aeruginosa, Proteus panneri and Alacaligenes faecalis*. In addition to imparting drought resistance, these bacteria also helped in enhancing the plant leaf area, shoot and root length and soil moisture. The bacteria also stimulated an increased synthesis in protein, sugar and water content (Naseem and Bano 2014). Under Indian conditions, numerous biological control microorganisms antagonistic to plant pathogens occurring in the semi-arid regions have been well documented. These include *Streptomyces, Trichoderma viride, Trichoderma harzianum, Trichoderma virens, Bacillus subtilis* and *Aspergillus niger* AN 27 showing inhibitory action against the pathogens *Fusarium solani* fsp. Solani, *F. oxysporum* f.sp. *ciceri* affecting chickpea (Gopalakrishnan et al. 2011; Singh 2014; Mahbobeh et al.2016), *Streptomyces griseus, Streptomyces* sp., *T. viride, T. harzianum* and *Pseudomonas fluoroscens* effective against the pathogen *Sclerotium rolfsii* causing collar rot of chickpea (Sreevidya and Gopalakrishnan 2013; Singh and Gaur 2016), *T. harzianum, Alcaligenes xylosoxydans, T. viride, T. harzianum, Pantoea dispersa, B. subtilis* AF1, *P. fluoroscens* and *Bacillus subtilis* are effective against *Fusarium udum* causing Fusarium wilt of pigeonpea (Siddiqui et al. 1998; Manjula and Podile 2001; Gounder and Srikanth 2002; Prasad et al. 2002; Vaidya et al. 2003; Mishra et al. 2018).

5 Role of Water Scarcity as a Cause of Plant Disease

Drought-like situations are highly favourable for developing of dry root rot and powdery mildew disease in legume crops (Gautam et al. 2013). The pathogen *Xyllela fastidiosa* is a fastidious xylem limted vascular bacteria causing Pierces disease of grapes. The effects of the disease are manifested in the form of clogged xylem vessels and reduced sap flow through plant system. Moreover, water-stressed conditions makes the plants more vulnerable to the attack of the pathogen (Mc Elrone et al.; Baccari and Lindow 2011; Choat et al. 2009). The arthropod vector *Homoladisca vitripennis* transmitting the pathogen was found to acquire the bacteria at a faster rate when they were subjected to infected plants under water-stressed conditions. However when a choice of water-stressed infected and well-watered uninfected plants was presented to the vectors, the transmission efficiency rate of pathogen by the vectors decreased with increasing water stress (Cid et al. 2018). Chickpea and pigeonpea are predominantly grown in semi-arid tropics. The soils in these regions are characterized as nutrient deficient and also receive scanty rainfall, which leads to impediments of proper crop growth. Some plants growing in semi-arid regions have evolved mechanisms to overcome such conditions. This includes the development of regulatory mechanisms to control the movement of stomata during a period of water stress and pathogen attack (Schroeaden et al. 2001; Arnaud and Hwang 2015). Under such harsh conditions the production of Reactive oxygen species is triggered in the apoplast leading to stomatal closure (Qi et al. 2017). In addition to the above, plants possess a complex innate immune system characterised by the recognition of Pathogen Associated Molecular Patterns (PAMPs) by Pattern Recognition Receptors (PRR) in the plant system which leads to the activation of PAMP-triggered immunity. Such a kind of response leads to heightened levels of cytosolic Ca^{2+}, stomatal closure and enhanced expression of pathogenesis-related genes which ultimately leads to the restriction of the pathogen (Lu et al. 2010; Zhang et al. 2010).

6 Management of Water Resources to Alleviate Plant Diseases

In the case of potato, the incidence of common scab disease caused by *Streptomyces scabies* is associated with water stress during the initial growth stages post-planting of potato tubers. This calls for maintaining optimum moisture levels in the soil for managing the disease. In order to curtain the disease overhead irrigation was suggested which minimized the disease incidence by 50%. Management of certain soil borne diseases is achieved by irrigating in surplus. Using this strategy flooding the soil was found to control several soil-borne pathogens like *Fusarium oxysporum* fsp. *Cubense* (Zentmeyer and Bald, Damping off disease caused by oomycete pathogens was found to be controlled when the quantity of flow irrigation was minimised thereby preventing injury to root system (Elmer et al. 2012). The necrotrophic fungus

Rhizoctonia bataticola causes dry root rot of chickpea. High temperature and low soil moisture are predisposing factors for this disease. Therefore optimal irrigation to prevent occurrence of drought is highly essential (Sinha et al. 2021).

The intensity of anthracnose disease of watermelon caused by *Colletotrichum gleosporoides* fsp. Cucurbitae was reduced when the method of irrigation was shifted to furrow method from sprinkler method of irrigation. This change was attributed to reduction in the time period of foliar wetness resulting in inhibition of pathogen invasion. In contrast sprinkler method of irrigation is often detrimental to plant health as it encourages foliar wetting and humidity buildup within the plant system allowing buildup of pathogens. An alternative method to this includes surface irrigation methods like furrow or flood irrigation which results in wetting only the soil regions thereby preventing foliar wetting (Rotem and Palti 1969). Conventional surface irrigation methods like flood and furrow are slowly being replaced by the drip irrigation systems keeping in view plant health. The drip irrigation method helps in the conservation of 50% of water and also decreases incidence of weeds, pests and diseases (Manda et al. 2021). The drip irrigation method is also highly essential for better plant nutrient utilization which helps it in generating resistance against major plant diseases. An appreciable increase in yield components in groundnut was observed when plant micronutrients were applied to groundnut through drip irrigation (Singh et al., 2021). However micronutrient irrigation through the drip method often leads to salt precipitation as a result of problems in components of the drip irrigation system. Such kind of prolonged precipitation leads to fluctuation of soil pH values which leads to a drastic change in the activity of plant pathogens and other beneficial microorganisms in the soil (Lucas and Davis 1961; Capra and Scicolone 1998; Stark et al. 2012). Such fluctuations in pH values lead to higher incidence of diseases in many crops. A low soil pH value of 5.5 enhances the disease incidence of bacterial wilt of tobacco (Li et al. 2017). Similarly, The soil-borne pathogen *Sclerotium rolfsii*, was able to colonize and invade the groundnut plants effectively in a soil pH range of 5–6 when compared with alkaline soil (Shim and Starr 1997).

Fusarium wilt disease of banana affects the vascular region of banana plant and with increased progression of the disease death of tissues occur due to dominance of necrotrophic phase of the pathogen (Dita et al. 2018; Ploetz 2019). The pathogen is perpetuated effectively by chlamydospores present in the soil which are stimulated by host nutrients (Dita et al. 2018). The upsurge of the disease generally occurs when banana plantations are irrigated using water from contaminated dams or rivers or when flood water encountering the pathogenic spores arrives in the cropped area (Stover 1962; Su et al. 1986). Therefore, irrigation water received as a result of surface run off from diseased plantations should be avoided. Moreover, its advised to use a flotation inlet for the irrigation system (BIPB 1989). Overall, excessive irrigation is to be avoided in cases when presence of excess water around the plant root zone makes the plant more susceptible to diseases. An increase in the level of soil moisture activates the zoospores of many oomycete plant pathogens. These zoospores can move through the film of water in the soil thereby infecting neighbouring crops with the disease (Agrios 1997). When the soil is at field capacity, some of the pathogens exhibit their fullest potential to cause infection which includes the soil-borne fungi

Phytophthora, Sclerotinia and Rhizoctonia, bacteria like *Pseudomonas* and *Erwinia* and the majority of the nematodes. Most of the bacterial plant pathogens and fungal pathogens attacking the plants during their juvenile stages prefer the presence of a high amount of soil moisture (Agrios 1997). Excessive moisture reduces the capacity of the host plant to protect itself from certain diseases. This is mainly attributed to lower amount of oxygen and lower temperature of soil under such situations (Hiltunen and White 2002).

The water exuded from industrial wastewater plants can be made fit for irrigation purposes once treated thereby helping in recycling the water (Pedrero et al. 2010). Using waste water for irrigation is mostly observed in water deficient areas (Jesse et al. 2019; Soller et al. 2017). Around twenty mha of land receives irrigation in the form of treated and untreated wastewater on a global scale which is estimated to increase in the future (Medoza-Espinoza et al. 2019). In recent days the handling and care of wastewater generated due to rapid urbanization is a challenge as untreated waste water tends to pollute water resources and leads to the spread of water borne diseases (Menegassi et al. 2020; Petousi et al. 2019). However certain problems may arise in the use of recycled water like deterioration of soil quality, toxicity to crops due to heavy metals and plant pathogens present in recycled water (Redekar et al. 2019; Rattan et al., 2005). In the point of view of plant health, recycled irrigation water is a harbouring area of many plant pathogens especially *Phytophthora* and *Pythium* (Redekar et al. 2019). Detection of plant pathogens in recycled irrigation water was tested using filters and bait methods. The filter method helped in detecting a large oomycete diversity whereas leaf baits showed specificity in capturing plant-associated oomycete pathogens (Redekar et al.2019). Chlorination was devised as an effective management strategy for the elimination of such oomycete plant pathogens from recycled irrigation water and was well established from previous studies.

A zero valent iron filtration was tested as a method for treating water before releasing it for irrigating crops. The zero valent iron consists of a metallic nano particle that serves as a filtering material. This filter helped in effectively reducing the amount of *E. coli* bacteria in irrigation water obtained after filtering compared with unfiltered water. In addition, the filtered water also had an increased pH (Anderson Coughlin et al. 2021). Analysis of plant pathogenic bacteria in irrigation water prior to their release for watering is highly essential. Some of the studies used to date for the purpose include immunofluorescence colony staining (Riffaud and Morris 2002), DAS-ELISA (Caruso et al. 2005) and quantitative-real time PCR (Cottyn et al. 2011) among which the real-time PCR method is highly efficacious and sensitive (Lamichhane and Bartoli 2015). More recently, the concept of organic application of nutrients is slowly taking shape where the nutrients are added in the form of manures, compost, residues of crops and legumes. The release of nutrients from such sources relies upon the natural processes in the soil system (Stockdale et al. 2002). The soils receiving the organic form of nutrients are known to have a better water-holding capacity which serves as an ideal feature to plants in the semi-arid conditions (Lampkin 1999).

7 Conclusion

The semi-arid areas involve a large proportion of the world's arable land. Therefore proper water management in such areas with the use of irrigation methods capable of conserving water without compromising on crop moisture requirements is essential. Such an attempt will help to prevent many crop diseases generated as a result of improper irrigation practices and help alleviate drought stress, thereby preventing plant pathogens favored by drought like situations. Most plant pathogenic fungi and bacteria depend on free water for their survival and dissemination. Therefore the irrigation methods should not lead to the accumulation of free water on the surface of foliage. Micronutrient application by the method of drip irrigation leads to reduced disease as well as helps in conserving water in semi-arid regions. A brief knowledge of the ecology and epidemiology of plant pathogens in irrigated and water-scarce areas is essential to help devise strategies for the management of pathogens in such specific locations. Technologies for rapidly detecting of plant pathogens in water resources meant for irrigation are highly peremptory. Beneficial soil microbe interactions are to be identified to determine the disease control and drought resistance imparting potential of soil microorganisms.

References

Agrios GN (1997) Plant pathology. Academic Press, San Diego

Aleandri MP, Chilosi G, Bruni N, Tomassini A, Vettraino AM, Vannini A (2015) Use of nursery potting mixes amended with local *Trichoderma* strains with multiple complementary mechanisms to control soilborne diseases. Crop Prot 67:269–278

Anderson-Coughlin BL, Litt PK, Kim S, Craighead S, Kelly AJ, Chiu P, Kniel KE (2021) Zero-valent iron filtration reduces microbial contaminants in irrigation water and transfer to raw agricultural commodities. Microorganisms 9(10):2009

Arnaud D, Hwang I (2015) A sophisticated network of signaling pathways regulates stomatal defenses to bacterial pathogens. Mol Plant 8:566–581

Atmatjidou VP (1991) Dissemination and transmission of *Xanthomonas campestris* pv. begoniae in an ebb and flow irrigation system. Plant Dis 75:1261–1265

Aw TG, Howe A, Rose JB (2014) Metagenomic approaches for direct and cell culture evaluation of the virological quality of wastewater. J Virol Methods 210:15e21. https://doi.org/10.1016/j.jviromet.2014.09.017

Baccari C, Lindow SE (2011) Assessment of the process of movement of *Xylella fastidiosa* within susceptible and resistant grape cultivars. Phytopathology 101:77–84

Badawi FSF, Biomy AMM, Desoky AH (2011) Peanut plant growth and yield as influenced by co-inoculation with Bradyrhizobium and some rhizo-microorganisms under sandy loam soil conditions. Ann Agric Sci 56:17–25

Barnawal D, Bharti N, Pandey SS, Pandey A, Chanotiya CS, Kalra A (2017) Plant growth-promoting rhizobacteria enhance wheat salt and drought stress tolerance by altering endogenous phytohormone levels and TaCTR1/TaDREB2 expression. Physiol Plant 161:502–514

Bhatti MA, Kraft JM (1992) Influence of soil moisture on root rot and wilt of chickpea. Plant Dis 76:1259–1262

BIPB (1989) Recommendations for dealing with panama disease. Queensland: Banana Industry Protection Board 4

Café-Filho AC, Lopes CA, Rossato M (2019) Management of plant disease epidemics with irrigation practices. Irrigation Agroecosyst 123

Capra A, Scicolone B (1998) Water quality and distribution uniformity in drip/trickle irrigation systems. J Agric Eng Res 70(4):355–365

Caruso P, Palomo JL, Bertolini E, Alvarez B, Lopez MM, Biosca EG (2005) Seasonal variation of *Ralstonia solanacearum* biovar 2 populations in a Spanish river: recovery of stressed cells at low temperatures. Appl Environ Microbiol 71:140–148

Chawla I, Karthikeyan L, Mishra AK (2020) A review of remote sensing applications for water security: Quantity, quality, and extremes. J Hydrol 585:124826

Choat B, Gambetta GA, Wada H, Shackel KA, Matthews MA (2009) The effects of Pierce's disease on leaf and petiole hydraulic conductance in Vitis vinifera cv. Chardonnay. Physiol Plant 136:384–394

Chowdhury SP, Schmid M, Hartmann A, Tripathi AK (2009) Diversity of 16S-rRNA and *nifh* genes derived from rhizosphere soil and roots of an endemic drought tolerant grass, *Lasiurus sindicus*. Eur J Soil Biol 45:114–122

Clover GRG, Smith HG, Azam-Ali SN, Jaggard KW (1999) The effects of drought on sugar beet growth in isolation and in combination with Beet yellows virus (BYV) infection. J Agric Sci 133:251–261

Colica G, Li H, Rossi F, Li D, Liu Y, De Philippis R (2014) Microbial secreted exopolysaccharides affect the hydrological behavior of induced biological soil crusts in desert sandy soils. Soil Biol Biochem 68:62–70

Cottyn B, Baeyen S, Pauwelyn E (2011) Development of a real-time PCR assay for *Pseudomonas cichorii,* the causal agent of midrib rot in greenhouse-grown lettuce, and its detection in irrigating water. Plant Pathol 60:453–461

Del Cid C, Krugner R, Zeilinger AR, Daugherty MP, Almeida RP (2018) Plant water stress and vector feeding preference mediate transmission efficiency of a plant pathogen. Environ Entomol 47(6):1471–1478

Dita M, Barquero M, Heck D, Mizubuti ESG, Staver CP (2018) Fusarium wilt of banana: current knowledge on epidemiology and research needs toward sustainable disease management. Front Plant Sci 9:1468

Djikeng A, Kuzmickas R, Anderson NG, Spiro DJ (2009) Metagenomic analysis of RNA viruses in a fresh water lake. PLoS ONE 4:e7264. https://doi.org/10.1371/journal.pone.0007264

El-Komy MH, Hassouna MG, Abou-Taleb EM, Al-Sarar AS, Abobakr Y (2020) A mixture of *Azotobacter*, *Azospirillum*, and *Klebsiella* strains improves rootrot disease complex management and promotes growth in sunflowers in calcareous soil. Eur J Plant Pathol 156:713–726

El-Komy MH, Al-Qahtani RM, Ibrahim YE, Almasrahi AA, Al-Saleh MA (2022) Soil application of Trichoderma asperellum strains significantly improves Fusarium root and stem rot disease management and promotes growth in cucumbers in semi-arid regions. Eur J Plant Pathol 1–17

Elmer WH, Gent MPN, McAoy RJ (2012) Partial saturation under ebb and flow irrigation suppresses Pythium root rot of ornamentals. Crop Prot 33:29–33

El-Sayed WS, Akhkha A, El-Naggar MY, Elbadry M (2014) In vitro antagonistic activity, plant growth promoting traits and phylogenetic affiliation of rhizobacteria associated with wild plants grown in arid soil. Front Microbiol 5

Fernandez-Cassi X, Timoneda N, Martinez-Puchol S, Rusinol M, RodriguezManzano J, Figuerola N, Bofill-Mas S, Abril JF, Girones R (2018) Metagenomics for the study of viruses in urban sewage as a tool for public health surveillance. Sci Total Environ 618:870e880. https://doi.org/10.1016/j.scitotenv.2017.08.249

Ganeshamurthy AN, Kalaivanan D, Rupa TR, Raghupathi HB (2020) Groundwater decline and prolonged drought could reduce vigour, enhance vulnerability to diseases and pests and kill perennial horticultural crops: needs urgent policy intervention. J Horticult Sci 15(1):9–16

Gautam HR, Bhardwaj ML, Kumar R (2013) Climate change and its impact on plant diseases (2013)

Gent DH, Lang JM, Schwartz HF, Bartolo ME (2005) Inoculum sources and survival of *Xanthomonas axonopodis* pv. *allii* in Colorado. Plant Dis 89:507–514

Gilbert RG, Nakayama FS, Bucks DA, French OF, Adamson KC (1981) Trickle irrigation: emitter clogging and other flow problems. Agric Water Manage 3(3):159–178

Gopalakrishnan S, Pande S, Sharma M, Humayun P, Kiran BK, Sandeep D, Vidya MS, Deepthi K, Rupela O (2011) Evaluation of actinomycete isolates obtained from herbal vermicompost for biological control of Fusarium wilt of chickpea. Crop Prot 30:1070–1078

Goudar SB, Srikanth K (2002) Bioassay of antagonist against Fusarium udum—causal agent of pigeonpea wilt. Karnataka J Agric Sci 13:64–67

Haramoto E, Kitajima M, Hata A, Torrey JR, Masago Y, Sano D, Katayama H (2018) A review on recent progress in the detection methods and prevalence of human enteric viruses in water. Water Res 135:168e186. https://doi.org/10.1016/j.watres.2018.02.004

Hentati O, Chaker S, Wali A, Ayoub T, Ksibi M (2014) Effects of long-term irrigation with treated wastewater on soil quality, soil-borne pathogens, and living organisms: case study of the vicinity of El Hajeb (Tunisia). Environ Monit Assess 186:2671–2683

Hiltunen LH, White JG (2002) Cavity spots of carrot (Daucus carota). Ann Appl Biol 141:201–223

Hjelmsø MH, Hellmér M, Fernandez-Cassi X, Timoneda N, Lukjancenko O, Seidel M, Elsässer D, Aarestrup FM, Löfström C, Bofill-Mas S, Abril JF, Girones R, Schultz AC (2017) Evaluation of methods for the concentration and extraction of viruses from sewage in the context of metagenomic sequencing. PloS One 12:1e17. https://doi.org/10.1371/journal.pone.0170199

Hoitink HAJ, Fynn RP, McMahon RW, Atmatjidou V (1991) Transmission of plant pathogens in an ebb and flood system. Ohio Florists' Assoc Bull 742:5–9

Hong CX, Moorman GW (2005) Plant pathogens in irrigation water: challenges and opportunities. Crit Rev Plant Sci 24(3):189–208

Jesse SD, Zhang Y, Margenot AJ, Davidson PC (2019) Hydroponic lettuce production using treated post-hydrothermal liquefaction wastewater (PHW). Sustainability 11:1–16

Kaushal M, Wani SP (2015) Plant-growth-promoting rhizobacteria: drought stress alleviators to ameliorate crop production in drylands. Ann Microbiol 66:35–42

Kitajima M, Iker BC, Pepper IL, Gerba CP (2014) Relative abundance and treatment reduction of viruses during wastewater treatment processes — Identification of potential viral indicators. Sci Total Environ 488–489:290–296

Kuroda K, Nakada N, Hanamoto S, Inaba M, Katayama H, Do AT, Nga TTV, Oguma K, Hayashi T, Takizawa S (2015) Pepper mild mottle virus as an indicator and a tracer of fecal pollution in water environments: comparative evaluation with wastewater-tracer pharmaceuticals in Hanoi, Vietnam. Sci Total Environ 506:287–298

Lamichhane JR, Bartoli C (2015) Plant pathogenic bacteria in open irrigation systems: what risk for crop health? Plant Pathol 64(4):757–766

Lampkin N (1999) Organic farming. Farming Press, Ipswich, pp 214–271

Lee KJ, Kang JY, Lee DY, Jang SW, Lee S, Lee BW, Kim KS (2016) Use of an empirical model to estimate leaf wetness duration for operation of a disease warning system under a shade in a ginseng field. Plant Disease 100(1):25–31

Lonsdale D, Gibbs JN (1996) Effects of climate change on fungal diseases of trees. In: Frankland et al. (eds) Fungi and environmental change. Cambridge University Press, Cambridge, p 1019

Lopes CA, Marouelli WA, Café Filho AC (2006) Associação da irrigação com doenças de hortaliças. Revisão Anual De Patologia De Plantas 14:151–179

Lu DP, Wu SJ, Gao XQ, Zhang YL, Shan LB, He P (2010) A receptor like cytoplasmic kinase, BIK1, associates with a flagellin receptor complex to initiate plant innate immunity. Proc Natl Acad Sci USA 107:496–501

Lucas RE, Davis J (1961) Relationships between pH values of organic soils and availabilities of 12 plant nutrients. Soil Sci 92:177–182

Mahboobeh S, Meysam SN, Shahidi GHB (2016) Application of soil-borne actinomycetes for biological control against Fusarium wilt of chickpea (Cicer arietinum) caused by Fusarium solani f. sp. pisi. J Phytopathol 164:967–978

Manda RR, Addanki VA, Srivastava S (2021) Role of drip irrigation in plant health management, its importance and maintenance. Plant Archives 21(1):1294–1302

Manjula K, Podile AR (2001) Chitin supplemented formulations improve biocontrol and plant growth promoting efficiency of Bacillus subtilis AF1. Can J Microbiol 47:618–625

Mayak S, Tirosh T, Glick BR (2004) Plant growth-promoting bacteria that confer resistance to water stress in tomatoes and peppers. Plant Sci 166(2):525–530

Mendoza-Espinosa LG et al (2019) Reclaimed water for the irrigation of vineyards: mexico and South Africa as case studies. Sustain Cities Soc 51:101769

Menegassi LC, Rossi F, Dominical LD, Tommaso G, Montes CR, Gomide CA, Gomes TM (2020) Reuse in the agro-industrial: Irrigation with treated slaughterhouse effluent in grass. J Clean Prod 251:119698

Mishra RJ, Bohra A, Kamaal N, Kumar K, Gandhi K, Sujayanand GK, Saabale PR, Satheesh Naik SJ, Sarma BK, Kumar D, Mishra M, Srivastava DK, Mishra NPS (2018) Utilization of biopesticides as sustainable solutions for management of pests in legume crops: achievements and prospects. Egypt J Biol Pest Control 28:3

Mirza MS, Mehnaz S, Normand P et al (2006) Molecular characterization and PCR detection of a nitrogen-fixing Pseudomonas strain promoting rice growth. Biol Fertil Soils 43:163–170

Mohiuddin M, Schellhorn HE (2015) Spatial and temporal dynamics of virus occurrence in two freshwater lakes captured through metagenomic analysis. Fronti Microbiol 6:960

Monteil CL, Lafolie F, Laurent J et al (2014) Soil water flow is a source of the plant pathogen *Pseudomonas syringae* in subalpine headwaters. Environ Microbiol 16:2038–2052

Naseem H, Bano A (2014) Role of plant growth-promoting rhizobacteria and their exopolysaccharide in drought tolerance of maize. J Plant Interact 9:689–701

Nene YL, Reddy MV, Haware MP, Ghanekar AM, Amin KS (1991) Field diagnosis of chickpea diseases and their control. In: Information bulletin, vol 28. International Crops Research Institute For The Semi-Arid Tropics, Patancheru

Obrêpalska-Stêplowska A, Renaut J, Planchon S, Przybylska A, Wieczorek P, Barylski J, Palukaitis P (2015) Effect of temperature on the pathogenesis, accumulation of viral and satellite RNAs and on plant proteome in peanut stunt virus and satellite RNA-infected plants. Front Plant Sci 6: 903. https://doi.org/10.3389/fpls.2015.00903

Ondrasek G (2014) Water scarcity and water stress in agriculture. In: Physiological mechanisms and adaptation strategies in plants under changing environment. Springer, New York, NY, pp 75–96

Pande CB, Moharir KN (2023) Application of hyperspectral remote sensing role in precision farming and sustainable agriculture under climate change: A review. In: Pande CB, Moharir KN, Singh SK, Pham QB, Elbeltagi A (eds) Climate change impacts on natural resources, ecosystems and agricultural systems. Springer Climate. Springer, Cham. https://doi.org/10.1007/978-3-031-19059-9_21

Pande S, Sharma M (2011) Climate change and changing scenario of plant diseases in semi arid tropics

Pavlou GC, Vakalounakis DJ (2005) Biological control of root and stem rot of greenhouse cucumber, caused by Fusarium oxysporum f. sp. radicis-cucumerinum, by lettuce soil amendment. Crop Prot 24(2):135–140

Pedrero F, Kalavrouziotis I, Alarcón JJ, Koukoulakis P, Asano T (2010) Use of treated municipal wastewater in irrigated agriculture—review of some practices in Spain and Greece. Agric Water Manag 97:1233–1241

Penrose DM, Glick BR (2003) Methods for isolating and characterizing ACC deaminase-containing plant growth-promoting rhizobacteria. Physiol Plant 118:10–15

Petousi I, Daskalakis G, Fountoulakis MS, Lydakis D, Fletcher L et al (2019) Effects of treated wastewater irrigation on the establishment of young grapevines. Sci Total Environ 658:485–492

Ploetz RC (2019) "Fusarium wilt." In: Abacá, Ense, Jones D (eds) Handbook of diseases of banana. CABI Publishing, Wallingford, pp 207–228

Prasad M, Srinivasan R, Chaudhary M, Choudhary M, Jat LK (2019) Plant growth promoting rhizobacteria (PGPR) for sustainable agriculture. In: PGPR amelioration in sustainable agriculture. Elsevier BV, Amsterdam,The Netherlands, pp 129–157

Prasad RD, Rangeshwaran R, Anuroop CP, Rashni HJ (2002a) Biological control of wilt and root rot of chickpea under field conditions. Ann Plant Prot Sci 10:72–75

Qi J, Wang J, Gong Z, Zhou JM (2017) Apoplastic ROS signaling in plant immunity. Curr Opin Plant Biol 38: 92–100

Rashid MI, Mujawar LH, Shahzad T, Almeelbi T, Ismail IMI, Oves M (2016) Bacteria and fungi can contribute to nutrients bioavailability and aggregate formation in degraded soils. Microbiol Res 183:26–41

Rattan RK, Datta SP, Chhonkar PK, Suribabu K, Singh AK (2005) Long-term impact of irrigation with sewage effluents on heavy metal content in soils, crops and groundwater—a case study. Agric Ecosyst Environ 109:310–322

Redekar NR, Eberhart JL, Parke JL (2019) Diversity of Phytophthora, Pythium, and Phytopythium species in recycled irrigation water in a container nursery. Phytobiomes J 3(1):31–45

Riffaud CMH, Morris CE (2002) Detection of *Pseudomonas syringae* pv. aptata in irrigation water retention basins by immunofluorescence colony-staining. Eur J Plant Pathol 108:539–545

Ritter L, Solomon K, Sibley P, Hall K, Keen P, Mattu G, Linton B (2002) Sources, pathways, and relative risks of contaminants in surface water and groundwater: a perspective prepared for the Walkerton inquiry. J Toxicol Environ Health 65:1–142

Rosario K, Morrison CM, Mettel KA, Betancourt Q (2019) Novel circular repencoding single-stranded DNA viruses detected in treated waste water 9e10. https://doi.org/10.1128/MRA.00318-19

Roozen NJM, (1990) Besmetting van Erwinia-vrij Pootgoed Vanuit Divese Bronnen. Een Literatu-uroverzicht. Lelystad, Netherlands: PAGV

Rotem J, Palti J (1969) Irrigation and plant diseases. Annu Rev Phytopathol 7:267–288

Schroeder JI, Kwak JM, Allen GJ (2001) Guard cell abscisic acid signalling and engineering drought hardiness in plants. Nature 410:327–330

Shim MY, Starr JL (1997) Effect of soil pH on sclerotial germination and pathogenicity of *Sclerotium rolfsii*. Peanut Sci 24:17–19

Siddiqui ZA, Mahmood I, Hayat S (1998) Biocontrol of Heterodera cajani and Fusarium udum on pigeonpea using Glomus mosseae, Paecilomyces lilacinus and Pseudomonas fluorescens. Thai J Agric Sci 31:310–321

Sinha R, Irulappan V, Mohan-Raju B, Suganthi A, Senthil-Kumar M (2019) Impact of drought stress on simultaneously occurring pathogen infection in field-grown chickpea. Sci Rep 9:1–15

Singh HB (2014) Management of plant pathogens with microorganisms. Proc Indian Natl Sci Acad 80:443–454

Singh SP, Gaur R (2016) Evaluation of antagonistic and plant growth-promoting activities of chitinolytic endophytic actinomycetes associated with medicinal plants against Sclerotium rolfsii in chickpea. J Appl Microbiol 121:506–518

Sinha R, Irulappan V, Patil BS, Reddy PCO, Ramegowda V, Mohan-Raju B, Senthil-Kumar M (2021) Low soil moisture predisposes field-grown chickpea plants to dry root rot disease: evidence from simulation modeling and correlation analysis. Sci Rep 11(1):1–12

Sreevidya M, Gopalakrishnan S (2013) Bacillus and Streptomyces spp. as potential biocontrol agents to control soil-borne pathogens of chickpea and sorghum. Acta Phytopathol Sin 43:520

Soller JA, Eftim SE, Warren I, Nappier SP (2017) Evaluation of microbiological risks associated with direct potable reuse. Microb Risk Anal 5:3–14

Stark S, Eskelinen A, Männistö MK (2012) Regulation of microbial community composition and activity by soil nutrient availability, soil pH, and herbivory in the tundra. Ecosystems 15:18–33

Steadman JR, Maier CR, Schwartz HF, Kerr ED (1975) Pollution of surface irrigation waters by plant pathogenic organisms. J Am Water Resour Assoc 11:796–804

Stockdale EA, Shepherd MA, Fortune S, Cuttle SP (2002) Soil fertility in organic farming systems—fundamentally different? Soil Use Manage 18:301–308

Stover RH (1962) Fusarial wilt (panama disease) of bananas and other musa species. Surrey: The Commonwealth Mycological Institute Kew 117

Su HJ, Hwang SC, Ko WH (1986) Fusarial wilt of 'Cavendish' bananas in Taiwan. Plant Dis 70:814–818. https://doi.org/10.1094/PD-70-814

Thebo AL, Drechsel P, Lambin EF, Nelson KL (2017) A global, spatially-explicit assessment of irrigated croplands influenced by urban wastewater flows. Environ Res Lett 12(7):074008

Timilsina S, Pereira-Martin JA, Minsavage GV, Iruegas-Bocardo F, Abrahamian P, Potnis N, Jones JB (2019) Multiple recombination events drive the current genetic structure of *Xanthomonas perforans* in Florida. Front Microbiol 10:448

Timmusk S, Wagner EGH (1999) The plant-growth-promoting rhizobacterium Paeni *Bacillus* polymyxa induces changes in *Arabidopsis thaliana* gene expression: a possible connection between biotic and abiotic stress responses. Mol Plant Microbe Interact 12(11):951–959

Vaidya RJ, Macmil SL, Vyas PR, Ghetiya LV, Thakor KJ, Chhatpar HS (2003) Biological control of Fusarium wilt of pigeonpea Cajanus cajan (L.) Millsp with chitinolytic Alcaligenes xylosoxydans. Indian J Exp Biol 41:1469–1472

Wang C, Guo Y, Wang C, Liu H, Niu D, Wang Y, Guo J (2012) Enhancement of tomato (*Lycopersicon esculentum*) tolerance to drought stress by plant-growth-promoting rhizobacterium (PGPR) *Bacillus* cereus AR156. J Agric Biotechnol 20:1097–1105

Wichelns D, Qadir M (2015) Achieving sustainable irrigation requires effective management of salts, soil salinity, and shallow groundwater. Agric Water Manage 157:31–38

Wohanka W (1995) Disinfection of recirculating nutrient solution by slow sand filtration. 4th international symposium on soil and substrate infestation and disinfestation. Acta Horticulurae 382:246–255

Wotton HR, Strange RN (1987) Increased susceptibility and reduced phytoalexin accumulation in drought-stressed peanut kernels challenged with Aspergillus flavus. Appl Environ Microbiol 53(2):270–273

Yadav SK (2010) Cold stress tolerance mechanisms in plants. A review. Agron Sustain Dev 30:515–527

Yang J, Kloepper JW, Ryu C-M (2009) Rhizosphere bacteria help plants tolerate abiotic stress. Trends Plant Sci 14:1–4

Yu Z, Wang Z, Zhang Y, Wang Y, Liu Z (2021) Biocontrol and growth-promoting effect of Trichoderma asperellum TaspHu1 isolate from Juglans mandshurica rhizosphere soil. Microbiol Res 242:126596

Zentmeyer GA, Bald JG (1977) Management of the environment. In: Horsfall JG, Cowling EB (ed) How disease is managed. Plant pathology-an advanced treatise, vol 1. Academic Press, London. Chapter 7, p 122–144

Zhang J, Li W, Xiang TT, Liu ZX, Laluk K, Ding XJ, Zou Y, GaoMH, Zhang XJ, Chen S, Mengiste T, Zhang YL, Zhou JM (2010) Receptor-like cytoplasmic kinases integrate signaling from multiple plant immune receptors and are targeted by a Pseudomonas syringae effector. Cell Host Microbe 7:290–301

Evaluation of Guelph Permeameter for Measuring Saturated Hydraulic Conductivity on Semi-arid Agricultural Catchment

B. B. Prajwal and P. N. Chandramouli

Abstract In an agricultural field, soil texture, cropping or tillage techniques, landscape orientations, and growing seasons have an impact on the spatial variability of saturated hydraulic conductivity (K_s). Regardless of the technique employed to measure K_s, the variability and heterogeneity of the soils have an impact on its measurements. This study's objective was to compare K_s at a depth of 15 cm in a semi-arid agricultural catchment using a variety of techniques, double ring infiltrometer (DRI) and guelph permeameter (GP) at the site, constant and falling head permeameters (CHPT and FHPT) at the laboratory and Rawls and Brakensiek regression equation based on the soil texture and porosity. Comparisons of K_s values of GP with all the other methods and evaluating the performance of the Rawls and Brakensiek regression equation in relation to the GP's K_s. Results of this study shows that mean K_s value obtained from five different methods, DRI, GP, CHPT, FHPT and Rawls-Brakensiek regression equation for selected agricultural catchment with varying soil type were differ from each other. Mean K_s measured using DRI were higher than any other methods. As CHPT and FHPT measures K_s vertically and ramming of cores into soil could disturb the soil's structure results in lesser K_s values when compared with field (DRI and GP) K_s. Rawls and Brakeinsiek regression equation agreed favorably and showed good performance with GP. The primary source of knowledge for choosing the right approach for certain conditions and soil properties is comparison of several methods.

Keywords Saturated hydraulic conductivity · Double ring infiltrometer · Guelph permeameter · Laboratory permeameters · Textural based model

B. B. Prajwal (✉) · P. N. Chandramouli
Department of Civil Engineering, The National Institute of Engineering, Mysuru 570008, India
e-mail: prajwalbb333@gmail.com

P. N. Chandramouli
e-mail: pncmouli@nie.ac.in

C. B. Pande et al. (eds.), *Surface and Groundwater Resources Development and Management in Semi-arid Region*, Springer Hydrogeology,
https://doi.org/10.1007/978-3-031-29394-8_12

1 Introduction

The design and monitoring of irrigation and drainage systems, canals and reservoirs, ponds and percolation pits, landfills, water storage tanks, ground water, septic tank systems, and many other agricultural, industrial, and environmental installations depend critically on the water transmission characteristics of the unsaturated zone (Reynolds and Elrick 1986; Bouwer and Jackson 1974; and others). To accurately anticipate how much water will flow through a soil profile, it is vital to know the saturated hydraulic conductivity (K_S) of the soil. K_S is an important parameter in governing a variety of hydrological processes, including rainfall, infiltration and surface runoff. K_S is defined as the rate of infiltration at steady state when the hydraulic gradient is equal to unity. In an agricultural field, soil texture, cropping or tillage techniques, landscape orientations, and growing seasons have an impact on the spatial variability of K_S. It is also one of the main input variables in the majority of physically based watershed models that are used to assess the effects of various land uses and management strategies on the dynamic behavior of soil and water. Regardless of the technique employed to measure K_S, the variability and heterogeneity of the majority of soils have an impact on its measurements.

Field experiments, such as guelph infiltrometer (GI), double ring infiltrometer (RI), rainfall stimulator (RS), and guelph permeameter (GP) to compare hydraulic conductivity (K_S) were conducted by Gupta et al. (1993). The results revealed that K_S values produced by RI and GP techniques were statistically equivalent, but were lower when compared to values obtained by RS and GI. Additionally, it was noted that the values acquired using GP and GI methods had high degrees of variability and required a larger number of observations to arrive at a mean value of K_S that had a similar standard error of estimation to that of the values obtained using RI and RS methods. Gallichand et al. (1990) compared the Guelph permeameter, falling head permeameter, and single auger hole method to determine the hydraulic conductivity. According to the results, the falling head permeameter had a lower K_S value than the other two methods. The existence of macropores, fewer sampling volumes, soil anisotropy, and disturbance of the cores all served as explanations for the lower hydraulic conductivity. Kanwar et al. (1990) conducted laboratory studies utilizing the constant head permeability approach as well as field experiments at a number of depths and sites using guelph permeameters and velocity permeameters to assess the effectiveness of two methods for determining hydraulic conductivity. The results indicate remarkably similar values for the two methods and due macropore flow, laboratory results demonstrated a greater K_S than guelph permeameter approach. Mohanty et al. (1994) performed analysis to evaluate the impact of four in situ saturated hydraulic conductivity measuring methods comprising of guelph permeameter, velocity permeameter, disk permeameter and double-tube methods at different field conditions. The K_S value was also calculated in the laboratory using undisturbed soil cores collected from all four sites and depths. The results revealed minimal variation, probably due to large sample sizes in K_S values obtained by the disc permeameter and double-tube methods and lowest K_s values due to small pore size for guelph permeameter method and presence or absence of open-ended macropores may have caused the maximum variation in K_s values for soil cores at shallow depths. Jačka et al. (2014) studied the Guelph permeameter, single ring infiltrometer, and lab permeameter as three saturated hydraulic conductivity evaluation methods in the shallow subsurface layer. This study produced significantly higher mean values of lab permeameters than the other two approaches, the presence of macropores and full saturation of the soil sample during lab permeameter technique resulted in higher K_s values. Mahapatra and Jha (2019) investigated the values of K_s by four pedotransfer functions, including Campbell, Rawls-Brakensiek, Cronican-Gribb, and Rosetta Model 2 and 3 with minimal data available on layered vadose zones and assessed saturated hydraulic conductivity in the field using a Guelph permeameter. The analysis revealed that the Models 2 and 3 of the Rosetta had K_s values that were, to some extent, equivalent, and that the Rawls-Brakwseink equation produced good K_s values in the deeper lateritic strata of the vadose zone with finer sandy clay loam texture.

The comparison of K_s from field, laboratory and texture-based models has only been studied in a small number of research. There were also surprisingly little studies conducted on K_s investigations in semi-arid agricultural catchments, particularly in southern Karnataka. This study's objective was to compare the saturated hydraulic conductivity (K_s) at a depth of 15 cm in a semi-arid agricultural catchment having varying soil types using a variety of techniques, double ring infiltrometer (DRI) and guelph permeameter (GP) at the site, constant and falling head permeameters at the laboratory and Rawls and Brakensiek regression equation based on the soil texture and porosity. Comparing K_s values of GP with all the other methods and evaluating the performance of the Rawls and Brakensiek regression equation in relation to GP's K_s.

2 Materials and Methods

2.1 Study Area Description

The experimental area selected as shown in Fig. 1 was an agricultural catchment near Baradanapura village, Mysore district, Karnataka. It covers an area of about 1.873 km^2 and is located between 12°13′2.72″ N latitude and 76°32′59.51″ E longitude. The catchment shows topographic variation with elevations ranging from 650 to 728 m (from 12.5 m ALOS DEM). The catchment comes under semi-arid climatic conditions with varying temperature of 24–29 °C during summer and 18–22 °C during winter. The evaporation ranges from 6.5–7.5 mm/day. Average annual rainfall over last 10 years is 852 mm and it is maximum during June to October. The catchment has a wide range of soil textures as well as geological formations that are heterogeneous. Agriculture is the most important socio-economic activity in Baradanapura

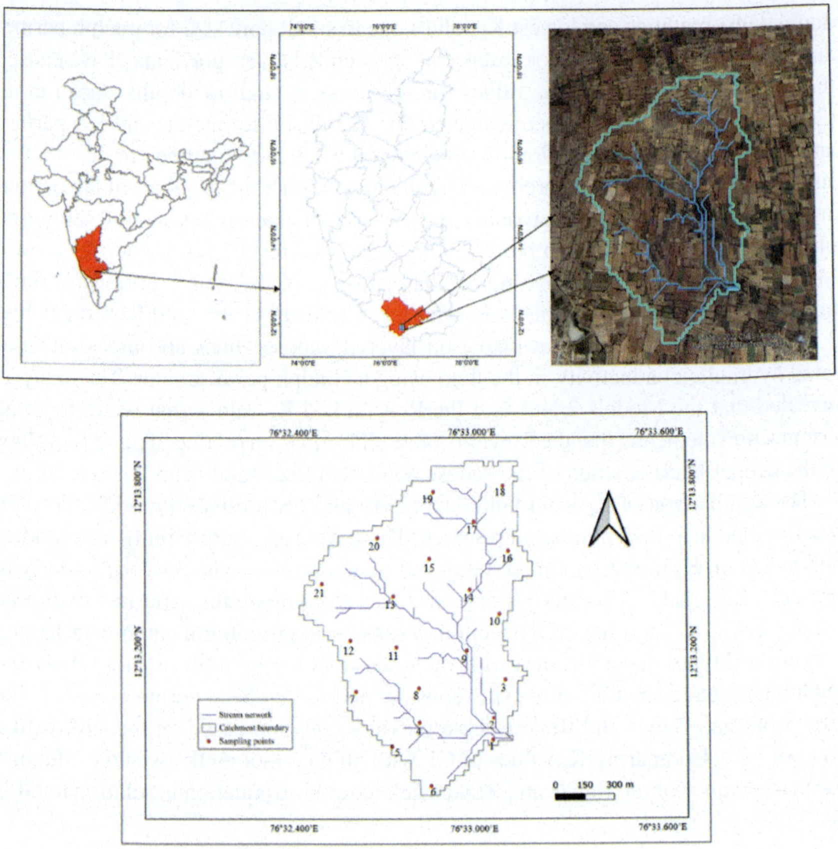

Fig. 1 Study area map

Catchment and agricultural land makes up roughly 90% of overall catchment area. Rainfall, tanks and ground water are the major irrigational sources for agriculture.

2.2 *Site Investigation*

At 15 cm deep, 21 distinct locations within the selected agricultural catchment have been studied and the tests were carried out in the months of March and April, 2022. Infiltration rate and hydraulic conductivity measurements at the sites were performed using Double Ring Infiltrometer and Guelph Permeameter and at the same time soil samples including core cutters were collected to carry out laboratory tests. The next sections provide a brief explanation on the theoretical aspects of the process used to calculate K_s using each technique.

2.3 Determination of Physical Properties of Soil

As shown in Fig. 2, undisturbed soil samples were collected for the determination of physical properties of soil such as texture, water content, bulk density, particle density followed by porosity. Wet sieve analysis and hydrometer analysis were used to determine the particle size distribution of the soil samples obtained from each location, and particle size distribution curves were plotted. As per United States Department of Agriculture (USDA) soil texture classification system, the soil samples were classified into several textural classes based on the percentages of Sand, Silt, and Clay so acquired. Core cutter method and density bottle method, respectively, were used in the laboratory to assess bulk density and particle density.

Bulk density (γ_b) (g/cm^3), Dry density (γ_d) (g/cm^3), Particle density and Porosity ($\varnothing$) (volume fraction) were calculated using following expressions:

$$\gamma_b = \frac{W_2 - W_1}{V} \tag{1}$$

$$\gamma_d = \frac{\gamma_b}{1 + w} \tag{2}$$

$$Particle\ density = \frac{Mass\ of\ soil\ sample}{Volume\ of\ soil\ sample} \tag{3}$$

$$Porosity = 1 - \frac{Bulk\ Density}{Particle\ Density} \tag{4}$$

Where, W_1 = weight of core cutter (g); W_2 = weight of core cutter with soil sample (g); V = volume of core cutter (cm^3); w = water content.

Fig. 2 Collection of soil samples at site

2.4 *Estimation of Saturated Hydraulic Conductivity by Various Method*

Hydraulic conductivity has been measured using a variety of instruments, including the Double Ring Infiltrometer (DRI), Guelph Permeameter (GP), Constant Head Permeability Test (CHPT), Falling Head Permeability Test (FHPT), and Rawls-Brakensiek regression equation (Estimated) based on the textural classification and soil porosity.

2.4.1 Double Ring Infiltrometer

Two concentric metal rings with diameters of 30 and 45 cm make up the double ring infiltrometer. As shown in Fig. 3 to a depth of 15 cm, these rings were carefully inserted into the ground. The rings were filled with water, and during the experiment, a steady head of water was maintained in the inside ring. Additionally, water was held in the outer ring to keep water from flowing laterally beneath the inner ring and maintain one-dimensional flow conditions. Following the recording of the observations of cumulative infiltration versus time, the infiltration rates for elapsed times from the start of each experiment were calculated. The modified Philip's infiltration equation was used to examine these infiltration data in order to calculate K_s values.

Philip (1958) presented that one-dimensional infiltration under ponded conditions could be defined by a simple and rapidly converging power series in $t^{1/2}$. In the current investigation, the first two terms of Philip's modified solution were used in

Fig. 3 Double ring infiltrometer at site

the following equation:

$$v(t) = \frac{1}{2}St^{\frac{1}{2}} + A \tag{5}$$

Where, $v(t)$= infiltration rate (cm/h); A= parameter known as transmissivity factor (cm/h); S= sorptivity (cm/h$^{1/2}$); t= elapsed time since the start of infiltration (h).

The saturated hydraulic conductivity (K_s) is functionally connected to both the A and S parameters and by analysing the experimental data and fitting the above equation to it, these parameters were established. The A values were converted to K_s estimates by using a multiplying factor of 2/3 (Youngs 1968).

2.4.2 Guelph Permeameter

The Mariotte principle is used by the Guelph permeameter, a constant head permeameter shown in Fig. 4 that relies on the assumption of steady-state flows from a 3 cm cylindrical auger hole of 15 cm depth in the soil (Reynolds and Elrick 1985). The instrument measures the steady-state flow rate at predetermined depths to determine both the horizontal and vertical K_s. Prior to taking the measurements, the well hole was first augured to a depth of 15 cm. Two heads (5 and 10 cm) for the same auger hole were used in the trials, which were run at the same depth until a steady state was reached. The field K_s is calculated by following equations:

Fig. 4 Guelph permeameter at site

$$K_s = G_2Q_2 - G_1Q_1 \quad (6)$$

$$Q_1 = R_1x35.39 \quad (7)$$

$$Q_2 = R_2x35.39 \quad (8)$$

$$G_1 = \frac{H_2C_1}{\pi(2H_1H_2(H_2 - H_1) + a^2(H_1C_2 - H_2C_1)} \quad (9)$$

$$G_2 = \frac{H_1C_2}{\pi(2H_1H_2(H_2 - H_1) + a^2(H_1C_2 - H_2C_1)} \quad (10)$$

Where, H_1 and H_2= head maintained (5 and 10 cm); a= radius of well; C_1 and C_2 = dimensionless shape factor that depends on H/a ratio and soil texture-structure category (Reynolds and Elrick 1986; Zhang et al. 1998).

2.4.3 Constant and Falling Head Permeability Test

Klute and Dirksen (1986) discussed the constant head and falling head methods for determining the saturated hydraulic conductivity of soil in a laboratory setting. With the dolly in position, the core cutter is rammed into the ground vertically at a depth of 15 cm. After it has been driven all the way through, the cutter is removed from the ground and both ends are trimmed. Undisturbed soil sample collected from the field is enclosed in the permeameter and soil sample is continuously subjected to water flow until it gets completely saturated. After reaching a steady state flow rate, the following Darcy's law formulae are used in laboratory testing as shown in Fig. 5 to calculate the saturated hydraulic conductivity (K_s).

$$k = \frac{q}{iA} \quad (11)$$

$$k = 2.3\frac{aL}{A(t_2 - t_1)}log_{10}\frac{h_1}{h_2} \quad (12)$$

Where, k= coefficient of permeability (cm/h) or saturated hydraulic conductivity (K_s); q= discharge of water flowing through soil sample (cm^3/h); i= hydraulic gradient; A= cross- sectional area of permeameter (cm^2); a= cross- sectional area of stand pipe (cm^2); L= length of permeameter (cm); $t_2 - t_1$= time interval (h); h_1 and h_2= head causing flow (cm).

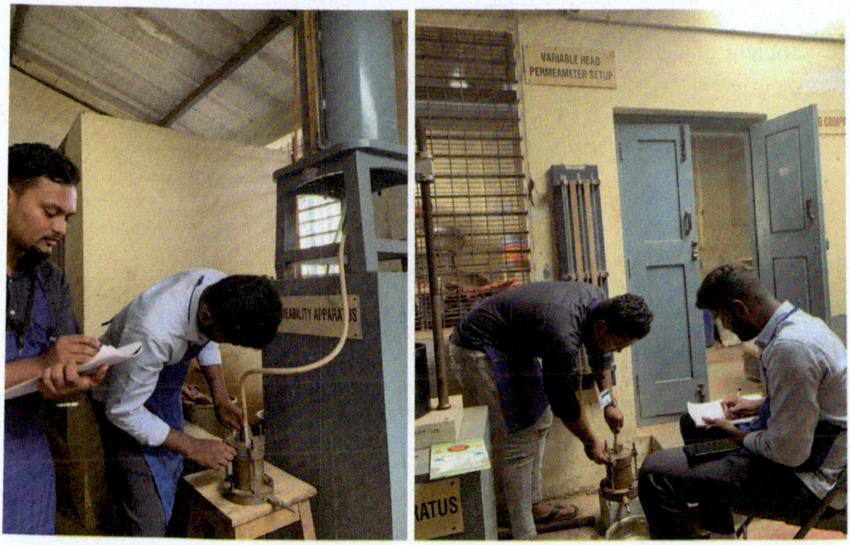

Fig. 5 Saturated hydraulic conductivity tests at laboratory

2.4.4 Rawls and Brakensiek Regression Equation

Rawls and Brakensiek (1989) published regression equations for the Brooks and Corey (1964) soil water retention and hydraulic conductivity parameters as a function of soil characteristics and also considered the porosity property which affects the infiltration rates. The formulae hold true for clay percentages more than 5% and below 60% as well as for sand percentages greater than 5% and below 70%. Equation for saturated hydraulic conductivity (K_s) (cm/h) for the USDA soil texture classes is given below:

$$K_s = e^{\left[\begin{array}{l} 19.52348(\varnothing) - 8.96847 - 0.028212(C) + 0.00018107\left(S^2\right) \\ -0.0094125\left(C^2\right) - 8.395215\left(\varnothing^2\right) + 0.077718(S)(\varnothing) - 0.00298\left(S^2\right)\left(\varnothing^2\right) \\ -0.019492\left(C^2\right)\left(\varnothing^2\right) + 0.0000173\left(S^2\right)(C) + 0.02733\left(C^2\right)(\varnothing) \\ +0.001434\left(S^2\right)(\varnothing) + 0.0000035(C^2)(S) \end{array}\right]} \tag{13}$$

Where, $\varnothing$= porosity (vol fraction), C= percentage of clay; S = percentage of sand.

2.5 *Performance Assessment of Rawls and Brakensiek Regression Equation*

The Performance assessment of K_s determined using Rawls-Brakensiek regression equation with K_s of GP was compared using the following equation:

$$\text{Coefficient of determination } (R^2) = \left[\frac{N\sum XY - \sum X \sum Y}{\sqrt{N\sum X^2 - \sum X^2}\sqrt{N\sum Y^2 - \sum Y^2}}\right]^2 \tag{14}$$

$$\text{Root Mean Square Error(RMSE)} = \sqrt{\frac{1}{N}\sum_{i=1}^{N}(X_i - Y_j)^2} \tag{15}$$

$$\text{Mean Absolute Error(MAE)} = \frac{1}{N}\left(\sum_{i=1}^{N}|X_i - Y_j|\right) \tag{16}$$

$$\text{Nash Sutcliffe Efficiency(NSE)} = 1 - \frac{\sum_{i=1}^{N}(X_i - Y_j)^2}{\sum_{i=1}^{N}(X_i - \overline{X})^2} \tag{17}$$

Where, X or X_i= Estimated K_s; Y or Y_j= Observed K_s; $\overline{X}$= mean of estimated K_s; N= number of measured values.

3 Results and Discussion

3.1 Determination of Physical Properties of Soil

In the laboratory, soil samples collected throughout the catchment are analysed and classified in accordance with the USDA (United States Department of Agriculture) soil textural classification. Texture of soil is typically coarser towards the surface and gradually finer as it descended. Table 1 presents the findings from the analysis of particle size and porosity for each site at a depth of 15 cm. It was found that the experimental site contains four main types of soils i.e., loamy sand, sandy loam, sandy clay loam and silty clay loam and the percentage of porosity ranged from 40.40 to 56.56.

3.2 Determination of K_s Obtained from Field, Laboratory and Estimated Values by Rawls-Brakensiek Regression Equation

Table 2 and Fig. 6 provides an overview of raw K_s results from five different methods, DRI, GP, CHPT, FHPT and Rawls and Brakensiek regression equation (Estimated) on four distinct soils. Because of its spatial variability, K_s is considered to have a log-normal distribution. Unless otherwise specified, the geometric mean is taken into account in the study (Bouwer and Jackson 1974; Lee et al. 1985). The measured mean

Table 1 Physical properties of soil at Baradanapura agricultural catchment

Site	Clay (C) (%)	Silt (%)	Sand (S) (%)	Porosity (∅) (%)	Textural class
1	28.00	48.00	24.00	50.82	Silty clay loam
2	37.00	39.00	24.00	51.95	Silty clay loam
3	27.00	47.00	26.00	56.56	Silty clay loam
4	27.00	47.00	26.00	52.87	Silty clay loam
5	27.00	47.00	26.00	56.12	Silty clay loam
6	26.00	48.00	26.00	52.15	Silty clay loam
7	28.00	18.00	54.00	46.32	Sandy clay loam
8	21.00	23.00	56.00	43.40	Sandy clay loam
9	29.00	18.00	53.00	45.40	Sandy clay loam
10	23.00	19.00	58.00	43.70	Sandy clay loam
11	23.00	19.00	58.00	43.16	Sandy clay loam
12	14.00	24.00	62.00	41.13	Sandy loam
13	16.00	22.00	62.00	41.70	Sandy loam
14	12.00	26.00	62.00	40.40	Sandy loam
15	13.00	24.00	63.00	42.48	Sandy loam
16	13.00	24.00	63.00	42.48	Sandy loam
17	6.00	25.00	69.00	41.57	Loam sand
18	6.00	26.00	68.00	47.90	Loam sand
19	7.00	27.00	66.00	47.59	Loam sand
20	8.00	24.00	68.00	42.54	Loam sand
21	7.00	26.00	67.00	43.59	Loam sand

K_s values (in cm/h) from five methods range from 0.39 to 0.48 for silty clay loam, 1.57 to 1.84 for sandy clay loam, 2.63–2.88 for sandy loam, and 5.35–7.03 for loamy sand. Also K_s values (cm/h) obtained from 21 sites using DRI, GP, CHPT, FHPT and regression equations ranges from 0.25–10.25, 0.25–8.42, 0.24–8.25, 0.25–8.31 and 0.22–8.28 respectively.

Due to the fact that DRI is a flooding-type infiltrometer and considers larger surface area for sampling than the other methods, its K_s values were generally higher at all of the sites. Because of unanticipated anthropogenic factors that lead to changes in pore structure at the ground surface, mean K_s values measured from the field (DRI and GP) are greater than the laboratory K_s (CHPT and FHPT). However, because an infiltrometer depends on water penetration at the soil surface for its readings, these factors may have a greater effect on those measurements (Ghosh et al. 2019). There was a slight difference in K_s measured by laboratory methods (CHPT and FHPT). The mean estimated K_s value derived from the regression equation, with the exception of silty clay loam soil, falls within the range of field and laboratory K_s and it is almost comparable with that of the field or lab K_s at every site.

Table 2 Comparison of K_s for different type of soils

Soil type	Site	K_s (cm/h)				
		DRI	GP	CHPT	FHPT	Estimated
Silty clay loam	1	0.25	0.25	0.24	0.26	0.30
	2	0.27	0.26	0.27	0.25	0.22
	3	1.04	0.83	0.73	0.65	0.91
	4	0.41	0.40	0.39	0.34	0.51
	5	0.78	0.77	0.79	0.76	0.85
	6	0.41	0.39	0.36	0.34	0.48
	Mean K_s (cm/h)	0.46	0.43	0.42	0.39	0.48
Sandy clay loam	7	1.87	1.68	1.44	1.43	1.75
	8	1.76	1.52	1.45	1.39	1.64
	9	1.32	1.30	1.13	1.27	1.25
	10	2.32	2.29	2.26	2.17	1.98
	11	2.11	1.83	1.81	1.73	1.79
	Mean K_s (cm/h)	1.84	1.69	1.57	1.57	1.66
Sandy loam	12	2.84	2.83	2.82	2.81	2.35
	13	2.68	2.54	2.50	2.27	2.51
	14	2.47	2.44	2.66	2.40	2.12
	15	3.10	3.07	2.99	2.95	3.18
	16	3.38	3.26	3.08	3.13	3.18
	Mean K_s (cm/h)	2.88	2.81	2.80	2.69	2.63
Loamy sand	17	4.76	4.46	4.30	4.16	4.22
	18	10.25	8.42	8.25	8.31	8.28
	19	8.74	7.49	6.07	6.46	7.12
	20	5.77	4.51	4.06	3.87	4.69
	21	6.96	6.27	5.98	5.06	4.80
	Mean K_s (cm/h)	7.03	6.03	5.54	5.35	5.62

3.3 Comparison of K_s Measured from GP with K_s of DRI, CHPT, FHPT and Rawls-Brakensiek Regression Equation

Figures 7, 8, 9 and 10 show comparisons of K_s obtained via GP with those obtained from other methods for silty clay loam, sandy clay loam, sandy loam, and loamy sand, respectively. K_s measured from GP demonstrated good agreement with all four approaches, as evidenced by the four methods' R^2 values which vary from 0.85 to 0.99 which is plotted against GP. As a result of its greater sampling area, different experimental installation depths, and reliance on water entry at the soil surface, DRI's K_s values were often higher than those produced by GP. In contrast, GP readings

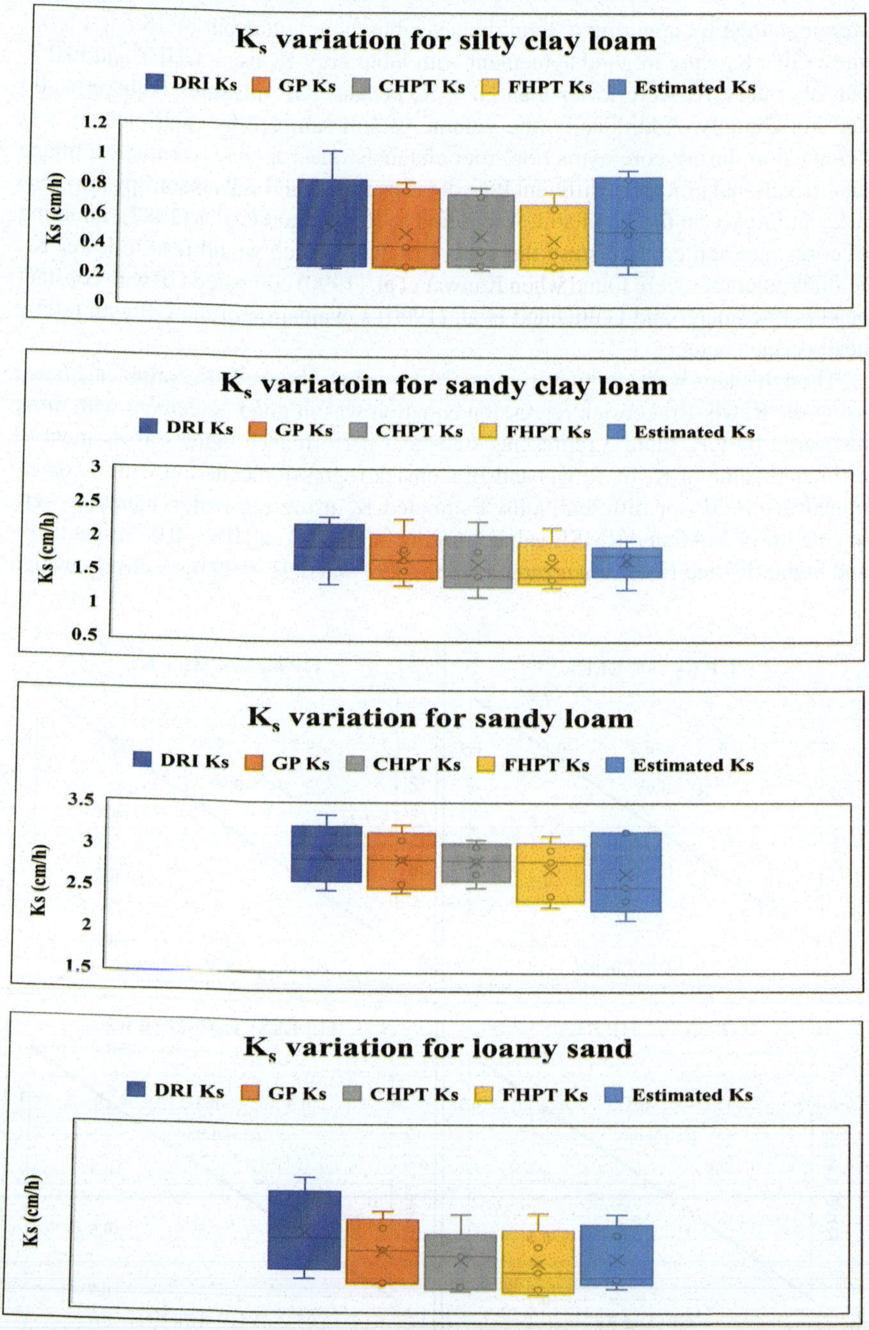

Fig. 6 Box plots for comparison of K_s obtained from various methods for different type of soils

were obtained by auguring a 3 cm diameter borehole at a depth of 15 cm. For all sites, GP's K_s were in good agreement with laboratory K_s using CHPT and FHPT, but laboratory K_s were lower than GP's K_s because GP measures both vertically and horizontally. Additionally, the volume of soil sampled by each method, soil compaction during core extraction, root channels, macropores, wormholes, tillage and irrigational practices, different type of crops grown and soil anisotropy may also have an impact on the K_s of soil. According to Rogers and Carter (1987), ramming of cores into soil could disturb the soil's structure, which would lead to lower K_s. Similar outcomes were found when Kanwar et al. (1990) compared GP with constant head permeameter and Gallichand et al. (1990)'s evaluation of the GP with falling head permeameter.

Though there were slight variations in K_s value obtained, K_s value calculated using the Rawls-Brakensiek regression equation was in good agreement with those measured by GP. Table 3 represents statistical performance indices assessment of estimated value of K_s by Rawls and Brakensiek regression equation with K_s determined using GP for different soils. Estimated K_s using regression equation were having lower MAE and RMSE values ranging 0.01–0.03 and 0.03–0.08 respectively and higher R^2 and NSE values ranging 0.85–0.93 and 0.91–0.99 respectively, which

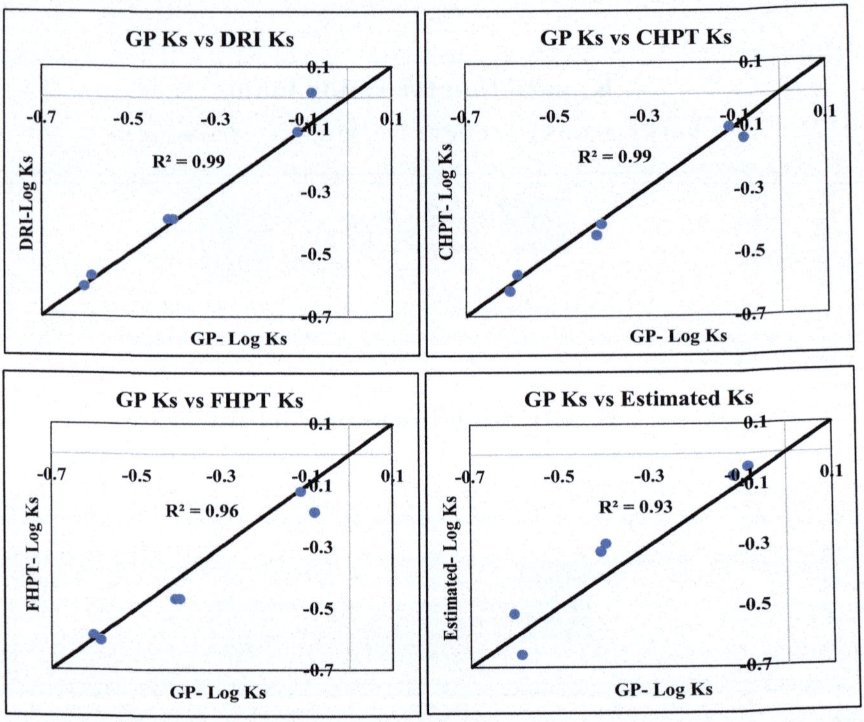

Fig. 7 Comparison of K_s obtained from GP with other methods for silty clay loam

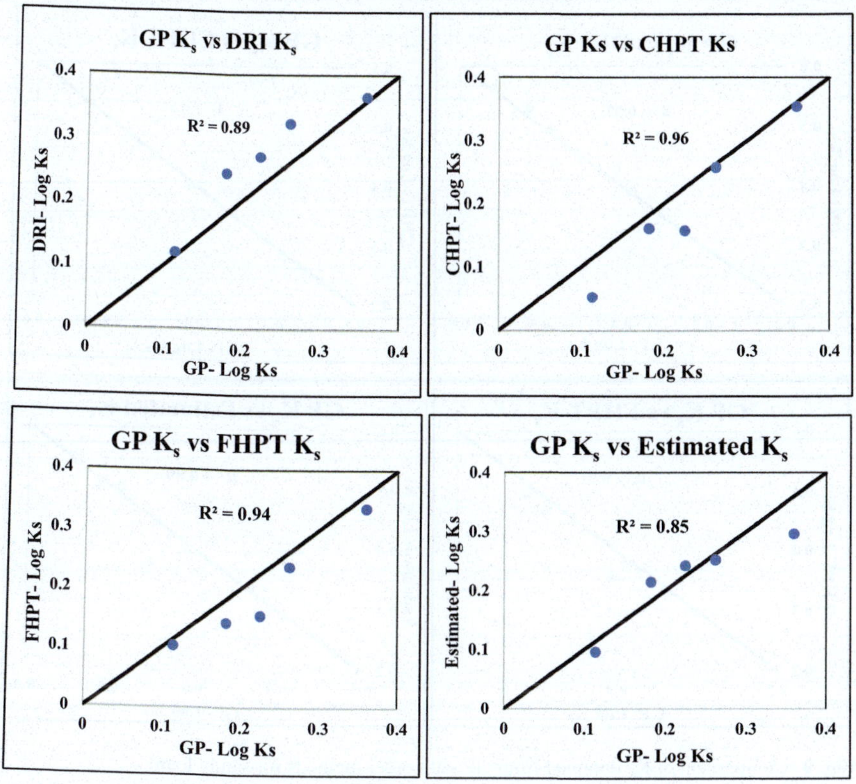

Fig. 8 Comparison of K_S obtained from GP with other methods for sandy clay loam

indicates the good performance of regression equation with GP. Several assumptions and considerations made to carry out each experiments led to variations in K_S measurements. Field K_s measured using GP is on basis of steady-state infiltration rate (Reynolds et al. 1993) while estimated K_S using regression equations is based on soil's porosity and percentage of sand and clay present in the soil (Rawls and Brakensiek 1989).

4 Conclusions

K_S is one of the hardest soil properties to measure, as it is variable in both space and time. Because of this, estimating K_S is a challenging undertaking involving testing, measurement, and judgement. Results of this study shows that mean K_S value obtained from five different methods, DRI, GP, CHPT, FHPT and Rawls-Brakensiek regression equation for selected agricultural catchment with varying soil type were differ from each other. Mean K_S measured using DRI were higher than any other

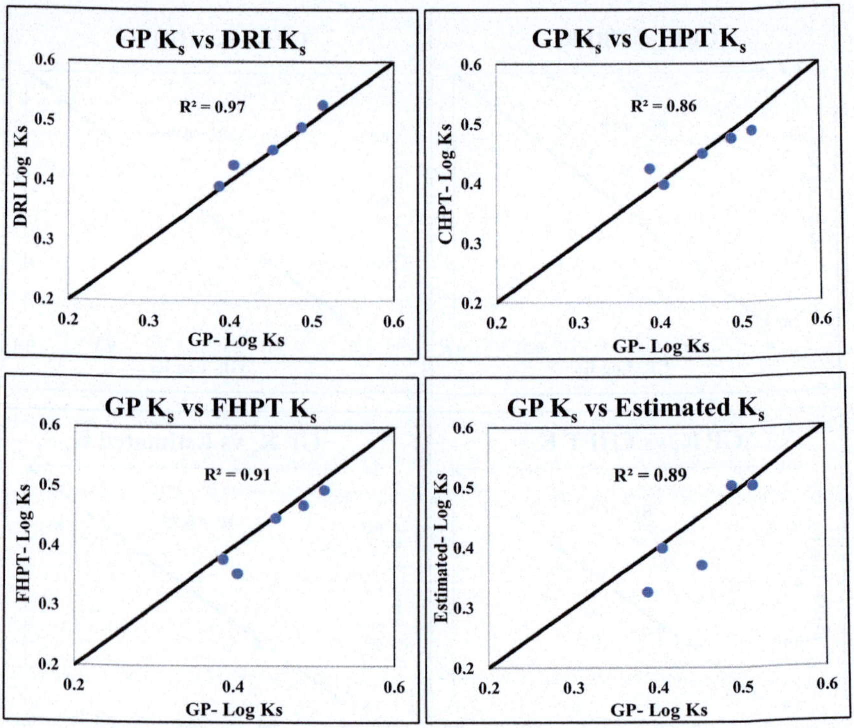

Fig. 9 Comparison of K_S obtained from GP with other methods for sandy loam

methods as it considers, greater surface areas for sampling, dependency on water penetration at the soil surface for its readings and difference in experimental installations depth. Since K_S obtained GP is by auguring a 3 cm diameter borehole mean K_S of GP were less than DRI. There was slight variations in mean K_S measured by two laboratory methods. As CHPT and FHPT measures K_S vertically and ramming of cores into soil could disturb the soil's structure results in lesser K_S values when compared with field (DRI and GP) K_S. However there is slight variation in K_S measure by GP, showed good accordance with DRI, laboratory methods and regression equations. Also irrespective of several assumptions and considerations made to carry out both experiments, regression equation agreed favourably and showed good performance with GP and it estimated results that are almost identical to field mean K_S. Although there are several methods and models for K_S estimate, each one has its own applications and limitations. DRI requires large volume of water in permeable soils while GP considers smaller diameter area for sampling, which requires many measurements to ensure accuracy. The primary source of knowledge for choosing the right approach for certain conditions and soil properties is comparison of several methods. By taking into account various methods' constraints and expectations, the findings of this study can aid in choosing which methods to use for K_S measurement in the agricultural catchment under consideration. Further researches can be done by

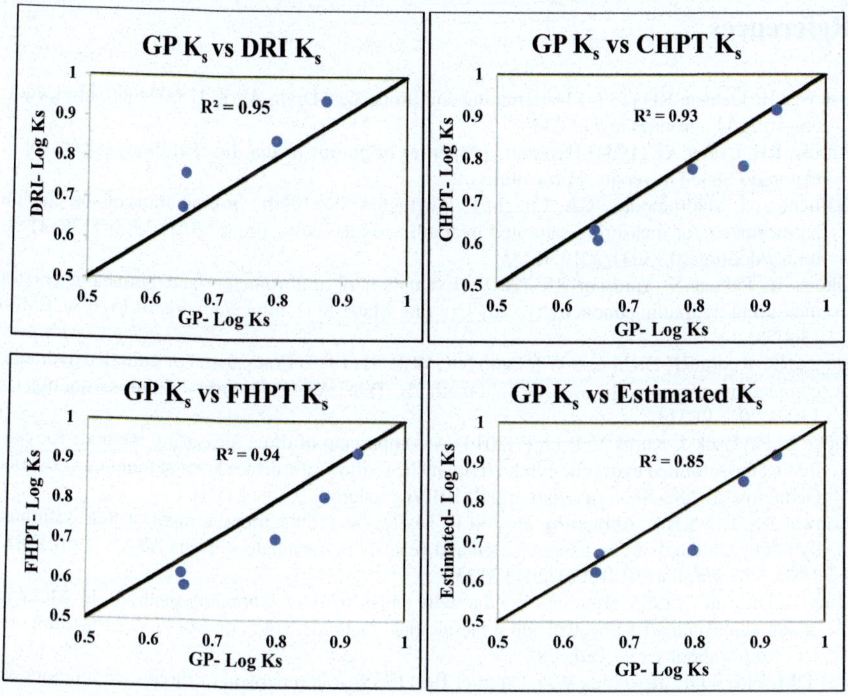

Fig. 10 Comparison of K_S obtained from GP with other methods for loamy sand

Table 3 Statistical performance indices assessment for GP's K_S and Estimated value of K_S by Rawls-Brakensiek regression equation

Type of soil	MAE	RMSE	R^2	NSE
Silty clay loam	0.03	0.08	0.93	0.91
Sandy clay loam	0.01	0.03	0.85	0.96
Sandy loam	0.01	0.05	0.89	0.97
Loamy sand	0.01	0.05	0.85	0.99

applying the present study to larger agricultural catchment and at different depths. Advanced field investigations are required to increase the sensitivity of the parameters of the soil and also further researches are required to study about the other pedotransfer functions based on soil properties other than soil texture and porosity.

References

Bouwer H, Jackson RD (1974) Determining soil properties. Drain Agric 17:609–666. https://doi.org/10.2134/agronmonogr17.c33

Brooks RH, Corey AT (1964) Hydraulic properties of porous media. In: Hydrology paper, vol 3. Colorado State University, Fort Collins

Gallichand J, Madramootoo CA, Emight P, Barrington SF (1990) An evaluation of the Guelph permeameter for measuring saturated hydraulic conductivity. Trans ASAE 33(4):1179–1184. https://doi.org/10.13031/2013.31455

Ghosh B, Pekkat S, Yamsani SK (2019) Evaluation of infiltrometers and permeameters for measuring hydraulic conductivity. Adv Civ Eng Mater 8(1). https://doi.org/10.1520/ACEM20180056

Gupta RK, Rudra RP, Dickinson WT, Patni NK, Wall GJ (1993) Comparison of saturated hydraulic conductivity measured by various field methods. Trans ASAE 36(1):51–55. https://doi.org/10.13031/2013.28313

Jačka L, Pavlásek J, Kuráž V, Pech P (2014) A comparison of three measuring methods for estimating the saturated hydraulic conductivity in the shallow subsurface layer of mountain podzols. Geoderma 219:82–88. https://doi.org/10.1016/j.geoderma.2013.12.027

Kanwar RS, Rizvi HA, Ahmed M, Horton R, Marley SJ (1990) Measurement of field-saturated hydraulic conductivity by using Guelph and velocity permeameters. Trans ASAE 32(6):1885–1890. https://doi.org/10.13031/2013.31239

Klute A, Dirksen C (1986) Hydraulic conductivity and diffusivity: laboratory methods. In: Methods of soil analysis: part 1 physical and mineralogical methods, vol 5, pp 687–734. https://doi.org/10.2136/sssabookser5.1.2ed.c28

Lee DM, Elrick DE, Reynolds WD, Clothier BE (1985) A comparison of three field methods for measuring saturated hydraulic conductivity. Can J Soil Sci 65(3):563–573. https://doi.org/10.4141/cjss85-060

Mahapatra S, Jha MK (2019) On the estimation of hydraulic conductivity of layered vadose zones with limited data availability. J Earth Syst Sci 128(3):1–17. https://doi.org/10.1007/s12040-019-1101-1

Mohanty BP, Kanwar RS, Everts CJ (1994) Comparison of saturated hydraulic conductivity measurement methods for a glacial-till soil. Soil Sci Soc Am J 58(3):672–677. https://doi.org/10.2136/sssaj1994.03615995005800030006x

Philip JR (1958) The theory of infiltration: 7. Soil Sci 85(6):333–337

Rawls WJ, Brakensiek DL (1989) Estimation of soil water retention and hydraulic properties. In: Unsaturated flow in hydrologic modelling. Springer, Dordrecht, pp 275–300. https://doi.org/10.1007/978-94-009-2352-2_10

Reynolds WD, Elrick DE (1985) In situ measurement of field-saturated hydraulic conductivity, sorptivity, and the α-parameter using the Guelph permeameter. Soil Sci 140(4):292–302

Reynolds WD, Elrick DE (1986) A method for simultaneous in situ measurement in the vadose zone of field-saturated hydraulic conductivity, sorptivity and the conductivity-pressure head relationship. Groundw Monit Remediat 6(1):84–95. https://doi.org/10.1111/j.1745-6592.1986.tb01229.x

Reynolds WD, Elrick DE, Topp GC (1983) A re-examination of the constant head well permeameter method for measuring saturated hydraulic conductivity

Rogers JS, Carter CE (1987) Soil core sampling for hydraulic conductivity and bulk density. Soil Sci Soc Am J 51(5):1393–1394. https://doi.org/10.2136/sssaj1987.03615995005100050057x

Youngs EG (1968) An estimation of sorptivity for infiltration studies from moisture moment considerations. Soil Sci 106(3):157–163

Zhang ZF, Groenevelt PH, Parkin GW (1998) The well-shape factor for the measurement of soil hydraulic properties using the Guelph Permeameter. Soil Till Res 49(3):219–221. https://doi.org/10.1016/S0167-1987(98)00174-3

Applicability of Geospatial Technology for Drainage and Hypsometric Analysis of Koyna River Basin, India

Tarate Suryakant Bajirao, Anuradha Kumari, and Nitin Madan Changade

Abstract In this study, drainage and hypsometric analysis of the semi-arid Koyna River basin of India was carried out using remote sensing (RS) and Geographical Information System (GIS) technology. The results obtained in this study will be useful for the management of soil and water resource. In this study, different sub-watersheds were prioritized using two methods i.e. flash flooding and soil erosion separately. This information is useful in the assessment of the risk of flash flooding and soil erosion for better watershed planning and management. Different morphometric parameters of the study area were analyzed for the estimation of flash flood proneness of different sub-watersheds. A parallel drainage pattern was found to be dominant in the study area. Hypsometric analysis was used to find out the sub-watershed-wise stage of geomorphic development or erosion cycle. Sub-watershed-wise hypsometric curve (HC) and hypsometric integral (HI) were derived in hypsometric analysis. Here, it was observed that different sub-watershed comes under the young and mature stage of geologic development. No, even a single sub-watershed was found under the monadnock/old stage of the erosion cycle. It was revealed that the HI values corresponding to different sub-watersheds vary from 0.4867 to 0.7533. Higher HI values represent the young stage of geologic development which indicates the high risk of soil erosion. Then sub-watershed-wise priority ranks were assigned based on value of morphometric parameters and HI for flash flood and soil erosion, respectively. Here, it was observed that there is no exact similarity between the priority ranks of both flash flood and soil erosion determined using drainage and hypsometric analysis, respectively. This study could be helpful to plan sub-watershed-wise soil and water conservation measures on a priority basis.

T. S. Bajirao (✉) · N. M. Changade
Department of Soil Science and Agriculture Chemistry, School of Agriculture, Lovely Professional University, Phagwara 144411, India
e-mail: taratesuryakant01@gmail.com

A. Kumari
Department of Soil and Water Conservation Engineering, G. B. Pant University of Agriculture and Technology, Pantnagar 263145, India

C. B. Pande et al. (eds.), *Surface and Groundwater Resources Development and Management in Semi-arid Region*, Springer Hydrogeology,
https://doi.org/10.1007/978-3-031-29394-8_13

Keywords Watershed management · Morphometry · HC · HI · Watershed prioritization

1 Introduction

Watershed development projects must give priority to the conservation of soil and water resources (Bajirao and Kumar 2021). These natural resources are severely endangered and damaged as a result of their careless use. Watershed management is essential for soil and water conservation to achieve sustainable development. Reduced vegetative cover and shifting agriculture are responsible for the increase in runoff per unit area due to low infiltration (Muhammad et al. 2015). To maximize productivity in sustainable agriculture, watershed management entails the prudent use of the natural resources that are already present (Nigam et al. 2017; Dumka and Kumar 2021). The term morphometry indicates taking measurements and analysis of the dimensions of the landforms, features, and surface of the earth. Prioritization of watersheds, investigation of flood frequency, assessment of drainage basins, and conservation and management of natural resources all depend heavily on morphometric analysis. Understanding the drainage basin's lithology, drainage pattern, topography, and erosional state is crucial for creating efficient watershed management and development strategies (Lole et al. 2016).

Flash floods are considered the most catastrophic as well as very dangerous natural hazards because of their sudden occurrences and unpredictability which lead to damage to infrastructure and threaten human life (Toduse et al. 2020). There are very few options for forecasting flash floods since they have a convective origin, happen locally in watersheds of less than 1,000 km^2, have short response times of a few hours or minutes, and have complex orography (Destro et al. 2018). The most significant factors affecting the severity of floods include the intensity of falling rainfall, rainfall duration, rainfall magnitudes, evaporation rate, water infiltration rate, drainage characteristics, environmental processes and anthropogenic activities (Jodar-Abellan et al. 2019). To protect people's lives and their property against flash floods, it is crucial to demarcate these critical regions (Ali et al. 2020). Watershed management attempts to reduce erosion, reduce dangerous runoff, and use it for productive purposes. It also aims to increase groundwater storage.

Soil erosion is one of the primary reasons for land degradation. Land degradation has an impact on a variety of things, including the quality of surface and groundwater, the operation of reservoirs due to siltation, the environment, and human survival (Ostovari et al. 2017). Soil erosion removes the land's top layer of fertile soil, which has an impact on soil productivity (Pham et al. 2018). For the conservation of land and water resources, sub-watersheds within a basin must be prioritized based on morphometric analysis. Morphometric analysis has been widely used to evaluate how susceptible watersheds are to natural disasters like flash floods and soil erosion (Aher et al. 2014; Shivhare et al. 2018; Asfaw and Workineh 2019; Hussein et al. 2019; Alam et al. 2020). Morphometric parameters provide very simple approaches that may be employed to investigate the geological and geomorphic history of a hydrological basin (Strahler 1952). Since the morphometric characteristics of watersheds are a significant factor that influences the intensity of flash floods and soil erosion, morphometry offers crucial insights into the hydrological response to rainfall (Borga et al. 2008). Morphometric parameters like linear, areal and relief parameters can be used in a variety of investigations like analysis and protection of natural resources as well as assessment of environmental hazards (Charizopoulos et al. 2019). Worldwide flash flood susceptibility has been successfully mapped using morphometric analysis (Alam et al. 2020; Das 2020; Pan et al. 2020).

Langbein (1947) first described the hypsometric analysis, which shows the overall slope and form of a watershed. Further, it is extended by Strahler (1952) and he included percent HC and HI. The HC gives the idea about the process of erosional pattern inside a basin. HI is an important terrain analysis factor that reflects what is the stage of the erosion cycle of a given watershed. A hypsometric analysis is the recognized component in the morphometric analysis of any watershed (Luo et al. 2018). The horizontal cross-sectional area is related to its relative elevation above the basin mouth by the area altitude curve, or HC (Strahler 1952). The shape of the HC represents how the slope of the basin has been changed temporally (Strahler 1952). At the early geomorphic stage of basin development, the change in the shape of the HC is frequent. But, the change in the shape of the HC becomes minimum when it attains a mature stage of geomorphic development over time. A true understanding of past soil movement can be derived by comparing the HC of various sub-watersheds. HC and HI are the two important parameters that indicate what is the health of a basin (Mehar et al. 2018) HI is an indicator of the "cycle of erosion" which represents how much time will be needed to remove the whole land mass to the base level. HC and HI are the two important indicators that state what is the evolutionary stage of a given watershed (Ritter et al. 2002; Singh et al. 2008). The age of the watershed can be assessed based on the shape of HC (Singh and Singh 2018; Tamilarasan and Brema 2019). HC indicates how much soil erosion is already taken place in a watershed while HI represents the erosion cycle of a watershed (Pande et al. 2021).

Watershed prioritization is the allotment of ranks to different sub-watersheds of the basin as per the order in which they should be treated for natural resources and hazard management (Puno and Puno 2019; Obeidat et al. 2021). Watershed management has recently benefited from the effective application of geospatial technology (RS and GIS). The effectiveness of the robust GIS tool allowed several researchers to evaluate drainage basins and accurately analyze their features. It has been further strengthened by the availability of a high-quality resolution digital elevation model (DEM). This advancement made it possible to use and apply morphometric analysis with the help of GIS tools in a variety of research areas like prioritizing sub-watersheds considering their vulnerability to flash floods and soil erosion (Obeidat et al. 2021). Geospatial technology, such as remote sensing and GIS, is advantageous for morphometric analysis since satellites can provide a synoptic image with a wide area range (Nigam et al. 2017). Traditional approaches can be used to quantify the drainage characteristics of various watersheds, but they are time- and labor-intensive (Kar et al. 2009). When compared to traditional approaches, the digital elevation model (DEM) and GIS make it simple to determine the morphometric parameters (Aher et al. 2014). GIS is a powerful tool for assessing the morphological features of drainage basins due to its flexibility in manipulating spatial data of multiple morphometric parameters that vary spatially and temporally within the basin (Aparna et al. 2015; Malik et al. 2019). GIS is user-friendly and offers better results with a high degree of accuracy (Chougale and Jagdish 2017). Since all lithologic and hydrological processes take place at the sub-watershed level, morphometric characteristics at this level can be used to learn more about how landform processes occur (Kaushal and Singh 2013; Pande and Moharir 2017).

Biswas et al. (1999) proposed that remote sensing and GIS are commonly used to study the morphometry of the basin. Nine sub-watersheds of West Bengal of India have been studied to analyze different morphometric parameters and prioritization of different sub-watershed had been done based on this analysis. Thakkar and Dhiman (2007) prioritized eight mini watersheds in the Kheda district of Gujarat State, India by employing geospatial technology based on morphometric analysis. Javed et al. (2009) utilized morphometric and land use change analysis for the prioritization of seven sub-watersheds in Madhya Pradesh using geospatial technology. Kar et al. (2009) utilized multi-spectral satellite data along with ground truth data for the management of rainfed watersheds in sustainable development in eastern India. Ahmed et al. (2010) derived different morphometric parameters using different input data like satellite data and available topographic maps. They proposed that satellite data gives more accurate and detailed information about the terrain characteristics. Pareta and Pareta (2011) studied morphometry and concluded that dendritic and radial drainage patterns of the Karawan watershed. Turkan and Bekir (2011) studied drainage nature, landform processes, flooding and erosional status in sub-basins of Central Anatolia. Aravinda and Balakrishna (2013) used Strahler's stream ordering technique for making a stream order map of the Vrishabhavathi watershed of Arkavathi river basin, Bangalore, and observed dendritic drainage pattern in the study area. Kaushal and Singh (2013) studied quantitative geomorphological properties of micro watersheds of the Ghataprabha river sub-basin in the Karnataka state of India. Aher et al. (2014) employed geospatial technology for the identification of critical areas in the semi-arid water-scarce region of India in the Ahmednagar district of Maharashtra. Bharadwaj et al. (2014) studied the drainage pattern of the Adyar watershed in the Chennai basin and concluded that this watershed represents low flood potential. Gajbhiye et al. (2014) proposed that the geomorphometric features of a watershed are important to develop a regional-based hydrological model for an ungauged watershed in the case of inadequate data availability. Yahya et al. (2015) carried out a hypsometric integral analysis to understand the geomorphologic development and geomorphic evolution stage of watersheds in Southern Jordan using DEM data. They concluded that soil erosion is a major concern in the study area. Nigam et al. (2017) studied the morphometry of a Kharun watershed located in the Seonath sub-basin. They found a homogeneous texture and less structural control by considering drainage analysis. Salvi et al. (2017) reviewed twenty-eight different literatures for watershed planning, management, and prioritization by considering morphometric analysis with the help of geospatial technology and proposed that morphometric analysis could be used for planning and construction of recharge shafts, check dams, and percolation tanks. Umrikar (2017) analyzed the morphometry of the drought-prone Andhale watershed of Maharashtra by employing geospatial technology and found that this watershed area is underlain by impervious rocks which would cause high runoff. Yahya (2017) conducted a study to investigate the drainage features of the Wadi Wala sub-basins. Adhikary and Dash (2018) found that the flow direction and the entire drainage network is controlled by local lithological and geomorphological structure in the Katra watershed. Kibate and Gessesse (2018) studied the hydro-geomorphology of the Dhidhessa river basin, a part of the Blue Nile basin via morphometric analysis and observed that the rate of soil erosion is severe. Zainab (2018) revealed the morphometric characteristics of the Khulgad watershed of the Kosi River and a GIS-based morphometric study revealed a sixth-order watershed mainly drained by a dendritic type of drainage pattern. Malik et al. (2019) carried out morphometric analysis using geospatial technology for studying the morphometry of the Naula watershed of India. Krishnan et al. (2021) used hypsometric analysis for studying the spatial variation of erosion in the Limbang River Basin. Prashanth et al. (2022) used hypsometric analysis for studying erosion proneness in North India. Shekhar and Mathew (2022) used hypsometric analysis for studying the erosion stages of the Bagh River basin.

Prioritization is important in a basin to initiate soil conservation practices from critically endangered sub-areas by considering different input resource limitations. The hypsometric analysis is tedious and time-consuming under traditional methods of data acquisition and analysis. However, the estimation procedure is now more accurate and less laborious as a result of the advent of geospatial technology (Singh et al. 2008; Malik and Kumar 2019). In this study, we have prioritized different sub-watershed based on flash floods and soil erosion separately for the Koyna river basin of India. The novelty of this study is that there is no exact similarity in priority ranks of different sub-watersheds between flash-flood and soil erosion determined using drainage and hypsometric analysis. This kind of prioritization using both drainage and hypsometric analysis for flash floods and soil erosion, respectively is not found previous literatures. Hence, this study was undertaken with the following objectives: (a) To study morphological parameters of the study area; (b) To prioritize different sub-watersheds for management of flash floods. (c) To study the hypsometry (HC and HI) of the study area and (b) To prioritize different sub-watersheds for management of soil erosion based on hypsometric analysis.

2 Materials and Methods

2.1 Study Area

The origin of the Koyna River is Western ghat, Satara district of Maharashtra state, India. The catchment area of the Koyna river basin is 1917 km^2 on the Deccan plateau. The catchment area of this river comes under semi-arid climatic conditions. Monsson rainfall is dominant in the study area. The topography of the study area is varying from gentle to steep slopy. The shape of the Koyna river basin is elongated. Due to the steep slope, the soil and runoff losses are dominant (Bajirao et al. 2021). The geographical location of different sub-watersheds of the study area is shown in Fig. 1.

2.2 Data Collection

ASTER DEM remote sensing free of cost data having 30 m spatial resolution available at the USGS earth explorer website (http://earthexplorer.usgs.gov) were collected for this analysis. This DEM data was used for the generation of stream network, delineation of sub-watershed boundaries, flash flood mapping, and hypsometric analysis of the Koyna river basin.

2.3 Methodology

The entire study area was divided into 15 sub-watersheds using ASTER DEM data in Arc GIS 10.2.2 software (as shown in Fig. 1). In this study, for the prioritization of sub-watersheds, two different methods were used. In the first method, prioritization was carried out for flash flooding based on drainage analysis. In the second method, prioritization was carried out for soil erosion based on hypsometric analysis. The flowchart presenting the methodology utilized for this analysis is shown in Fig. 2.

2.4 Drainage Analysis

For the purpose of creating the drainage network, hypsometric analysis, and delineating the sub-watersheds, ASTER DEM data were employed. Firstly, the DEM data was mosaiced for the delineation of basin boundaries. Different sub-watershed boundaries were delineated using collected DEM data in ArcGIS 10.2.2 environment. Different important steps are performed one by one as shown in Fig. 2. Creation of depression less DEM is the first step to fill the sink which helps the water flow to find the flow path through different cells. The filling can be done by either cutting off tall cells or filling the sink. To avoid discontinuity in the drainage flow sinks were removed. Flow direction denotes the direction where each cell will drain in the landscape; only one neighboring cell receives water from a certain cell in the direction of the steepest descent. Flow accumulation is the next step of hydrological modelling which determines the flow path of the drainage network through different cells over the area. To create a drainage network, a flow accumulation map and a

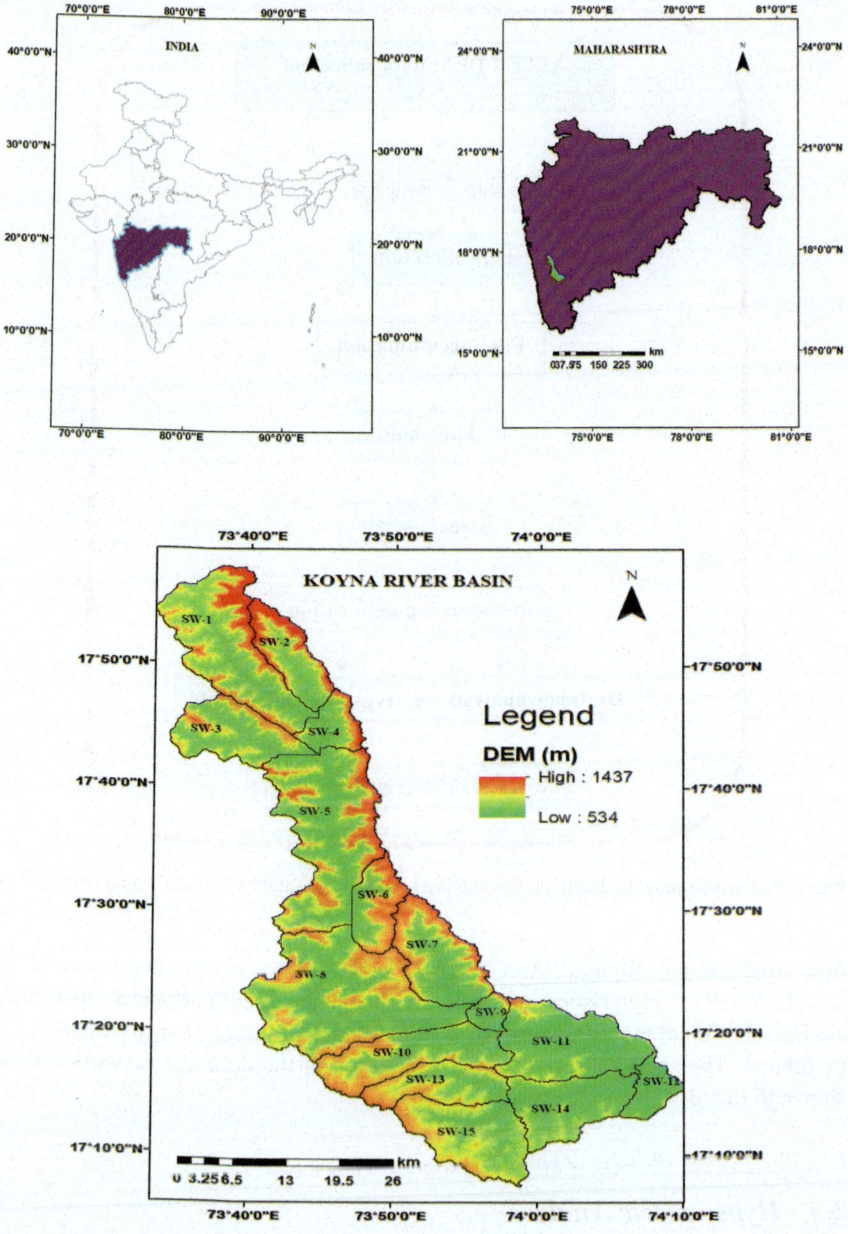

Fig. 1 Location of a study area

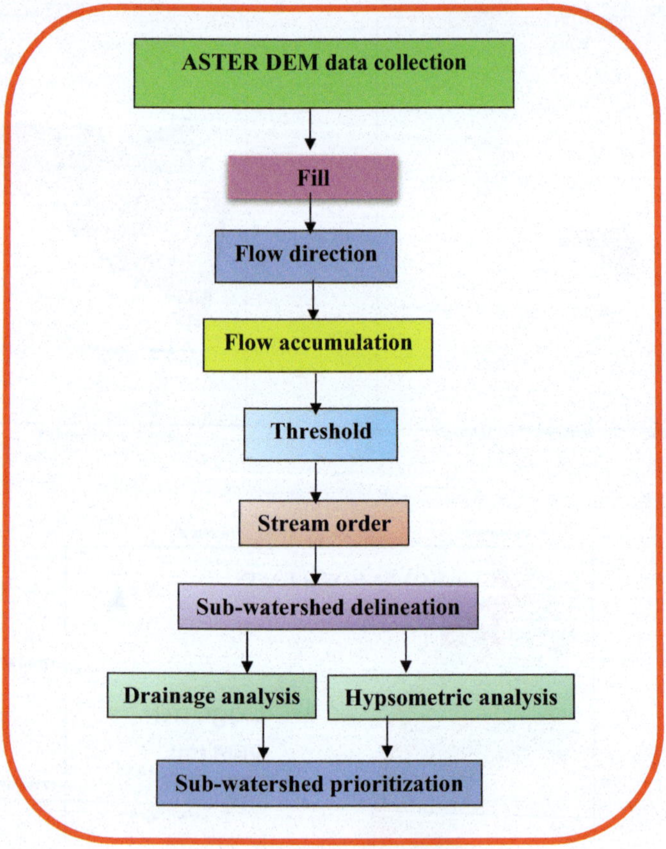

Fig. 2 Flowchart showing methodology adopted for drainage and hypsometric analysis

flow direction map are used. ArcGIS 10.2.2 software was used to determine several morphometric characteristics such as linear, shape, and relief parameters in drainage analysis. Different morphometric parameters used in drainage analysis are presented in Table 1. The steps followed for the generation of the drainage network map are shown in Fig. 3.

2.5 *Hypsometric Analysis*

2.5.1 Hypsometric Curve (HC)

Strahler (1952) classified different geomorphic evolutionary stages/erosion cycles of watersheds into (i) youth stage (convex upward HC and HI $\geq$ 0.60) in which the sub-watershed is extremely susceptible to the erosion process; (ii) equilibrium or

Table 1 Standard formulae for determination of morphometric parameters used in drainage analysis

Morphometric parameters	Formula/method	References
Linear		
Stream order	Hierarchical order	Strahler (1964)
Stream length (km)	Length of all streams	Horton (1945)
Drainage density (D_d), km/km^2	$D_d = \frac{L}{A}$ where, L = Sum of the length of all streams; A = watershed area	Horton (1945)
Texture ratio (T), number/km	$T = \frac{N_1}{P}$ where, N_1 is the number of streams of 1st order stream; P is the sub-watershed perimeter	Horton (1945)
Length of overland flow (L_{ovf}), km	$L_{ovf} = \frac{1}{2D_d}$	Horton (1945)
Stream frequency (F_s), number/km^2	$F_s = \frac{N}{A}$ where, N is the number of streams of all orders	Horton (1945)
Constant of channel maintenance (C_m)	$Cm = \frac{1}{D_d}$	Horton (1945)
Infiltration number (I_n)	$I_n = D_d \times F_s$	Faniran (1968)
Drainage texture (D_t), number/km	$D_t = \frac{N}{P}$	Horton (1945)
Shape		
Form factor (R_f)	$R_f = \frac{A}{L_{sw}^2}$ where, L_{sw} is the length of sub-watershed	Horton (1945)
Compactness coefficient (C_c)	$C_c = \frac{0.2821P}{A^{0.5}}$	Adhikary and Dash (2018)
Circulatory ratio (R_c)	$R_c = \frac{4\pi A}{P^2}$	Miller (1953)
Elongation ratio (R_e)	$R_e = \frac{2\sqrt{(A/\pi)}}{L_{sw}}$	Schumm (1956)
Relief		
Sub-watershed relief (H), m	The elevation difference between the highest and lowest points of sub-watershed	Schumm (1956)
Relief ratio (R_h)	$R_h = \frac{H}{L_{sw}}$ where, L_{sw} is the length of sub-watershed	Schumm (1956)
Relative relief (R_r)	$R_r = \frac{H}{P}$	Melton (1957)
Ruggedness number (R_n)	$R_n = H \times D_d$	Schumm (1956)

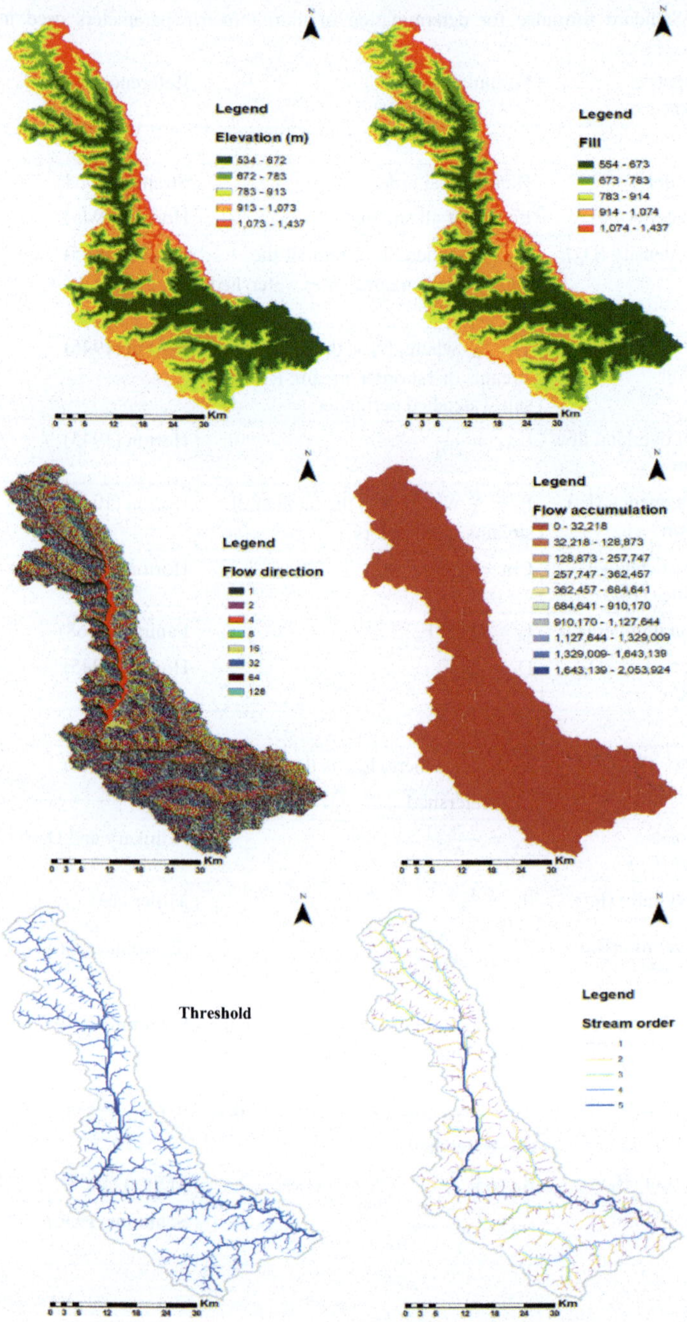

Fig. 3 Steps followed for preparation of drainage network map

mature stage (S-shaped HC which is concave upward at high elevation and convex downward at low elevation and $0.30 \leq HI \leq 0.60$) and (iii) old or monadnock stage (concave upward HC and $HI \leq 0.30$) which is fully stabilized.

2.5.2 Hypsometric Integral (HI)

The HI is a geomorphological parameter that classifies any watershed development into different geologic stages. It is most important to estimate the erosional stage of a watershed and for prioritization to start conservation practices for the conservation of natural resources on a priority basis. HI is an indicator used for the study of the 'cycle of erosion' at any given watershed (Strahler 1952). The time needed to bring the land topological unit down to its base level is known as the cycle of erosion which is shown in Fig. 4. The HI can be stated as a percentage, indicating how much land mass (volume) is still present in the basin today relative to its original volume (Sarangi et al. 2001; Ritter et al. 2002). HI will describe how several hydrological processes contributed to the erosion that has occurred in the watershed.

In this study, different HC curves were plotted concerning relative height (h/H) and relative area (a/A) corresponding to different sub-watersheds as shown in Fig. 7. Then, to describe the equation of the curve, a trend line (dotted line) was fitted to the plotted HCs of different sub-watersheds of the study area (Fig. 7). The best-fit equation was developed for the highest value of the coefficient of determination (R^2). In order to estimate the area under the curve, this equation was then further integrated between 0 and 1 (due to the graph's non-dimensionality). Thus, the HI value of each sub-watershed is determined by the anticipated area under the curve. The HC and HI values were estimated for individual sub-watersheds.

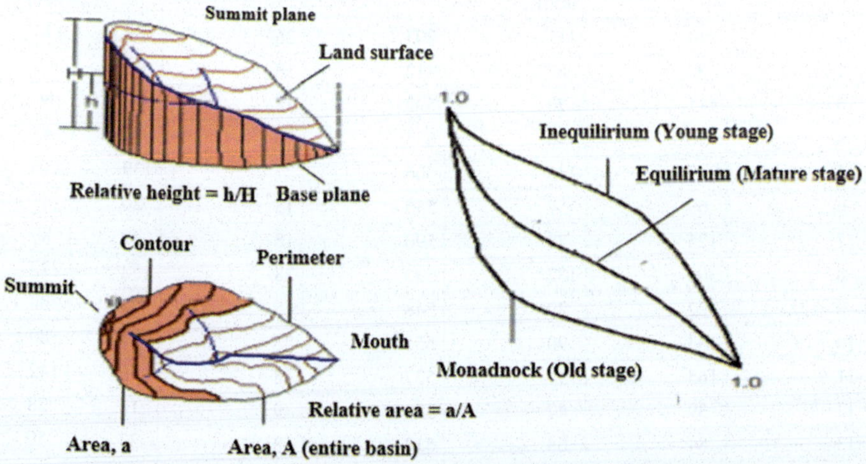

Fig. 4 Hypsometric analysis and hypsometric curve (Ritter et al. 2002; Mandi and Soren 2016)

2.6 *Prioritization of Sub-watershed Using Drainage Analysis and Hypsometric Analysis*

Different morphometric parameters namely D_d, F_s, T, D_t, I_n, R_c, R_f, R_e, R_h, R_r, R_n, L_{ovf}, C_m and C_c were used for prioritizing different sub-watershed for flash-flooding in drainage analysis. Those parameters which are directly proportional to flash flooding are assigned with higher priority and vice versa.

In hypsometric analysis, priority ranks were assigned based on hypsometric integral (HI) values. The highest HI value sub-watersheds were assigned the highest (first) priority and vice versa.

3 Result and Discussion

The results obtained after carrying out morphometric analysis are described below. Basic features corresponding to different 15 sub-watersheds as per morphometric analysis are presented in Table 2. The Koyna River basin is a fifth-order basin as per Strahler's stream ordering method. Stream numbers corresponding to different stream order and their corresponding stream length are presented in Table 3.

A highly important hydrological element is the watershed area since it controls how much runoff water will flow as a result of rainfall. The correlation between

Table 2 Basic features of different sub-watersheds in the study area

Sub-watershed (SW)	Sub-watershed area (km^2)	Sub-watershed perimeter (km)	Sub-watershed relief (m)	Sub-watershed length (km)	Stream numbers	Stream length (km)
1	214	111	803	26	81	143
2	82	67	812	18	36	58
3	111	70	592	18	53	86
4	46	46	635	10	24	42
5	263	135	606	27	139	216
6	69	57	556	13	26	47
7	125	78	590	18	49	89
8	308	132	533	28	152	222
9	22	29	548	5	11	22
10	133	90	551	23	61	89
11	165	94	537	18	76	142
12	39	47	388	9	16	28
13	84	68	513	18	37	50
14	118	79	408	16	64	96
15	140	83	507	14	70	96

Table 3 Variation of a stream number and stream length corresponding to different sub-watersheds

SW	Number of streams					Stream length (km)				
	I	II	III	IV	V	I	II	III	IV	V
1	66	14	1	–	–	78	28	37	–	–
2	29	4	2	1	–	32	11	15	0	–
3	37	10	3	2	1	42	17	15	10	2
4	18	4	1	1	–	24	11	1	6	–
5	106	27	4	1	1	106	59	21	0	30
6	18	5	2	1	–	28	10	10	0	–
7	40	8	1	–	–	52	20	17	–	–
8	129	20	2	–	1	121	52	12	–	37
9	8	2	–	–	1	15	1	–	–	6
10	51	8	1	–	1	40	22	25	–	2
11	59	13	2	1	1	84	23	7	2	26
12	12	3	–	–	1	16	6	–	–	6
13	31	5	1	–	–	21	11	17	–	–
14	51	9	2	1	1	50	20	6	20	1
15	57	9	3	1	–	52	19	18	7	–

sub-watershed area and sub-watershed perimeter, sub-watershed length and sub-watershed slope, sub-watershed length and stream length, stream order and stream number are shown in Fig. 5. There was a significant correlation found to be between the sub-watershed area and sub-watershed perimeter ($R^2 = 0.96$). Basin length plays an important role in the formation of surface runoff. Longer streams will have flatter gradients (Obeidat et al. 2021). A negative correlation ($R^2 = -0.60$) was observed between the sub-watershed length and the sub-watershed slope in this study. A strong positive correlation ($R^2 = 0.71$) was observed between sub-watershed length and stream length. When compared to watersheds with low stream numbers, those watersheds with high stream numbers have more runoff and faster peak flows (Bhatt and Ahmed 2014). There is a negative correlation ($R^2 = -0.65$) between a stream number and stream order. The correlation between stream order and stream length, sub-watershed area and stream length, drainage density and sub-watershed relief, sub-watershed relief and ruggedness number is shown in Fig. 6. Number of streams of the first-order stream was found to be more and vice versa. The relation between stream length and stream order was observed to be strongly negative ($R^2 = -0.77$). In contrast, a strong positive correlation ($R^2 = 0.97$) was found to be in between stream length and sub-watershed area. However, in this study, no relation was observed between sub-watershed relief and drainage density. Additionally, there is a positive correlation ($R^2 = 0.61$) found between sub-watershed relief and ruggedness number.

Different linear, shape and relief parameters corresponding to different sub-watersheds are presented in Table 4.

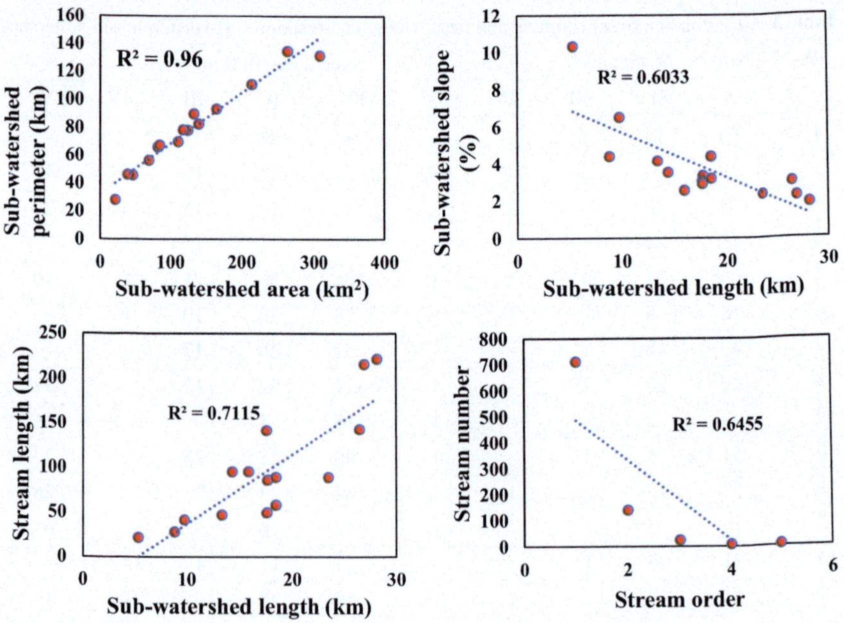

Fig. 5 Relationship between **a** Sub-watershed area and Sub-watershed perimeter **b** Sub-watershed length and Sub-watershed slope **c** Sub-watershed length and stream length **d** Stream order and stream number

4 Linear Factors

4.1 *Drainage Density (D_d)*

It can suggest the closed spacing between different channels (Horton 1945). It will be used to study landforms and dissection complexity. Its lower value represents a coarser drainage network, and permeable topography while the higher value represents weak strata, mountainous areas and impervious land (Nigam et al. 2017). This parameter is most important to study the drainage capacity of any watershed. Chougale and Jagdish (2017) categorized D_d into five classes like very coarse, coarse, moderate, fine and very fine if the D_d is <2, 2–4, 4–6, 6–8 and <8 km/km^2, respectively. If the infiltration is high the D_d will be low (Chandrashekar et al. 2015). High D_d is responsible for quick peak runoff flow and flash flooding. The values of D_d corresponding to different sub-watersheds like sub-watershed (SW) 1, 2, 3, 4, 5, 6, 7, 8, 9, 10, 11, 12, 13, 14 and 15 was found to be 0.67, 0.71, 0.78, 0.9, 0.82, 0.68, 0.72, 0.72, 1.03, 0.67, 0.86, 0.73, 0.59, 0.82, 0.68 km/km^2, respectively. The higher value of D_d was observed in sub-watershed 9 (SW-9) which is highly susceptible to flash flooding.

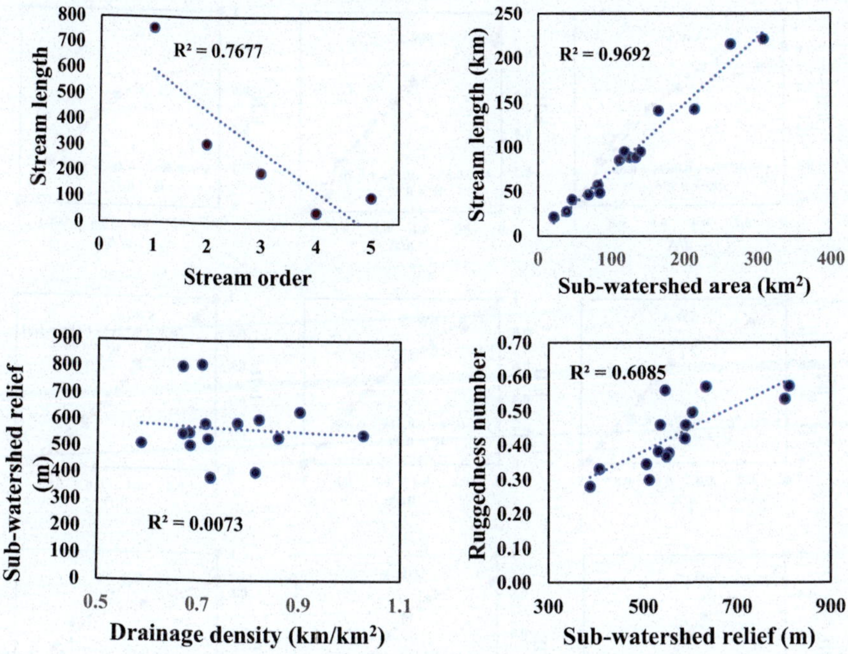

Fig. 6 Relationship between **a** Stream order and stream length **b** Sub-watershed area and stream length **c** Drainage density and sub-watershed relief **d** sub-watershed relief and ruggedness number

4.2 *Stream Frequency (F_S)*

Its value can be affected by lithology, climatic factors, vegetative cover, rainfall amount, rainfall pattern, soil permeability and topography of the watershed. Its higher value is responsible for quick runoff, less infiltration, peak water flow and flooding watershed. Its lower value indicates relatively pervious soil strata and flat topography. The values of Fs corresponding to different sub-watersheds like sub-watershed (SW) 1, 2, 3, 4, 5, 6, 7, 8, 9, 10, 11, 12, 13, 14 and 15 was observed to be 0.38, 0.44, 0.48, 0.52, 0.53, 0.38, 0.39, 0.49, 0.51, 0.46, 0.46, 0.41, 0.44, 0.55, 0.5 streams/km^2, respectively. The higher value of Fs was observed to be in sub-watershed 14 (SW-14) which is highly susceptible to flash flooding.

4.3 *Texture Ratio (T)*

It is an important morphometric indicator for studying the drainage of the watershed. It depends on climatic conditions, rainfall conditions, vegetation health, underlying lithology, soil permeability and relief of the watershed (Schumm 1956). Its higher value represents higher runoff and vice versa. Its lower value represents a

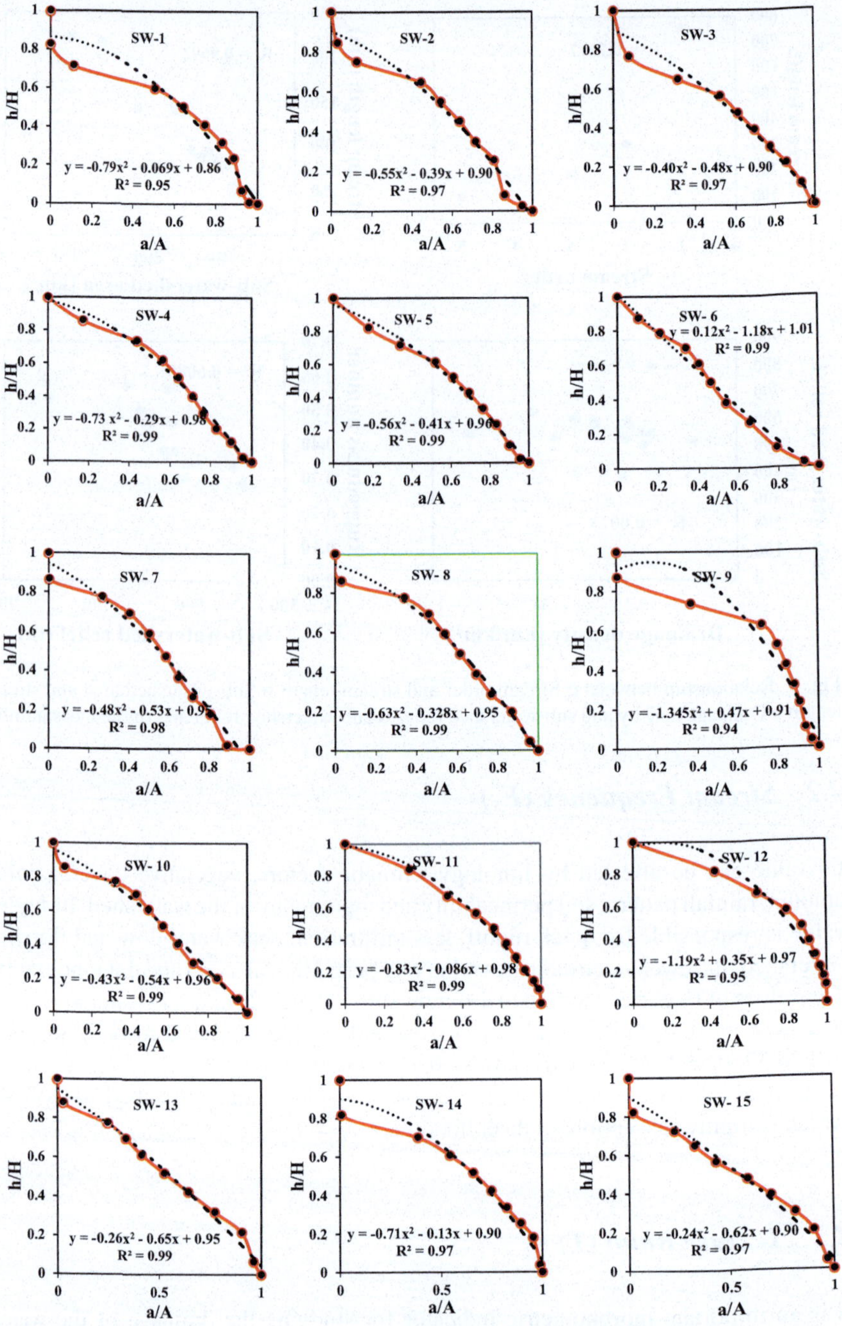

Fig.7 HCs of different 15 sub-watersheds of the study area

Table 4 Variation of different morphometric parameters in different sub-watersheds

SW	D_d	F_s	T	D_t	I_n	L_{ovf}	C_m	R_c	C_c	R_f	R_e	R_h	R_r	R_n
1	0.67	0.38	0.59	0.73	0.25	0.75	1.49	0.22	2.15	0.31	0.62	31	7	538
2	0.71	0.44	0.44	0.54	0.31	0.71	1.41	0.23	2.08	0.24	0.55	45	12	577
3	0.78	0.48	0.53	0.76	0.37	0.64	1.28	0.28	1.89	0.36	0.67	33	8	462
4	0.9	0.52	0.39	0.52	0.47	0.55	1.11	0.28	1.91	0.49	0.79	64	14	572
5	0.82	0.53	0.79	1.03	0.43	0.61	1.22	0.18	2.35	0.36	0.68	22	4	497
6	0.68	0.38	0.32	0.46	0.26	0.73	1.47	0.27	1.92	0.39	0.7	43	10	378
7	0.72	0.39	0.51	0.63	0.28	0.7	1.39	0.25	1.98	0.37	0.68	33	8	425
8	0.72	0.49	0.98	1.15	0.35	0.69	1.39	0.22	2.12	0.39	0.7	19	4	384
9	1.03	0.51	0.28	0.38	0.53	0.49	0.97	0.33	1.74	0.76	0.99	110	19	564
10	0.67	0.46	0.57	0.68	0.31	0.75	1.49	0.21	2.21	0.24	0.55	24	6	369
11	0.86	0.46	0.63	0.81	0.40	0.58	1.16	0.24	2.06	0.53	0.82	30	6	462
12	0.73	0.41	0.26	0.34	0.30	0.69	1.37	0.23	2.11	0.51	0.8	43	8	283
13	0.59	0.44	0.46	0.54	0.26	0.85	1.69	0.23	2.08	0.27	0.59	29	8	303
14	0.82	0.55	0.65	0.81	0.45	0.61	1.22	0.24	2.05	0.47	0.77	26	5	335
15	0.68	0.5	0.69	0.84	0.34	0.73	1.47	0.26	1.98	0.68	0.93	36	6	345

relatively flat area. The values of T corresponding to different sub-watersheds like sub-watershed (SW) 1, 2, 3, 4, 5, 6, 7, 8, 9, 10, 11, 12, 13, 14 and 15 was found to be 0.59, 0.44, 0.53, 0.39, 0.79, 0.32, 0.51, 0.98, 0.28, 0.57, 0.63, 0.26, 0.46, 0.65 and 0.69 streams/km, respectively. The T value of sub-watershed 8 (SW-8) was found to be more causing high runoff or flash floods.

4.4 Drainage Texture (D_t)

It gives an idea about the closeness of channels in the watershed. It depends on lithology, rock type, relief, climate and vegetation. It can be classified as very coarse, coarse, moderate, fine and very fine when the value of drainage texture is <2, 2–4, 4–6, 6–8 and >8, respectively (Chougale and Jagdish 2017). The values of D_t with respect to different sub-watersheds like sub-watershed (SW) 1, 2, 3, 4, 5, 6, 7, 8, 9, 10, 11, 12, 13, 14 and 15 was found to be 0.73, 0.54, 0.76, 0.52, 1.03, 0.46, 0.63, 1.15, 0.38, 0.68, 0.81, 0.34, 0.54, 0.81 and 0.84 streams/km, respectively. D_t of sub-watershed 8 (SW-8) was found to be more causing high runoff or flash floods.

4.5 *Infiltration Number (I_n)*

It is simply a multiplication of D_d and F_s. It is used to study the infiltration potential of the watershed area. It is inverse to that of the infiltration rate of land. Its higher value represents more runoff water flow and vice versa (Strahler 1964; Pareta and Pareta 2011; Nigam et al. 2017). The values of I_n with respect to different sub-watersheds like sub-watershed (SW) 1, 2, 3, 4, 5, 6, 7, 8, 9, 10, 11, 12, 13, 14 and 15 was found to be 0.25, 0.31, 0.37, 0.47, 0.43, 0.26, 0.28, 0.35, 0.53, 0.31, 0.40, 0.30, 0.26, 0.45 and 0.34/km^3, respectively. The I_n of sub-watershed 9 (SW-9) was found to be more causing high runoff or flash floods.

4.6 *Length of Overland Flow (L_{ovf})*

It is a parameter that represents erosion, runoff flow and infiltration of the area and its Spatio-temporal variation (Sahu et al. 2016). Its high value indicates relatively low relief and hence high infiltration and less runoff while a low value indicates high relief and hence less infiltration and more runoff (Bharadwaj et al. 2014). A shorter L_{ovf} is responsible for very quick peak runoff flow. The values of L_{ovf} with respect to different sub-watersheds like sub-watershed (SW) 1, 2, 3, 4, 5, 6, 7, 8, 9, 10, 11, 12, 13, 14 and 15 was found to be about 0.75, 0.71, 0.64, 0.55, 0.61, 0.73, 0.7, 0.69, 0.49, 0.75, 0.58, 0.69, 0.85, 0.61 and 0.73 km, respectively. L_{ovf} of sub-watershed 9 (SW-9) was found to be less which is responsible for high runoff or flash floods.

4.7 *Constant of Channel Maintenance (C_m)*

It is reciprocal to D_d. It is an important parameter to study landform runoff flow and erosion (Schumm 1956). It is dependent on factors like rainfall amount, vegetation health, soil permeability, rock strata, and degree of erosion. Its value will be low in the area of close dissection, steep slope and high relief (Chandrashekar et al. 2015). Its lesser value represents more runoff flow and vice versa (Schumm 1956). Its value corresponding to different sub-watersheds like sub-watershed (SW) 1, 2, 3, 4, 5, 6, 7, 8, 9, 10, 11, 12, 13, 14 and 15 was found to be about 1.49, 1.41, 1.28, 1.11, 1.22, 1.47, 1.39, 1.39, 0.97, 1.49, 1.16, 1.37, 1.69, 1.22 and 1.47 km, respectively. Its value is lesser in sub-watershed 9 (SW-9) which is responsible for high runoff or flash floods.

5 Shape Factors

5.1 Circulatory Ratio (R_c)

Different factors like geologic, climatic factors, vegetative and topographic factors affect it (Miller 1953). Its value lies between 0 (straight line) and 1 (perfect circle). Its higher, medium and lower values represent the old, mature, and young stages. Its lower value denotes a basin with a strong elongation and a highly permeable substrata. R_c and flash flood or runoff is directly correlated. Its higher value represents less time available for infiltration and thereby causing more runoff or flash floods. Its value corresponding to different sub-watersheds like sub-watershed (SW) 1, 2, 3, 4, 5, 6, 7, 8, 9, 10, 11, 12, 13, 14 and 15 was found to be about 0.22, 0.23, 0.28, 0.28, 0.18, 0.27, 0.25, 0.22, 0.33, 0.21, 0.24, 0.23, 0.23, 0.24, and 0.26, respectively. Its value is more in sub-watershed 9 (SW-9) which is responsible for high runoff or flash floods.

5.2 Compactness Coefficient (C_c)

It is a parameter whose dependence on the watershed's slope rather than its size is important (Horton 1945). For a perfect circle, its value is 1 and increases with respect to the length of a watershed. This is a factor used to study the degree of elongation of the basin. Circular watersheds are causing more runoff than elongated watersheds (Singh and Singh 1997). Its minimum value represents more runoff and vice versa. Its value corresponding to different sub-watersheds like sub-watershed (SW) 1, 2, 3, 4, 5, 6, 7, 8, 9, 10, 11, 12, 13, 14 and 15 was found to be about 2.15, 2.08, 1.89, 1.91, 2.35, 1.92, 1.98, 2.12, 1.74, 2.21, 2.06, 2.11, 2.08, 2.05 and 1.98, respectively. Its value is less in sub-watershed 9 (SW-9) which is responsible for high runoff or flash flooding.

5.3 Form Factor (R_f)

It is a factor that depends upon the shape of the watershed and affects runoff formation. Its smaller value represents a highly elongated watershed while a higher value indicates a circular watershed. Circular watersheds with a higher value of R_f are always responsible for higher peak flow (Horton 1932). This factor is used to study flood intensity of different watersheds (Bharadwaj et al. 2014). Its value corresponding to different sub-watersheds like sub-watershed (SW) 1, 2, 3, 4, 5, 6, 7, 8, 9, 10, 11, 12, 13, 14 and 15 was found to be about 0.31, 0.24, 0.36, 0.49, 0.36, 0.39, 0.37, 0.39, 0.76, 0.24, 0.53, 0.51, 0.27, 0.47 and 0.68, respectively. Its value is more in sub-watershed 9 (SW-9) which is responsible for high runoff or flash flooding.

5.4 Elongation Ratio (R_e)

It is a factor used to study the relief and circularity of watersheds. Its value will vary between 0 (straight line) to 1 (perfect circle). In different climatic and geological conditions, its value mostly lies between 0.4 to 1. If its value is closer to 1 then it indicates a circular watershed with low relief while its value equal to 0.4 indicates an elongated watershed with high relief (Strahler 1964). Biswas et al. (1999) classified watersheds based on its value as: 0.9 to 1 (circular watershed), 0.8–0.9 (oval watershed), 0.7–0.8 (less elongated watershed), 0.5–0.6 (elongated watershed), and <0.5 (more elongated). Its higher value indicates more runoff and vice versa. Its value corresponding to different sub-watersheds like sub-watershed (SW) 1, 2, 3, 4, 5, 6, 7, 8, 9, 10, 11, 12, 13, 14 and 15 was found to be about 0.62, 0.55, 0.67, 0.79, 0.68, 0.7, 0.68, 0.7, 0.99, 0.55, 0.82, 0.8, 0.59, 0.77 and 0.93, respectively. Its value is more in sub-watershed 9 (SW-9) which is responsible for high runoff or flash floods.

6 Relief Factors

6.1 Relief Ratio (R_h) and Relative Relief (R_r)

These factors are crucial for the formation of landforms, drainage, surface water movement, subsurface flow, permeability, and erosional status of the land (Magesh et al. 2011). Its higher value indicates lower infiltration and higher surface runoff flow conditions and vice versa (Obeidat et al. 2021). If the watershed R_h and R_r are higher then it will cause a sudden occurrence of peak runoff and flash flooding in the watershed area (Ameri et al. 2018). The R_h with respect to different sub-watersheds like sub-watershed (SW) 1, 2, 3, 4, 5, 6, 7, 8, 9, 10, 11, 12, 13, 14 and 15 was found to be 31, 45, 33, 64, 22, 43, 33, 19, 110, 24, 30, 43, 29, 26 and 36, respectively. R_r corresponding to different sub-watersheds like sub-watershed (SW) 1, 2, 3, 4, 5, 6, 7, 8, 9, 10, 11, 12, 13, 14 and 15 was found to be 7, 12, 8, 14, 4, 10, 8, 4, 19, 6, 6, 8, 8, 5 and 6, respectively. Both R_h and R_r are more in sub-watershed 9 (SW-9) which is responsible for high runoff or flash flooding.

6.2 Ruggedness Number (R_n)

This factor is used to study unevenness, smoothness, or roughness of the terrain. Its higher value represents steep slopy terrain and is responsible for the quick formation of runoff and flash flooding (Patton and Baker 1976; Obeidat et al. 2021). Its value corresponding to different sub-watersheds like sub-watershed (SW) 1, 2, 3, 4, 5, 6, 7, 8, 9, 10, 11, 12, 13, 14 and 15 was found to be 538, 577, 462, 572, 497, 378,

425, 384, 564, 369, 462, 283, 303, 335 and 345, respectively. Its value is more in sub-watershed 2 (SW-2) which is responsible for high runoff or flash flooding.

6.3 *Prioritization of Different Sub-watersheds for Flash Flooding Using Morphometric Analysis*

D_d, F_s, T, D_t, I_n, R_c, R_f, R_e, R_h, R_e, and R_n have a positive relation with flash flood or runoff; the greater the values of these parameters, the greater the likelihood that flooding may occur. (Bajirao and Kumar 2021). On the contrary, L_{ovf}, C_m and C_c have a negative relationship to flash floods or runoff. It implies that the likelihood of flooding is increased by lower values of these parameters (Bajirao and Kumar 2021; Obeidat et al. 2021). The sub-watershed causing more runoff/flash flood was assigned with first priority and vice versa. On the basis of all these factors, sub-watershed-wise priority ranks were assigned from 1 to 15 as presented in Table 5. Total rank is calculated by adding the ranks of all morphometric parameters for a given sub-watershed as presented in Table 5.

Table 5 Final priority rankings for flash flooding based on morphometric parameters

SW	D_d	F_s	T	L_{ovf}	C_m	I_n	D_t	R_c	R_f	R_e	C_c	R_h	R_r	R_n	Total rank	Final priority rank
1	14	15	6	14	14	15	7	13	12	12	13	9	9	4	157	XIII
2	10	11	11	10	10	10	11	11	15	15	10	3	3	1	131	X
3	6	7	8	6	6	6	6	3	11	11	2	8	8	7	95	V
4	2	3	12	2	2	2	12	3	5	5	3	2	2	2	57	II
5	5	2	2	5	5	4	2	15	11	10	15	14	15	5	110	VII
6	12	15	13	12	12	14	13	4	8	8	4	5	4	10	134	XII
7	9	13	9	9	9	12	9	6	9	10	6	8	8	8	125	IX
8	9	6	1	8	9	7	1	13	8	8	12	15	15	9	121	VIII
9	1	4	14	1	1	1	14	1	1	1	1	1	1	3	45	I
10	14	9	7	14	14	10	8	14	15	15	14	13	12	11	170	XIV
11	3	9	5	3	3	5	5	8	3	3	8	10	12	7	84	III
12	7	12	15	8	7	11	15	11	4	4	11	5	8	15	133	XI
13	15	11	10	15	15	14	11	11	13	13	10	11	8	14	171	XV
14	5	1	4	5	5	3	5	8	6	6	7	12	13	13	93	IV
15	12	5	3	12	12	8	3	5	2	2	6	6	12	12	100	VI

6.4 Prioritization of Different Sub-watersheds Based on Soil Erosion Using Hypsometric Analysis

The hypsometric analysis of different sub-watersheds was carried out in the ArcGIS 10.2 environment. The hypsometric curves corresponding to different sub-watersheds were generated using ASTER DEM data as shown in Fig. 7. As per the shapes of HCs of different sub-watersheds, it was observed that the areas of the Koyna river basin come under the young and mature stages of geologic development. Here, it was observed that different sub-watersheds represent convex and S-type hypsometric curves. The convex shape of any sub-watershed indicates more hazard due to soil erosion/loss while the concave shape represents less hazard to soil resources. The S-shaped curve represents a mature stage of geologic development. In this study, it was observed that most of the sub-watersheds come under the convex type of hypsometric curve.

Sub-watershed-wise minimum, mean and maximum elevation, HI, geologic stage and final prioritization rank of different sub-watersheds are presented in Table 6. Based on HI value, different sub-watersheds like sub-watershed (SW) 1, 2, 3, 4, 5, 6, 7, 8, 10, 13 and 15 comes under the mature/equilibrium stage of geologic development. Different sub-watersheds like sub-watershed (SW) 9, 11, 12 and 14 come under the young/inequilibrium stage of geologic development. Here, not even a single watershed came under the monadnock/old stage of geologic development. The Monadnock stage represents a very low erosion hazard, the mature stage represents a low to moderate rate of soil erosion while the young stage represents a severe rate of soil erosion. HI represents how much percent landmass still exists in the watershed.

In this study, priority rankings to different sub-watersheds were given by considering HI values. The greatest (first) priority was given to the sub-watershed with the highest HI score, and vice versa. The final priority ranks corresponding to different sub-watersheds are presented in Table 6 based on HI value. Sub-watershed SW-12 got the highest priority, while sub-watershed SW-6 got the lowest priority for conservation purpose.

Here it was observed that there is a mismatch of final priority ranks of different sub-watersheds for flash floods and soil erosion risk assessment except for sub-watersheds (SW) 11 and 14. Here, it was found that both methods are providing different final priority ranks for different sub-watersheds. Hence, prioritization should be done by considering flash flood and soil erosion risk issues separately. Different sub-watershed were categorized into three priority zones namely low, medium and high priority zones as shown in Fig. 8. Those sub-watersheds received final priority ranks from I to V categorized under high priority zones. The sub-watershed that comes under final priority ranking between VI to X are categorized under medium priority zone while the remaining sub-watershed showing final priority rank between XI to XV have been categorized under low priority zones in both flash flood as well as soil erosion analysis (Fig. 8).

Table 6 Final priority rankings for soil erosion based on hypsometric analysis

Sub watershed	Min. elevation (m)	Max. elevation (m)	Mean elevation (m)	HI by integration method	Geologic stage	Final priority rank
SW-1	624	1427	862.00	0.5659	Mature	IX
SW-2	625	1437	926.24	0.5251	Mature	XII
SW-3	626	1218	788.31	0.5733	Mature	VI
SW-4	628	1263	805.72	0.5935	Mature	V
SW-5	626	1232	807.17	0.567	Mature	VIII
SW-6	626	1182	893.51	0.4867	Mature	XV
SW-7	559	1149	815.53	0.5166	Mature	XIII
SW-8	556	1089	771.43	0.5702	Mature	VII
SW-9	534	1082	659.83	0.6946	Young	II
SW-10	555	1106	797.85	0.5496	Mature	X
SW-11	542	1079	656.98	0.6635	Young	III
SW-12	539	927	607.02	0.7533	Young	I
SW13	596	1109	837.78	0.5382	Mature	XI
SW-14	554	962	674.80	0.601	Young	IV
SW-15	595	1102	809.87	0.5158	Mature	XIV

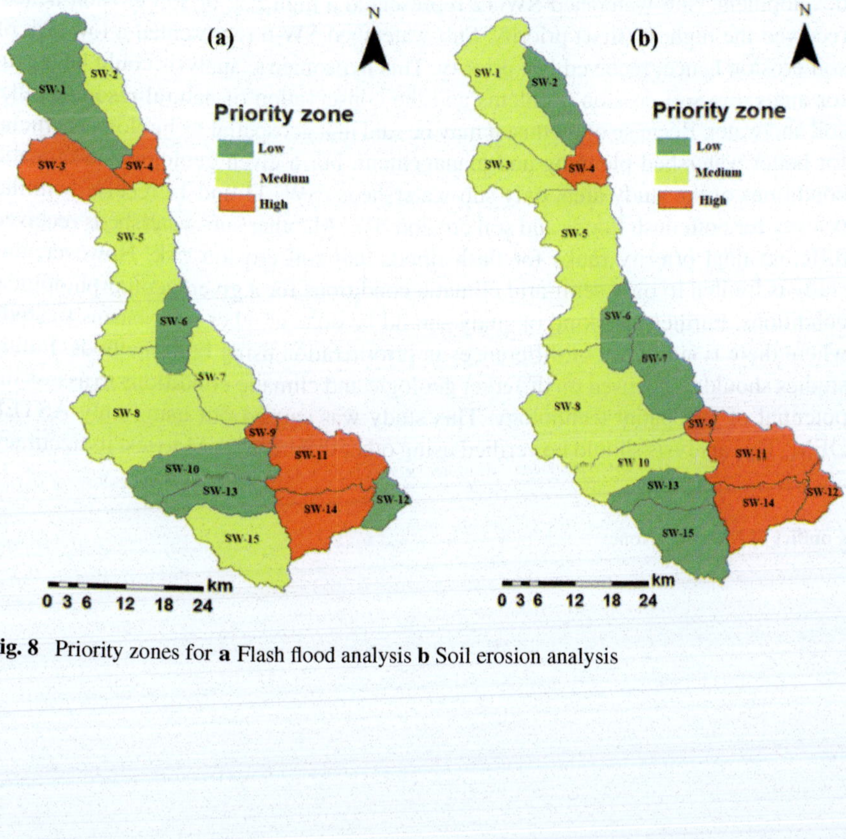

Fig. 8 Priority zones for **a** Flash flood analysis **b** Soil erosion analysis

7 Conclusions

Being rapid and unexpected, one of the most dangerous and destructive natural disasters is flash flooding, endangering both human lives and property, and destroying infrastructure. Watershed management entails the management of water resources in light of potential climate change scenarios. Prioritization of sub-watersheds is required for the development of land and water conservation measures. In this study, two different methods namely drainage and hypsometric analysis were used for the prioritization of different sub-watersheds for flash flood and soil erosion estimation, respectively. Different morphometric parameters were used for the assessment of flood potential from different sub-watersheds. Those sub-watersheds which are more prone to flash flood formations must be given higher priority for preventing flash floods. Sub-watershed SW-9 represented a high risk of flash floods; hence, received the highest (first) priority. Sub-watershed SW-13 represented a low risk of flash floods hence, received last priority. In the second method, prioritization is carried out using hypsometric analysis for assessing the soil erosion risk of different sub-watersheds. The main purpose of a hypsometric analysis is to investigate the erosional stage of different sub-watersheds. In this research, the erosion risk of various sub-watersheds in the Koyna river basin of India was evaluated using HC and HI. Here, it was observed that HI values vary from 0.4867 to 0.7533 corresponding to different sub-watersheds. The study area comes under the mature and young stages of geomorphic development. Sub-watershed SW-12 represented a high risk of soil erosion hence, received the highest (first) priority. Sub-watershed SW-6 represented a low risk of soil erosion hence, received last priority. This hypsometric analysis could be useful for analyzing soil erosion problems and the conservation of natural resources like soil and water. Because of all this, it may be said that geospatial technology is crucial for better watershed planning and management. For a given geologic and climatic conditions of the study area, only sub-watersheds (SW) 11 and 14 received similar priority for both flash floods and soil erosion risk. All other sub-watersheds received different final priority ranks for flash floods and soil erosion risk. However, this study is limited to only semi-arid climatic conditions for a given geomorphological conditions. Further, this kind of study should be done at other river basins to study where there is similarity or differences in prioritization using both methods. Future studies should be focused on different geologic and climatic conditions to assess the potential of geospatial technology. This study was carried out using only ASTER DEM. This analysis should be verified using other different DEM to test its accuracy.

Conflict of Interest None.

References

Adhikary PP, Dash CJ (2018) Morphometric analysis of Katra watershed of Eastern Ghats: a GIS approach. Int J Curr Microbiol App Sci 7:1651–1665

Aher PD, Adinarayana J, Gorantiwar SD (2014) Quantification of morphometric characterization and prioritization for management planning in semi-arid tropics of India: a remote sensing and GIS approach. J Hydrol 511:550–560

Ahmed SA, Chandrashekarappa KN, Raj SK, Nischitha V, Kavitha G (2010) Evaluation of morphometric parameters derived from ASTER and SRTM DEM—a study on Bandihole sub-watershed basin in Karnataka. J Indian Soc Remote Sens 38:227–238

Alam A, Ahmed B, Sammonds P (2020) Flash flood susceptibility assessment using the parameters of drainage basin morphometry in SE Bangladesh. Quatern Int. https://doi.org/10.1016/j.quaint.2020.04.047

Ali SA, Parvin F, Pham QB, Vojtek M, Vojtekova J, Costache R, Ghorbani MA (2020) GIS-based comparative assessment of flood susceptibility mapping using hybrid multi-criteria decision making approach, nave Bayes tree, bivariate statistics and logistic regression: a case study of Topl'a basin, Slovakia. Ecol Indic 117:106620. https://doi.org/10.1016/j.ecolind.2020.106620

Ameri AA, Pourghasemi HR, Cerda A (2018) Erodibility prioritization of sub-watersheds using morphometric parameters analysis and its mapping: a comparison among TOPSIS, VIKOR, SAW, and CF multi-criteria decision making models. Sci Total Environ 613–614:1385–1400

Aparna P, Nigee K, Shimna P, Drissia TK (2015) Quantitative analysis of geomorphology and flow pattern analysis of Muvattupuzha River basin using Geographic Information System. Aquatic Proc 4:609–616

Aravinda PT, Balakrishna HB (2013) Morphometric analysis of Vrishabhavathi watershed using remote sensing and GIS. Int J Res Eng Technol 2(8):514–522

Asfaw D, Workineh G (2019) Quantitative analysis of morphometry on Ribb and Gumara watersheds: implications for soil and water conservation. Int Soil Water Conserv Res 7:150–157

Bajirao TS, Kumar P, Kumar M, Elbeltagi A, Kuriqi A (2021) Superiority of hybrid soft computing models in daily suspended sediment estimation in highly dynamic rivers. Sustainability 13:542

Bajirao TS, Kumar P (2021) Geospatial technology for prioritization of Koyna River basin of India based on soil erosion rates using different approaches. https://doi.org/10.1007/s11356-021-13155-7

Bharadwaj AK, Pradeep C, Thirumalaivasan D, Shankar CP, & Madhavan N (2014) Morphometric analysis of Adyar watershed. IOSR J Mech Civ Eng 71–77

Bhatt S, Ahmed SA (2014) Morphometric analysis to determine floods in the Upper Krishna basin using Cartosat DEM. Geocarto Int 29(8):878–894

Biswas S, Sudhakar S, Desai VR (1999) Prioritisation of sub watersheds based on morphometric analysis of drainage basin: a remote sensing and GIS approach. J Indian Soc Remote Sens 27:155–166

Borga M, Gaume E, Creutin JD, Marchi L (2008) Surveying flash floods: gauging the ungauged extremes. Hydrol Process 22:3883–3885

Chandrashekar H, Lokesh KV, Sameena M, Jyothi R, Ranganna G (2015) GIS-based morphometric analysis of two reservoir catchments of Arkavati River, Ramanagaram District, Karnataka. Aquatic Proc 4:1345–1353

Charizopoulos N, Mourtzios P, Psilovikos T, Psilovikos A, Karamotsou L (2019) Morphometric analysis of the drainage network of Samos Island (northern Aegean Sea): insights into tectonic control and flood hazards. Compes Rendus Geoscience 351:375–383

Chougale SS, Jagdish BS (2017) Morphometric analysis of Kadvi River basin, Maharashtra using geospatial techniques. Curr World Environ 12:635–645

Das S (2020) Flood susceptibility mapping of the Western Ghat coastal belt using multi-source geospatial data and analytical hierarchy process (AHP). Remote Sens Appl Soc Environ 20:100379. https://doi.org/10.1016/j.rsase.2020.100379

Destro E, Amponsah W, Nikolopoulos EI, Marchi L, Marra F, Zoccatelli D, Borga M (2018) Coupled prediction of flash flood response and debris flow occurrence: application on an alpine extreme flood event. J Hydrol 558:225–237

Dumka BB, Kumar P (2021) Modeling rainfall-runoff using artificial neural network (ANNs) and wavelet based ANNs (WANNs) for Haripura dam, Uttarakhand. Indian J Ecol 48(1):271–274

Faniran A (1968) The index of drainage intensity—a provisional new drainage factor. Aust J Sci 31:328–330

Gajbhiye S, Mishra SK, Pandey A (2014) Prioritizing erosion-prone area through morphometric analysis: an RS and GIS perspective. Appl Water Sci 4:51–61

Horton RE (1932) Drainage basin characteristics. Trans Am Geophys Union 13:350–361

Horton RE (1945) Erosional development of streams and their drainage basins. Geol Soc Am Bull 56:275–370

Hussein S, Abdelkareem M, Hussein R, Askalany M (2019) Using remote sensing data for predicting potential areas to flash flood hazards and water resources. Remote Sens Appl Soc Environ 16:100254. https://doi.org/10.1016/j.rsase.2019.100254

Javed A, Khanday MY, Ahmed R (2009) Prioritization of sub-watersheds based on morphometric and land use analysis using remote sensing and GIS techniques. J Indian Soc Remote Sens 37:261–274

Jodar-Abellan A, Valdes-Abellan J, Pla C, Gomariz-Castillo F (2019) Impact of land use changes on flash flood prediction using a sub-daily SWAT model in five Mediterranean ungauged watersheds (SE Spain). Sci Total Environ 657:1578–1591

Kar G, Kumar A, Singh R (2009) Spatial distribution of soil hydrophysical properties and morphometric analysis of a rainfed watershed as a tool for sustainable land use planning. Agric Water Manag 96:1449–1459

Kaushal N, Singh S (2013) Quantitative analysis of drainage system is an important aspect of prioritisation of watersheds. Int J Sci Technol Manag 2:39–50

Kibate G, Gessesse B (2018) Hydro-geomorphological characterization of Dhidhessa River basin, Ethiopia. Int Soil Water Conserv Res 6:175–183

Krishnan MVN, Prasanna MV, Vijith H (2021) Modelling spatial variation in erosional status and geomorphic evolution of the Limbang river basin in Northern Borneo through hypsometric analysis. Geocarto Int. https://doi.org/10.1080/10106049.2021.2002428

Langbein WB (1947) Topographic characteristics of drainage basins. Water Supply Paper 968C:127–157

Lole AA, Koren SB, Sagar RS, Dhokare AS, Bagade AS, Londhe SS (2016) Morphometric analysis of Herle Nala basin, Kolhapur district, Maharashtra, India. Int Res J Eng Technol 3:1771–1775

Luo M, Xu Y, Mu K, Wang R, Pu Y (2018) Spatial variation of the hypsometric integral and the implications for local base levels in the Yanhe River, China. Arab J Geosci 11:366

Magesh NS, Chadrasekar N, Soundranagyagam JP (2011) Morphometric evaluation of Papanasam and Manimuthar watersheds, part of Western Ghats. Tirunelveli District, Tamil Nadu, India: a GIS approach. Environ Earth Sci 64:374–381

Malik A, Kumar A (2019) Use of GIS for hypsometric (area-elevation) analysis of Gagas watershed (Uttarakhand). Indian J Ecol 46(3):481–485

Malik A, Kumar A, Kandpal H (2019) Morphometric analysis and prioritization of sub-watersheds in a hilly watershed using weighted sum approach. Arab J Geosci 12:118

Mandi S, Soren K (2016) Hypsometric analysis of Bunbuni River, Chotanagpur Plateau, in India. Int J Sci Res 5(8):1698–1701

Mehar R, Verma MK, Tripathi RK (2018) Hypsometric analysis of Sheonath River Basin, Chhatisgarh, India: a remote sensing and GIS approach. Int J Eng Res Technol 7(10):116–121

Melton MA (1957) An analysis of the relations among elements of climate, surface properties and geomorphology. Technical Report Number 11, Project NR 389-042, Columbia University, Department of Geology, Office of Naval Research, New York, USA

Miller VC (1953) A quantitative geomorphic study of drainage basin characteristics in the Clinch Mountain area, Virginia and Tennessee. Technical Report Number 3, Project NR 389–402, Columbia University, Department of Geology, ONR, New York, USA

Muhammad T, Yogrema SP, Noorlaila H (2015) The utilization of global digital elevation model for watershed management a case study: Bungbuntu Sub Watershed, Pamekasan. Procedia Environ Sci 24:297–302

Nigam GK, Tripathi MP, Ambast SK, Kumar L, Khalkho D (2017) Morphometric analysis of drainage basin using aerial photographs: a case of Karun watershed of Seonath sub-basin of Chhattisgarh. Int J Adv Biotechnol Res 7:623–629

Obeidat M, Awawdeh M, Al-Hantouli F (2021) Morphometric analysis and prioritisation of watersheds for flood risk management in Wadi Easal Basin (WEB), Jordan, using geospatial technologies. J Flood Risk Manag e12711. https://doi.org/10.1111/jfr3.12711

Ostovari Y, Ghorbani-Dashtaki S, Bahrami H, Naderi M, Dematte JAM (2017) Soil loss estimation using RUSLE model, GIS and remote sensing techniques: a case study from the Dembecha Watershed, Northwestern Ethiopia. Geoderma Reg 11:28–36

Pan N, Dubey RK, Bhatt A, Rai SP, Semwal P, Mishram S (2020) Soil erosion and flood hazard zonation using morphometric and morphotectonic parameters in Upper Alaknanda river basin. Nat Hazards 103:3263–3301

Pande CB, Moharir K (2017) GIS based quantitative morphometric analysis and its consequences: A case study from Shanur River Basin, Maharashtra India. Appl Water Sci 7:861–871. https://doi.org/10.1007/s13201-015-0298-7

Pande C, Moharir K, Pande R (2021) Assessment of morphometric and hypsometric study for catchment development using spatial technology-a case study of Wardha river basin in Maharashtra, India. Int J River Basin Manage, 19(1):43–53. https://doi.org/10.1080/15715124.2018.1505737

Pareta K, Pareta U (2011) Quantitative morphometric analysis of a watershed of Yamuna basin, India using ASTER (DEM) data and GIS. Int J Geomatics Geosci 2:248–269

Patton PC, Baker VR (1976) Morphometry and floods in small drainage basins subject to diverse hydrogeomorphic controls. Water Resour Res 12:941–952

Pham TG, Degener J, Kappas M (2018) Integrated universal soil loss equation (USLE) and geographical information system (GIS) for soil erosion estimation in a Sap basin: Central Vietnam. Int Soil Water Conserv Res 6:99–110

Prashanth M, Kumar A, Dhar S, Verma O, Gogoi K (2022) Hypsometric Analysis for Determining Erosion Proneness of Dehar Watershed, Himachal Himalaya, North India. J Geosci Res 7(1):86–94

Puno GR, Puno RCC (2019) Watershed conservation prioritization using geomorphometric and land use-land cover parameters. Glob J Environ Sci Manag 5(30):279–294

Ritter DF, Kochel RC, Miller IR (2002) Process geomorphology. McGraw Hill, Boston

Sahu N, Reddy GPO, Kumar N, Nagaraju MSS, Srivastava R, Singh SK (2016) Morphometric analysis in basaltic terrain of central India using GIS techniques: a case study. Appl Water Sci 7:2493–2499

Salvi SS, Mukhopadhyay SD, Ranade AR (2017) Morphometric analysis of river drainage basin/watershed using GIS and RS: a review. Int J Res Appl Sci Eng Technol 5:503–508

Sarangi A, Bhattacharya AK, Singh A, Singh AK (2001) Use of geographic information system (GIS) in assessing the erosion status of watersheds. Indian J Soil Conserv 29:190–195

Schumm SA (1956) Evolution of drainage systems and slopes in badlands at Perth Amboy, New Jersey. Geol Soc Am Bull 67:597–646

Shekhar PR, Mathew A (2022) Evaluation of morphometric and hypsometric analysis of the Bagh RiverBasin using remote sensing and geographic information system techniques. Energy Nexus 7:100104

Shivhare N, Rahul AK, Omar PJ, Chauhan MS, Gaur S, Dikshit PKS, Dwivedi SB (2018) Identification of critical soil erosion prone areas and prioritization of microwatersheds using geoinformatics techniques. Ecol Eng 121:26–34

Singh S, Singh MC (1997) Morphometric analysis of Kanhar River Basin. Nat Geogr J India 43:31–43

Singh O, Sarangi A, Sharma MC (2008) Hypsome-tric integral estimation methods and its relevance on erosion status of north western lesser Himalayan watershed. Water Resour Manag 22:1545–1560

Singh V, Singh SK (2018) Hypsometric analysis using microwave satellite data and GIS of Naina–Gorma River Basin (Rewa district, Madhya Pradesh, India). Water Conserv Sci Eng. https://doi.org/10.1007/s41101-018-0053-7

Strahler AN (1952) Hypsometric (area-altitude) analysis of erosional topography. Bull Geol Soc Am 63:1117–1142

Strahler AN (1964) Quantitative geomorphology of drainage basin and channel network. Handbook of applied hydrology. McGraw Hill Book Company, New York, pp 39–76

Tamilarasan A, Brema J (2019) Assessment of soil erosion in Siruvani watersheds based on USLE and hypsometric curve methods. Int J Innov Technol Explor Eng 8(6S4):1070–1073

Thakkar AK, Dhiman SD (2007) Morphometric analysis and prioritization of miniwatersheds in Mohr Watershed, Gujarat using remote sensing and GIS techniques. J Indian Soc Remote Sens 35:313–321

Toduse NC, Ungurean C, Davidescu S, Clinciu I, Marin M, Nita MD, Davidescu A (2020) Torrential flood risk assessment and environmentally friendly solutions for small catchments located in the Romania Natura 2000 sites Ciucas, Postavaru and Mare. Sci Total Environ 698:134271. https://doi.org/10.1016/j.scitotenv.2019.134271

Turkan BA, Bekir NA (2011) Drainage morphometry and its influence on landforms in volcanic terrain, Central Anatolia, Turkey. Procedia Soc Behav Sci 19:732–740

Umrikar BN (2017) Morphometric analysis of Andhale watershed, taluka Mulshi, district Pune, India. Appl Water Sci 7:2231–2243

Yahya F (2017) Morphometric assessment of Wadi Wala watershed, Southern Jordan using ASTER (DEM) and GIS. J Geogr Inf Syst 9:158–190

Yahya F, Ali A, Omar E, Nisrin AS (2015) Quantitative analysis of geomorphometric parameters of Wadi Kerak, Jordan, using remote sensing and GIS. J Water Resour Prot 7:456–475

Zainab F (2018) Morphometric analysis of Khulgad watershed Almora, Uttarakhand. Int J Mod Trends Eng Res 5:162–173

Exploring the Suitability of Groundwater for Domestic Water Quality and Irrigation Purpose in Dindigul District, Tamil Nadu

Rajee Radhakrishnan and Saravanabavan Vaithialingam

Abstract The quality of groundwater is exaggerated by a wide range of natural and anthropogenic influences. Whenever there is a limited supply of water and it must be used as efficiently as possible, their influence is typically greater. The study area has lot of leather industries, cement industries which affects the excellence of groundwater. This leads to realize the chemical characteristics of groundwater based on BIS standards. The piper plot indicates that the cations and anions are in the order of $Na^+ > Mg^+ > Ca^+$ and $Cl^- > HCO^3 > SO_4{}^2$ and the samples fallen are under the characteristics of alkaline earth exceeds alkalies. Weighted overlay techniques in GIS and multivariate statistical technique such as cluster analysis (CA) were used to identify the quality indexes. Derived results state that 12% of the samples are fit in both post and pre monsoon season and unfit for drinking purposes in DGWI. IWQ indicate that 56.08% of area in POM (post monsoon) and 70.31% of area in PRM (pre monsoon season) is suitable for Irrigation purposes. Integrated groundwater quality shows that 78.16% and 40.26% of area is undesirable for domestic and irrigation purposes. The outputs of each chemical parameter are cartographically and spatially visualized as a map using GIS Techniques.

Keywords Domestic groundwater quality index · IWQ · Cluster analysis · GIS · Inverse distance weight · BIS

1 Introduction

Groundwater is the strategic resource for basic necessities, primary, secondary and quaternary activities (UNEP 1999). The quality of groundwater affects the urban expansion, industrial and agricultural activities in both developed and developing

---

R. Radhakrishnan (✉) · S. Vaithialingam
Department of Geography, Madurai Kamaraj University, Tamil Nadu, Madurai 625 021, India
e-mail: rajee.14mku@gmail.com

S. Vaithialingam
e-mail: vsaravanabavan@gmail.com

C. B. Pande et al. (eds.), *Surface and Groundwater Resources Development and Management in Semi-arid Region*, Springer Hydrogeology,
https://doi.org/10.1007/978-3-031-29394-8_14

countries directly or indirectly (Singh and Chandel, 2006; Saleem, 2007; Gupta et al. 2008; Srinivasamoorthy et al. 2011). The decrease in quality of groundwater that makes contamination has introduced to the environment due to human activities. Certain groundwater constituents have high concentration due to natural processes.

The color and taste in the groundwater decide the suitability of drinking water rather considering the chemical and heavy metal contamination (Kumar et al. 2007). About 80% of the diseases prevalent are because of the contaminated water. The concern on water scarcity and water quality for agriculture and other sectors is more sensitive. According to Omran et al. (2014), severe water shortage is experienced over the driest part of the globe. The groundwater quality degrades due to pressure formed over hydrological and hydro-geologic systems as an impact of dynamic nature of the climate (Gurdak et al. 2012; Bondu et al. 2016). The quality of the groundwater depends on the structural and chemical composition of rocks as well as on a number of hydrological elements. The solubility, complexity, sorption, and exchange processes of minerals are significantly influenced by water quality (Apambire et al. 1997; Raju et al. 2009). Hydro geochemical parameter is to be understood to expedicate differences in groundwater resources (Bozdag and Gocmez 2013; Pande and Moharir 2022).

There are number of studies related to groundwater quality assessment carried through out the country. Sadashivaiah et al. (2008) studied Tumkur Taluk, Karnataka and assess the water scarcity and quality problems. Water quality for Bhadravathi taluk using GIS has been analysed for both POM and PRM (Raikar and Sneha 2012). Reddy (2013) studied the potential of groundwater for irrigation using APHA method for Bhaskar Rao Kunta watershed, Nalgonda. Venkateswaran and Vediappan (2013) assessed quality of the same for irrigation in Lower Bhavani Sub Basin, Cauvery River, Tamil Nadu. Samahajira and Annal (2017a) collected and analyzed groundwater samples from Rediyarchatram block for irrigation practices located in Dindigul district. The suitability of groundwater has been evaluated by applying multiple techniques (Ahmadi and sedghamiz 2007; Nas and Berktay 2010; Samson et al. 2010; Ketata-Rokbani et al. 2011; Ordookhani et al. 2012) for domestic as well as irrigation purposes. Vinothkanna et al. (2020a, b) assessed the quality of water on behalf of irrigation for Dindigul district.

The area considered for study consists of many of industries like leather industries, spinning mills, cement and chemical industries and heavy fertilizer used in the agriculture making groundwater unfit for domestic and agriculture purposes. Based on the above aspect, a GIS based MCDSS (multi criteria decision support system) is a sound platform for decision makers not only to integrate the chemical parameters but also for easy visual interpretation (Simsek and Gunduz 2007; Pande et al. 2020). Geographic Information system (GIS) is an effective platform to monitor the groundwater and to create maps based on chemical values present in the water (Krishnaraj et al. 2015; Lozano et al. 2012; Vinothkanna et al. 2020a, b). Pollution in water can be effectively monitored and managed using GIS (Balakrishnan et al. 2011). Along with GIS, cluster analysis were used in number of studies to identify pollutants (Lin et al. 2017; Misaghi et al. 2017; Zamani-Ahmadmahmoodi et al. 2017). For sustainable use the quality of groundwater should be asses and monitor properly (Selvam and

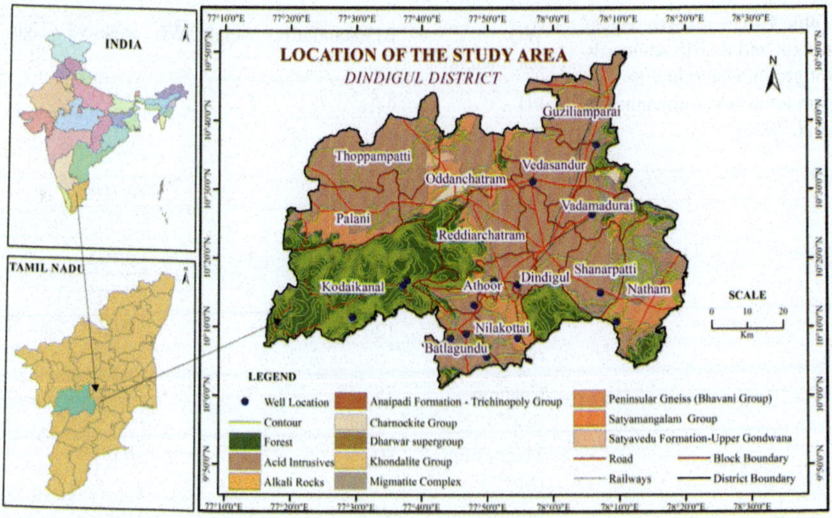

Fig. 1 Study area location

Sivasubramanian 2012; Singaraja et al. 2012). Vinothkanna (2019) used correlation techniques to identify the suitability of groundwater in Dindigul district.

Geographically, Dindigul is bounded by Erode, Karur and Tiruchirappalli districts on the north; a part of Madurai and Tiruchirappalli districts on the east; a part of Madurai and Theni districts on the south and Kerala State and Coimbatore district on the west and it locates in the latitudinal extent of 10°05' and 10° 09' north and longitudinal extent of 77° 30' to 78°20' east (Fig. 1). Total geographical area of the district is 6036.11 km.

Dominant soil type is Alfisols, occupying 38.53% of Dindigul. Having a tropical climate, the district is geologically flanked by acid intrusive in the north, east and south. Another important geological feature present in the area is peninsular gneiss (Fig. 1).

2 Material and Methods

This study purely used secondary data for checking groundwater quality and was collected from Central Groundwater Board, Tharamani, and Chennai. For this study the mean values from 2008 to 2017 were considered to analyze the groundwater quality for domestic and irrigation. The number of wells varied from one year to another year because it may close or dry up. So, common wells for all the years for both POM and PRM are extracted using excel software. Finally, 17 common wells have been derived to study the groundwater quality based on BIS standards (Table 1). The base spatial administrative unit is prepared from SOI (Survey of India

Table 1 The relative weight, weight and its BIS standards for groundwater quality parameters (Vinothkanna et al. 2022)

WQ parameters	BIS standards	Weight (Wi)	Relative weight
pH	8.5	3	0.103
Total dissolved solids (TDS)	500	5	0.172
Sulphates ($So4^{2-}$)	200	3	0.103
Chloride (Cl^-)	250	3	0.103
Sodium (Na^+)	200	3	0.103
Potassium (K^+)	12	1	0.034
Calcium (Ca^{2+})	75	2	0.069
Magnesium (Mg^{2+})	30	2	0.069
Fluoride (F^-)	1	5	0.172
		$\Sigma wi = 27$	$\Sigma Wi = 1.00$

Toposheets) in the scale of 1: 50,000. Location of each well has pointed and the result of each chemical parameter of water has been joined to the point as an attribute and final geo data base was created using Arc GIS software (Pande et al. 2018). The spatial analysis tool has been used extensively; mainly interpolation technique such as the Inverse Distance Weight (IDW) method is used for generating the surfaces (Karunanidhi et al. 2013). Cluster analysis by SPSS has been used to find out the similarities in the data. Hence, in groundwater, there are multiple types of clusters to identify the physic-chemical character (Troiano et al. 1994; Farnham et al. 2000).

2.1 *Domestic Water Quality Index (DWQI)*

The parameters to understand domestic water quality index are, pH, Total dissolved solids (TDS), Sulphates (So_4^{2-}), Chloride (Cl^-), Sodium (Na^+), Potassium (K^+), Calcium (Ca^{2+}), Magnesium (Mg^{2+}) and Fluoride (F-). The assigning of weight has been done based on the relative importance on the water quality parameters. Vinothkanna et al. (2016), applied the relative weight using a weighted arithmetic index method (WAI) (Table 1).

DWQI = Σqiwi.

Where qi (parameter rating) = 100 X (Va-Vi) / (Vs-Vi),

When Va = actual value.

Vi = Desired value (null for all; except pH (7.0) and DO (14.6 mg/l).

Vs = standard value.

Wi (unit weight) = K/S_n.

Where K indicates constant = 1.
1/Vs1 + 1/Vs2 + 1/Vs3 + 1/Vs4…….. + 1/Vsn.
S_n = 'n' number of standard values.

2.2 *Irrigation Ground Water Quality Index (IWQI)*

The indices such as SAR, EC, and RSC, % Na, PI, MR, KR and SSP are mainly concerned as irrigation parameters for assessing the IWQ index (Vinothkanna et al. 2021a, b). The equations given in Table 2 is used as an input for IWQI assessment.

There may be a radical difference in geographical settings with different soil conditions and cropping pattern. Based on the physical aspect the decision maker can easily change the procedure and alternate weight can assign for parameters used. In addition to weighing coefficient rating of each parameter has also worked out to calculate the irrigation water quality index. The IWQ index is calculated using the formula,

$$\text{IWQ index} = \sum_{i-1}^{8} G_i$$

where I is an index and G is contribution of each categories important to assess the quality of irrigation water resources.

$$\text{Gi} = W_1 \times R_1$$

Where W is the weight and r is the ranking value of the parameter.

Table 2 Equation for calculation of water quality for irrigation

Parameter	Equation	Author
SAR	= Na/ √Ca + Mg/2	Richards (1954)
RSC	= (CO3 + HCO3) − Ca + Mg)	Richards (1954)
Na %	= (Na + K)/(Ca + Mg + Na + K) * 100	Wilcox (1955)
PI	= {(Na + HCO3)/[√(Ca + Mg + Na)]} * 100	Doneen (1964)
MR	= Mg * 100/(Ca + Mg)	Raghunath (1987)
KR	= Na/(Ca + Mg)	Kelly's (1963)
SSP	= (Na + K)*100/Ca + Mg + Na + K	Todd (1980)

2.3 Cluster Analysis

Cluster analysis (CA) is a statistical method used to classify variables based on the similarities and to identify the homogeneity dendograms were used to perform CA.

3 Result and Discussion

The descriptive statistics such as, minimum, maximum, mean and Standard deviation values for both post and pre monsoon for domestic groundwater quality index are shown in the Table 3. Classification of chemical parameters based on BIS standards and their suitability classes for post monsoon (POM) and pre monsoon season (PRM) are shown in Table 4.

3.1 Groundwater Chemistry Mechanism

Several factors affect the hydro-geochemical compositions of groundwater in the aquifer, and the Piper tri-linear diagram can be used to study these issues more thoroughly (Fig. 2).

Piper plot indicates analytical value of both cations and anions. The cations is dominated by sodium and potassium ions ($Na^+ + K^+$) followed by no dominant type and the anions is dominated by chloride ions (Cl^-). The concentration of positive and negative ions are in the order of $Na^+ > Mg^+ > Ca^+$ and $Cl^- > HCO^3 > SO_4{}^2$. Mixed type water is highly dominated in the study area followed by magnesium bi-carbonate type in both monsoon seasons. The samples present in the study area demonstrates that weak acids exceed strong acids followed by strong acids exceeds weak acids. Based on the graph it clearly illustrates that the alkaline earth exceeds alkalies. During pre-monsoon season almost all the samples are found between the percentages of 40 to 60. In post monsoon season most of the samples are found between 40 and 60%

3.2 Domestic Ground Water Quality

The easy ways of transferring diseases to humans is water. Water consists of charged particles, soluble gases and micro-livings. When these exceed particular extent; it becomes unfit for domestic purposes. The maximum value for individual chemical parameter is fixed and suggested by many governments and non-government organization such as, WHO and BIS to determine the usability of universal solvent. To

Table 3 Minimum and maximum value of domestic ground water quality parameter (mg/l)

Parameters	Characteristics	POM		PRM		Mean		SD	
		Min	Max	Min	Max	POM	PRM	POM	PRM
General	pH	7.75	8.38	7.94	8.21	8.07	8.09	0.16	0.08
	Total dissolved solids (TDS)	303.40	1441.73	101.40	1407.30	697.4	667.76	431.96	412.74
Major Cations	Magnesium (Mg^{2+})	30.83	142.53	30.90	124.27	52.49	52.9	36.03	33.67
	Potassium (K^{+})	15.20	71.60	5.20	85.00	15.51	15.92	18.01	18.5
	Calcium (Ca^{2+})	26.60	125.27	20.60	124.40	54.72	44.78	29.78	27.83
	Sodium (Na^{+})	6.00	254.00	32.90	257.60	114.07	116.37	81.15	80.04
Major Anions	Sulphates(So_4^{2-})	22.30	233.73	23.60	274.00	62.31	60.83	57.17	62.52
	Chloride (Cl^{-})	65.70	551.00	48.80	524.50	202.84	197.49	167.72	165.2
	Fluoride (F^{-})	0.16	1.33	0.22	1.3	0.63	0.66	0.37	0.4

Table 4 Classification of chemical parameters based on BIS level for POM and PRM

Chemical parameter	BIS standards	Suitability class	POM		PRM	
			%	No. well	%	No. well
pH	<8.5	Desirable	100	17	100	17
	>8.5	Permissible	–	Nil	–	Nil
TDS (mg/L)	<500	Desirable	35.29	6	41.17	7
	500–1000	Permissible	35.29	6	35.29	6
	>1000	Not permissible	29.41	5	23.52	4
Mg (mg/L)	<30	Desirable	23.52	4	23.52	4
	30–100	Permissible	64.70	11	64.70	11
	>100	Not permissible	11.76	2	11.76	2
K (mg/L)	<20	Permissible	82.35	14	88.23	15
	>20	Not permissible	17.64	3	11.76	2
Ca (mg/L)	<75	Desirable	82.35	14	88.23	15
	>75	Permissible	17.64	3	11.76	2
Na (mg/L)	<200	Permissible	76.47	13	88.23	15
	>200	Not permissible	23.52	4	11.76	2
So_4 (mg/L)	<200	Desirable	94.11	16	94.11	16
	>200	Permissible	5.88	1	5.88	1
Cl (mg/L)	<200	Desirable	64.70	11	70.58	12
	>200	Permissible	35.29	6	29.41	5
F (mg/L)	<1	Desirable	82.35	14	76.47	13
	>1	Permissible	17.64	3	23.52	4

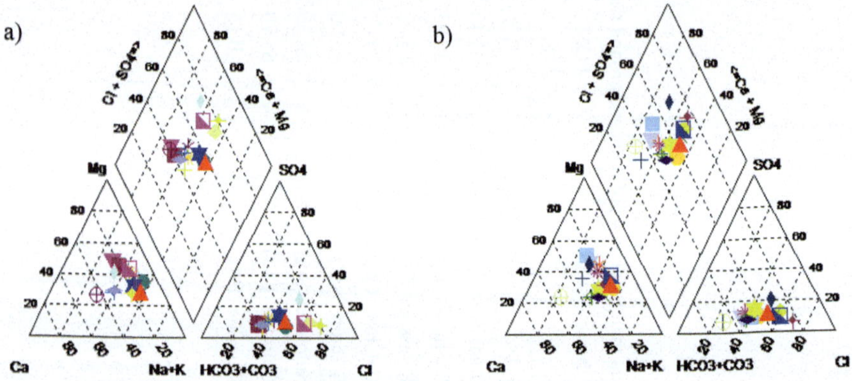

Fig. 2 **a** PRM and **b** POM samples (Piper Plot)

examine the suitability of subsurface water for drinking purpose chemical parameters used will be pH, TDS, Mg^{2+}, K^+, Ca^{2+}, Na^+, $So4^{2-}$, $Cl^-$and F^-.

3.2.1 PH

pH determines the acidic and alkaline nature of water. This acidic and alkaline nature can be evaluated with the interaction of rock and other materials to water (Hem 1985). The mean pH in the collected samples during post monsoon season ranging from 7.75 to 8.38 mg/l and in pre monsoon it is 7.94 to 8.21 mg/l in the study. The samples are under the permissible limit as prescribed for the drinking water (BIS 2012).

3.3 *Total Dissolved Solids (TDS)*

The preferred limit for TDS is 1000 mg/l (WHO 2004 and BIS 2012). Nearly 71 and 76% of groundwater representative is under permissible limit for drinking during post and pre-monsoon respectively. About 29 and 24% of samples found in the study locations were above the permissible level prescribed by the BIS and considered as not suitable for drinking purposes during post and pre monsoon season respectively. High values are due to inappropriate method of sewage disposals.

3.3.1 Magnesium (Mg^{2+})

For enzyme activation, magnesium is one of the important components. If the content of magnesium is high it is considered as laxative agent. For human beings, the structural and functional changes may occur due to the efficiency of magnesium. During the post monsoon season mean values range from 30.83 to 142.53 mg/l and in pre-monsoon season it is from 30.90 to 124.27 mg/l. There are only 2 samples exceeds the BIS limit (100 mg/l). Excess Magnesium content in the ground water leads to unpleasant taste (Ramesh and Elango 2012).

3.3.2 Potassium (K^+)

The potassium level below 20 is permissible in the groundwater as per BIS standards and its values ranges from 15.20 to 71.60 mg/l as well as between 5.20 and 85.00 mg/l in POM and PRM season respectively. About 82% (14 representatives) in POM and 88% (15 representatives) in PRM are under permissible limit.

3.3.3 Calcium (Ca^{2+}) and Sodium (Na^{+})

For the growth of bone calcium is one of the important parameters. It is found alkaline in nature. Water hardness is depending on dissolved ca^{2+} and Mg^{2+} in water (Rabeiy 2018). The mean level of calcium ranges from 26.60 to 125.27 mg/l and 20.60 to 124.40 mg/l in post and pre monsoon season in the study area.

Sodium is a common chemical element present in drinking water. However, intolerable taste occurs, if it exceeds the allowable extent. The acceptable limit for sodium is 200 mg/l (BIS 2012). 13 samples (76%) and 15 samples (88%) during POM and PRM are under the range prescribed by the BIS.

3.3.4 Sulphate (So_4)

Due to the application of sulfate in agricultural field and the occurrence of sulphide minerals leads to the presence of Sulphate in the groundwater. It increased because of industrial waste. The Sulphate concentration is varying from 22.30 to 244.73 mg/l in the POM and in PRM between 23.60 and 274 mg/l. The desirable level for sulfate is 200 mg/l (BIS 2012). Approximately, all the samples in the location are under desirable limit except one sample.

3.3.5 Chloride (Cl^{-}) and Fluoride (F^{-})

The permissible limit of chloride is 200 mg/l (BIS 2012) and it is considered as the significant chemical parameter in assessing the quality of water. High chloride concentration indicates high constituents of biotic contaminators (Singh and Khan 2011). 6 samples (35%) and 5 samples (29%) in post and pre monsoon season exceeds the permissible limit as set by the BIS. During the post and pre monsoon season, the mean fluoride concentration in the groundwater samples ranges from 0.16 to 1.33 mg/l and 0.22 to 1.3 mg/l respectively. All the samples are under permissible limit in the district. Fluoride is one of the significant health hazards in ground water resources (Vinothkanna et al. 2021a, b).

3.3.6 Water Quality Index (WQI)

The WQI is a mathematical method of showing groundwater quality and the index value is changing from one standard method to another depend upon the physical nature of the study undertaken. BIS standard been adopted to classify the values into 5 categories (Table 5). The resulted WQI shows that, about 12% of the representative values are excellent and good in both POM and PRM. Also, 53% of post monsoon samples and 53% of pre monsoon samples are under poor to very poor water. 12% of water is unfit for drinking purposes (Fig. 3).

Table 5 Water quality index—Pre monsoon and Post monsoon

Sl. No	WQI	Water class	Pre monsoon season		Post monsoon season	
			Total well	%	Total well	%
1	<25	Excellent	2	11.76	2	11.76
2	25–50	Good	2	11.76	3	11.64
3	50–75	Poor	4	23.52	4	23.52
4	75–100	Very poor	6	35.29	5	29.41
5	>100	Unfit for drinking	3	11.64	3	11.64

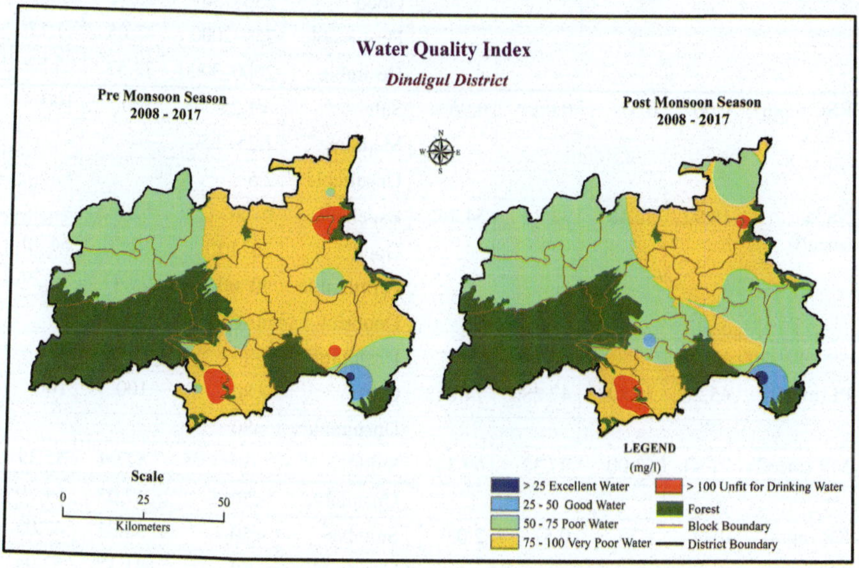

Fig. 3 Spatial distribution of WQI for pre and post monsoon seasons

3.4 *Irrigation Water Quality*

Standard of groundwater is important for plant growth apart from soil. The salt present in the water influences permeability along with soil structure. Plant growth also affects indirectly (Karunanidhi et al. 2013). The productivity of crop is also affected if the water is in poor quality (Iqbal et al. 2012). To identify the groundwater quality condition for irrigation, eight quality indices were calculated. The results of each index are shown as map (Table 6).

Table 6 Irrigation ground water quality parameter suitability classes

Irrigation parameters	Post monsoon season		Pre monsoon season		Suitability class		Percentage	
	Min	Max	Min	Max			POM	PRM
SAR (epm)	0.36	4.59	0.60	5.23	Excellent	<10	100	100
					Good	10–18	–	–
					Doubtful	18–26	–	–
					Unsuitable	>26	–	–
EC (μS/cm)	128	2613.6	167	2421	Excellent	<250	5.88	5.88
					Good	250–750	23.52	29.41
					Permissible	750–2000	47.05	41.17
					Desirable	2000–3000	23.52	23.52
RSC (epm)	0.00	0.57	0.00	0.96	Safe	<1.25	100	100
					Medium	1.25–2.5	–	–
					Unsuitable	2.5	–	–
Na % (meq/l)	18.92	52.48	23.74	54.74	Excellent	0–20	5.88	–
					Good	20–40	64.70	64.70
					Permissible	40–60	29.41	35.29
					Doubtful	60–80	–	–
					Unsuitable	>80	–	–
PI (meq/l)	45.28	70.89	47.280	74.082	Good	<80	100	100
					Unsuitable	>80	–	–
MR (meq/l)	21.41	63.00	27.32	69.34	Suitable	>50	52.94	35.29
					Unsuitable	<50	47.05	64.70
KR (epm)	0.32	1.77	0.51	2.03	Suitable	<50	58.82	47.05
					Unsuitable	>50	41.17	52.94
SSP (meq/l)	30.65	64.98	38.74	69.42	Suitable	<200	100	100
					Unsuitable	>200	–	–

3.4.1 Sodium Adsorption Ratio (SAR) and Electrical Conductivity (EC)

The amount of salt absorbed by the soil and the irrigation water utilised have a strong relationship (Bhunia et al. 2018). The irrigation suitability can be identified by using SAR as one parameter and helps to determine the presence of alkaline and earth alkaline in the water. The sodium concentration is due to the presence of Ca and Mg in the subsurface water (Iqbal et al. 2012). Richard(1954) proposed equations are used to calculate the SAR. The SAR values represents that all the samples (both post and pre monsoons) presents in the district are suitable for doing agricultural activities (Figs. 4, 5a).

The 29 and 35% of representatives during both seasons are under excellent and good category respectively (Fig. 5b). EC value ranges from 2000 to 3000 μmhos/cm

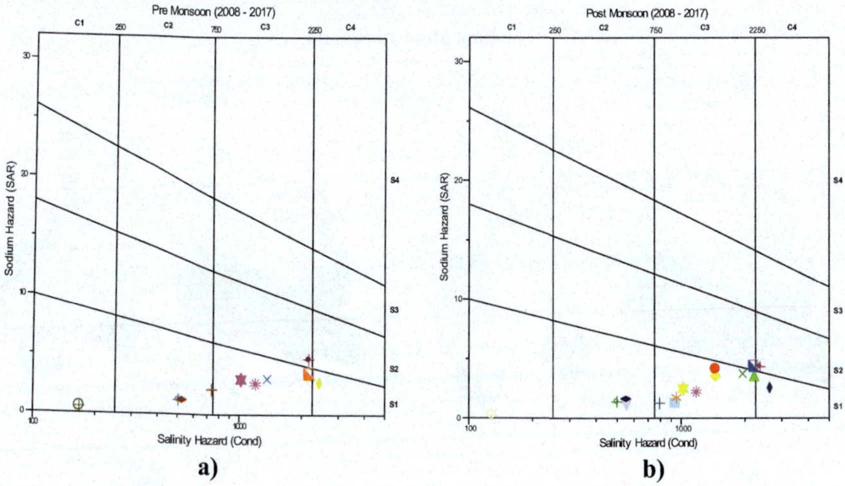

Fig. 4 Graphical representation of Wilcox diagram

are 24% during post and pre monsoon season indicates medium salt enhancement (Sarath Prasanth et al. 2012).

The classifications of irrigation water are graphically plotted using Wilcox diagram (SAR versus EC). Only one sample in both seasons comes under C1S1 (Low salt—low sodium type). Under C2S1 (medium saline-low sodic type), 5 samples in pre and 4 samples in post are categorized as good salinity. Dominant samples are under C3S1 (High saline-low sodic type) and are equally distributed in post and pre monsoon season. Only few samples are in C4S1 (Very high saline-low sodic type) such as 1 and 2 in pre and post monsoon respectively. The medium saline and low sodium groundwater in both monsoon conditions are clearly portrays using plots (Table 7; Fig. 4).

3.4.2 Residual Sodium Carbonate (RSC)

The alkalinity risk for soil is determined by the irrigation water's residual sodium carbonate (RSC) index. The water with RSC greater than 2.5epm is well thought-out as unsuitable for irrigation (Richard's 1954). Almost, 100% of the ground water samples (17 wells) in post and pre monsoon are suitable for agriculture.

3.4.3 Percent Sodium (%Na)

The soil permeability can be reduced due to the presence of high amounts of sodium in irrigation water and arrest the growth of plants (Joshi et al. 2009). Na% present in the groundwater samples of >20 meq/l is excellent for irrigation and it is found only

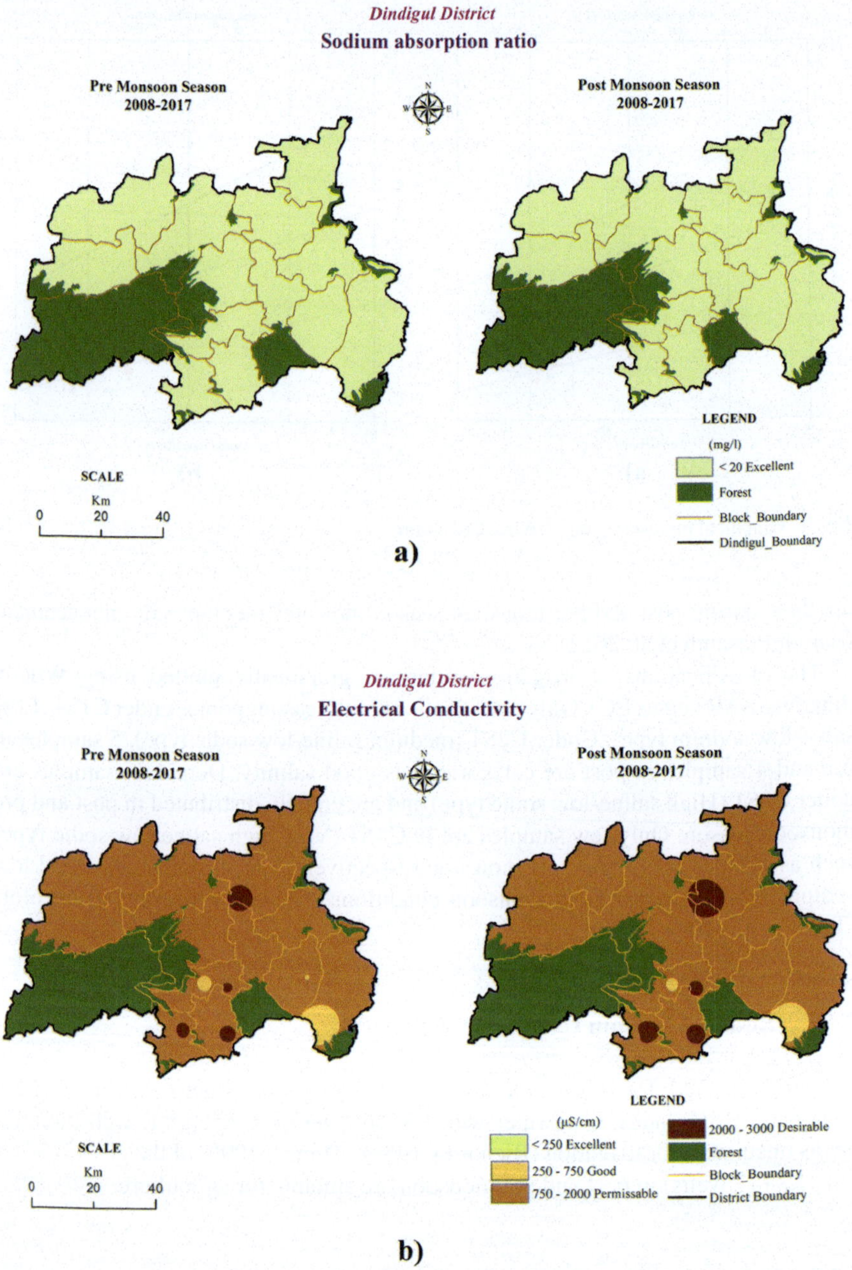

Fig. 5 **a** and **b** Geographical extension of SAR and Electrical Conductivity for both seasons

Table 7 Salinity hazards

Salinity hazards class	Sodium hazard class	SAR in equivalents per mole	EC in (micromohs/cm)	Quality	Pre monsoon samples (µS/cm)	No of sample	Post monsoon samples (µS/cm)	No of sample
C1	S1	<10	100–250	Excellent	160	1	128	1
C2	S2	10–18	250–750	Good	294 to 736	5	256 to 550	4
C3	S3	18–26	750–2250	Doubtful	787 to 2174	10	789to2207	10
C4 and C5	S4 and S5	>26	> 2250	Unsuitable	2421	1	2350 to 2613.64	2

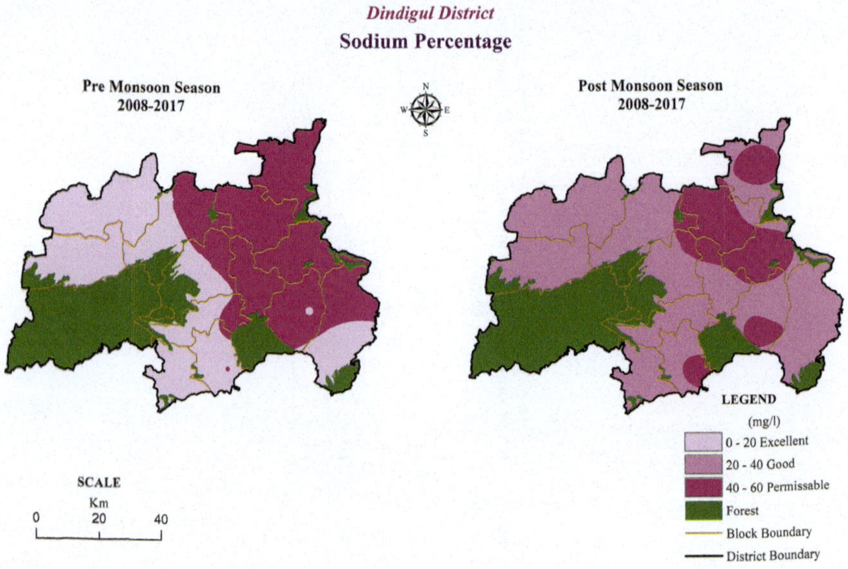

Fig. 6 NA% for pre and post monsoon seasons

in post monsoon season (6%). About 65% of test variables samples in the selected spatial extent are fit for cultivation. The test variables are under doubtful as well as unsuitable classes are not found in any of the samples during both seasons (Fig. 6).

3.4.4 Permeability Index (PI)

The suitability of water for irrigation can be assessed using Permeability Index (Doneen 1964). PI values ranges from 18.92 to 52.48 meq/l and 47.28 to 74.28 meq/l in post and pre monsoon season respectively. PI values specify that all the 17 ground water samples in both PRM and POM are suitable for irrigation.

3.4.5 Magnesium Ratio

Larger amount of magnesium in ground waters disturbs the soil quality in the way of alkaline conversion inversely affects crop production (Paliwal 1972). If the MR is <50 meq/l considered as suitable and if MR is above 50 meq/l, it is considered as unsuitable for agriculture. About 53 and 35% of collected representatives in both seasons are excellent in behalf of cultivation (Fig. 7). Samples unsuitable for agriculture are 65% over the study area. The high Magnesium ratio in the samples is due to interactions of surface and subsurface water with the country rock (Pandian and Sankar 2007).

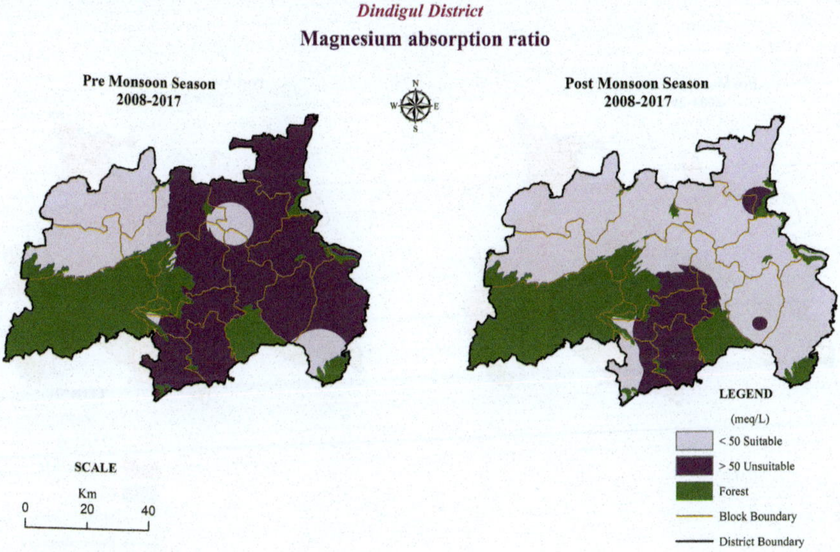

Fig. 7 MAR for seasons

3.4.6 Kelly's Ratio

The sodium problem in irrigation water can be work easily with the help of Kelly's ratio (Kelley et al. 1940). A KR value indicates that 59% (10 samples in POM) and 47% (8 samples in PRM) during seasons are suitable (Fig. 8).

Soluble Sodium Percentage (SSP)

The excess of sodium concentration reduces the crop production and plant growth (Iqbal et al. 2012). In the pre monsoon season in the SSP values varies from 41.05 to 69.42 meq/l and in post monsoon it is 37.98 to 64.98 meq/l. The SSP values in entire groundwater test variables are suitable for agriculture in both seasons.

Irrigation Water Quality Index

To evaluate the suitability of water for irrigation purpose the chemical parameters are ranked and weighted based on its importance (Table 8). The weight has assigned equally due to their magnitude. The outputs of samples are classified as suitable and unsuitable.

In post monsoon season, the irrigation water quality index indicates that 56.08% of area was under suitable classes and 43.91% area is found in the unsuitable classes

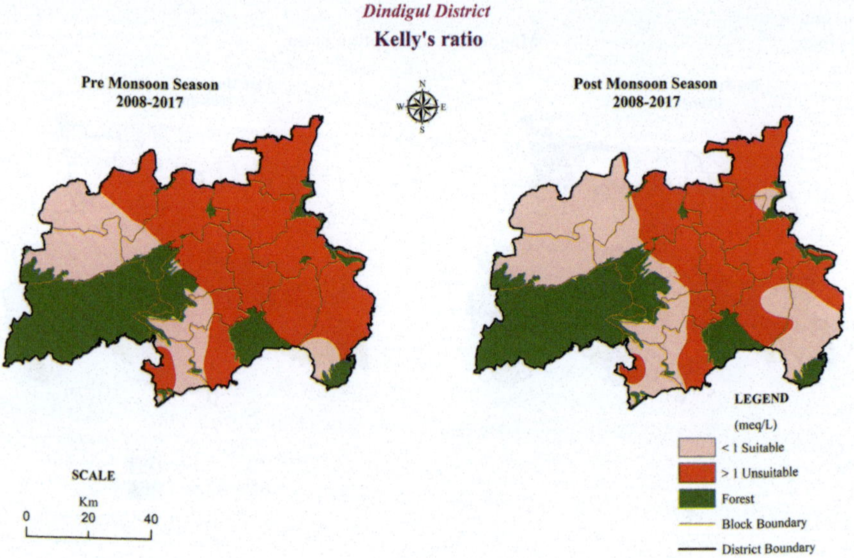

Fig. 8 KR for seasons

(Fig. 9). In pre-monsoon season 70.31% of area is under suitable classes and nearly 29.68% are under unsuitable classes (Table 9).

3.5 *Integrated Ground Water Quality*

The main intention of doing integrated ground water assessment on quality is to squashed drinking and irrigation groundwater quality index in a single map. Weighted overlay analysis in GIS has performed to integrate the multiple layers into single layers and grouped in to four categories. They are undesirable for domestic and irrigation, moderate desirable for irrigation, desirable for irrigation, desirable for basic as well as irrigation purpose. Table 10 contains calculated area and spatial distributions are shown as a map (Fig. 10).

Table 8 Irrigation zones identifying ground water suitability of water quality parameters ratings

Parameter	Classes	Ratings
SAR	Excellent	6
	Good	3
	Doubtful	1
	Unsuitable	0
EC	Excellent	4
	Good	3
	Permissible	2
	Doubtful	1
RSC	Safe	6
	Medium	4
	Unsuitable	0
%Na	Excellent	5
	Good	3
	Permissible	1
	Doubtful	1
	Unsuitable	0
PI	Suitable	6
	Moderate Suitable	4
MR	Suitable	10
	Unsuitable	0
KR	Suitable	10
	Unsuitable	0
SSP	Suitable	10
	Unsuitable	0

In pre-monsoon season, the desirable for domestic and irrigation classes covering an area of 381.61 sq. km and in post monsoon season, the area is 208.93 sq. km. About 624.88 sq. km (POM) and 567.26 sq. km (PRM) of geographical extent is desirable for agriculture. In study area, moderate desirable for irrigation is found to be 9% of area in POM and 43% of area in PRM. The area undesirable for domestic and irrigation is 78% in POM and 40% in PRM.

The dendrogram resulted that only 3 clusters are identified over post monsoon season irrigation and all other pre monsoon season and post–pre DWQI have only 2 clusters during the study area. Even though there are 2 clusters the sample location is not common for all and it's varied from one monsoon season to another (Fig. 11).

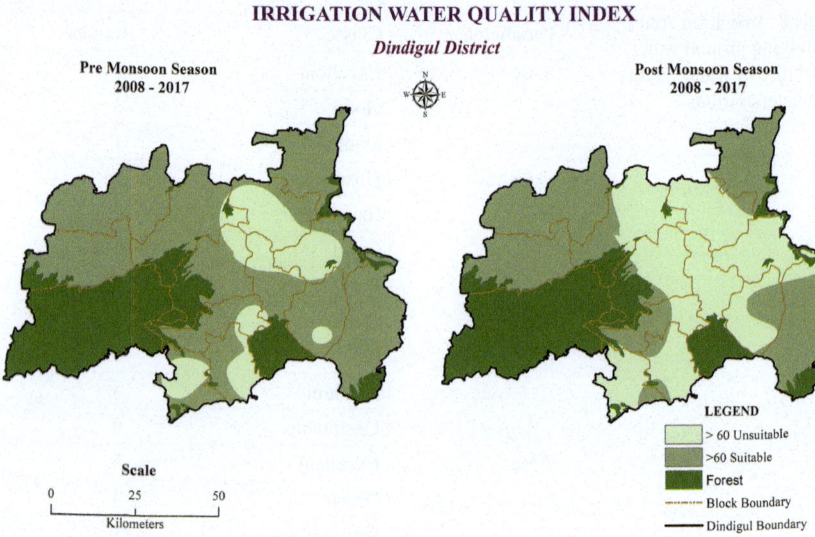

Fig. 9 Irrigation Water Quality Index

Table 9 Irrigation water quality index

Class	Area in sq.km		In percentage	
	POM	PRM	POM	PRM
Suitable	3385.27	4244.00	56.08	70.31
Unsuitable	2650.84	1792.10	43.91	29.68
Total	**6036.11**	**6036.11**	**100**	**100**

Table 10 Integrated water quality index

Class	Area in sq.km		In %	
	POM	PRM	POM	PRM
Desirable for domestic and irrigation	208.93	381.61	3.46	6.32
Desirable for irrigation	567.26	624.88	9.39	10.35
Moderate desirable for irrigation	541.50	2598.92	8.97	43.05
Undesirable for domestic and irrigation	4718.40	2430.69	78.16	40.26
Total	**6036.11**	**6036.11**	**100**	**100**

4 Conclusion

The earthly presence of groundwater as well as its usability is crucial. For the optimum and proactive use of water, the timely monitoring of groundwater quality is mandatory. The present study examines the aptness of groundwater for domestic and

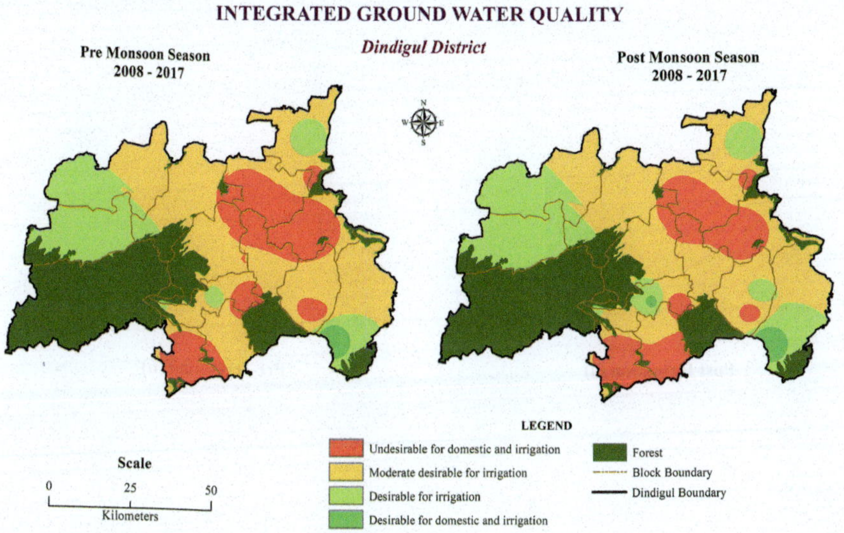

Fig. 10 Integrated Ground Water Quality for seasons

irrigation using water quality indices for Dindigul district, Tamil Nadu. General characteristics, major positive and negative ions in the groundwater are direct indicators of quality indices based on BIS standards. The water quality index was calculated for both post and pre monsoon seasons. Geographical variations of each chemical parameter reveal the nature of groundwater in the study area. DWQI reveals that 53 and 59% of groundwater samples are poor water quality in post and pre monsoon season respectively. 12% of samples are unfit for drinking purposes in both the seasons. About, 3385.27 and 4244.00sq.km area is suitable for watering in agricultural fields in POM and PRM season based on IWQ indices. Assessing the integrated water quality index reveals only 3% in post monsoon and 6% in pre monsoon season are desirable for both basic and agricultural activities. This type of research helps realizing the quality of the water for drinking and agricultural purposes as well as to develop appropriate management practice needed in the contaminated groundwater environment. BIS, WHO, and numerous other standards are followed widely for the evaluation of water quality, and variation in each parameter's range is a limitation for water quality studies. It is challenging to understand the water status of a certain place when applying various types of water quality standards because they commonly show the permissible and non-permissible range at the same place. So, using the local geologic and lithological parameters will help in deriving more accurate results. Majority of the area's water quality is affected by improper sewage treatment, contaminated water supplies, and the mixing of industrial wastes, which in turn affects the local population and environment. The water quality checking is incomplete unless otherwise contaminated water is treated and used accordingly. Considering the study area, the major sources of water contamination are agricultural fields in the form of pesticides and fertilizers along with leather industries. So the

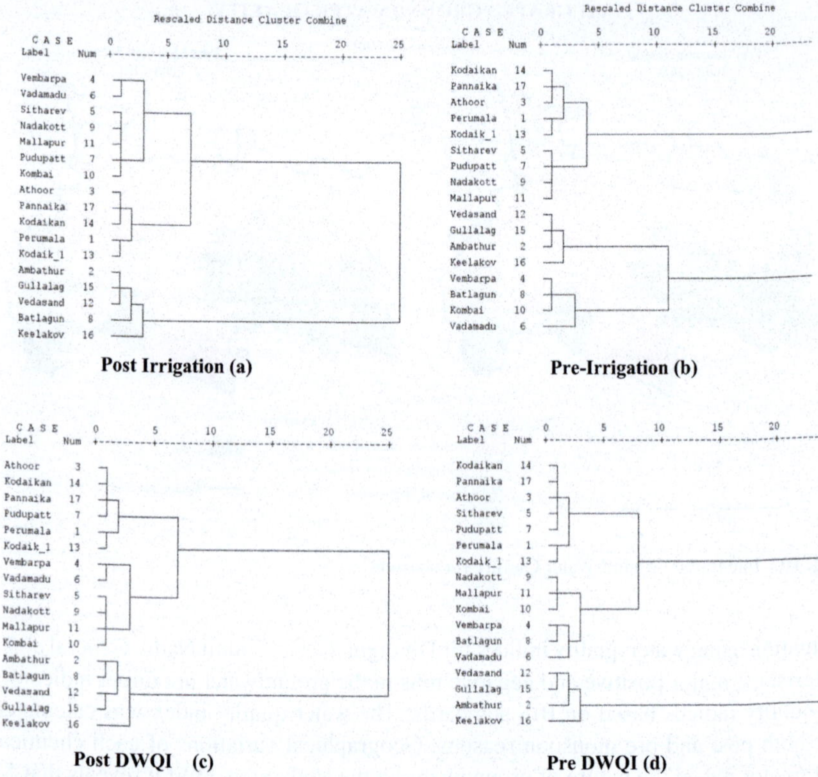

Fig. 11 Dendrogram of water sampling locations during Post and Pre monsoon season for Irrigation **a** and **b**, **c** and **d** represents water sampling location dendrogram for post and pre monsoon season for domestic groundwater quality

quality of water can be improved by promoting organic farming and proper monitoring of waste management by the industries. Safe water influences the health and living conditions of the people, thus affecting the social life of public.

References

Ahmadi SH, Sedghamiz A (2007) Geostatistical analysis of spatial and temporal variations of groundwater level. Environ Monit Assess 129(1):277–294. https://doi.org/10.1007/s10661-006-9361-z

Apambire WB, Boyle DR, Michel FA (1997) Geochemistry, genesis, and health implications of fluoriferous groundwaters in the upper regions of Ghana. Environ Geol 33(1):13–24. https://doi.org/10.1007/s002540050221

Balakrishnan P, Saleem A, Mallikarjun ND (2011) Groundwater quality mapping using geographic information system (GIS): a case study of Gulbarga City, Karnataka, India. Afr J Environ Sci Technol 5(12):1069–1084. https://doi.org/10.5897/AJEST11.134ISSN

Bhunia GS, Keshavarzi A, Shit PK, Omran ESE, Bagherzadeh A (2018) Evaluation of groundwater quality and its suitability for drinking and irrigation using GIS and geostatistics techniques in semiarid region of Neyshabur, Iran. Appl Water Sci 8(6):1–16. https://doi.org/10.1007/s13201-018-0795-6

Bondu R, Cloutier V, Rosa E, Benzaazoua M (2016) A review and evaluation of the impacts of climate change on geogenic arsenic in groundwater from fractured bedrock aquifers. Water Air Soil Pollut 227(9):1–14. https://doi.org/10.1007/s11270-016-2936-6

Bozdağ A, Göçmez G (2013) Evaluation of groundwater quality in the Cihanbeyli basin, Konya, Central Anatolia, Turkey. Environ Earth Sci 69(3):921–937. https://doi.org/10.1007/s12665-012-1977-4

Bureau of Indian Standards (BIS) (2012) Indian standard drinking water specification (second revision) BIS 10500:2012, New Delhi

Doneen LD (1964) Water quality for agriculture, Department of Irrigation. University of California, Davis 48

Farnham IM, Stetzenbach KJ, Singh AK, Johannesson KH (2000) Deciphering groundwater flow systems in Oasis Valley, Nevada, using trace element chemistry, multivariate statistics, and geographical information system. Math Geol 32(8):943–968. https://doi.org/10.1023/A:1007522519268

Gupta S, Mahato A, Roy P, Datta JK, Saha RN (2008) Geochemistry of groundwater, Burdwan District, West Bengal, India. Environ Geol 53(6):1271–1282. https://doi.org/10.1007/s00254-007-0725-7

Gurdak JJ, McMahon PB, Bruce BW (2012) Vulnerability of groundwater quality to human activity and climate change and variability, High Plains aquifer, USA. In: Treidel H, Martin-Bordes JL, Gurdak JJ (eds) Climate change effects on groundwater resources-a global synthesis of findings and recommendations. Taylor & Francis Group, pp 145–168

Hem JD (1985) Study and interpretation of the chemical characteristics of natural water, vol 2254. Department of the Interior, US Geological Survey

Iqbal H, Inam A, Bakhtiyar Y, Inam A (2012) Effluent quality parameters for safe use in agriculture. Water Qual Soil Manag Irrig Crops. https://doi.org/10.5772/31557

Joshi DM, Kumar A, Agrawal N (2009) Assessment of the irrigation water quality of river Ganga in Haridwar district. Rasayan J Chem 2(2):285–292

Karunanidhi D, Vennila G, Suresh M, Subramanian SK (2013) Evaluation of the groundwater quality feasibility zones for irrigational purposes through GIS in Omalur Taluk, Salem District, South India. Environ Sci Pollut Res 20(10):7320–7333. https://doi.org/10.1007/s11356-013-1746-2

Kelley WP, Brown SM, Liebig GF Jr (1940) Chemical effects of saline irrigation water on soils. Soil Sci 49(2):95–108

Kelley WP (1963) Use of saline irrigation water. Soil Sci 95(6):385–391

Ketata-Rokbani M, Gueddari M, Bouhlila R (2011) Use of geographical information system and water quality index to assess groundwater quality in El Khairat deep aquifer (Enfidha, Tunisian Sahel). Iran J Energy Environ 2(2):133–144

Krishnaraj S, Kumar S, Elango KP (2015) Spatial analysis of groundwater quality using geographic information system–a case study. IOSR J Environ Sci, Toxicol Food Technol 9(2):1–6

Kumar M, Kumari K, Ramanathan AL, Saxena R (2007) A comparative evaluation of groundwater suitability for irrigation and drinking purposes in two intensively cultivated districts of Punjab, India. Environ Geol 53(3):553–574. https://doi.org/10.1007/s00254-007-0672-3

Lin Y, Han P, Huang Y, Yuan GL, Guo JX, Li J (2017) Source identification of potentially hazardous elements and their relationships with soil properties in agricultural soil of the Pinggu district of Beijing, China: multivariate statistical analysis and redundancy analysis. J Geochem Explor 173:110–118. https://doi.org/10.1016/j.gexplo.2016.12.006

Lozano R, Naghavi M, Foreman K, Lim S, Shibuya K, Aboyans V, Remuzzi G (2012) Global and regional mortality from 235 causes of death for 20 age groups in 1990 and 2010: a systematic analysis for the global burden of disease study 2010. Lancet 380(9859):2095–2128

Misaghi F, Delgosha F, Razzaghmanesh M, Myers B (2017) Introducing a water quality index for assessing water for irrigation purposes: a case study of the Ghezel Ozan River. Sci Total Environ 589:107–116

Nas B, Berktay A (2010) Groundwater quality mapping in urban groundwater using GIS. Environ Monit Assess 160:215–227

Omran ESE, Ghallab A, Selmy S, Gad AA (2014) Evaluation and mapping water wells suitability for irrigation using GIS in Darb El-Arbaein, South Western Desert, Egypt. Int J Water Resour Arid Environ 3(1):63–76

Ordookhani K, Amiri B, Saravani A (2012) Assessment of ground water quality using geographic information system in Parsian plain. In: International conference on environment, agriculture and food sciences (ICEAFS'2012), August 11–12, 2012 Phuket (Thailand)

Paliwal KV (1972) Irrigation with saline water. IARI Monograph no 2 (New series), New Delhi, p 198

Pande CB, Moharir K (2018) Spatial analysis of groundwater quality mapping in hard rock area in the Akola and Buldhana districts of Maharashtra, India. Appl Water Sci 8:106. https://doi.org/10.1007/s13201-018-0754-2

Pande CB, Moharir KN (2022) Estimation of groundwater quality parameters for drinking purpose using IDW, GIS and statistical analysis methods: A case study of basaltic rock in Mahesh river basin, Akola and Buldhana districts (MS), India. In: Yadav S, Negm AM, Yadava RN (eds) Water quality, assessment and management in India. Earth and Environmental Sciences Library. Springer, Cham. https://doi.org/10.1007/978-3-030-95687-5_16

Pande CB, Moharir KN, Singh SK et al (2020) Groundwater evaluation for drinking purposes using statistical index: study of Akola and Buldhana districts of Maharashtra, India. Environ Dev Sustain 22:7453–7471. https://doi.org/10.1007/s10668-019-00531-0

Pandian K, Sankar K (2007) Hydrogeochemistry and groundwater quality in the Vaippar River basin Tamil Nadu. J-Geol Soc India 69(5):970

Rabeiy RE (2018) Assessment and modeling of groundwater quality using WQI and GIS in Upper Egypt area. Environ Sci Pollut Res 25(31):30808–30817. https://doi.org/10.1007/s11356-017-8617-1

Raghunath HM (1987) Ground water: hydrogeology, ground water survey and pumping tests, rural water supply and irrigation systems. New Age Int

Raikar RV, Sneha MK (2012) Water quality analysis of Bhadravathi taluk using GIS–a case study. Int J Environ Sci 2(4):2443–2453. https://doi.org/10.6088/ijes.00202030125

Raju NJ, Dey S, Das K (2009) Fluoride contamination in groundwaters of Sonbhadra district, Uttar Pradesh, India. Curr Sci 979–985

Ramesh K, Elango L (2012) Groundwater quality and its suitability for domestic and agricultural use in Tondiar river basin, Tamil Nadu, India. Environ Monit Assess 184(6):3887–3899. https://doi.org/10.1007/s10661-011-2231-3

Reddy KS (2013) Assessment of groundwater quality for irrigation of Bhaskar Rao Kunta watershed, Nalgonda District, India. Int J Water Resour Environ Eng 5(7):418–425. https://doi.org/10.5897/IJWREE2012.0375

Richard LA (1954) Diagnosis and Improvement of Saline and Alkaline Soils. Agricultural Handbook No. 60. US Department of Agricultue. https://www.ars.usda.gov/pacificwest-area/riverside-ca/us-salinity-laboratory/docs/handbook-no-60/. (accessed 26.27. 2017)

Sadashivaiah CRRC, Ramakrishnaiah CR, Ranganna G (2008) Hydrochemical analysis and evaluation of groundwater quality in Tumkur Taluk, Karnataka State, India. Int J Environ Res Public Health 5(3):158–164. https://doi.org/10.3390/ijerph5030158

Saleem R (2007) Groundwater management—emerging challenges. Water Digest 1:12–28

Samahajira S, Annal CF (2017a) Evaluation of groundwater quality for block of Dindigul distr. Evaluation 6(5):34–39

Samahajira S, Florence AC (2017b) Evaluation of groundwater quality for irrigation purpose in Rediyarchatram block of Dindigul district Tamil Nadu, India. Int Res J Environ Sci 6(5):34–39

Samson M, Swaminathan G, Kumar NV (2010) Assessing groundwater quality for potability using a fuzzy logic and GIS–a case study of Tiruchirappalli city–India. Comput Model New Technol 14(2):58–68

Sarath Prasanth SV, Magesh NS, Jitheshlal KV, Chandrasekar N, Gangadhar KJAWS (2012) Evaluation of groundwater quality and its suitability for drinking and agricultural use in the coastal stretch of Alappuzha District, Kerala, India. Appl Water Sci 2(3):165–175. https://doi.org/10.1007/s13201-012-0042-5

Selvam S, Sivasubramanian P (2012) Groundwater potential zone identification using geoelectrical survey: a case study from Medak district, Andhra Pradesh, India. Int J Geomat Geosci 3(1):55

Simsek C, Gunduz O (2007) IWQ index: a GIS-integrated technique to assess irrigation water quality. Environ Monit Assess 128(1):277–300

Singaraja C, Chidambaram S, Anandhan P, Prasanna MV, Thivya C, Thilagavathi R (2012) A study on the status of fluoride ion in groundwater of coastal hard rock aquifers of south India. Arab J Geosci. https://doi.org/10.1007/s12517-012-0675-6

Singh V, Chandel CS (2006) Analysis of wastewater of Jaipur City for agricultural use. Res J Chem Environ 10(1):30–33

Singh P, Khan IA (2011) Ground water quality assessment of Dhankawadi ward of Pune by using GIS. Int J Geomat Geosci 2(2):688–703

Srinivasamoorthy K, Nanthakumar C, Vasanthavigar M, Vijayaraghavan K, Rajivgandhi R, Chidambaram S, ..., Vasudevan S (2011). Groundwater quality assessment from a hard rock terrain, Salem district of Tamilnadu, India. Arabian J Geosci 4(1):91–102. https://doi.org/10.1007/s12517-009-0076-7

Todd DK, Mays LW (1980) Groundwater Hydrology. John Willey & Sons. Inc., New York, 535

Troiano J, Johnson BR, Powell S, Schoenig S (1994) Use of cluster and principal component analyses to profile areas in California where ground water has been contaminated by pesticides. Environ Monit Assess 32(3):269–288

UNEP (1999) Global environment outlook 2000. Earthscan, London

Venkateswaran S, Vediappan S (2013) Assessment of groundwater quality for irrigation use and evaluate the feasibility zones through geospatial technology in Lower Bhavani Sub Basin, Cauvery River, Tamil Nadu, India. Int J Innov Technol Explor Eng 3(2):180–187

Vinothkanna S, Rajee R, Senthilraja K (2020a) Assessing ground water quality for the suitability of irrigation in Dindigul District, Tamil Nadu India. Indian J Ecol 47(1):23–29

Vinothkanna S, Rajee R, Senthilraja K (2020b) Assessing the domestic groundwater quality of Dharmapuri District, Tamil Nadu. Res Journey Int E-Res J 249:238–245

Vinothkanna S, Rajee R, Senthilraja K (2021a) Fluoride health hazard assessment in ground water resources of Tiruchirappalli district Tamil Nadu. Indian J Ecol 48(2):496–502

Vinothkanna S, Rajee R, Senthilraja K (2021b) Determining ground water quality for irrigation in Varahanadhi River Basin, Tamil Nadu, India. Indian J Ecol 48(3):842–849

Vinothkanna S, Rajee R, Senthilraja K (2022) GIS-based assessment of groundwater quality for drinking in Perambalur District of Tamil Nadu. Indian J Ecol 49(1):149–154

Vinothkanna S, Emayavaramban V, Kannadasan K, Senthilraja K (2016) Assessing groundwater quality in Namakkal district, Tamil Nadu—a GIS approach. In: Angadi DP (eds), Proceedings of the UGC sponsored 4th international conference on remote sensing and GIS applications on coastal management, February 16–17, 2016. Mangalore University, Mangalore, India pp 87–98

Vinothkanna S (2019) A Statistical Approach for Determining Ground water quality in Dindigul district, Tamil Nadu. International Journal of Research in Social Sciences 9(7):546–554

Wilcox L (1955) Classification and use of irrigation waters (No. 969). US Department of Agriculture

World Health Organization (2004) Guidelines for drinking-water quality: recommendations, vol 1. World Health Organization

Zamani-Ahmadmahmoodi R, Jafari A, Alibeygi-Beni H (2017) Potential ecological risk assessment, enrichment, geoaccumulation, and source identification of metals in the surface sediments of Choghakhor Wetland, Iran. Environ Earth Sci 76(11):1–9. https://doi.org/10.1007/s12665-017-6718-2

Evaluation of Morphometric Parameters of Drainage Networks Derived from Topographic Maps and DEM Using Geographical Information System—A Study on Semi-Arid River Basin, India

B. Y. Chinmayi and H. Ramesh

Abstract Agriculture plays a vital role in India's economy and there is a prodigious need for efficient utilization of available water resources. Effective management of any river basin necessitates a precise delineation of the watershed as it is crucial to determining stream flow paths and the contributing areas, which are the main factors in a river basin's morphometric characteristics. In the current study, an effort has been made to evaluate various morphometric characteristics using geographic information system (GIS) and remote sensing (RS) techniques, including linear aspects, relief aspects, and aerial aspects. The GIS platform is used for the preprocessing of the data and the delineation of the watershed. The current study focuses on the upper Cauvery River basin i.e., the Hemavathi subbasin in Karnataka, which is classified as semi-arid. The Hemavathi River runs for about 202 km and drains about 5,575 square kilometers. A total of 5 sub-basins are created using a hydrological model for the current study. The morphometric parameters evaluated for the study are derived using the Survey of India topographic map (1:50,000) and SRTM (Shuttle Radar Topographic Mission) Digital Elevation Model. The systematic analysis of morphometric parameters within the drainage network provides significant insight for hydrologists and decision-makers in understanding and framing efficient water and soil conservation practices.

Keywords Drainage network · Morphometry · SRTM

B. Y. Chinmayi (✉) · H. Ramesh
Department of Water Resources and Ocean Engineering, National Institute of Technology Karnataka, Surathkal Mangaluru-575025, India
e-mail: chinmayi.yogendra@gmail.com

C. B. Pande et al. (eds.), *Surface and Groundwater Resources Development and Management in Semi-arid Region*, Springer Hydrogeology,
https://doi.org/10.1007/978-3-031-29394-8_15

1 Introduction

Morphometric analysis is defined as quantitative and mathematical analysis and measurements of the earth's landforms to have a significant understanding of the geological and hydrological characteristics of a drainage basin (Asfaw and Workineh 2019). Drainage characteristics of a river basin are often studied using conventional methods (Athol 1984). More methodical approaches to studying the basin morphometric characteristics have been presented in recent studies (Mahala 2020; Pande et al. 2021a). The development of morphometric analysis follows that of fundamental basin parameters like catchment area, perimeter, stream length, maximum and minimum elevation, and stream orders. Additionally, derived and shape parameters are obtained using these parameters. In order to study different morphometric parameters, drainage networks' pattern, texture, and shape have been quantitatively analyzed in a significant amount of research (Ahmed et al. 2010; Pande et al. 2020).

Traditional techniques, such as field observations and the use of topographic maps, can also be used to identify the drainage networks that exist within the basin (Pande et al. 2021b). Due to the various constraints involved in the field survey, it is practically not feasible to analyze every stream network (Maidment 2002) so in the current study, the drainage network is created using Digital Elevation Map (DEM) under Geographical Information System (GIS) platform (Pande et al. 2017). The generated stream networks are subsequently validated by superimposing them on topographical maps procured from the Govt. of India. Drainage networks extracted from DEM use the concept of water flowing under gravity from higher elevation to lower elevation under a few assumptions that there are no interception, evaporation, or other losses (Ahmed et al. 2010; Prabhakaran and Jawahar Raj 2018). In the current work, a semi-arid basin located in the southern part of India is selected for the study.

2 Materials and Methods

2.1 *Description of the Study Area*

The Hemavathi river originates in the Western Ghats at a height of approximately 1,869 m close to Ballalarayanadurga in the Chikkamagalur district of the State of Karnataka, which falls in the upper Cauvery basin. It flows through Hassan district where it is joined by its chief tributary, the Yagachi River, and then into Mandya district before joining the Cauvery river near Krishnarajasagara. Hemavathi river traverses a length of approximately 202 km and has a drainage area of about 5,575 km^2. Figure 1 shows the location of the study area (Fig. 2).

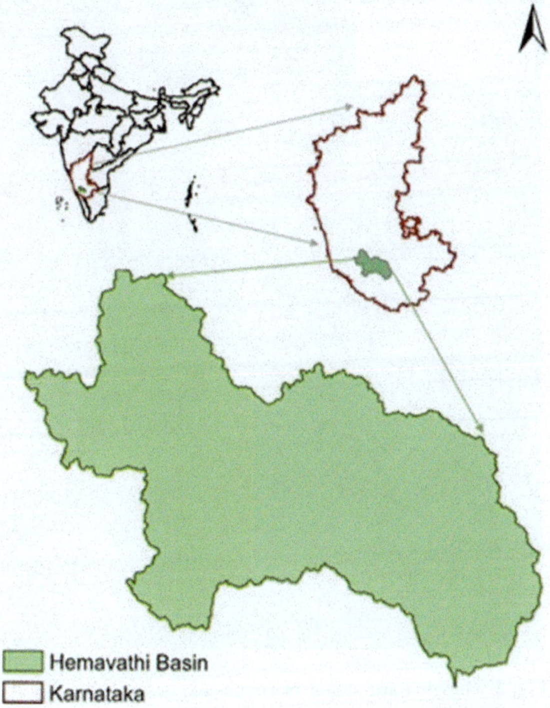

Fig. 1 Location map of the study area

3 Methodology

The study area for the current study is delineated using 30 m resolution Shuttle Radar Topographic Mission (STRM) DEM procured from USGS which is projected in Universal Transverse Mercator (UTM) -WGS 84, Zone 43N. The DEM must be created without sinks or depressions, as these features could result in the generation of discontinuous drainage networks, which is the first crucial step before creating drainage networks (Pande et al. 2018). Sinks are DEM cells that are lower in elevation than the cells around them. As a result, when creating stream networks, the flow tends to congregate at the depressions or sinks in the DEM and cannot move elsewhere. Using Arc Hydro in ArcGIS, the sinks were located and eliminated in the current work.

The next step in the analysis is to determine the flow direction. The basin landscape is analyzed using the eight direction pour point model, also known as the D-8 algorithm, to determine the direction of each cell. The cells are checked for the surrounding elevation of 8 cells and drained to the cell with the least elevation assuming gravity flow. Similarly, the entire cells in the area of interest are examined, and a flow direction map with a predefined threshold is generated (shown in Fig. 3a).

The accumulation of the flow is the following step in the modelling process. It is crucial to create the stream with the ultimate flow path now that the model is aware

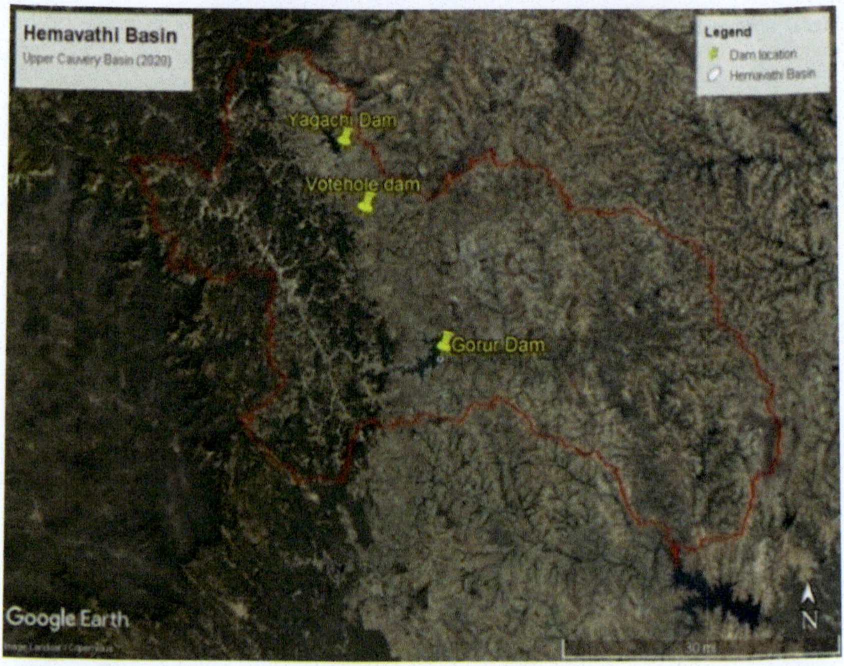

Fig. 2 Google Earth image of Hemavathi basin in the year 2022

of the direction of the flow in each cell. The flow is accumulated in each cell of a raster that is made in this step (shown in Fig. 3b).

The study area for the current analysis is characterized into 5 sub-basins based on the 1st order of streams and major dams located in the study area. The first subbasin is taken up to upstream of the Yagachi dam built across the Yagachi river in the Chikkamagalur district. The second subbasin is considered up to upstream of the Gorur dam built across the Hemavathi river in Hassan district.

The third subbasin is considered till the point where the Yagachi river confluences the Hemavathi river. The fourth subbasin is considered at the confluence point of Hebba Halla and Hemavathi rivers basically to reduce the geographical study area of the last subbasin. Lastly, the fifth subbasin up to entire subbasin outlet of the study area as shown in Fig. 4.

The streams generated were then overlaid on topographical maps to ensure that they were accurate before being used to create the sub-basins for the current study. Figure 5a, b show the reaches overlayed on topographical maps in GIS.

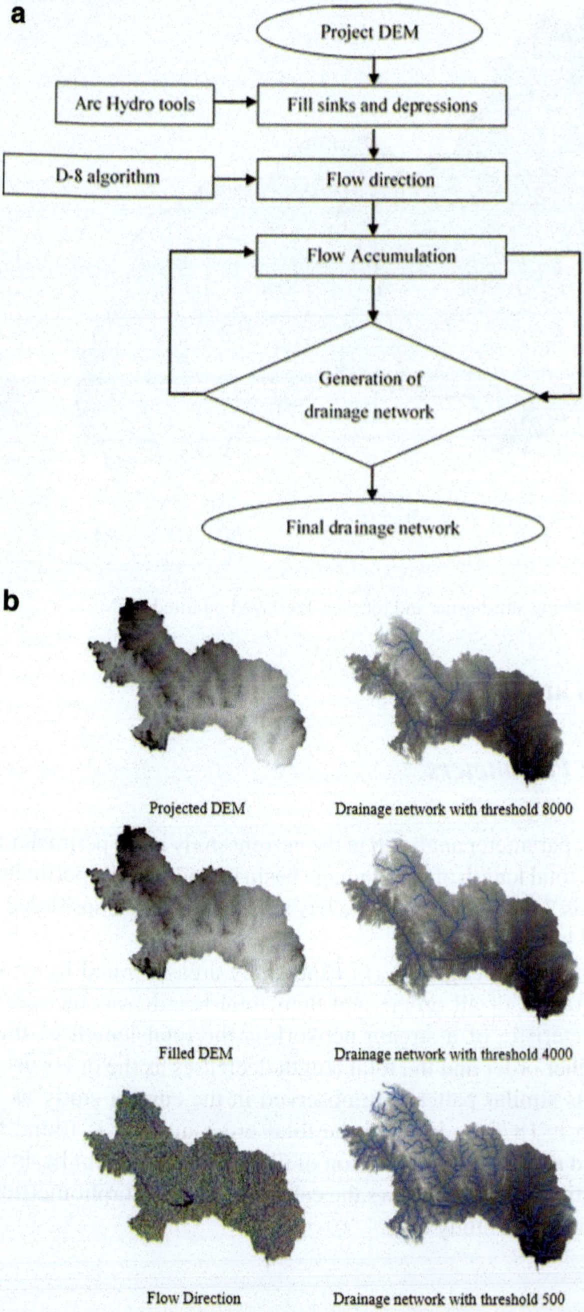

Fig. 3 **a** Flowchart for extracting the drainage network from the DEM, **b** Extracting drainage network from DEM

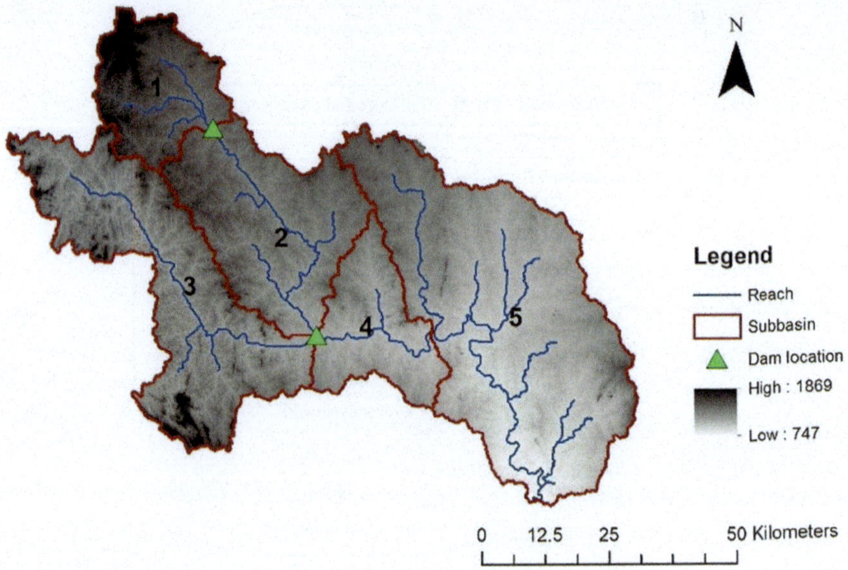

Fig. 4 Map showing sub-basins and reaches overlayed on filled DEM

4 Results and Discussion

4.1 Basic Parameters

The first basic parameter analyzed in the current study is the perimeter (P) of the basin defined as the total length of the drainage basin boundary. The perimeter derived from SRTM DEM is 745 km for the entire basin. The total area considered for the present study is 5575 km^2.

The total stream length (L_u) of the study area is found by counting the total number of streams of all orders and their total length was measured in GIS. The typical characteristic of a stream network is the total length of the stream is the longest in higher order and the total length decreases as the order decreases (Biswas 2016) and this similar pattern are observed in the current study as well. The total stream length is 1878.89 km and the total order number is found to be 636. The maximum and the minimum elevation of the entire Hemavathi basin are 1869 m and 747 m respectively. Table 2 shows the calculated basic morphometric parameters for each sub-basin in the study area.

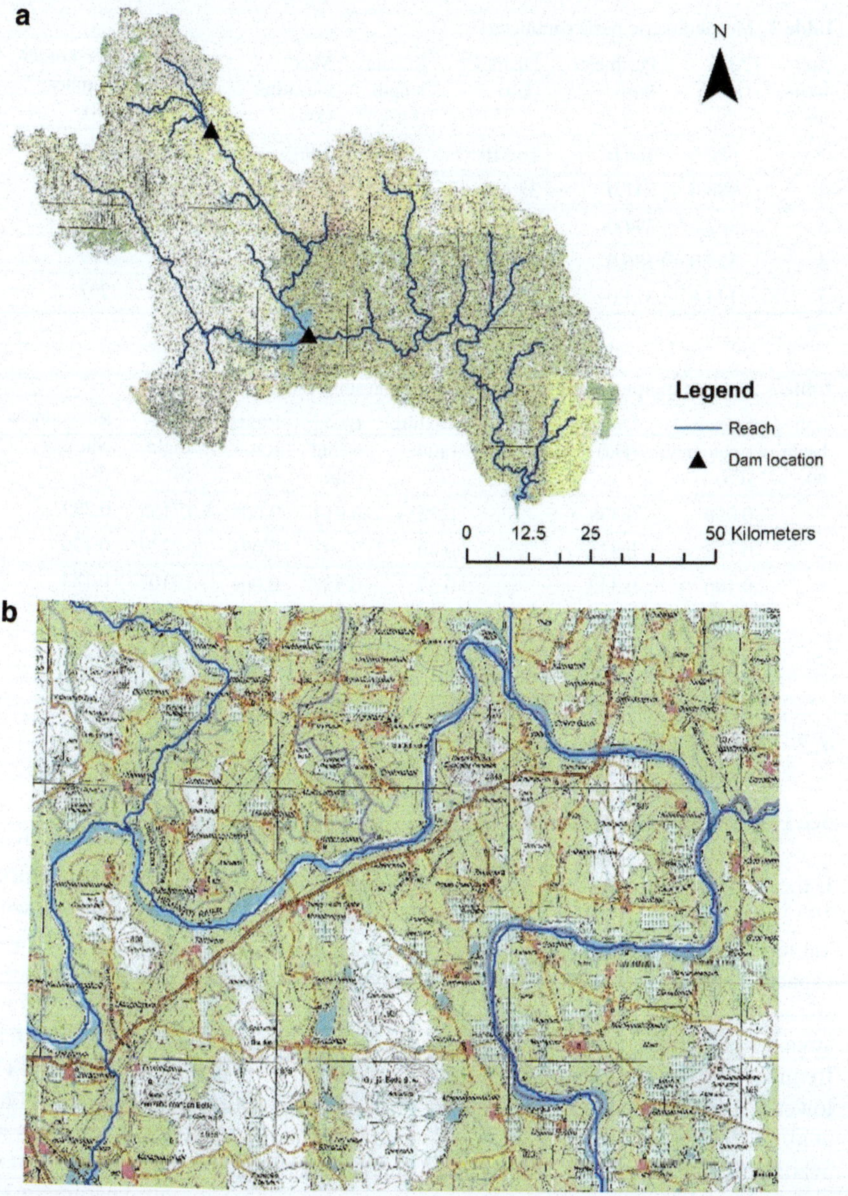

Fig. 5 **a** Map showing reaches overlayed on topographical maps, **b** Map showing the enlarged view of the reach in the fourth subbasin overlayed on the topographical map

Table 2 Morphometric basic parameters

Sub-basin no	Area (km²)	Perimeter (km)	Diameter (km)	Stream length (km)	Max elevation (km)	Min elevation (km)	Total order number (N)
1	552.0	168.0	26.510	181.00	1.867	0.954	54
2	988.0	241.0	35.467	327.27	1.279	0.886	111
3	1362.0	394.0	41.643	466.13	1.423	0.886	145
4	529.0	186.0	25.952	186.11	1.066	0.832	69
5	2144.0	374.0	52.247	718.37	1.191	0.747	257

Table 3 Derived morphometric parameters of Hemavathi basin

Sub-basin no	Stream frequency (FS)	Drainage Density (Dd)	Texture ratio (T)	Basin Relief (Bh)	Relief Ratio (Rr)	Slope Angle (S)	Ruggedness Number (Rn)
1	0.098	0.328	0.32	0.913	0.005	0.7165	0.299
2	0.112	0.331	0.46	0.393	0.001	0.2229	0.130
3	0.106	0.342	0.37	0.537	0.001	0.3107	0.184
4	0.130	0.352	0.37	0.234	0.001	0.1718	0.082
5	0.120	0.335	0.69	0.444	0.001	0.1217	0.149

4.2 *Derived Parameters*

4.2.1 Stream Frequency (S_f)

Is the total number of stream segments of all orders per unit area of the basin (Horton 1932). The stream frequencies calculated individually for each sub-basin are shown in Table 3. The stream frequency is directly related to the slope of the terrain, infiltration capacity, and permeability of the basin and provides drainage basin responses to stream processes. Recent studies have shown that it is possible to have different stream frequencies though the drainage density is the same (Mahala 2020). Stream frequency in the Hemavathi basin ranges between 0.09 and 0.13 (units/km^2). This lower Stream frequency represents a poor drainage network in the basin (Thomas et al. 2010). The lower values of Sf being reported in the study area represents less available relief indicative of lesser slope and permeability.

4.2.2 Drainage Density (D_d)

Is a measure of the closeness of spacing of streams in the basin according to Horton (1945). Dd is the ratio between the total stream length of all orders to the area of the basin. Dd also provides the mathematical extent of runoff potentiality of the basin irrespective of stream order. Dd is considered an important parameter for studying the travel time of water. Dd for the Hemavathi basin is found to be 0.32 to 0.33 km/km^2 indicating low Dd. The Dd values for each of the sub-basins are tabulated in Table 3.

4.2.3 Texture ratio (T)

Is defined as the ratio between the total number of streams of all orders to the perimeter of the basin. According to Horton (1945) drainage texture lesser than 2 indicates poor drainage texture ratio and the infiltration capacity of the basin acts directly on the texture ratio values as recognized by Horton. In the current study, the drainage texture ratio values range between 0.3 and 0.6 units/km.

4.2.4 Basin Relief (B_h)

Is defined as the maximum vertical distance between the lowest and the highest points of a sub-basin. B_h is a general relief morphometric parameter used to understand the slope of a basin as a hydrological parameter. The general relation of B_h with the slope is that with the increasing slope the relief ratio increases thereby resulting in higher stream gradients resulting in a lower time of concentration of runoff thereby increasing the flood peaks and sediment transport in the basin. B_h values are dependent on various underlying geology and geomorphology of the basin. B_h can be used as the best indicator to determine the floodplain zones and to identify the stages of the flood (Sreekumar and Aslam 2016). In the current work, the B_h values range between 0.2 and 0.9 km. The values are lesser in the downstream of the basin indicating the presence of relatively plain terrain while 0.9 value at the upstream subbasin 1 indicates mountainous terrain with relatively higher B_h values in the study area. The values of B_h in each subbasin are tabulated in Table 3.

4.2.5 Relief Ratio (R_r)

Is defined as the ratio of the maximum vertical distance between the lowest and the highest points of a sub-basin to the basin length. Also, it is defined as the ratio between the relief of the basin to the longest dimension parallel to the highest order stream in the basin. R_r gives the calculation of the elevation difference between the source of the stream and the confluence point downstream divided by the total length of the stream. Normally R_r is used to understand the average drop in elevation per unit length of the stream considered. R_r increases with decreasing drainage area and the study area considered for the current study is relatively large hence there could be a dip in R_r values. The study reveals that the R_r of the Hemavathi basin ranges between 0.001 and 0.005 (refer to Table 3).

4.2.6 Slope Angle (S)

Is defined as the inverse tangent value for the ratio between the maximum vertical distance between the lowest and the highest points of a sub-basin to the basin length. Normally slope angle is considered with hydrological relevance. Steeper slopes generate higher runoff values and transport higher loads of sediment to downstream sub-basins. As slope governs the relationship between the infiltration and runoff, it is imperative to study the slope of the area of interest. The slope angle of the Hemavathi subbasin is shown in Table 3.

4.2.7 Ruggedness Number (R_n)

Is a measure that indicates the structural complexity of the terrain. R_n is a dimensionless number obtained by the product of the basin relief and its drainage density and the basin relief is the maximum vertical distance between the lowest and the highest points of a sub-basin. In the study area, R_n is found to range between 0.13 and 0.29. Studies suggest that the mountainous basin will have higher R_n when compared to the rest of the terrain in the basin (Ansari et al. 2012). Here in the current study, subbasin 1 has the highest R_n followed by subbasin 2 and 3 downstream of subbasin 1 (refer to Table 3).

4.3 Shape Parameters

4.3.1 Elongation Ratio (R_e)

Is defined as the ratio between the diameter of a circle with the same area as that of the basin and the maximum length of the basin (Schumm 1956). Studies have revealed that circular basins discharge the runoff relatively more efficiently compared to elongated basins (Ahmed et al. 2010). The values of R_e generally range between 0.1 and 0.9. These values can be grouped as circular when R_e is > 0.8, as oval when R_e ranges between 0.6 and 0.8 and the basin is grouped as less elongated when the Re values are < 0.6 (Schumm, 1956). In the current study, it is observed that the R_e values for the Hemavathi basin fall under a less elongated region as shown in Table 4.

4.3.2 Circularity Index (R_c)

Is defined as the ratio of basin area to the area of a circle having the same perimeter as the basin. R_c is often influenced by the length, frequency of streams, geology, relief, and slope of the basin (Sukristiyanti et al. 2018). Basins with R_c values greater than 0.5 is considered to be more or less circular in shape with moderate to high drainage

Table 4 Morphometric shape parameters of Hemavathi basin

Sub-basin no	Elongation ratio (R_e)	Circularity index (R_c)	Form factor (F_f)
1	0.146	0.246	0.026
2	0.108	0.214	0.014
3	0.089	0.110	0.010
4	0.139	0.192	0.024
5	0.073	0.193	0.006

and relief ratio. R_c values indicate how well the basin is structurally controlled concerning the drainage network (Das et al. 2022). The Hemavathi basin is having R_c values ranging between 0.11 to 0.24 and thus indicates having very less R_c. The R_c values are tabulated in Table 4.

4.3.3 Form Factor (F_f)

Is defined as the ratio of the basin area to the square of the basin length. F_f is also an indicator of the morphometric shape parameter. The higher the value of F_f more circular in shape is the basin (Javed et al. 2011). The Ff values are tabulated in Table 4 for the sub-basins in the present study. The values indicate very low ranges of F_f indicating the basin is relatively elongated and less circular.

5 Conclusions

Morphometric parameters of a watershed can be calculated by field visits and other conventional methods. Systematic analysis and a significant understanding of the basin morphometric parameters demand finer data and advanced tools like GIS and remote sensing data. The current study analyses the basin morphometric parameters of the Hemavathi river basin using DEM and also topographical maps. The morphometric shape parameters analyzed in the current study indicate that the basin is less circular and more elongated in shape which may further be affecting its surface runoff and sediment transport characteristics.

The present study also concludes that the basin is having low stream frequency values and low drainage density. However, stream frequency and drainage density are greatly dependent on the total area of the basin, so researchers must consider the effect of the size of the basin on the morphometrical parameters analyzed. Similarly, from the current study, it is suggested that care must be taken in the selection of a threshold value for the flow accumulation of the drainage network. Since stream order and stream density are major inputs for most of the derived morphometric parameters, it is vital to generate realistic streams for the study before beginning the analysis. In the current study, the applicable threshold value was decided by superimposing the stream networks generated using DEM on the topographical maps for validating the stream network generated.

The present study aims to interpret the basin-related information such as its form, relief, shape, geology, and hydrologic processes in the 5 subbasin of the Hemavathi basin to stakeholders and decision-makers. In addition to this, a systematic understanding of the variations in form and process in the basin is necessary for efficient policy making and serves as an important input in the prioritization of sub-basins for management and conservational practices planning. The current work can also provide an insight for.

References

Ahmed SA, Chandrashekarappa KN, Raj SK, Nischitha V, Kavitha G (2010) Evaluation of morphometric parameters derived from ASTER and SRTM DEM—A study on Bandihole sub-watershed basin in Karnataka. J Indian Soc Remote Sens 38(2):227–238. https://doi.org/10.1007/s12524-010-0029-3

Ansari ZR, Rao LAK, Yusuf A (2012) Gis based morphometric analysis of yamuna drainage network in parts of fatehabad area of agra district. Uttar Pradesh. J Geol Soc India 79(5):505–514. https://doi.org/10.1007/s12594-012-0075-2

Asfaw D, Workineh G (2019) Quantitative analysis of morphometry on Ribb and Gumara watersheds: Implications for soil and water conservation. Int Soil Water Conserv Res 7(2):150–157. https://doi.org/10.1016/j.iswcr.2019.02.003

Abrhams AD (1984) Channel networks' a geomorphological perspectiv. Water Resour Res 20(2):161–168

Biswas SS (2016) Analysis of GIS based morphometric parameters and hydrological changes in parbati river basin, Himachal Pradesh, India. J Geogr Nat Disasters, 6(2). https://doi.org/10.4172/2167-0587.1000175

Das BC, Islam A, Sarkar B (2022) Drainage basin shape indices to understanding channel hydraulics. Water Resour Manage 36(1):2523–2547. https://doi.org/10.1007/s11269-022-03121-4

Horton R (1932) Drainage basin characteristics. Trans Am Geophys Union 13:350–361. https://doi.org/10.1029/TR013i001p00350

Horton R (1945) Erosional development of streams and their drainage basins: Hydro physical approach to quantitative morphology. Bulletin of the geological society of America. Prog Phys Geogr, 19(4), 533–554. https://doi.org/10.1177/030913339501900406

Javed A, Khanday MY, Rais S (2011) Watershed prioritization using morphometric and land use/land cover parameters: A remote sensing and GIS based approach. J Geol Soc India 78(1):63–75. https://doi.org/10.1007/s12594-011-0068-6

Mahala A (2020) The significance of morphometric analysis to understand the hydrological and morphological characteristics in two different morpho-climatic settings. Appl Water Sci, 10(1). https://doi.org/10.1007/s13201-019-1118-2

Manjare BS (2015) Prioritization of Sub-Watersheds for sustainable development and management of natural resources: An integrated approach using remote sensing, GIS techniques. In: 16th ESRI India User Conf., 1–13

Pande CB, Moharir K (2017) GIS based quantitative morphometric analysis and its consequences: a case study from Shanur River Basin, Maharashtra India. Appl Water Sci 7, 861–871. https://doi.org/10.1007/s13201-015-0298-7

Pande CB, Khadri SFR, Moharir KN et al (2018) Assessment of groundwater potential zonation of Mahesh River basin Akola and Buldhana districts, Maharashtra, India using remote sensing and GIS techniques. Sustain Water Resour Manag 4:965–979. https://doi.org/10.1007/s40899-017-0193-5

Pande CB, Moharir KN, Singh SK et al (2020) An integrated approach to delineate the groundwater potential zones in Devdari watershed area of Akola district, Maharashtra, Central India. Environ Dev Sustain 22:4867–4887. https://doi.org/10.1007/s10668-019-00409-1

Pande CB, Moharir KN, Khadri S (2021a) Watershed planning and development based on morphometric analysis and remote sensing and GIS techniques: a case study of semi-arid watershed in Maharashtra, India. In: Pande CB, Moharir KN (eds) Groundwater resources development and planning in the semi-arid region. Springer, Cham. https://doi.org/10.1007/978-3-030-68124-1_1

Pande C, Moharir K, Pande R (2021b) Assessment of morphometric and hypsometric study for watershed development using spatial technology–a case study of Wardha river basin in Maharashtra, India. Int J River Basin Manage 19(1):43–11

Prabhakaran A, Jawahar Raj N (2018) Drainage morphometric analysis for assessing form and processes of the watersheds of Pachamalai hills and its adjoinings, Central Tamil Nadu. India Appl Water Sci 8(1):1–19. https://doi.org/10.1007/s13201-018-0646-5

Schumm SA (1956) Evolution of drainage systems and slopes in badlands at perth amboy, New Jersey. Geol Soc Am Bull 67:597–646

https://doi.org/10.1130/0016-7606(1956)67[597:eodsas]2.0.co;2

Sreekumar KJBS, Aslam A (2016) Implication of drainage basin parameters of a tropical river basin of South India. Appl Water Sci, 67–75. https://doi.org/10.1007/s13201-014-0212-8

Strahler A (1957) Quantitative analysis of watershed geomorphology. Trans Am Geophys Union 38:913–920

https://doi.org/10.1029/TR038i006p00913

Sukristiyanti S, Maria R, Lestiana H (2018) Watershed-based Morphometric Analysis: A Review. In: IOP Conference Series: Earth and Environmental Science, 118(1). https://doi.org/10.1088/1755-1315/118/1/012028

Thomas J, Joseph S, Thrivikramaji KP (2010) Morphometric aspects of a small tropical mountain river system, the southern Western Ghats. India Int J Digit Earth 3(2):135–156. https://doi.org/10.1080/17538940903464370

Surface Water Quality Forecasting Using Machine Learning Approach

Ayushi Jha, Manojit Chowdhury, and Ajay N. Satpute

Abstract Surface water resources play a crucial role in drinking, industrial and agricultural domains. Since anthropogenic and environmental pollution sources affect surface water by threatening its ability to be used for industrial, agricultural, drinking, and other purposes. Estimating the contamination level of existing surface water is critical for effective water quality management. Evaluation of suitability of surface water for various purposes is an important task to establish the proper remedies or safety measures if it is unhealthy. However, conventional method of water quality (WQ) monitoring is generally expensive and tedious. Machine learning (ML) approach can be deployed to accurately assess the surface water quality and forecasting the suitability of water for various applications. In recent years, ML models have been extensively used in predicting the surface Water Quality Index (WQI) that depends upon the physical and chemical parameters of the water such as pH, turbidity, dissolved oxygen, total dissolved solid, biochemical oxygen demand, and other pollutants concentrations. A wide number of ML models such as linear regression (LR), decision tree, random forest (RF), naive bayes, k-nearest neighbour (KNN), support vector machine (SVM) and artificial neural network (ANN)-based algorithms have been used to predict and forecast surface water quality (SWQ). Analysing the efficiency statistics (R^2 and RMSE) can provide insight into the predictive performance of ML models. According to the comprehensive analysis, ML may be successfully implemented to water quality monitoring and forecasting from natural environmental processes and human impacts on an ecosystem.

A. Jha
Department of Soil and Water Engineering, Punjab Agricultural University, Ludhiana, Punjab 141004, India
e-mail: ayushijhagsm@gmail.com

M. Chowdhury (✉) · A. N. Satpute
Division of Agricultural Engineering, ICAR-Indian Agricultural Research Institute, New Delhi 110012, India
e-mail: manojitchowdhury13@gmail.com

A. N. Satpute
e-mail: ajaysatpute48@gmail.com

C. B. Pande et al. (eds.), *Surface and Groundwater Resources Development and Management in Semi-arid Region*, Springer Hydrogeology,
https://doi.org/10.1007/978-3-031-29394-8_16

Keywords Surface water quality · Water quality monitoring · Water quality forecasting · Machine learning · Water quality index

1 Introduction

Water is one of the most important resources for sustaining life on earth, essential for both human and the existence of the majority of other creatures (Aldhyani et al. 2020). There are water resources covering over 70% of the earth's surface (Mishra and Dubey 2015). Surface water and groundwater are the two different categories of water resources. Surface water can be found in places like rivers, lakes, reservoirs, and coastal areas (Mustafa et al. 2017). 65% of the water utilised in agriculture comes from rivers, with the remaining 35% going toward drinking, industrial purposes, and other human requirements. Surface water quality is affected by human factors such sewage, urbanisation, agricultural waste, and industrial pollution (Bhatti et al. 2019). Living things require water of a certain quality to survive. Only a certain limit of pollution level can be tolerated by aquatic creatures, other animals and human beings. It is crucial to monitor WQ to maintain a consistent and secure water supply (Hassan et al. 2021).

Increasing population and climate change have made water quality assessment challenging. Water scarcity is a problem brought on by population increase and inadequate infrastructure. The industrial revolution, extensive use of pesticides and fertilisers have had a negative impact on water quality ecosystems. People all throughout the world are impacted by the rise in global water usage that coincides with population growth. Individuals are forced to drink polluted water, which can cause several ailments for example cholera, diarrhoea, dysentery, and hepatitis (Kalaivanan and Vellingiri 2022).

The conventional approach of forecasting water quality involves manually gathering raw data at regular intervals and analysing it in laboratory. This method results in a tedious and perhaps risky process for creating water policy. The previous methods of data collection produced an imbalanced and noisy dataset. Water condition is determined by its physical, chemical, and biological characteristics (Bordalo et al. 2006). SWQ can be determined by a number of parameters, including electrical conductivity, dissolved oxygen, total dissolved solids, chemical oxygen demand, biological oxygen demand, turbidity, temperature, and pH (Tchobanoglous and Schroeder 1985; Nikoo et al. 2013). Another frequently used indicator of water quality that is necessary for time-consuming calculations involving large numbers of data and mathematical formulas is the water quality index (WQI). Based on the criteria defined by regulatory bodies in the study sector, WQI classifies water qualities into excellent, poor, and worst (Fernández et al. 2004).

Innovative methods for assessing and predicting WQ are essential. Predicting water quality trends requires an understanding of the temporal aspect of predicting WQ changes. In order to monitor and evaluate the WQ around the world, numerous strategies have been implemented, including the multivariate statistical method

(Alam et al. 2021), fuzzy inference (Oladipo et al. 2021), and the WQI (Wang et al. 2019). To forecast changes in water quality, a particular model variation performs better than a single model. A variety of methods are being suggested for forecasting and simulating water quality. Predictive algorithms, visual modelling, algorithm analysis, and statistical methods are frequently employed. To ascertain the correlation and association between various water quality indices, multivariate statistical methods were applied. Regression analysis, multivariate interpolation, and geostatistical methods were applied to transitional probability (Nair and Vijaya 2022).

Currently machine learning (ML) techniques are being used for rapid and precise prediction of SWQ based on training and testing of large datasets. The models for forecasting quality of models are predominantly capable of monitoring contaminants in surface water. Mechanism-oriented and no-mechanism-oriented models are used to model and predict WQ. The mechanism-oriented model is complex and uses cutting-edge system structure data to predict water quality; it is acknowledged as a multipurpose model that can be used to any aquatic body. Different ML models such as gradient boosting, decision tree, regression tree, random forest, group method of data handling (GMDH), support vector machine (SVM) and ANN have been used to forecast the surface WQ. The ML algorithms predict the long-term evolution of the WQ (Chen et al. 2018). Forecasting of WQ is essential for ecological sustainability, environmental monitoring, and human and animal health (Fijani et al. 2019). The purpose of this chapter is to provide an overview of the ML based approaches available for predicting and forecasting the surface water quality.

2 Parameters Associated to Surface Water Quality Estimation

WQ parameters are split into three categories such as physical, chemical and biological (Gray 2008; Spellman 2017), given in Table 1.

2.1 Physical Parameters

There are number of physical parameters that defines the quality of water such as:

2.1.1 Turbidity

Turbidity can be expressed as cloudiness or haziness of water. It is an assessment of capability of light to pass through water. Suspended material in water causes turbidity (Alley 2007). Higher turbidity elevates water temperatures because more

Table 1 Surface water quality indicating parameters

SWQ parameters		
Physical parameters	Chemical parameters	Biological parameters
TS	BOD	Bacteria
EC	COD	Viruses
Turbidity	DO	Protozoa
Temperature	pH	Algae
Taste and odor	Acidity	
	Alkalinity	
	Fluoride	
	Chloride	
	Hardness	
	Sulphate	
	Toxic organic and inorganic substance	
	Iron and manganese	

sun heat is absorbed by suspended particles, lowering the amount of food accessible. Because of this, heated water contains less oxygen than cold water (Cole et al. 2000). Nephelometric turbidimeter is used to measure the turbidity of water which represents turbidity in NTU or TU unit.

2.1.2 Temperature

Temperature influences chemical reactions, smells, solubility, viscosity, and taste of water. Temperature has an influential effect on the sedimentation and chlorination process, and BOD. It also impacts on the biosorption of water's dissolved heavy metals. Water tastes best when water temperature is in between 10 and 15 °C (APHA 2005; Davis 2010).

2.1.3 Taste and Odor

Foreign substances including organic and inorganic molecules, soluble gases can have an impact on the taste and odour of water. These materials may originate from residential, agricultural, or natural sources. The quantitative value of odor or taste is calculated by diluting a known volume of surface water sample with an odor free distilled water sample to prepare a 200 ml of total volume (Tchobanoglous et al. 1985). The odour or taste unit is denoted by the following expression:

$$\text{TON or TNN} = \frac{(\text{Sample A} + \text{Sample B})}{\text{Sample A}}$$

where;

TON and TNN are the threshold odor and taste number respectively.

Sample A is volume of the surface water sample.

Sample B is the volume of distilled water added.

2.1.4 Total Solid (TS)

Solids exist in water in two forms: dissolved and suspended. A glass fibre filter is used to distinguish the solids present in the surface water sample by passing the water sample through the filter. The surface of the filter retains the suspended solids while dissolved solids travel through the filter (APHA 2005). Total dissolved solid (TDS) is obtained when the filtered portion of surface water sample is stored in a dish and evaporated. Total solid present in the water sample is the TDS and total suspended solid (TSS). The procedure for acquiring TS is depicted in Fig. 1.

2.1.5 Electrical Conductivity (EC)

EC measures the capacity of a water sample to transmit electric current. Conductivity rises as ion concentration increases due to electrical current is transmitted by ions in solution. Therefore, it is one of the key variables used to determine irrigation water suitability. EC of pure water is always less than polluted water. The electrometric method is the most used way of measuring EC (APHA 2005).

2.2 Chemical Parameters

Important Chemical parameters representing surface water quality are discussed below.

2.2.1 pH

pH is among the most significant aspects of WQ. It is a dimensionless number and expressed as the hydrogen ion concentration's negative logarithm determining whether a solution is more acidic or basic. H^+ ions are more prevalent in acidic water, while basic water contains more OH^- ions (Alley 2007; Spellman 2017). The pH scale which ranges from 0 to 14, while 7 signifying neutral value. Acidic solutions

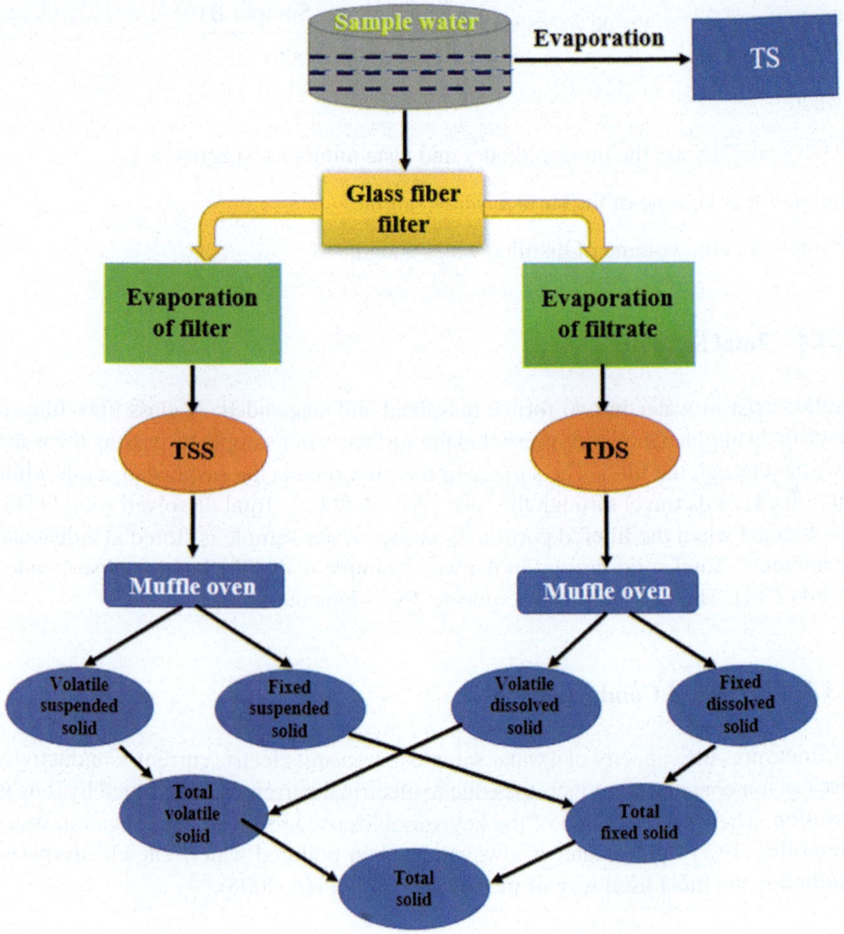

Fig. 1 Flow diagram of obtaining TS

have a pH under 7, while a pH more than 7 indicates a basic solution. Safest pH range for daily household purpose is 6.5–8.5 (APHA 2005). pH is measured with pH meter.

2.2.2 Dissolved Oxygen (DO)

DO is a major aspect of WQ estimation of surface water. It is used to represent the level of water contamination. More DO in water represents better quality of water. Oxygen is somewhat soluble in water and extremely temperature sensitive. The amount of oxygen dissolved in the surface water varies according on water temperature, pressure, and salinity. Laboratory based Winkler titration method is mostly used to determine DO (APHA 2005).

2.2.3 Biochemical Oxygen Demand (BOD)

BOD is a major parameter to represent the quality of water. Bacteria and other microbes feed on organic material in surface water. They consume oxygen as they decompose organic material. Energy released from the conversion of organic compounds to CO_2 and H_2O is utilised by the microorganisms for their growth and reproduction.

This process needs the dissolved oxygen of water to decompose the organic matter. This oxygen requirement is termed as BOD. It is the measure of pollution level of surface water as highly polluted water is having higher BOD. It is measured in the laboratory by titration process (Tchobanoglous et al. 2003; APHA 2005).

2.2.4 Chemical Oxygen Demand (COD)

COD is another essential parameter that represent the quality surface water. COD can be defined as quantity of oxygen needed to dissolve the organic matter in water. It is a chemical analysis that makes utilisation of heat, potassic dichromate, and sulfuric acid to obtain the outcome in less than two hours. The value of COD is always greater than BOD.

2.2.5 Acidity and Alkalinity

Acidity and alkalinity are another two parameters that give the idea about the SWQ. Acidity is a measurement of the presence of acids in the surface water from mineral acids, CO_2 and hydrolysed salts. Acidity level in surface water is measured by titrating a sample of the surface water with 0.02 N sodium hydroxide and phenolphthalein as an indicator in laboratory (APHA 2005).

The ability of water to neutralise acids, or its alkalinity, is determined by the sum of all titratable bases. The major contributors of alkalinity of water are the hydroxide (OH^-), bicarbonate (HCO^{3-}) and carbonate ($CO_3{}^{2-}$) ions, or a combination of two of these ions. Alkalinity is measured by titration process with 0.02 N H_2SO_4 solution and methyl orange or phenolphthalein (APHA 2005).

2.2.6 Chloride and Fluoride

Chloride is a naturally occurring mineral that can be found in surface water such as lakes and streams. However, relatively high chloride concentrations in freshwater may be a sign of wastewater contamination. Small amount of chloride content does not affect the water quality as it is required for cell functioning of animals and plants. Chloride is dangerous for human being and animals when it is consumed as sodium chloride in large amount (Chatterjee 2001).

Fluoride is also required in moderate amount in surface water to ensure good dental health of human being. Fluoride in excess amount causes discoloured teeth, a disease known as fluorosis so it should be managed in the surface water (Tchobanoglouset al. 1985; Davis 2010).

2.3 *Biological Parameters*

The biological parameters are the useful indicator of SWQ as the absence or presence of the parameters defines the WQ. Bacteria, viruses, algae, protozoa etc. are the biological parameters of water quality indication. Microorganisms can be found all around nature. The intestinal system of humans contains a normal population of germs, the majority of which are coliform bacteria. Although surface water is a storehouse to numerous bacteria per millilitre, but major population of them are harmless. The presence of bacteria in surface water is dangerous when it contains harmful bacteria and viruses from persons affected with diseases (APHA 2005; Abbas et al. 2014).

Traditional method of surface water quality estimation is a tedious and laborious process as it involves the laboratory-based analysis of the individual foresaid parameters. Furthermore, the traditional method cannot predict and forecast the SWQ over time. Due to the limitations of the traditional method of water quality estimation, ML based approach can be taken into consideration. ML based approach can rapidly and accurately forecast the surface water quality. A comprehensive overview of the ML based approaches has been discussed below.

3 Conspectus of ML

ML is a sub-part of artificial intelligence to evaluate data and look for patterns that could be used to forecast new information (Zhu et al. 2022; Alamet al. 2021; Oladipoet al. 2021). Machine learning is a novel approach to data analysis and processing. This approach has found broad application in numerous fields due to its high accuracy, adaptable customisation, and easy expansion (Wang et al. 2019). Machine learning allows managing complex and nonlinear datasets very easily, facilitating the identification of underlying mechanisms (Wang et al. 2019). Machine learning approach has proven its ability as a tool that has remarkable adaptability and excellent performance in environmental science as well as engineering areas (Simoeset al. 2008). Prior to application of machine learning, acquisition of data,

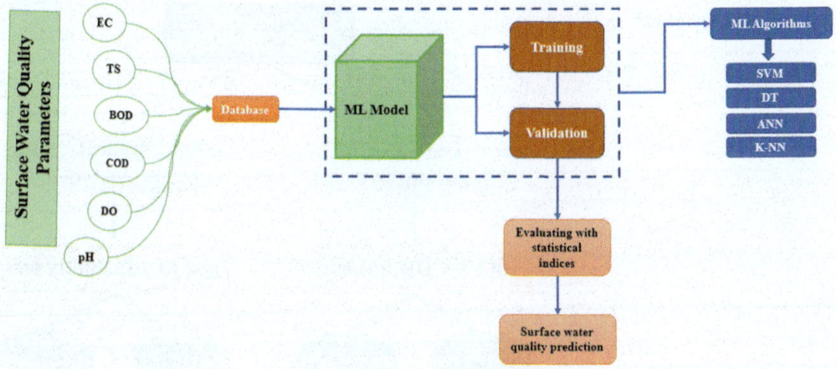

Fig. 2 Framework of ML in surface water quality forecasting

selection of appropriate algorithm, training of model, testing of model with predefined set of data needs to be conducted. Among all these, selection of appropriate algorithm is one of the crucial tasks (Ma et al. 2014). Application of machine learning system has a rich collection of algorithms that can serve several purposes and would analyse various water quality parameters as shown in Fig. 2. Machine learning algorithms mainly includes LR, ANN, DT, SVM, naive Bayes, KNN, RF.

3.1 *Linear Regression*

Linear regression executes the operation based on regression operation. The regression operation consists of dependent and independent variables, in which independent variables is used to determine the dependent variable to develop a model. It provides a relationship and determines how variables are related to each other. The controlling parameters in the regression models are the type and number of independent variables, type of relationship between the variables and other factors. In the figure below, dependent variable represents salinity of water and independent variable represents hardness of water. So, with increase in hardness of water, salinity increases. Thus, there exist a linear relationship between hardness and salinity which can be presented by linear regression model. The regression line is the best fit line for the model.

3.2 *Decision Tree (DT)*

Decision tree is a algorithm is used widely in developing machine learning model. Decision tree algorithm can be used solve both the regression and classification requiring queries, but classification is more frequently practices practically. It has tree structure classifier with three nodes, starting with initial node called root node.

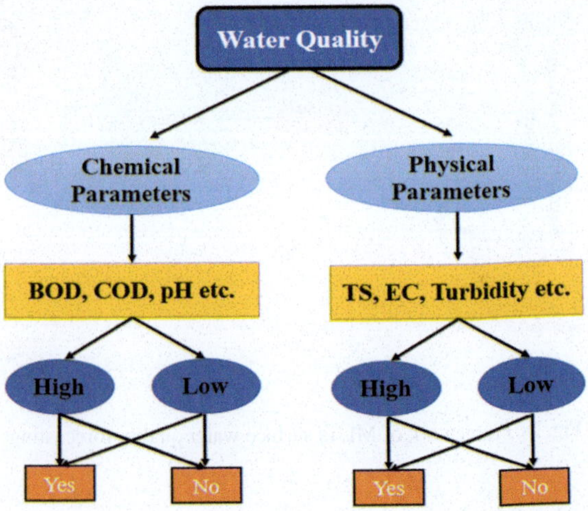

Fig. 3 Structure of DT

It consists of root node and branch nodes, where root node represents the complete sample. The root nodes are further divided into branches that are branch nodes or interior nodes. The branch nodes participate in for decision making and do the analysis using the codes embedded in the nodes, while the nodes as branches represents the nature of a data collection process. In the end of the process the output is reflected by the leaf nodes. In dealing with issues involving decisions, this algorithm might be helpful. A decision tree structure is shown in Fig. 3.

3.3 K-Nearest Neighbour (KNN)

KNN is considered amongst the simplest ML algorithms that has been used widely for refined analysis. The k-nearest neighbour algorithm is based on an assumption that is the present data case and new case is comparable and then when new data is fed then that data point is placed to the most similar category, that too rapidly and precisely into most suitable category. The classification of new data point is done on the basis of similarity, once all the existing data has been stored. k-nearest neighbour can also be used for regression problems, but mostly employed for classification problems. This algorithm stores the results during the train phase and learns from it later, so it is also called as lazy learner. The data set is the k number nearest neighbour chosen on the basis of Euclidean distance. The category with the highest number of neighbours receives the additional data points, and then the model is ready. Figure 4 depicts the KNN structure.

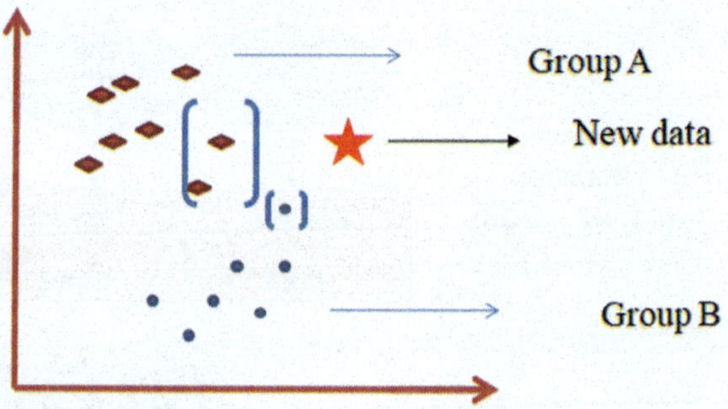

Fig. 4 KNN structure

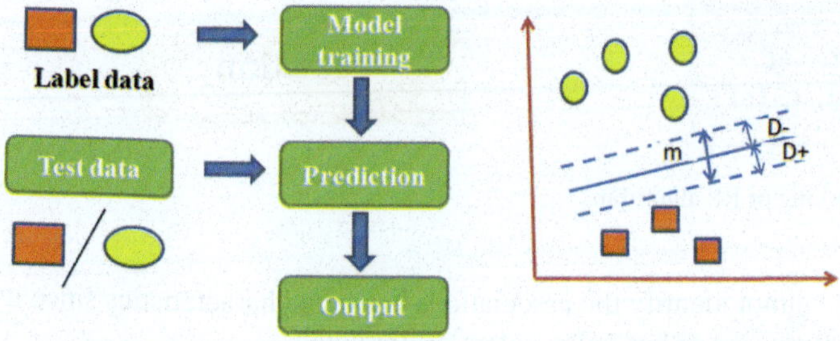

Fig. 5 Framework of SVM

3.4 *Support Vector Machine (SVM)*

SVM is an algorithm that can be utilized in solving classification problems and also used for regression problems. The decision making through SVM algorithm is based on decision line, this divides the n number of data into classes, and thus allows rapid classification of the new data points that are fed further. The decision boundary that is established is named as hyper plane. This hyper plane is created with the help of extreme vectors or points that is selected by SVM. The SVM technique is based on utilises support vectors that represent these extreme instances. The Fig. 5 below shows how hyperplane separates the data points into the classes.

3.5 *Naive Bayes*

The basis of this algorithm is the Bayes theorem, which is used to resolve classification issues. This algorithm categorizes the text categorization having a large training set. It helps in developing one the accurate working machine learning algorithm that would produce outputs quickly and accurately. As the analysis is based on the statistical theorem, so it can work as a probabilistic classifier and make the predictions on the basis of likelihood of occurrence of any event or say object. Naive Bayes

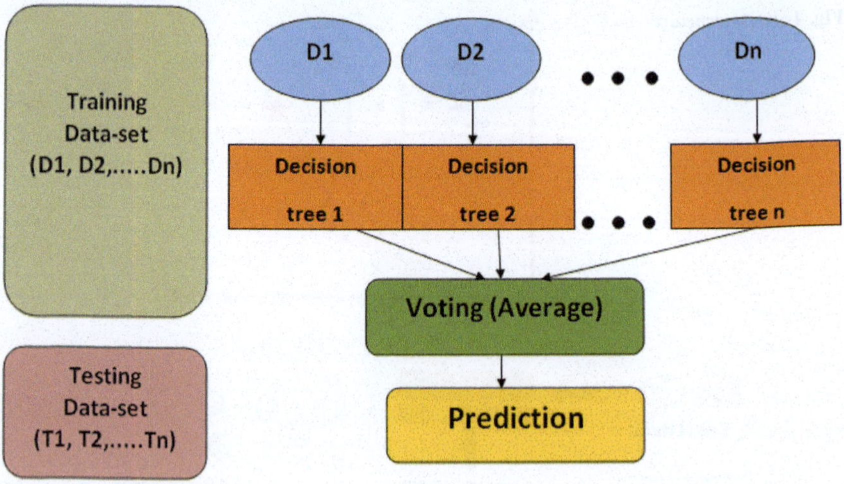

Fig. 6 Working of RF algorithm

algorithm cannot identify the associations between characteristics since it considers that all features are either independent or unrelated.

3.6 Random Forest (RF)

Random Forest can be applied to machine learning problems involving both classification and regression. The idea behind this algorithm ensemble learning that means learning altogether to produce desirable output, as it is a combination of various decision tree models. In this algorithm output of various classifiers are integrated, then difficult issues are addressed and model performance is improved. Random forest utilizes the result forecasted from different decision tree rather than relying on a single DT and forecasts the result on the basis of majority of prediction votes. Working of RF algorithm is shown in Fig. 6.

3.7 Artificial Neural Network (ANN)

ANN is an algorithm inspired by brain of human being and is used to forecast problems and model the complicated patterns feasibly. ANN is the result of efforts that has been made to simulate the human brain and its functions. The operations of ANN and neural networks present in human brains are very similar, although they are not exactly the same. ANN algorithm accepts only structured and numeric data. The architecture of ANN consists of inner layer, outer layer and a hidden layer (can be

Fig. 7 Architecture of ANN

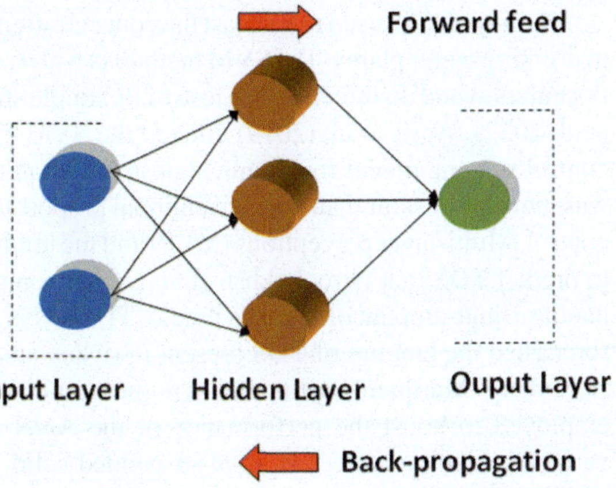

more than one). The hidden layer identifies some of the most important patterns in the inputs and sends them to the following layer for additional analysis. By selecting only, the most crucial information from the inputs and ignoring the redundant information, hidden layer speeds up and enhances the network's performance. Figure 7 represents the architecture of ANN.

4 An Overview on Application of ML for SWQ Analysis

This section offers a range of ML-based strategies for analysing WQ. Many researchers have attempted to evaluate water quality parameters by means of machine learning approach by using several algorithms available, integrating real-time observation, forecasting, tracking of pollutant sources, determination of pollutant concentrations, allocation of water resources, and improvement of water treatment technologies (Olyaie et al. 2017; Zhu et al. 2022). The main cause of the declining water quality in metropolitan areas is now human-generated municipal and industrial wastewater (Mohammadpour et al. 2015) and WQ is depleted is assessed by parameters mainly BOD, COD, pH, TDS etc. In surface water quality research, the use of machine learning, have become prominent topic (Tung and Yaseen 2020; Sharma et al. 2021). A study was performed by Ahmed and Shah (2017) and Khaled et al. (2018) to predict BOD using ANFIS (adaptive-network-based fuzzy inference system) and the study recommended that ANFIS technique could be effectively applied to develop models for predicting the river WQ (Deng et al. 2015; Khaled et al. 2018). SVM model have also given success in study of WQ parameters such as BOD (Deng et al. 2015; Noori et al. 2015), DO (Liu and Lu 2014; Li et al. 2018), COD (Kisi and Parmar 2016), Total Phosphate, and Total Nitrogen (Liu and Lu 2014). Arnon et al. (2019) utilized SVM to build a novel method for pollutant prediction under uncertain circumstances. The datasets acquired by the SVM, the detection rates were high and the error rates were low.

ML may also be used to forecast the concentrations of coagulants and disinfectants in drinking water plants. The SVM method is widely used in constructing schemes for flocculation and disinfection because of its simple structure and high robustness (Zhu et al. 2022). Wang et al. (2019) utilized the SVM model and suggested a predictive control scheme model for chemical dose based on the residual free chlorine, which was more successful than the conventional proportional-integral-derivative feedback control. Multi-layer perceptions (MLP) and the random forest were two ANN models to predict TDS that Niroobakhsh et al. (2012) compared (RF). The RF findings can handle a huge amount of data and predict TDS levels with accuracy. Tarke et al. (2016) forecasted the amount of TDS present in rivers, an ANN model was employed. The Levenberg–Marquardt optimization process and the back-propagation technique are employed to boost the performance of the ANN model. Using four independent variables—temperature, pH, total suspended solid, and total suspended—the MLR model is used to forecast both biochemical oxygen demand (BOD) and chemical oxygen demand (Zare 2014). With a correlation coefficient value of 0.5, the system promptly predicts BOD with a relatively good outcome. The Mathura River's DO concentration was predicted using a two-layer ANN model (Wang et al. 2021a, b), and the results of the experiment supported this claim. Different neural network types are examined by Maxwell (2015) for stream water temperature prediction. To anticipate the chemical oxygen requirement, Abyaneh (2014) has used machine learning techniques like ANN and regression (COD). Sakizadeh (2016) estimated the water quality index using ANN and Bayesian regularization (WQI). However, the prediction and categorization of water quality were done using the radial-basis-function (RBF), a form of ANN model (Yesilanacar et al. 2008; Bouamar and Ladjal 2008).

Defining the process of pollution transmission and assessing the components of water quality are two ways that are taken into consideration while evaluating the quality of water (Kashefipour 2002; Kashefipour and Falconer 2002; Nasseri and Kashefipour 2012; Qishlaqi et al. 2017). Emamgholizadeh et al. (2014) utilised multilayer perceptron (MLP), radial basis network (RBF) and an ANFIS for WQ components of Karoon River. Comparing the DT model to the Kappa statistical model, it was found that the Decision Tree model was more accurate and effective. The effectiveness of three ML classifiers—namely, PNN, SVM, and KNN—in classifying WQ metrics was assessed by Modaresi and Araghinejad (2014). As a result, SVM performed superbly in comparison to the other two classifiers. Danadeset al. (2016) assessed how well SVM and KNN models performed in categorising WQ indices. For the classification of five water quality metrics, Radhakrishnan and Pillai (2020) examined the effectiveness of three ML algorithms (SVM, DT, and Nave Bayes). In this study, categorization factors like as pH, DO, BOD5, and EC were taken into account. The classifiers performance results revealed the effectiveness machine learning algorithm.

To improve model performance, many ML models are increasingly being combined with big data and some other technologies such as IoT. The accuracy is related to model computation time, determine how well a prediction of water quality performs. The research found that the time required by ML algorithms to estimate SWQ is not becoming any shorter (Kalaivanan and Vellingiri 2022).

5 Comparison of Different Water Quality Prediction Models

A major flaw in the existing water quality prediction methods or models is their incompetence in considering the cross effects between two explanatory variables. Different variables are interdependent and hence there is a cross-correlation between them that should be given due attention (Hwang et al. 2016; Lintern et al. 2017). ML-based models take into account all cross-correlations between input parameters and therefore their prediction accuracy is improved. It is the main advantage over traditional statistical models (Li et al. 2015; Wang et al. 2021a, b).

Different ML algorithms have been applied to predict surface water quality. It includes regression tree analysis, decision trees, neural networks, byes function, etc. There exists a non-linear and very complex relationship between the different parameters water quality. The ML models like neural networks and regression tree have been found satisfactory in predicting the effect of land use characteristics on surface water quality (Castrillo and García 2020; Sajedi-Hosseini et al. 2018; Xu et al. 2020). The ML model developed and tested on a new dataset will help in predicting the surface water quality under different scenario of future land use. This will aid in policy decision-making (Schreiber et al. 2019).

Among all the ML algorithm, SVM is an extensively used to develop various models. It has been also used to predict the quality of surface water. In the studies carried out to predict surface water quality using ML-based models, it was found that SVM is the best model. In the study conducted by Babbar and Babbar (2017), SVM and Decision tree were the best classifiers. The error rate of 0% which is the lowest among all the used models, made them superior to ANN, K-NN, and Naive Bayes classifiers in classifying water quality. It was also found that if the data provided to models accurately represent the domain knowledge, then ML models can determine the WQ. The SVM model has better forecasting accuracy of WQ than the ANN model mainly due to its ability to optimize a lesser number of parameters. This structural risk minimization principle helps the SVM model to avoid the overtraining of the data and hence aid to have a better generalization ability (Giri and Singh 2014; Sengorur et al. 2015; Kalaivanan and Vellingiri 2022; Malek et al. 2022).

ANN is a very popular data-driven ML model. It offers solutions to both the linear and non-linear association between variables. Since the data associated with the water quality is highly nonlinear ANN model suits more than any other model. The output of the ANN model is affected if the initial weights assigned to all the input parameters have a similar value. The performance of the ANN models is impeded if the training data are imbalanced. Even though ANN models are widely used, they possess certain drawbacks. The main drawback arises due to input dataset size. If the size of the input dataset is small then the prediction power of the ANN models is lowered. This problem is further exaggerated if the data used for testing the model don't lie in the range of the training dataset (Zare 2014; Khosravi et al. 2018; Asadollah et al. 2021).

Traditional ML models like the Decision Tree model are often used to predict surface water quality and have been found to perform well. Among the decision tree models, there are different models that can be developed which include Gradient Boosting (GB), single decision tree, Random Forest (RF), Logistic Model Tree (LMT), Ordinary Decision Tree (ODT), etc. Here the ensemble models such as RF and GB are always superior to the models based on a single decision tree. This is attributed to the ability of these models to manage both regular attributes and data. Also, they are very efficient in the predictions and their performance is not influenced by the missing values in the data. (Gakii and Jepkoech 2019; Bui et al. 2020; Lu and Ma 2020; Wang et al. 2021a, b). The decision tree-based models have more calculation speed and are more suited to prediction of surface water quality at short-term scale. These models produce good results when continuous input datasets are used (Jeihouni et al. 2020).

Every ML model has its benefits and shortcomings for the prediction of surface water quality. Hence it is better to combine the benefits of such models to have better accuracy. So, the hybrid ML model approach is the best way ahead.

6 Limitations of ML Based Approach

Machine learning is a powerful tool which because it can be applied to various purposes in the water resources area such as optimizing the allocation of water resources, predicting water quality, managing the recurrent shortages of water resources, etc. Despite this, there are still numerous challenges that hinder the use of ML algorithms in this area to assess water quality. Those challenges are described below.

6.1 Availability of the Data

A large amount of high-quality data is a prerequisite for the use of ML algorithms for any specific purpose. The ML models are trained based on the data that was obtained. These models then detect the trend or pattern in the data provided to them. Such a pattern or trend is useful in predicting any variable. The models are then validated and when certain accuracy is obtained in the results then only these models are used for real field application. The cost involved in the data acquisition limits the use of ML algorithms in the water resources domain. The technological limitations in this case further aggravate the problem (Wang et al. 2021a, b; Zhu et al. 2022). Even if the data is available, it will always not be in the required resolution (Wang et al. 2021a, b).

6.2 *Limited Applicability*

Models developed using ML algorithms are often empirical in nature. They are suitable only for those areas for which they are developed. The conditions in real-world water bodies for which the water quality predictions have to be done are very different and extremely complex. Further, these conditions vary in every water body. So, the model developed can be applied to the specific system. This factor is a major obstacle in the wide-scale applicability of ML-based approaches. To overcome this problem, models are required to be calibrated every time they are put in use in an area different than the one for which they were developed. This process will additional expenditure of resources (Zhu et al. 2022). Therefore, it is important that one should avoid black-box solutions when applying a ML approach, as it doesn't provide process-based scientific insight (Schäfer et al. 2022).

6.3 *Selection of Appropriate Input Parameters*

The very first step in developing any ML based model is the selection appropriate input parameters. The water quality is influenced by number of factors which are presented in Table 1. It is imperative that out of all those factors only sufficient parameters which have enough underlying information to predict water quality are selected. This is very important because too many input variables will lead to redundancy and too less will cause inaccuracies. It is very hectic and complex to choose the right set of input variables because the extent to which each input parameter influences the water quality also depends upon the other variable. The proper selection of the variables can improve the accuracy of the model because improper selection leads to an undesirable impact on model performance. (Khoi et al. 2022). Also, every ML model will respond to different input variables and input data patterns in a different way. So, it is very essential to choose the input variables that enhance the performance of a particular ML model (Hussain and Khan 2020). There are various techniques available to assess and select the input parameter combinations, which mainly include partial autocorrelation function, autocorrelation function, correlation coefficient, and cross-correlation function (Asadollah et al. 2021).

6.4 *Requirement of Professional Background*

Since the ML based approach deals with the large-scale data, sophisticated models, statistical data analysis and AI, the approach demands the technical knowledge. The implementation of ML algorithms in applications demands professional background knowledge (Zhu et al. 2022). Otherwise, the development and application of such models will be limited.

7 Conclusion

Water is the most significant and crucial resource of the nature. Not only the quantity but also the quality of the water influences every aspect of its use. There is a certain limit to the level of deterioration in the water quality which can be tolerated depending upon the purpose of use. Exceeding these boundaries is also harmful to the organisms surviving in water. With the rapid growth of industrialization, the discharge of pollutant in the surface water bodies is increasing which further deteriorates the water quality. To mitigate this pollutant load it is essential to determine source areas and corresponding quantity. Therefore, it is imperative to continuously monitor the surface water quality. Traditionally the surface water quality is assessed by estimating different parameters (Table 1) from the water sample in the laboratory. But there are many constraints to use laboratory method which includes large time requirement, unable to predict future water quality, etc. Therefore, new methods should be developed to assess and monitor the surface water quality.

The machine learning offers a solution to the development of the model for monitoring and prediction of surface water quality. The ML based models can be trained and tested on the past data of surface water quality available. Different ML algorithms such as Linear regression, ANN, SVM, KNN, DT, RF, Naive Bayes, etc., can be used to develop a model that will forecast the surface water quality under various future scenarios. The ML-based models detect the trend in the data and then predict the output based on the trend. It offers many benefits over traditional approaches such as rapid and accurate estimation of multiple parameters, cost-effectiveness, non-laborious, and ability of surface water quality forecasting over time. The performance of these models is hampered by the availability of the large datasets, their empirical nature, requirement of the professional knowledge and numerous input variables. Nevertheless, ML based models have proved very efficient in predicting the surface water quality.

Future Scope

Now a days ML models are being integrated with IoT and big data to enhance their performance. The performance of ML models in surface water quality prediction is dependent on the accuracy of the detection. In the future following research gaps in the ML-based approach for surface water quality prediction can be explored.

1. New hybrid machine learning models should be introduced that can accurately predict surface water quality.
2. Optimization methods need to be improved by adding new algorithms to predict the surface water quality.
3. Advanced sensors have to be developed and used in surface water quality monitoring so that it will aid in the collection of more accurate data and facilitate ML based approach for the study.
4. The reliability and feasibility of the existing ML models need to be improved and more universal ML model should be constructed to suit the different requirements.

5. Multidisciplinary approach should be followed so that knowledge from different field can be integrated to develop the more sophisticated ML model could be developed.

With the advances in the machine learning studies, it is promising for future studies in surface water quality prediction.

References

Abbas SH, Ismail IM, Mostafa TM, Sulaymon AH (2014) Biosorption of heavy metals: a review. J Chem Sci Technol 3:74–102

Abyaneh HZ (2014) Evaluation of multivariate linear regression and artificial neural networks in prediction of water quality parameters. J Environ Health Sci Eng 12:1–8. https://doi.org/10.1186/2052-336X-12-40

Ahmed AM, Shah SMA (2017) Application of adaptive neuro-fuzzy inference system (ANFIS) to estimate the biochemical oxygen demand (BOD) of Surma River. J King Saud Univ Eng Sci 29(3):237–243

Alam R, Ahmed Z, Seefat SM, Nahin KTK (2021) Assessment of surface water quality around a landfill using multivariate statistical method, Sylhet, Bangladesh. Environ Nanotechnol Monit Manag 15:100422. https://doi.org/10.1016/j.enmm.2020.100422

Aldhyani TH, Al-Yaari M, Alkahtani H, Maashi M (2020) Water quality prediction using artificial intelligence algorithms. Appl Bionics Biomech 1–12. https://doi.org/10.1155/2020/6659314

Alley ER (2007) Water quality control handbook, vol 2. McGraw-Hill, New York

APHA (2005) Standard methods for the examination of water and wastewater, 21st edn. American Public Health Association, Washington, DC

Arnon TA, Ezra S, Fishbain B (2019) Water characterization and early contamination detection in highly varying stochastic background water, based on machine learning methodology for processing real-time UV-spectrophotometry. Water Res 155:333–342. https://doi.org/10.1016/j.watres.2019.02.027

Asadollah SBHS, Sharafati A, Motta D, Yaseen ZM (2021) River water quality index prediction and uncertainty analysis: a comparative study of machine learning models. J Environ Chem Eng 9(1):104599. https://doi.org/10.1016/j.jece.2020.104599

Babbar R, Babbar S (2017) Predicting river water quality index using data mining techniques. Environ Earth Sci 76(14):1–15. https://doi.org/10.1007/s12665-017-6845-9

Bhatti NB, Siyal AA, Qureshi AL, Bhatti IA (2019) Socio-economic impact assessment of small dams based on T-paired sample test using SPSS software. Civ Eng J 5(1):153–164. https://doi.org/10.28991/cej-2019-03091233

Bordalo AA, Teixeira R, Wiebe WJ (2006) A water quality index was applied to an international shared river basin: the case of the Douro River. Environ Manag 38(6):910–920. https://doi.org/10.1007/s00267-004-0037-6

Bouamar M, Ladjal M (2008) A comparative study of RBF neural network and SVM classification techniques performed on real data for drinking water quality. In: 2008 5th international multi-conference on systems, signals and devices. IEEE, pp 1–5. https://doi.org/10.1109/SSD.2008.4632856

Bui DT, Khosravi K, Tiefenbacher J, Nguyen H, Kazakis N (2020) Improving prediction of water quality indices using novel hybrid machine-learning algorithms. Sci Total Environ 721:137612. https://doi.org/10.1016/j.scitotenv.2020.137612

Castrillo M, García ÁL (2020) Estimation of high frequency nutrient concentrations from water quality surrogates using machine learning methods. Water Res 172:115490. https://doi.org/10.1016/j.watres.2020.115490

Chatterjee A (2001) Water supply waste disposal and environmental pollution engineering (including odour, noise and air pollution and its control), 7th edn. Khanna Publishers, Delhi

Chen S, Fang G, Huang X, Zhang Y (2018) Water quality prediction model of a water diversion project based on the improved artificial bee colony—backpropagation neural network. Water 10(6):806. https://doi.org/10.3390/w10060806

Cole S, Codling ID, Parr W, Zabel T, Nature E, Heritage SN (2000) Guidelines for managing water quality impacts within UK European Marine sites. Natura

Danades A, Pratama D, Anggraini D, Anggriani D (2016) Comparison of accuracy level K-nearest neighbor algorithm and support vector machine algorithm in classification water quality status. In: 2016 6th International Conference on System Engineering and Technology, IEEE, pp 137–141. https://doi.org/10.1109/ICSEngT.2016.7849638

Davis ML (2010) Water and wastewater engineering: design principles and practice. McGraw-Hill Education

Deng W, Wang G, Zhang X (2015) A novel hybrid water quality time series prediction method based on cloud model and fuzzy forecasting. Chemom Intell Lab Syst 149:39–49. https://doi.org/10.1016/j.chemolab.2015.09.017

Emamgholizadeh S, Kashi H, Marofpoor I, Zalaghi E (2014) Prediction of water quality parameters of Karoon River (Iran) by artificial intelligence-based models. Int J Environ Sci Technol 11(3):645–656. https://doi.org/10.1007/s13762-013-0378-x

Fernández N, Ramírez A, Solano F (2004) Physico-chemical water quality indices—a comparative review. Bistua: Revista de la Facultad de Ciencias Básicas, 2(1):19–30

Fijani E, Barzegar R, Deo R, Tziritis E, Skordas K (2019) Design and implementation of a hybrid model based on a two-layer decomposition method coupled with extreme learning machines to support real-time environmental monitoring of water quality parameters. Sci Total Environ 648:839–853. https://doi.org/10.1016/j.scitotenv.2018.08.221

Gakii C, Jepkoech J (2019) A classification model for water quality analysis using decision tree. Eur J Comput Sci Inf Technol 7:1–8

Giri A, Singh NB (2014) Comparison of artificial neural network algorithm for water quality prediction of river Ganga. Environ Res J 8(2):55–63. http://docsdrive.com/pdfs/medwelljournals/erj/2014/55-63.pdf

Gray NF (2008) Drinking water quality: problems and solutions, 2nd edn. Cambridge University Press, Cambridge

Hassan MM, Hassan MM, Akter L, Rahman MM, Zaman S, Hasib KM et al (2021) Efficient prediction of water quality index (WQI) using machine learning algorithms. Human Centric Intell Syst 1(3–4):86–97. https://doi.org/10.2991/hcis.k.211203.001

Hussain D, Khan AA (2020) Machine learning techniques for monthly river flow forecasting of Hunza River, Pakistan. Earth Sci Inform 13(3):939–949. https://doi.org/10.1007/s12145-020-00450-z

Hwang SA, Hwang SJ, Park SR, Lee SW (2016) Examining the relationships between watershed urban land use and stream water quality using linear and generalized additive models. Water 8(4):155. https://doi.org/10.3390/w8040155

Jeihouni M, Toomanian A, Mansourian A (2020) Decision tree-based data mining and rule induction for identifying high quality groundwater zones to water supply management: a novel hybrid use of data mining and GIS. Water Resour Manag 34(1):139–154. https://doi.org/10.1007/s11269-019-02447-w

Kalaivanan K, Vellingiri J (2022) Survival study on different water quality prediction methods using machine learning. Nat Environ Pollut Technol 21(3):1259–1267. https://doi.org/10.46488/NEPT.2022.v21i03.032

Kashefipour SM, Falconer RA (2002) Longitudinal dispersion coefficients in natural channels. Water Res 36(6):1596–1608. https://doi.org/10.1016/S0043-1354(01)00351-7

Kashefipour SM (2002) Modelling flow, water quality and sediment transport processes in reverine basins. Doctoral dissertation, Cardiff University

Khaled B, Abdellah A, Noureddine D, Salim H, Sabeha A (2018) Modelling of biochemical oxygen demand from limited water quality variable by ANFIS using two partition methods. Water Qual Res J 53(1):24–40. https://doi.org/10.2166/wqrj.2017.015

Khoi DN, Quan NT, Linh DQ, Nhi PTT, Thuy NTD (2022) Using machine learning models for predicting the water quality index in the La Buong River, Vietnam. Water 14(10):1552. https://doi.org/10.3390/w14101552

Khosravi K, Mao L, Kisi O, Yaseen ZM, Shahid S (2018) Quantifying hourly suspended sediment load using data mining models: case study of a glacierized Andean catchment in Chile. J Hydrol 567:165–179. https://doi.org/10.1016/j.jhydrol.2018.10.015

Kisi O, Parmar KS (2016) Application of least square support vector machine and multivariate adaptive regression spline models in long term prediction of river water pollution. J Hydrol 534:104–112. https://doi.org/10.1016/j.jhydrol.2015.12.014

Li Y, Li Y, Qureshi S, Kappas M, Hubacek K (2015) On the relationship between landscape ecological patterns and water quality across gradient zones of rapid urbanization in coastal China. Ecol Model 318:100–108. https://doi.org/10.1016/j.ecolmodel.2015.01.028

Li C, Li Z, Wu J, Zhu L, Yue J (2018) A hybrid model for dissolved oxygen prediction in aquaculture based on multi-scale features. Inf Process Agric 5(1):11–20. https://doi.org/10.1016/j.inpa.2017.11.002

Lintern A, Webb JA, Ryu D, Liu S, Bende-Michl U, Waters D, Leahy P, Wilson P, Western AW (2017) Key factors influencing differences in stream water quality across space. Wiley Interdiscip Rev Water 5(1):e1260. https://doi.org/10.1002/wat2.1260

Liu M, Lu J (2014) Support vector machine—an alternative to artificial neuron network for water quality forecasting in an agricultural nonpoint source polluted river? Environ Sci Pollut Res 21(18):11036–11053. https://doi.org/10.1007/s11356-014-3046

Lu H, Ma X (2020) Hybrid decision tree-based machine learning models for short-term water quality prediction. Chemosphere 249:126169. https://doi.org/10.1016/j.chemosphere.2020.126169

Ma C, Zhang HH, Wang X (2014) Machine learning for big data analytics in plants. Trends Plant Sci 19(12):798–808. https://doi.org/10.1016/j.tplants.2014.08.004

Malek NHA, Yaacob WFW, Nasir SAM (2022) Prediction of water quality classification using machine learning. Encyclopedia. https://encyclopedia.pub/entry/24076

Maxwell S (2015) One water: the need for more holistic thinking, analysis, and policymaking in water. J Am Water Works Ass 107(3):21–24. https://doi.org/10.5942/jawwa.2015.107.0048

Mishra RK, Dubey SC (2015) Freshwater availability and its global challenge. Int J Eng Sci Inven Res Dev 2(6):65–83

Modaresi F, Araghinejad S (2014) A comparative assessment of support vector machines, probabilistic neural networks, and K-nearest neighbor algorithms for water quality classification. Water Resour Manag 28(12):4095–4111. https://doi.org/10.1007/s11269-014-0730-z

Mohammadpour R, Shaharuddin S, Chang CK, Zakaria NA, Ghani AA, Chan NW (2015) Prediction of water quality index in constructed wetlands using support vector machine. Environ Sci Pollut Res 22(8):6208–6219. https://doi.org/10.1007/s11356-014-3806-7

Mustafa A, Sulaiman O, Shahooth S (2017) Application of QUAL2K for water quality modeling and management in the lower reach of the Diyala river. Iraqi J Civ Eng 11:66–80. https://doi.org/10.37650/ijce.2017.134910

Nair JP, Vijaya MS (2022) River water quality prediction and index classification using machine learning. J Phys Conf Ser IOP Publishing 2325:012011

Nasseri MM, Kashefipour SM (2012) Application of numerical modeling for solution of flow equations and estimation of water quality pollutants in rivers (Case study: Karkheh River), pp 51–60

Nikoo MR, Karimi A, Kerachian R, Poorsepahy-Samian H, Daneshmand F (2013) Rules for optimal operation of reservoir-river-groundwater systems considering water quality targets: Application of M5P model. Water Resour Manag 27(8):2771–2784. https://doi.org/10.1007/s11269-013-0314-3

Niroobakhsh M, Musavi-Jahromi SH, Manshouri M, Sedghi H (2012) Prediction of water quality parameter in Jajrood River basin: application of multi layer perceptron (MLP) perceptron and radial basis function networks of artificial neural networks (ANNs). Afr J Agric Res 7(29):4131–4139. https://doi.org/10.5897/AJAR11.1645

Noori R, Yeh HD, Abbasi M, Kachoosangi FT, Moazami S (2015) Uncertainty analysis of support vector machine for online prediction of five-day biochemical oxygen demand. J Hydrol 527:833–843. https://doi.org/10.1016/j.jhydrol.2015.05.046

Oladipo JO, Akinwumiju AS, Aboyeji OS, Adelodun AA (2021) Comparison between fuzzy logic and water quality index methods: a case of water quality assessment in Ikare community, Southwestern Nigeria. Environ Challenges 3:100038. https://doi.org/10.1016/j.envc.2021.100038

Olyaie E, Abyaneh HZ, Mehr AD (2017) A comparative analysis among computational intelligence techniques for dissolved oxygen prediction in Delaware River. Geosci Front 8(3):517–527. https://doi.org/10.1016/j.gsf.2016.04.007

Qishlaqi A, Kordian S, Parsaie A (2017) Hydrochemical evaluation of river water quality—a case study. Appl Water Sci 7(5):2337–2342. https://doi.org/10.1007/s13201-016-0409-0

Radhakrishnan N, Pillai AS (2020) Comparison of water quality classification models using machine learning. In: 2020 5th international conference on communication and electronics systems (ICCES). IEEE, pp 1183–1188. https://doi.org/10.1109/ICCES48766.2020.9137903

Sajedi-Hosseini F, Malekian A, Choubin B, Rahmati O, Cipullo S, Coulon F, Pradhan B (2018) A novel machine learning-based approach for the risk assessment of nitrate groundwater contamination. Sci Total Environ 644:954–962. https://doi.org/10.1016/j.scitotenv.2018.07.054

Sakizadeh M (2016) Artificial intelligence for the prediction of water quality index in groundwater systems. Model Earth Syst Environ 2(1):1–9. https://doi.org/10.1007/s40808-015-0063-9

Schäfer B, Beck C, Rhys H, Soteriou H, Jennings P, Beechey A, Heppell CM (2022) Machine learning approach towards explaining water quality dynamics in an urbanised river. Sci Rep 12(1):1–15. https://doi.org/10.1038/s41598-022-16342-9

Schreiber J, Jessulat M, Sick B (2019) Generative adversarial networks for operational scenario planning of renewable energy farms: a study on wind and photovoltaic. In: Artificial neural networks and machine learning—ICANN 2019: image processing, pp 550–564. https://doi.org/10.1007/978-3-030-30508-6_44

Sengorur B, Koklu R, Ates A (2015) Water quality assessment using artificial intelligence techniques: SOM and ANN—a case study of Melen River Turkey. Water Qual Expo Health 7(4):469–490. https://doi.org/10.1007/s12403-015-0163-9

Sharma N, Sharma R, Jindal N (2021) Machine learning and deep learning applications—a vision. Glob Transit Proc 2(1):24–28. https://doi.org/10.1016/j.gltp.2021.01.004

Simoes FDS, Moreira AB, Bisinoti MC, Gimenez SMN, Yabe MJS (2008) Water quality index as a simple indicator of aquaculture effects on aquatic bodies. Ecol Ind 8(5):476–484

Spellman FR (2017) The drinking water handbook, 3rd edn. CRC Press, Boca Raton

Tarke PD, Sarda PR, Sadgir PA (2016) Performance of ANNs for prediction of TDS of Godavari river, India. Int J Eng Res 5(2):115–118. https://doi.org/10.17950/ijer/v5s2/209

Tchobanoglous G, Schroeder EE (1985) Water quality: characteristics, modeling, modification

Tchobanoglous G, Peavy HS, Rowe DR (1985) Environmental engineering. McGraw-Hill Interamericana, New York

Tchobanoglous G, Burton FL, Stensel HD, Metcalf & Eddy (2003) Wastewater engineering: treatment and reuse, 4th edn. Tata McGraw-Hill Limited, New Delhi

Tung TM, Yaseen ZM (2020) A survey on river water quality modelling using artificial intelligence models: 2000–2020. J Hydrol 585:124670. https://doi.org/10.1016/j.jhydrol.2020.124670

Wang J, Fu Z, Qiao H, Liu F (2019) Assessment of eutrophication and water quality in the estuarine area of Lake Wuli, Lake Taihu, China. Sci Tot Environ 650:1392–1402. https://doi.org/10.1016/j.scitotenv.2018.09.137

Wang B, Wang Y, Wang S (2021a) Improved water pollution index for determining spatiotemporal water quality dynamics: case study in the Erdao Songhua River Basin, China. Ecol Indic 129:107931. https://doi.org/10.1016/j.ecolind.2021.107931
Wang R, Kim JH, Li MH (2021b) Predicting stream water quality under different urban development pattern scenarios with an interpretable machine learning approach. Sci Total Environ 761:144057. https://doi.org/10.1016/j.scitotenv.2020.144057
Xu T, Coco G, Neale M (2020) A predictive model of recreational water quality based on adaptive synthetic sampling algorithms and machine learning. Water Res 177:115788. https://doi.org/10.1016/j.watres.2020.115788
Yesilnacar MI, Sahinkaya E, Naz M, Ozkaya B (2008) Neural network prediction of nitrate in groundwater of Harran Plain, Turkey. Environ Geol 56(1):19–25. https://doi.org/10.1007/s00254-007-1136-5
Zare AH (2014) Evaluation of multivariate linear regression and artificial neural networks in prediction of water quality parameters. J Environ Health Sci Eng 12(1):1–8. https://doi.org/10.1186/2052-336X-12-40
Zhu M, Wang J, Yang X, Zhang Y, Zhang L, Ren H, Wu B, Ye L (2022) A review of the application of machine learning in water quality evaluation. Eco-Environ Health 1(2):107–116. https://doi.org/10.1016/j.eehl.2022.06.001

Assessment of Groundwater Prospects Zones Using RS, GIS, and MIF Methods

Dheeraj Mohan Gururani, Shekhar Singh, Himanshu Joshi, Yogendra Kumar, Anil Kumar, Manoj Singh Bohra, and Priyanka Mehta

Abstract The pivotal role of groundwater in managing domestic and industrial water demand and its current unsustainable rate of exploitation from subsurface reserves following anthropogenic and Climate Change impacts necessitates an efficient strategy in terms of resource identification, tenable extraction, and replenishment. The presented study is an attempt towards the same with an objective to identify groundwater potential zones for the Panna district in Madhya Pradesh, a semi-arid region under the Bundelkhand belt in Central India using an MCDM approach following a weighted overlay analysis integrated over RS and GIS platform. The study area encompasses a spatial extent of 7135 sq. km. The input layers used for the study include lithology, land use, slope, lineaments, rainfall, drainage density, and soil. Typically, these parameters first handily govern the subsurface hydrogeology of any region. The results from the overlay analysis suggested that 20.37, 59.42, 19.40, and 0.81% of the total geographical area of the district have very good, good, moderate, and poor potentiality for groundwater, respectively. Further, this contemplates the strong control of lithological distributions, land use, and slopes in the determination of groundwater potential zones. Moreover, the study also advocates for upscaling to other regions following its validation using some geophysical methods of subsurface exploration.

Keywords Groundwater potential · Remote sensing · GIS · MIF · Semi-arid region

D. M. Gururani
Department of Civil Engineering, MIET College, Institute of Meerut, Meerut, Uttar Pradesh 250005, India

S. Singh (✉) · H. Joshi · A. Kumar · M. S. Bohra
Department of Soil and Water Conservation Engineering, G.B. Pant University of Agriculture and Technology, Pantnagar, Udham Singh Nagar, Uttarakhand 263145, India
e-mail: shekharsingh861@gmail.com

Y. Kumar · P. Mehta
Deapartment of Irrigation and Drainage Engineering, G.B. Pant University of Agriculture and Technology, Pantnagar, Udham Singh Nagar, Uttarakhand 263145, India

C. B. Pande et al. (eds.), *Surface and Groundwater Resources Development and Management in Semi-arid Region*, Springer Hydrogeology,
https://doi.org/10.1007/978-3-031-29394-8_17

1 Introduction

Groundwater, the second largest repository of fresh water on Earth, plays a decisive role in the operation and management of freshwater supplies, either domestic or public (Nazaroff and Cohen 2013; Khan et al. 2020). Alone, it supports the potable water demand of over 33% of the global population (Nickson et al. 2005; Pande et al. 2021a). In India, greater than 85% of rural, 50% of urban, and over 50% of irrigation water demands are met from groundwater reserves (CGWB 2007). Typically, groundwater is stored in saturated geologic formations beneath the earth's surface, which are permeable enough to allow its flow fairly easily through it. These formations are termed an aquifer, which usually rests on the top of a relatively impermeable bed called, an aquitard or aquiclude, thus, restricting the movement of groundwater (Masters and Ela 2018). Naturally, aquifers are continuously recharged through precipitation water, which slowly percolates down through cracks and pores of overlaying soil and rock matrix, thus replenishing the subsurface reserves.

1.1 Statement of Problem

With the advent of substantial and water-intensive agriculture following the green revolution to feed this large population of the country followed industrialization and globalization, the dependency and withdrawal rate from available groundwater reserves has increased substantially to meet the overall demand of growing anthropogenic operations (Adeyeye et al. 2018). Oftentimes, the water retraction rate to meet the demand from the subsurface reserves becomes so immense that it exceeds the annual recharge rate, thus making the management of demand unsustainable. This scenario is further exacerbated by the Climate Change (CC) impacts on meteorological processes like erratic rainfall patterns manifesting as a decrease in the typical number of wet days over annual time scales, the occurrence of excessive intensity precipitation events, etc (Pande et al. 2023). These coupled together physically manifest as declining groundwater levels, a decrease in the resurgence of discharges from aquifers and springs, freshwater water shortages, widened gap between demand, availability, and supply, land subsidence, etc., thus placing a question over the future water resource sustainability. These exigencies for an efficient groundwater resource development plan and policy in terms of resource identification, withdrawal, and replenishment, especially for the semi-arid region which receives most of its precipitation during the summer half and where most of the precipitated water escapes as surface runoff due to non-arresting land use and soils. Concerning groundwater resource identification, the use of traditional/physical techniques like geophysical surveys using resistivity method, seismic refraction, and reflection, Ground Penetration Radar (GPR), etc. for groundwater exploration follows a heft cost and manpower resources, and in a resource-limited environ with large spatial extent these methods find little to limited use (Rani et al. 2022). Following the onset of Remote Sensing

(RS) and Geographic Information System (GIS), lately, the use of these technologies has greatly influenced hydrological explorations and are now extensively accepted in groundwater investigations as well (Hoffman and Sander 2007; Rodell et al. 2009; Das et al. 2018; Mallick et al. 2019; Ajay Kumar et al. 2020). GIS has now become a very nimble tool in the evaluation and management operations of various parameters/features that affects the availability and recharging processes of groundwater like slope, soil, lithology, lineaments, etc. (Ahmad et al. 2020; Roy et al. 2020; Jha et al. 2010; Pande et al. 2021b). Furthermore, incorporating the Multi-Criteria Decision Making (MCDM) approaches, which explicitly accounts for the inter-relational influence of one factor over the other and vice-versa, over the GIS platform for groundwater resource explorations has proved much laudable (Kadam et al. 2020). The presented work, therefore, is an attempt towards the identification of the groundwater potential zones for a semi-arid district in Central India following an MCDM technique of Multi Influencing Factors (MIF) over a GIS platform using facilely available geological and hydro-meteorological RS data. The study, therefore, is a contribution towards the identification of suitable sites for sustainable groundwater exploration and withdrawal along with the recharging of groundwater reserves via manmade interventions.

2 Study Area

The study area (Fig. 1), i.e., the Panna district, is a part of the Sagar division in the state of Madhya Pradesh in Central India. It spans from latitude 23°49′11″ N to 24°4′40″ N and longitude 79°44′18″ E to 80°40′36″ E, covering a geographical area of 7135 sq. km, with an absolute relief ranging from 659 to 99 m Above Mean Sea Level (AMSL), and an average elevation of 363.2 m AMSL. The district was constituted in 1950. Historically, the region was the part of Gond settlement until the thirteenth century and later served as the capital of Raja Chhatrasal Bundela. Also, the region is famous for its diamond mines and is the only producer of diamonds in the country. Further, the district is a biodiversity hotspot housing the renowned Panna Tiger Reserve. The primary river draining through the district is the Ken R., which is the major tributary of the river Yamuna. At annual timescales, the study area is a water-scarce region receiving most of its precipitation during the Monsoonal period and is the part of semi-arid Bundelkhand belt (Jana et al. 2017). The major problems associated with the region are depleting groundwater reserves, soil erosion, and infertile lands (Jain et al. 2020). This adversely affects the overall economy and Human Development Index (HDI) of the region making it one of the most economically backward districts of Madhya Pradesh.

The geology of the study area suggests the presence of rocks under the Bhander, Bijawar, Kaimur, Rewa, and Semri groups along with Bundelkhand granitoid complex and undifferentiated fluvial sediments. Whereas, the geomorphology of the region is dominated by homocline, pediplain, pediment, scarp, highly dissected

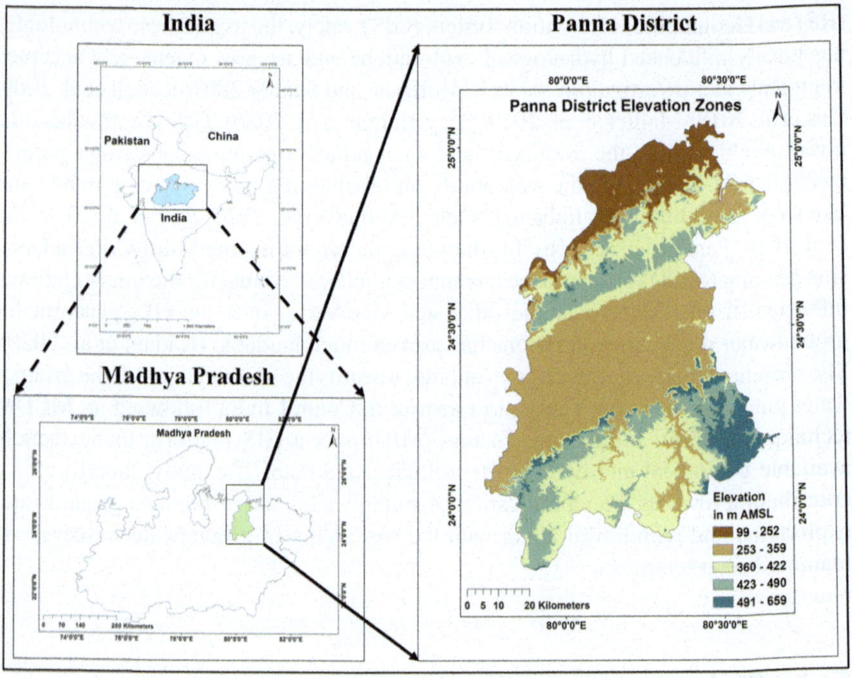

Fig. 1 Index map of the study area

hills and valleys, moderately dissected hills, valleys, and lower plateau, and waterbodies with traces of residual hill, residual capping, plateau top, and butte. Besides these, active flood and older alluvial plains, gullied tracts, and lateritic uplands are also observed in the mid-west and southeast portions of the study area, respectively.

3 Methodology

Figure 2, illustrates the methodology used for the presented investigation. The boundary of the Panna district was prepared based on the Survey of India (SoI) topographic and political maps (54P/12, 54P/14, 54P/15, 54P/16, 55M/13, 63C/12, 63C/4, 63C/8, 63D/1, 63D/11, 63D/12, 63D/2, 63D/3, 63D/4, 63D/5) of 1:50,000 scale. The drainage network and slope for the study area were developed from Advanced Spaceborne Thermal Emission and Reflection Radiometer (ASTER) Global Digital Elevation Model (GDEM) version 3, of spatial resolution 30 m using ArcGIS v 10.4.1. The lineament and lithology data were taken from the respective maps of scale 1:250,000 available at the Bhukosh portal of the Geological Survey of India (GSI). The line density tool under the Spatial Analyst Toolbox in ArcMap v 10.4.1 was then used for the generation of both drainage and lineament density maps. The soil data is taken from the Digital Soil Map of the World (DSMW) of the Food and Agriculture Organization (FAO), United Nations (UN) of scale 1:250,000 and Soil Water Assessment Tool (SWAT) 2012 'usersoil' database, while the land use map is developed using the Sentinel-2A data of 2021 of spatial resolution 10 m. For the development of the rainfall map, 22-year gridded precipitation data of resolution 0.5° × 0.625°, from 2000 to 2021, was taken from NASA CERES MERRA 2 (available from: https://power.larc.nasa.gov/data-access-viewer/). The average annual rainfall was then computed for various locations within the study area, which then was imported to ArcMap. This was followed by the use of Inverse Distance Weighing (IDW) for plotting the rainfall data for the area of interest. The generated maps were then transformed to raster (.tiff files), which were then resampled, for conforming the pixel/cell size, and reclassified, for assigning index values/sub-classes weights to developed classes under each layer. Finally, a weighted overlay analysis following a Multi-Influence Factor (MIF) technique was performed over the aforesaid layers for the generation of groundwater prospect zones. For the assessment of the influence/assignment of weights of the specific layer during the final overlay, the influence of each feature/input parameter over the hydrogeology of the study area was considered (Shaban et al. 2006). The identified classes from the developed overlay were then again reclassed under the poor, moderate, good, and very good groups following a quartile approach.

3.1 *MIF for Assessment of Sub-class Weights and Percentage Influence*

MIF, one of the MCDM techniques, is used to assess the influence of each input layer towards the groundwater prospects zonation of the study area. For this purpose, the inter-relationship between inputs factors is ascertained and is subdivided into two categories, namely, the major and minor influences as per the recommendations of Magesh et al. (2012; Pande et al. 2020). In the case of each major influence, a weight of unity is assigned, while for the minor effect, a weight of 0.5 is awarded. The final weight (W_L) of the layer is the sum of the major (W_{Maj}) and minor (W_{Min}) influence for the respective layer (as mentioned in Eq. 1).

$$W_L = \sum_{i=1}^{n} W_{Maj_i} + \sum_{j=1}^{m} W_{Min_j} \quad (1)$$

where, n and m are the number of major and minor relations, respectively. To perceive the percent influence (I_L) of each layer, the final weight of the individual layer is divided by the cumulative weights of all the input layers and then multiplied by 100 and rounded off to whole numbers (as mentioned in Eq. 2). This percent influence shall also provide the range of whole numbers which can be used to rate various sub-classes under the respective layer.

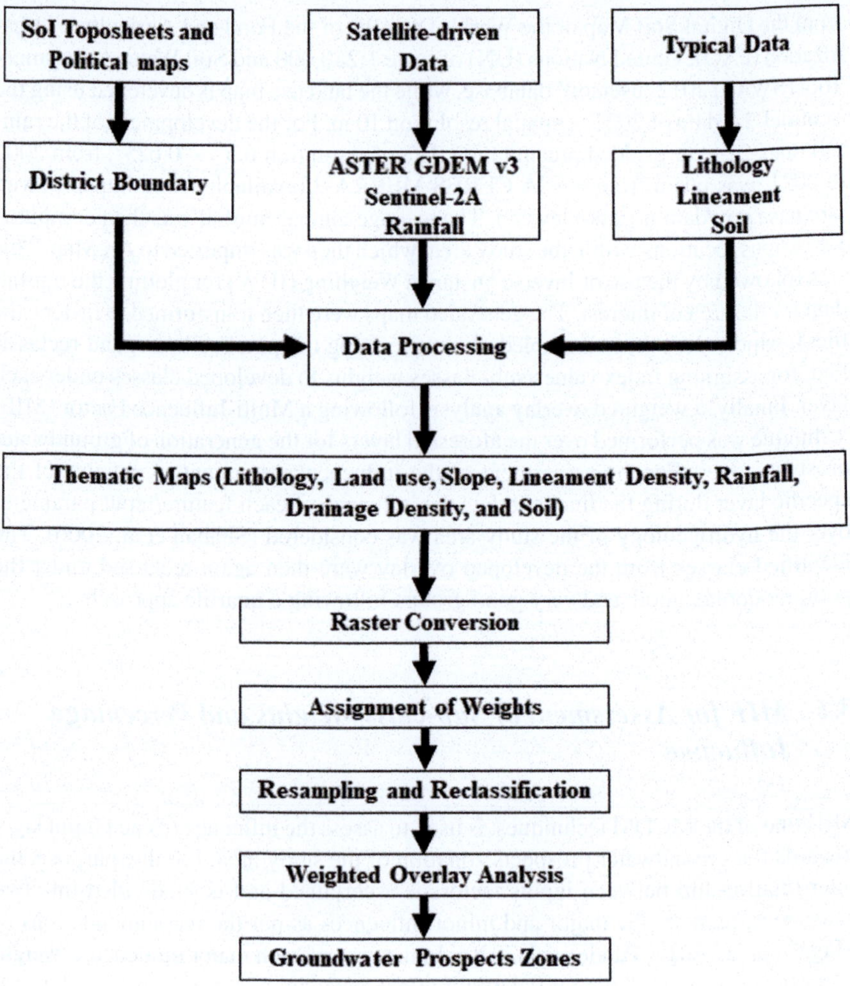

Fig. 2 Flowchart for the groundwater prospects zonation mapping

$$I_L = \left[\frac{W_{L_i}}{\sum_{i=1}^{p} W_{L_i}}\right] * 100 \quad (2)$$

Figure 3 depicts the major and minor influence relationship between the seven input parameters, while Table 1 illustrates the respective W_L and I_L for each input layer.

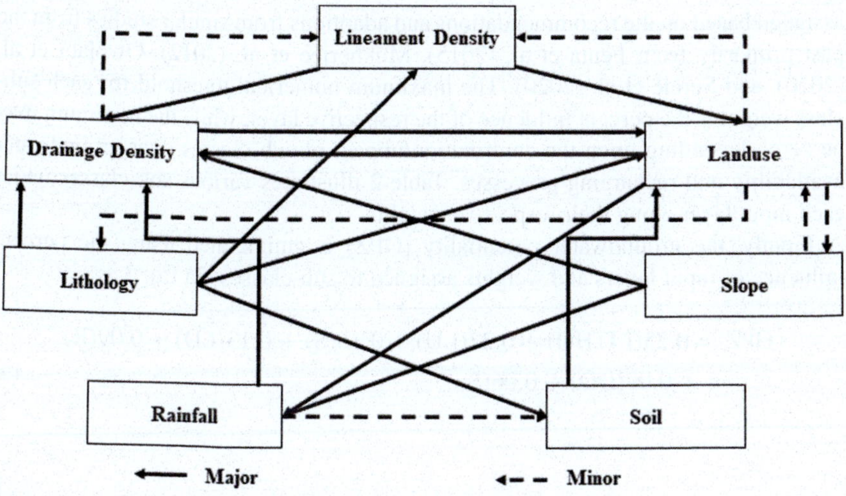

Fig. 3 Inter-relationship between various input layers for MIF

Table 1 Final weights and percent influence of each layer following MIF

Factor	Major (W_{Maj})	Minor (W_{Min})	Final weight (W_L)	Percent influence (I_L)
Lithology	1 + 1 + 1 + 1	0	4	25
Land use	1 + 1	0.5 + 0.5 + 0.5	3.5	22
Slope	1 + 1	0.5	2.5	16
Lineament density	1 + 1	0	2	13
Rainfall	1	0.5	1.5	9
Drainage density	1	0.5	1.5	9
Soil	1	0	1	6
			Σ16	Σ100

4 Results and Discussion

4.1 MIF and Sub-class Weights

The seven input layers, namely, lithology, land use, slope, lineament density, rainfall, drainage density, and soil, were overlayed for the development of final groundwater prospects zonation mapping of the Panna district. These chosen factors affect both, groundwater availability as well as recharging/replenishment potential (Rani et al. 2019). Hence understanding these parameters at the local level becomes a prerequisite for opportune groundwater prospects zonation. Following the computation of percent influence, the respective weights for each sub-classes under various input layers were assigned based on the recommendations and adaptions from similar studies from the past primarily from Fenta et al. (2015), Mukherjee et al. (2012), Owolabi et al. (2020), and Serele et al. (2020). The maximum numerical threshold for each sub-class weight is the percent influence of the respective layer, while the minimum may be zero, depending upon the contribution/impact of sub-classes over groundwater availability and recharging processes. Table 2 illustrates various sub-classes under each input layer along their respective weights.

Finally, the groundwater potentiality (GPZ) is enumerated using the percent influence of input layers and weights assigned to sub-classes via Eq. 3 as

$$\begin{aligned} \text{GPZ} = {} & 0.25(\text{LITHO}) + 0.22(\text{LU}) + 0.16(\text{S}) + 0.13(\text{LD}) + 0.09(\text{R}) \\ & + 0.09(\text{DD}) + 0.06(\text{So}) \end{aligned} \tag{3}$$

4.2 *Lithology*

Lithology governs the permeability as well as the porosity of the underlying formation of any drainage system (Tessema et al. 2014), hence plays a crucial role in the percolation of surface water to the groundwater table, thus affecting groundwater availability and recharging process (CGWB 2014). The presented work suggests that the lithology has major control over lineament density, soil, drainage density, and land use, hence is assigned the largest percent influence, i.e., 25%, in the context of groundwater potential zonation. A total of 22 lithologic sub-classes were identified for the study area, with alluvium/sand sub-classes with the highest weight (25) and shale sub-class with the lowest weight (1), respectively. The assigned weights were the function of the relatively easy/perviousness that the given lithologic class allows towards the movement of water in the subsurface matrix. The dominant lithologic classes are the sandstone and shale of Bhandar and Sumeri groups of the Neoproterozoic and Mesoproterozoic periods. These two cover 47.77 and 31.81% of the total spatial extent of the area of interest. While, granite gneiss, kimberlite, quartz reef, and ferruginous sandstone belonging to Bundelkhand Granitoid Complex and Bijawar group of Paleoproterozoic & Archaean-Paleoproterozoic periods are the least dominant, with cumulative coverage of 0.01% of the entire geographical area of the Panna district. Figure 4a illustrates the lithology map of the area of interest.

4.3 *Land Use*

The land use data from Sentinel-2A suggests the presence of six (06) major land use classes, namely, waterbodies, forest, cropland/arable land, built-up/settlement, bare ground, and grasslands with relative percentages of 0.73, 10.08, 45.30, 2.01, 0.01,

Table 2 Sub-class weights for input layers

Parameter with % influence	Sub-classes	Weight
Lithology (LITHO) (25%)	Alluvium/sand	25
	Basalt/meta basalt	21
	Grey sand, silt, and clay/oxidised silt–clay with kankar and micaceous sand	19
	Ferruginous sandstone/sandstone/sandstone and orthoquartzite	18
	Limestone	16
	Clay	15
	Chlorite schist/laterite/quartz reef	14
	Diamondiferous conglomerate	9
	Diorite/kimberlite	07
	Medium-grained granite	04
	Coarse-grained porphyritic granite/fine-grained granite/granite gneiss	03
	Shale	01
Land use (LU) (22%)	Waterbody	22
	Flooded vegetation	20
	Cropland/arable land	18
	Forest	14
	Pastures/grassland	09
	Bare ground	05
	Built-up/settlement	01
Slope (S), in degree (16%)	Less than 3	16
	3–5	12
	5–15	08
	15 30	04
	30–50	01
	Greater than 50	01
Lineament density (LD), in km/sq. km (13%)	0–1.39	01
	1.40–2.69	05
	2.70–3.99	09
	4.00–5.20	13
Drainage density (DD), in km/sq. km (9%)	0–1.79	09
	1.80–3.59	07
	3.60–5.29	05
	5.30–7	02
Rainfall (R), in mm/year (9%)	938–992	02

(continued)

Table 2 (continued)

Parameter with % influence	Sub-classes	Weight
	993–1045	05
	1046–1099	07
	1100–1153	09
Soil (So), as per FAO soil codes (6%)	SNUM 3772 (sandy-clay loam)	06
	SNUM 3780 (sandy-clay loam)	06
	SNUM 3861 (clay)	02

and 41.88% of the total geographical area of the Panna district, respectively. Another land use class, particularly, the flooded vegetation is also detected but has very low to nil relative contribution towards the overall land use of the study area. Land use defines the relative perviousness of the surface towards the incoming precipitation water and the overland flow, thus governing the surface runoff and the infiltration rate (Douglas and Ansari 2005), therefore are ranked according to the same. From Fig. 3, Table 1, and previously mentioned recommendations, the land use is said to have a major impact on rainfall and drainage density, and have minor influence over the slope, lineament density, and lithology, with a total percent influence of 22% in the context of groundwater potentiality. The land use which directly endows to the groundwater availability/recharge, i.e., waterbody, flooded vegetation, and cropland, were assigned higher weights, which are 22, 20, and 18, respectively. While, land use classes like built-up/settlement and bare ground including rock outcrops, were assigned lower weights, i.e., 1 and 5. The land use for the Panna district is represented in Fig. 4b.

4.4 Slope

Slope affects the groundwater flow paths at both local and regional scales (Gleeson and Manning 2008). With higher slopes the flow velocity of overland flow and the surface runoff increases which in turn curtails the contact time between the sheet of flowing water and the underlying soil, thus reducing the infiltration rates and vice versa. Also, in excessively steep slopes the meager portion of the water which gets infiltrated in the soil matrix escapes out as the prompt interflow, thus having a very insignificant contribution towards groundwater. For the presented work slope classification as recommended by Rani et al. (2019) is adapted. The slope map of the study area is sub-classified into six (06) categories as mentioned in Table 2. The average slope of the study area was estimated to be 4.1°, while the maximum slope was 53.72°. The first slope class, i.e., less than 3°, covers the maximum (51.93%) of the study area followed by classes 3° to 5° (22.83%), and 5° to 15° (21.62%), respectively. The slope has a major effect on rainfall and drainage density and a minor effect on land use and is assigned a percent influence of 16%. Higher slopes were

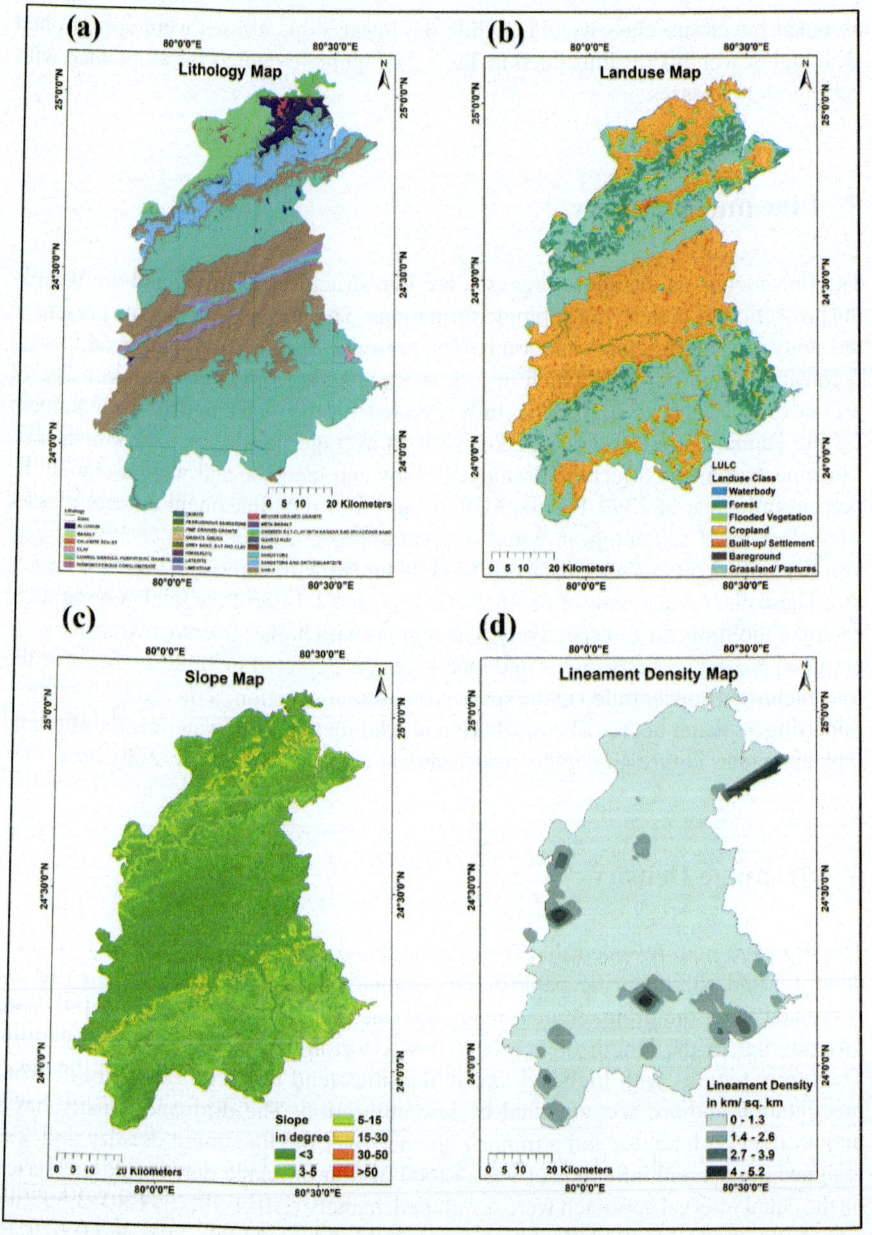

Fig. 4 **a** Lithology, **b** land use, **c** slope, and **d** lineament density map for the Panna district, Madhya Pradesh

assigned lower sub-class weights, while the lower slope classes were apportioned with higher weights (as illustrated in Table 2). Figure 4c depicts the slope map with various slope classes.

5 Lineament Density

The lineaments, as the name suggests, are line structures/features, and are usually the projection of underlying geologic formations. The primary lineaments present in the study area are fractures and joints. The presence of lineaments offers additional permeability and porosity which in turn augments the availability and recharge of groundwater (Dar et al. 2010). The study revealed that the total length of the lineament for the Panna district was 691.71 km with an average density of 0.44 km/sq. km. Lineaments majorly affect the drainage density and land use and were assigned the percent influence of 13%. For the MIF, in sum four (04) lineament density classes were developed following an equal interval approach, namely, (i) 0–1.39 km/sq. km, (ii) 1.40–2.69 km/sq. km, (iii) 2.70–3.99 km/sq. km, and (iv) 4.00–5.20 km/sq. km. These classes account for 88.46, 8.77, 1.63, and 1.14% of the total geographical extent of the study area, respectively. The regions with higher lineament density were assigned greater scores/weights and vice-versa (as depicted in Table 2). Most of the lineaments are concentrated in the southeast and south portion of the study area, thus, imparting a greater degree of contribution in relation to groundwater availability and replenishment. Figure 4d depicts the lineament density map for the study area.

6 Drainage Density

The drainage map revealed that the drainage network of the study area follows a dendritic and sub-dendritic pattern. The drainage density, as depicted in Fig. 5a is derived from the drainage map using the line density function, and is inversely proportional to the length of overland flow (Horton 1945) and the permeability. Therefore, regions with higher drainage densities tend to have higher runoff from precipitation and are accompanied by low infiltration. The drainage density has a major effect on land use and a minor influence over the lineament density and was assigned the percent influence of 9%. A total of four drainage density classes based on the equal interval approach were developed, namely (i) 0–1.79, (ii) 1.80–3.59, (iii) 3.60–5.29, and (iv) 5.30–7.00, which accounts for 52.22, 44.21, 2.93, and 0.64% of the spatial extent of the study area. The areas of low drainage density were assigned higher weights and vice versa (see Table 2). The average drainage density of the study area was estimated to be 1.83 km/sq. km.

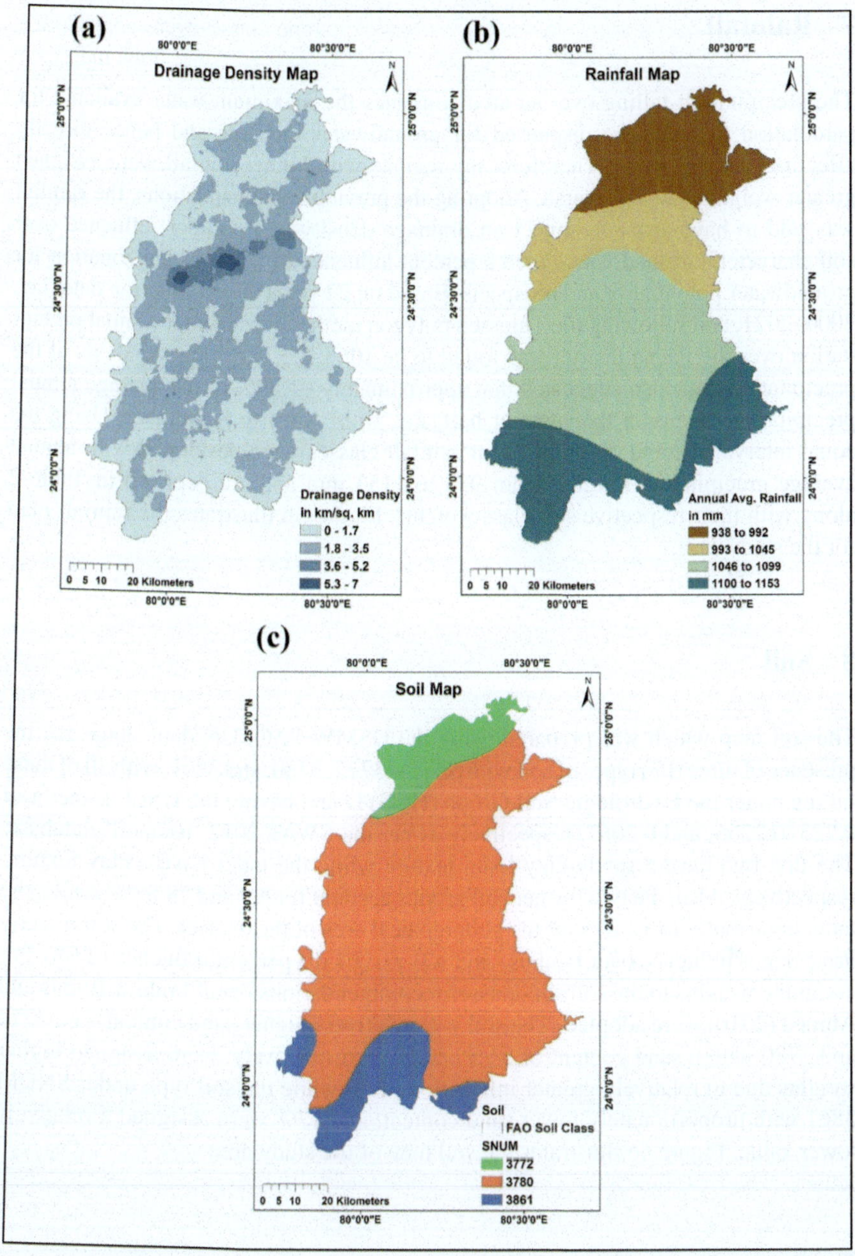

Fig. 5 **a** Drainage density, **b** rainfall, and **c** soil map for the study area

7 Rainfall

The precipitation falling over an area construes the maximum water available for percolation that may be dispensed for groundwater recharge and hence directly affecting its availability. Therefore, the regions with higher rainfall were assigned greater weights and vice-versa. Adopting the previous commendations, the rainfall was said to have a major impact on drainage density and a minor influence over soil characteristics and is assigned a percent influence of 9% over the zonation for groundwater potentiality and prospects. Based on 22 years of precipitation data, i.e., 2000–2021, and following the Thiessen polygon method, the average annual precipitation over the Panna district was found to be 1063.7(±51.8) mm. Analysis of the precipitation data also suggested that approximately 91.50% of the average annual precipitation occurs in the summer half, i.e., from April to September. Using the equal interval method, a total of four rainfall classes were developed with annual average precipitation ranging from 938 to 1153 mm and are depicted in Table 2 along with their respective sub-class weights. Figure 5b illustrates the rainfall map for the study area.

8 Soil

The soil map which was prepared using the DSMW FAO, UN data suggested the presence of three (03) types of soil with SNUM 3772, 3780, and 3861, with all of these falling under the Hydrologic Soil Group (HSG) D and having the USLE k-factor of 0.225, 0.2266, and 0.2067, respectively as per the SWAT 2012 'usersoil' database. The first two have a sandy-clay loam texture while the latter have a clay texture, respectively. Also, the two former soil groups account for 8.9 and 78.88%, while, the latter contributes to 12.22% of the geographical area of the district. The soil directly has a major influence over land use and was assigned a percent influence of 6%. For assigning weights to the soil sub-classes recommendations from Ibrahim-Bathis and Ahmed (2016) were adopted. The soil with relatively higher sand content, i.e., 3772 and 3780 with a sand content of 54.5 and 51%, respectively, were assigned higher weights due to relatively greater infiltration rates, while the soil type under SNUM 3861 with proportionately lower sand content, i.e., 17% were assigned a relatively lower value. Figure 5c illustrates the soil map of the study area.

9 Groundwater Potential Zones

For the development of the groundwater potential zonation map, the above-mentioned seven thematic layers were then overlayed using the percent influence of each layer over ArcMap 10.4.1 via the weighted overlay analysis tool. The raster generated after the overlay is then again reclassed to get four (04) groups for the final groundwater prospects zonation, namely, poor, moderate, good, and very good based on the quartile approach, i.e., raster index values less than the first quartile are under poor, between the first and the second quartile is in moderate, while the values in second and third quartile range, and over the third quartile are under good and very good classes, respectively. From the final groundwater potential zonation raster, it was concluded that approximately, 20.37 and 59.42% of the total geographical area of the Panna district was found to be under very good and good potentiality for groundwater, whereas, 19.40 and 0.81% have moderate and poor prospects for groundwater. The potential zones under the 'very good' class are chiefly concentrated near the northern and mid-reaches as well as in the southern and south-eastern portions of the area of interest in the form of non-continuous stretches. This suggests the higher control of lithologic groups, primarily alluvium, grey sand, sandstone, and limestone deposits along with the land use, which is dominated by grasslands, cropland, and forest, and gentler slopes in governing the groundwater potentiality for the study area. Figure 6 demonstrates the groundwater potential zones for the Panna district. The sites with very good and good groundwater prospects represent locations where the probability of finding groundwater is relatively higher and the sustainable extraction process is comparatively easier. Furthermore, these locations also act as opportune spots where site-suitable Groundwater Augmentation Measures (GAMs) can be implemented for the replenishment and rejuvenation of the subsurface water reserves. The study suggests the development of contour trenches and recharge pits for different cross-sectional shapes in areas of very good and good potentiality. The purpose of these measures is to increase the cumulative amount of infiltrated water towards recharging the subsurface reserves along with soil conservation. In the areas of moderate potentiality, implementation of check dams like masonry dams, live check dams, brush-wood check dams, etc. is suggested over second and lower-order streams towards arresting overland flow, surface runoff, and excessive channel flow during the Monsoonal period. Finally, in the areas of low potentiality, bunding, stone pitching, storage ponds, plantation of adaptable flora including hardwood and grasses, etc. should be done for enhancing soil moisture at one end and abbreviating soil erosion over the other.

10 Conclusions and Future Scope

The RS and GIS-based approaches in coherence with the use of the MCDM technique have proved frugal, cost-effective, and nimble propositions towards the delineation of groundwater potential zones for the semi-arid Panna district of Madhya Pradesh. The study took into consideration seven thematic layers, namely, lithology, land use, slope, lineament density, rainfall, drainage density, and soil, as these parameters first-handily affect the hydrogeology and groundwater regime at both local and regional scales. These layers were then overlayed after assigning suitable percent influence and sub-class weights following the MIF method. Following this approach, a total of

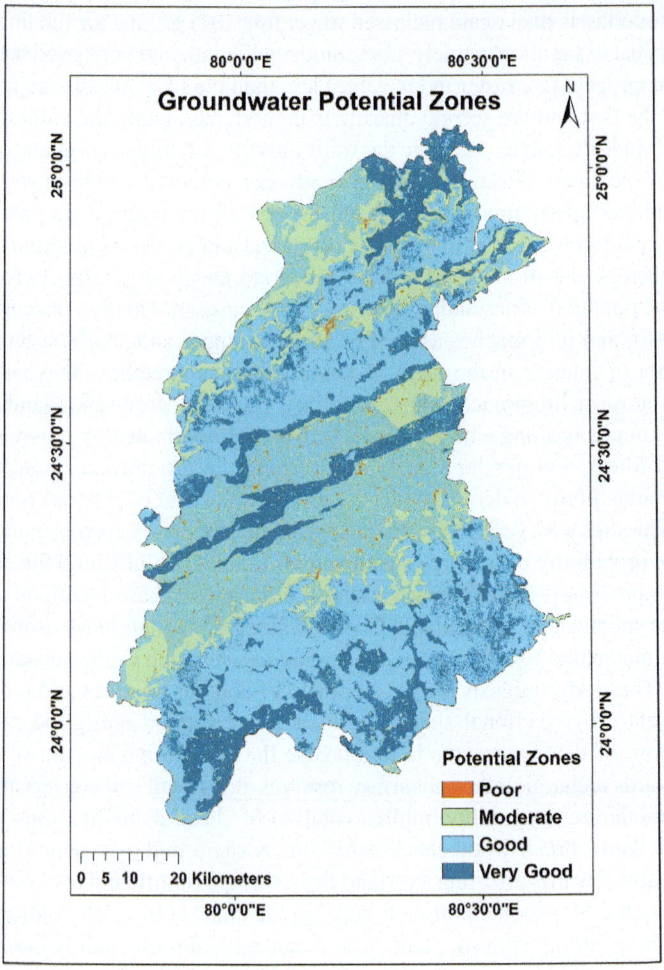

Fig. 6 Groundwater prospects zones for the Panna district

four qualitative zones for groundwater potentiality and recharge, viz. poor, moderate, good, and very good, were developed for the study area. This shall contribute towards sustainable water resource management, especially groundwater when it comes to the identification of potential sites for tenable extraction as well as for the replenishment of the same via site-suitable artificial augmentation structures. The study also suggests opportune GAMs for the identified sites of various degrees of potentialities in order to rejuvenate and provisioning of subsurface water reserves and for soil conservation. The MCDM technique followed for the quantitative assessment and dependence of various input influencers, i.e., MIF is relatively uncluttered, but leaves scope for more refined prospecting via the application of MCMD approaches which accounts for checking of mathematical consistencies in the assignment of weights along with uncertainty assessment like the Analytical Hierarchical Process (AHP) and Fuzzy AHP. Also, the addition of other influencing parameters like Topographical Proximity Index (TPI), curvature, roughness, geomorphology, etc. in the final overlay after assigning suitable weights and use of more refined and fine thematic data may bring the map depicted potential zones much closer to the on-ground scenario. Validation of the developed map via ground truthing remains open and can be done via geo-tagging followed by recording water levels for pre and post-monsoonal periods from various borewells distributed across the study area and then plotting the water level fluctuation data and matching the same with the developed results. Another approach is to go for the geophysical investigation like the resistivity method, GPR, etc. for a few locations dispersed along the study area but that may prove manpower and financially intensive. Further, an improved understanding of the relationship of known influencers along identification of site-specific indices that governs the subsurface hydrology along with addressing their cumulative impact on the groundwater regime under the dynamic land use and CC scenario is also to be advocated.

References

Adeyeye O, Ikpokonte E, Arabi S (2018) GIS based groundwater potential mapping within Dengi area, North Central Nigeria. Egypt J Remote Sens Space Sci 22(2):175–181. https://doi.org/10.1016/j.ejrs.2018.04.003

Ahmad I, Dar MA, Andualem TG, Teka AH (2020) GIS-based multi-criteria evaluation of groundwater potential of the Beshilo River basin. Ethiopia J Afr Earth Sci 164:103747. https://doi.org/10.1016/j.jafrearsci.2019.103747

Ajay Kumar V, Mondal NC, Ahmed S (2020) Identification of groundwater potential zones using RS, GIS and AHP techniques: a case study in a part of deccan volcanic province (DVP), Maharashtra, India. J Indian Soc Remote Sens 48(3):497–511. https://doi.org/10.1007/s12524-019-01086-3

CGWB. 2007. Manual on Artificial Recharge of Groundwater. Ministry of Water Resource, Government of India, New Delhi, India. 185 p

CGWB (2014) Groundwater scenario of Himalayan region. Ministry of Water Resource, Government of India, India, 226 p

Dar MA, Sankar K, Dar IA (2010) GW prospects evaluation-based on hydro-geomorphological mapping: a case study in Kancheepuram district, Tamil Nadu. Ind J Remote Sens 38(2):333–343

Das S, Pardeshi SD, Kulkarni PP, Doke A (2018) Extraction of lineaments from different azimuth angles using geospatial techniques: a case study of Pravara basin, Maharashtra, India. Arab J Geosci 11:160. https://doi.org/10.1007/s12517-018-3522-6

Douglas C, Ansari SA (2005) Estimating ground water recharge from topography, hydrogedogy and land cover. Ground Water 43(1):102–112

Fenta AA, Kifle A, Gebreyohannes T, Hailu G (2015) Spatial analysis of groundwater potential using remote sensing and GIS-based multi-criteria evaluation in Raya valley, north Ethiopia. Hydrogeol J 23:195–206

Gleeson T, Manning AH (2008) Regional groundwater flow in mountainous terrain: three-dimensional simulations of topographic and hydrogeologic controls. Water Resour Res 44:W10403

Hoffmann J, Sander P (2007) Remote sensing and GIS in hydrogeology. Hydrogeol J 15:1–3

Horton RE (1945) Erosion development of streams and their drainage basins: hydro-physical approach to quantitative morphology. Geol Soc Am Bull 56:275–370

Ibrahim-Bathis K, Ahmed SA (2016) Geospatial technology for delineating groundwater potential zones in Doddahalla watershed of Chitradurga district, India. Egypt J Remote Sens Space Sci 19(2):223–234

Jain R, Chand P, Rao S, Agarwal P (2020) Crop and soil suitability using multi-criteria decision making in drought prone semi-arid tropics in India. J Soil Water Conserv 19(3):271–283

Jana C, Alam NM, Mandal D, Shamim M, Kaushal R (2017) Spatio-temporal rainfall trends in the twentieth century for Bundelkhand region, India. J Water Clim Chang 8(3):441–455

Jha MK, Chowdary VM, Chowdhury A (2010) Groundwater assessment in salboni block, West Bengal (India) using remote sensing, geographical information system and multi-criteria decision analysis techniques. Hydrogeol J 18:1713–1728

Kadam AK, Umrikar BN, Sankhua RN (2020) Assessment of recharge potential zones for groundwater development and management using geospatial and MCDA technologies in semiarid region of Western India. SN Appl Sci 2:312

Khan MYA, ElKashouty M, Bob M (2020) Impact of rapid urbanization and tourism on the groundwater quality in Al Madinah city, Saudi Arabia: a monitoring and modeling approach. Arab J Geosci 13(18):1–22. https://doi.org/10.1007/s12517-020-05906-6

Magesh NS, Chandrashekar N, Soundranayagam JP (2012) Delineation of groundwater potential zones in Theni district, Tamil Nadu using remote sensing, GIS, and MIF techniques. Geosci Front 3(2):189–196

Mallick J, Khan RA, Ahmed M, Alqadhi SD, Alsubih M, Falqi I, Hasan MA (2019) Modeling groundwater potential zone in a semi-arid region of aseer using fuzzy-ahp and geoinformation techniques. Water (Switz) 11(12):2656. https://doi.org/10.3390/W111226562656

Masters GL, Ela WP (2018) Introduction to environmental engineering and science. Pearson India Education Services Pvt. Ltd., Noida, Uttar Pradesh, India, p 692

Mukherjee P, Singh CK, Mukherjee S (2012) Delineation of groundwater potential zones in arid regions of India—a remote sensing and GIS approach. Water Res Manag 26:2643–2672

Nazaroff WW, Cohen LA (2013) Environmental engineering and science. Wiley India Pvt. Ltd., New Delhi, p 690

Nickson RT, McArthur JM, Shrestha B, Kyaw-Nyint TO, Lowrt D (2005) Arsenic and other drinking water quality issues, Muzaffargarh District, Pakistan. Appl Geochem 20:55–68

Owolabi ST, Madi K, Kaulamba AM, Orimoloye IR (2020) A groundwater potential zone mapping approach for semi-arid environments using remote sensing (RS), geographic information system (GIS), and analytical hierarchical process (AHP) techniques: a case study of Buffalo catchment, Eastern Cape, South Africa. Arab J Geosci 1184

Pande CB, Moharir KN, Singh SK et al (2020) An integrated approach to delineate the groundwater potential zones in Devdari watershed area of Akola district, Maharashtra, Central India. Environ Dev Sustain 22:4867–4887 (2020). https://doi.org/10.1007/s10668-019-00409-1

Pande CB, Moharir KN, Khadri S (2021a) Watershed planning and development based on morphometric analysis and remote sensing and GIS techniques: a case study of semi-arid watershed in Maharashtra, India. In: Pande CB, Moharir KN (eds) Groundwater resources development and planning in the semi-arid region. Springer, Cham. https://doi.org/10.1007/978-3-030-68124-1_11

Pande CB, Moharir KN, Panneerselvam B et al (2021b) Delineation of groundwater potential zones for sustainable development and planning using analytical hierarchy process (AHP), and MIF techniques. Appl Water Sci 11:186. https://doi.org/10.1007/s13201-021-01522-1

Rani M, Pande A, Kumar K, Joshi H, Rawat DS, Kumar D (2022) Investigation of groundwater recharge prospect and hydrological response of groundwater augmentation measures in Upper Kosi watershed, Kumaun Himalaya, India. Groundw Sustain Dev 16:100720

Rani M, Joshi H, Kumar K, Pande A, Rawat DS (2019) Development of recharge conservation site suitability model for groundwater retrieval and evaluation of artificial recharge potential in a complex hydro-geological spring-fed river basin. Arab J Geosci 12

Rodell M, Velicogna I, Famiglietti JS (2009) Satellite-based estimates of groundwater depletion in India. Nature 460:999–1002

Roy S, Hazra S, Chanda A, Das S (2020) Assessment of groundwater potential zones using multi-criteria decision-making technique: a micro-level case study from red and lateritic zone (RLZ) of West Bengal, India. Sustain Water Resour Manag 6(1):1–14. https://doi.org/10.1007/s40899-020-00373-z

Serele C, Perez-Hoyos A, Kayitakire F (2020) Mapping of groundwater potential zones in the drought-prone areas of south Madagascar using geospatial techniques. Geosci Front 11(4):1403–1413

Shaban A, Khawlie M, Abdallah C (2006) Use of remote sensing and GIS to determine recharge potential zone: the case of Occidental Lebanon. Hydrogeol J 14:433–443

Tessema A, Nzotta U, Chirenje E (2014) Assessment of groundwater potential in fractured hard rocks around Vryburg, North West Province, South Africa, WRC Project No: K5/2055

Impact of Surface Temperature on Soil Chemical Properties Using Coupled Approach of Satellite Imagery, Gamma Test and Regression Based Models in Semi-arid Area

Vijay Kant Singh, Ram Prakash, and Daniel Prakash Kushwaha

Abstract Temperature of land surface has crucial effect on soil natural environment by controlling soil pH, soil water retention, organic content, physical and microbiological forms of soil. In the present study, semi-arid Guhla and Kaithal blocks of Kaithal district, Haryana state, India, have been selected for the assessment of effect of land surface temperature (LST) on the soil chemical properties. Soil sampling was done from twenty one random sites of this blocks in which nine belong to Guhla block and twelve belong to Kaithal block on 11 June 2015. Total eighteen soil chemical properties viz. soil saturation, cation exchange capacity, organic carbon, calcium carbonate, N, P, K, exchangeable sodium percentage, electrical conductivity, pH, water soluble anions viz. carbonate, bicarbonate, chloride, sulphate and water soluble cations viz. calcium, magnesium, sodium, and potassium have been determined for each sample. Spilt window technique has been used for LST determination by utilizing weather data of the location. Multiple linear regression (MLR) and multiple non-linear regression (MNLR) analysis have been used for modeling LST and soil chemical properties. Due to no correlation among the variables, gamma test has been used for selecting best input structure. Coefficient of multiple determination, multiple correlation coefficient and root mean square error (RMSE) between remote sensing (RS) based LST and soil chemical parameters, were found as 0.367, 0.606 and 2.276 °K, respectively, in MLR model. In MNLR model, dataset length was divided into 70% for training and 30% for testing. Coefficient of determination (R^2) and RMSE, between RS based LST and MNLR based LST, were found as 0.861 and 3.333 °K, respectively during testing period. MNLR model has given

V. K. Singh
Department of Soil Science, Punjab Agricultural University, Ludhiana 141004, Punjab, India

R. Prakash
Department of Soil Science, Chaudary Charan Singh Haryana Agricultural University, Hisar 125004, Haryana, India

D. P. Kushwaha (✉)
College of Agriculture Engineering, Ara (Bhojpur), Under Bihar Agricultural University, Sabour, Bhagalpur 813210, Bihar, India
e-mail: danielprakash1991@gmail.com

C. B. Pande et al. (eds.), *Surface and Groundwater Resources Development and Management in Semi-arid Region*, Springer Hydrogeology,
https://doi.org/10.1007/978-3-031-29394-8_18

better results in terms of coefficient of determination in comparison to MLR analysis, with overestimated values of LST.

Keywords Land surface temperature · Soil chemical properties · Gamma test · Remote sensing and GIS · Multiple regression analysis · Regression based residual study

Abbreviations

CEC	Cation exchange capacity
ECe	Electrical conductivity of a saturation soil paste
EDTA	Ethylene diamine tetra acetic
ESP	Exchangeable sodium percentage
FVC	Fractional vegetation cover
GT	Gamma test
K	Available potassium
LSE	Land surface emissivity
LST	Land surface temperature
MLR	Multiple linear regression
MNLR	Multiple non-linear regression
N	Available nitrogen
OLI	Operational land imager
P	Available phosphorous
Q-Q	Quantity versus quantity
RH	Relative humidity
RSC	Residual sodium carbonate
TIR	Thermal infrared
TOA	Top of atmosphere
WS	Water soluble

1 Introduction

Surface temperature not only influences the plant growth but also effects the natural environment of soil by influencing the soil respiration process of micro-organisms in soil ecosystem. Surface temperature of soil is responsible for maintaining desired depth of soil profile temperature by conduction mode of heat transfer. Role of micro-organisms is important to retain the balanced amount of nutrients in the soil for proper and healthier growth in soil ecosystem. Some studies state that temperature along the any vertical profile of soil decreases from top to bottom, especially if bottom profile is approaching towards a confined or unconfined aquifer (Singh et al. 2018b).

Many researchers have determined the importance of soil temperature on physical, chemical and biological phenomena (Toselli et al. 1999; Lahti et al. 2002; Weih and Karlsson 1999; Lahti et al. 2002; Repo et al. 2004; Probert 2000; Jebamalar et al. 2012; Tarate et al. 2017; Davidson and Janssens 2006; Buchas 2001) occurring in the crop root zone. Lehnert (2013) noted that inter-spherical processes which occur between soil and atmosphere by means of gas exchange is also influenced by LST. Soil respiration by means of micro-organisms and its temperature are positively correlated (Lloyd and Taylor 1994; Pajari 1995; Singh and Gupta 1977; Rajesh et al. 2023; Gadekar et al. 2023).

In recent years, the satellite remote sensing (RS) and geographic information system (GIS) have been used in many fields such as agriculture, hydrology, soil erosion, river network analysis (Kushwaha et al. 2019), change in earth surface features and many more (Jiang and Eastman 2000; Malczewski 2006; Chatterjee et al. 2017; Tabari et al. 2011; Kushwaha et al. 2017a). Along with this, RS combined with crucial soil properties through statistical modeling is another idea and no such study has been done till now. The soil temperature at earth surface which is called LST is crucial for air temperature, top of atmosphere (TOA), spectral radiance, brightness temperature, land surface emissivity, relative humidity (RH), and atmospheric water vapour studies. In satellite remote sensing, sensor's images do not provide LST directly but requires some band mathematics which is based on pixel values of sensor's images and metadata file information associated with that image.

Modeling between soil and crop parameters is longstanding. Too many models have been developed in this context by many researchers (Reimer and Shaykewich 1980; Gupta et al. 1981; Ghuman and Lal 1982; Manrique 1990; Kushwaha et al. 2021; Rehman et al. 2022). Langholz (1989) performed his study in bare soil condition and correlated this with natural vegetation. Kang et al. (2000) and Paul et al. (2004) studied in natural forest conditions and Parton (1984) shown the same study for pasture lands. Selection of modeling techniques is also a part of research and always been a challenge for researchers. There are many statistical technique for linear and non-linear modeling for these parameters viz. simple linear regression, multiple linear regression (MLR) and multiple non-linear regression (MLNR), artificial neural network (ANN), random forest (RF) data mining and other advanced techniques (Kushwaha et al. 2017b; Kushwaha and Kumar 2017a, b, c and 2021; Malik et al. 2019; Kumar et al. 2020; Anh et al. 2023). Simple linear regression can-not solve the problems if input variables are large.

Many agriculturist have stated that regression based models are better than complex data driven techniques (Navale et al. 2018; Zheng, et al. 2020, Tamta, et al. 2023; Mairizal, et al. 2020). Rastgou et al. (2020) have found that MLNR technique is better than RF data mining technique in estimating the soil water retention curve. Stangierski et al. (2019) have investigated MLR and ANN techniques results and stated that both model are good to predict the overall desirability of spreadable Gouda cheese during storage. Pahlavan-Rad et al. (2020) have found that MLR results are similar to RF technique in terms of root mean square error and mean absolute error. Complex non-linear data driven techniques do not provide equation for prediction. MLR and MNLR techniques help us to include potentially important variables in a single model. The benefits of these approaches are that this may lead to a more accurate and precise understanding of the addition of each individual input variables with the output through a single regression equation (Marill 2004; Kaup 2020).

The selection and assortment of inputs is one of the important criteria in modeling (Noori et al. 2011). Some of the studies suggest for the trial and error process for selecting appropriate input variables but time consumption cannot be avoided. In this context, gamma test (GT) has been proven by many meteorologists and hydrologists to generate a smooth model (Moghaddamnia et al. 2009; Stefansson et al. 1997; Remesan et al. 2008; Moghaddamnia et al. 2009; Lafdani et al. 2013; Singh et al. 2017; Singh et al. 2018a, b). GT gives the certain statistical values, which helps us to select and assort the effective inputs for modeling output data.

As per author's knowledge, no such studies have been carried that provides exact relationship by means of relating satellite data to identify soil chemical composition and its fertility status anywhere in the location under highly variable climatic conditions. This study can be a benchmark for future research. In this study, following objectives have been resolved: (i) to estimate LST over Guhla and Kaithal block of Kaithal district by using split-window technique, (ii) to construct a most efficient combination of input variables using gamma test technique to predict LST, (iii) to evaluate results of MLR analysis and MNLR analysis to predict remote sensing based LST associated with soil chemical parameters, (iv) to study the residuals associated with each output values of LST obtained by using MLR technique.

2 Material and Methods

2.1 *Description of Study Area*

Kaithal district is semi-arid region, located in north eastern part of Haryana state, India. It is located between 76°10′-76°42′ E longitudes and 29°31′-30°12′ N latitudes. The district has been divided into six administrative blocks viz. Kaithal, Guhla, Kalayat, Pundari, Rajound and Siwan. Guhla and Kaithal block have been selected in this study (Fig. 1). The geographical area of Kaithal and Guhla bocks are 1280 and 553 km^2, respectively. Soil of these regions is irrigated with all identified categories of groundwater. Mean temperature of these regions is 40 °C in summer (May and June) which is mainly dry and hot and 7 °C in winder in the month of January. Climatic condition is tropical steppe, semiarid and hot with four seasons in a year. These regions are nearly flat with gentle slope in the direction of south west. Soil of this district was found as sandy to sandy loam in texture.

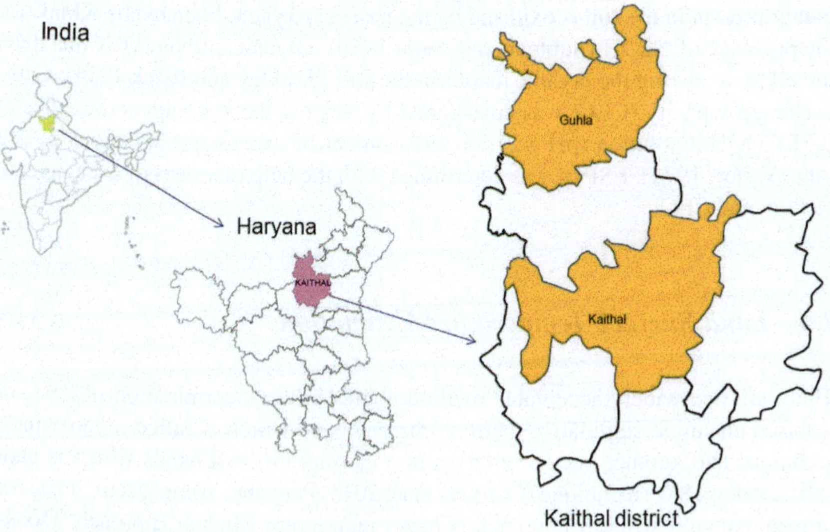

Fig. 1 Location map of Guhla and Kaithal blocks

2.2 Collection of Soil Samples

Twenty one soil samples have been taken from random locations of these blocks in which nine belong to Guhla block and twelve belong to Kaithal block. Selection of random locations was also dependent on the spatially classified ground water map, because it was found that ground water has the great impact on soil properties in these regions (Singh et al. 2017; Singh et al. 2018a, b). All the soil samples have been collected with auger from the depths of 0–15 cm. Top 15 cm soil is more responsible for crop growth and it was considered that this layer is also highly affected by surface temperature (Singh et al. 2020). Sampling was done before the showers of monsoon on 11 June 2015. The location of soil sampling in the form of longitudes and latitudes was recorded by hand held global positioning system and given in Table 3.

2.3 Analysis of Soil Samples

All collected soil samples have been air dried in shed for few days because of initial moisture present in the soil samples. Subsequently, grinding and sieving was done. Wooden mortar and plastic hammer have been used for grinding of soil and 2 mm sieve has been used to separate gravel particles from the samples. The soil samples were analyzed for pH, ECe, N, P and K, $CaCO_3$, water soluble cations (Ca^{2+}, Mg^{2+}, Na^+ and K^+) and water soluble anions ($CO_3{}^{2-}$, $HCO_3{}^-$, Cl^- and $SO_4{}^{2-}$) as per standard procedures described in USDA handbook No. 60 (Richads 1954). The organic matter present in the soil is oxidized by the nascent oxygen, liberated by $KMnO_4$, in the presence of NaOH (Subbiah and Asija 1956). Organic carbon (OC) was determined by oxidizing the organic matter in the soil (Walkley and Black 1934). Cation exchange capacity (CEC) was determined by treating the soil with sodium acetate (CH_3COONa) solution (pH 8.2) for replacement of exchangeable cations by Na^+ ions (Varley, 1971). ESP of was determined with the help of values of exchangeable sodium and CEC.

2.4 Land Surface Temperature Estimation

There are two widely acceptable methodologies in the determination of LST; one is based on single cumulative thermal infrared band which is called mono-window technique and another one is based on two thermal infrared bands which is called split window (SW) technique (Gabriele et al. 2015; Peng and Weng 2016). It has been proven that split-window approach is better than mono window approach (Dousset and Gourmelon, 2003; Sobrino et al. 2004, 2016; Weng et al. 2004; Mallick et al. 2008; Chatterjee et al. 2017; Kumar et al. 2017; Kushwaha et al. 2017a; Singh et al. 2018a, b). In this study, split-window approach has been used with Landset-8 imagery of 11 June 2015. This approach uses the following algorithm:

$$LST = T_{B10} + C_1(T_{B10} - T_{B11}) + C_2(T_{B10} - T_{B11})^2 + C_0 + (C_3 + C_4e_w)(1 - \varepsilon) + (C_5 + C_6e_w)\Delta\varepsilon \quad (1)$$

where *LST* (°K), T_{B10} and T_{B11} (brightness temperatures; °K), ε (mean LSE of TIR bands), e_w (atmospheric water vapor content; g/cm^2) and $\Delta\varepsilon$ (difference in LSE of TIR bands). All *C* notations are the split window coefficients (Sobrino et al. 1991, 2006; Shaouhua Zhao et al. 2009). To find $\Delta\varepsilon$, ε, T_{B10} and T_{B11}, steps have been given in Table 1.

Where, L_λ (TOA spectral radiance; watts per (m^2 × srad × μm)), M_L (radiance multiplicative band value), A_L (radiance additive band value), K_1 and K_2 (thermal conversion constants), *NIR* (near infrared band, *R* (red band), ε_s and ε_v (soil and vegetative emissivity values of the respective bands), B (band value) and suffices *s, v, 10 and 11* indicate soil, vegetation, band 10 and band 11, respectively. Values of M_L, A_L, K_1 and K_2 can be directly obtained from the metadata file of the image provided by the agency. To get the value of e_w, it is necessary to determine equilibrium vapor pressure ($e*_w$) (Eq. 2).

$$e_w^* = \left(1.0007 + 3.46 \times 10^{-6}P\right) \times (6.1121)e^{\left(\frac{17.502*T}{240.97+T}\right)} \quad (2)$$

$$e_w = RH \times e*_w \quad (3)$$

Table 1 Systematic calculations of difference in land surface emissivity ($\Delta\varepsilon$), mean land surface emissivity (ε), brightness temperatures of band 10 (T_{B10}) and band 11 (T_{B11})

Steps	Factor	Generalized formula	
		Band 10	Band 11
1	L_λ	$L_{\lambda,10} = M_L \times B_{10} + A_L$	$L_{\lambda,11} = M_L \times B_{11} + A_L$
2	T_B	$T_{B10} = \frac{K_{2,10}}{Ln\left\{\left(\frac{K_{1,10}}{L_{\lambda,10}}\right)+1\right\}}$	$T_{B11} = \frac{K_{2,11}}{Ln\left\{\left(\frac{K_{1,11}}{L_{\lambda,11}}\right)+1\right\}}$
3	*NDVI*	((NIR) – (R))/((NIR) + (R))	
4	*FVC*	$FVC = \frac{NDVI-NDVIs}{NDVIv-NDVIs}$	
5	*LSE*	$LSE_{10} = \varepsilon_{s,10}\ (1\text{-}FVC) + \varepsilon_{v,10} \times FVC;$	$LSE_{11} = \varepsilon_{s,11}\ (1\text{-}FVC) + \varepsilon_{v,11} \times FVC$
6	ε	$(LSE_{10} + LSE_{11})/2$	
7	$\Delta\varepsilon$	$(LSE_{10} - LSE_{11})$	

where T (dry bulb temperature; °C), P (absolute pressure; milli-bar), e^*_w and e_w (milli-bar) and RH (fraction).

Due to lack of weather data of Guhla and Kaithal blocks individually, lumped values of relative humidity (RH), absolute pressure and dry bulb temperature of whole Kaithal district have been taken, of given day and given time when satellite passed over the district. Soil samples were taken on the same day when LANDSAT-8 passed through the location. Work of soil sampling was started early morning and completed till the sunset on 11 June 2015. LANDSAT-8 imagery of the same day has been used, which was found cloud free fortunately. LANDSAT-8 imagery is having Cirrus band (i.e. band-9; wavelength is from 1.360 to 1.390 μm) which shows the high-altitude cloud contamination and this contamination is not visible in other spectral bands. Cirrus clouds appear bright in color whereas most land surfaces appear dark through the cloud-free atmospheres which containing water vapor. In spite of this, atmospheric correction has also been done in "Semi-Automatic Classification Plugin" in QGIS software and by using "Build Virtual Raster" processing tool.

2.5 *Gamma Test (GT)*

Gamma test helps us to select appropriate inputs for modeling the given output in terms of establishing a smooth model (Lafdanı et al. 2013; Singh et al. 2018a, b). It has the non-linear relationships that help us to use low and high variable data both in input selection. With the help of GT we can identify which input and what combination of it, should actually be used in a model (Changa and Heinemann 2018; Agalbjorn et al. 1997; Tsui et al. 2002). In this study, gamma values and variance ratio values have been determined in winGamma software. Lowest gamma and variance ratio values have been considered as best input/inputs (Jones et al. 2002).

2.6 Multiple Linear Regression Analysis (MLR)

In multiple regression analysis, we try to find the joint association of several independent variables with the dependent variable. We can estimate the value of dependent variable for any give value of independent variable. It is applied when we have multivariate data set. The general form of MLR is as follows:

$$Y = \alpha + \beta_1 X_1 + \beta_2 X_2 + \cdots \beta_k X_k \tag{4}$$

$$\hat{Y} = \alpha + \beta_1 X_1 + \beta_2 X_2 + \cdots \beta_k X_k + \epsilon \tag{5}$$

$$(Residual)_i = Y_i - \hat{Y}_i \tag{6}$$

where, Y is a dependent variable, $\hat{Y}$ is fitted value, $X_1, X_2, \ldots, X_k$ are independent variables, k is number of independent variable, i shows the ith observation, ϵ is error term or residual, α is intercept and $\beta_1, \beta_2, \ldots, \beta_k$ are partial regression coefficients. In this study, LST is dependent variable and soil chemical parameters are independent variables. The residual data of MLR model is the difference between the observed data of the dependent variable LST (i.e. LST_i) and the fitted values of LST (i.e. $\widehat{LST}_i$) (Eq. 6).

2.7 Multiple Non-linear Regression Analysis (MNLR)

It is applied when dependent variable has no linear relationship with independent variables in multivariate data set (Tabari et al. 2010; Malik and Kumar 2018). In this study, a special form of non-linear relationship has been used, which best suits for non-linear relationship between LST and soil chemical parameters (Eq. 7).

$$Y = \alpha \beta_1^{X_1} \beta_2^{X_2} \ldots \beta_k^{X_k} \tag{7}$$

where Y is a dependent output and $X_1, X_2, \ldots, X_k$ are independent inputs, k is number of independent variable, α is intercept and $\beta_1, \beta_2, \ldots, \beta_k$ are partial regression coefficients of MNLR equation.

2.8 Performance Evaluation of Models

Performance measures are used to indicate how well a model performs its tasks. The performance of the model can be measured qualitatively and quantitatively. Karl Pearson's correlation coefficient (r), multiple correlation coefficient (MCC),

Table 2 Quantitative performance evaluation indices

Parameter	Relationship	Range
Root mean square error (RMSE)	$\sqrt{\frac{\sum_{i=i}^{N}(LST_i - LST_{ci})^2}{N}}$	0 to ∞
Correlation coefficient (r)	$\frac{\sum_{i=1}^{N}(LST_i - LST_m)\sum_{i=1}^{N}(LST_{ci} - LST_{cm})}{\sqrt{\sum_{i=1}^{N}(LST_i - LST_m)^2}\sqrt{\sum_{i=1}^{N}(LST_{ci} - LST_{cm})^2}}$	−1 to +1
Multiple correlation coefficient (MCC)	$\sqrt{\frac{\sum_{i=1}^{N}(\widehat{LST}_i - LST_m)^2}{\sum_{i=1}^{N}(LST_i - LST_m)^2}}$	−1 to +1

coefficient of determination (R^2), coefficient of multiple determination (CMD) and root mean square error (RMSE) have been used as quantitative performance evaluating indices for predicting the effectiveness of the models and given in Table 2. MCC indicates the joint association of several independent variables with the dependent variable. Coefficient of determination and coefficient of multiple determination help us to find out percent of variation in dependent variable due to independent variable/variables; both parameters are very close to the square of r and multiple correlation coefficient, respectively. In this study, multiple correlation coefficient, coefficient of multiple determination and root mean square error were estimated between remote sensing based LST and soil chemical parameters in multiple linear regression whereas r, R^2 and root mean square error (RMSE) were estimated between multiple non-linear regression model computed LST and remote sensing based LST.

Where, LST_{ci} and LST_i are the model computed LST and remote sensing based LST for i^{th} observation, respectively, LST_{cm} and LST_m are the mean of model computed and mean of remote sensing based LST, respectively, N is the total number of observations and $\widehat{LST}_i$ is fitted values of LST.

3 Results and Discussion

3.1 *Land Surface Temperature Estimation and Statistical Analysis of Data*

LST were determined using split-window approach (Fig. 2) and values of LST were interpolated at 21 locations of soil sampling and given in Table 3. From Fig. 2, green colour shows most of the water bodies and vegetation, whereas red colour shows concrete structure, upland plains, built up lands etc. 71.43% sampling was done in the LST range 307.357 to 316.482 °K whereas 28.57% sampling was done in LST range 316.483 to 318.74 °K.

Coefficient of variation (C_V) is very useful statistical technique to compare variations in the parameters (Table 4). Calcium carbonate has highest coefficient of variation (i.e. 135.385%) whereas LST has lowest C_V (i.e. 0.637%). Saturation and

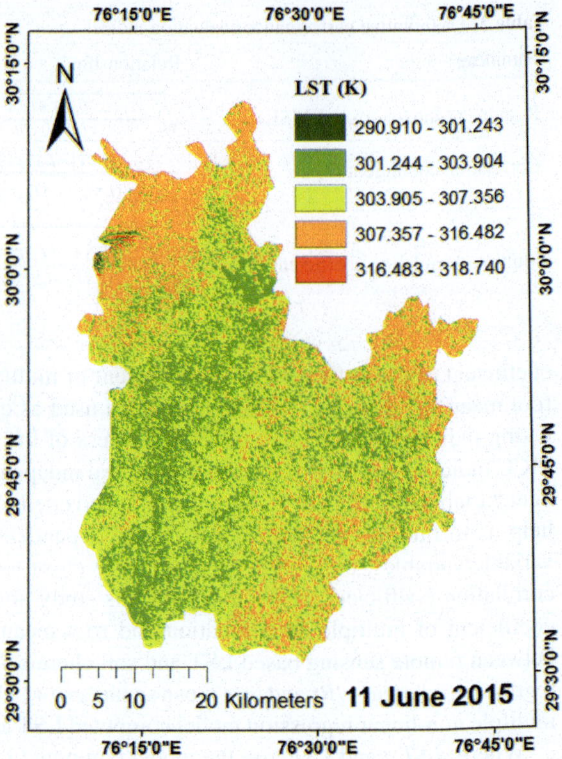

Fig. 2 Land surface temperature of Kaithal district using split-window technique

cation exchange capacity have C_V value less than 18% whereas organic carbon and pH have C_V value less than 10%. In available N, available P and available K, available P has highest C_V i.e. 17.993% and available N has lowest C_V i.e. 11.955%. Exchangeable sodium percentage and electrical conductivity have C_V value 37.732% and 34.858%, respectively. Among all the anions, chloride has highest 71.171% C_V and Bicarbonate has lowest 28.395% C_V. Among all the cations, calcium has highest value of C_V i.e. 65.53% and potassium has lowest value of C_V i.e. 22.365%. C_V from highest to lowest can be arranged in following manner; calcium carbonate > chloride > calcium > magnesium > carbonate > sulphate > exchangeable sodium percentage > sodium > electrical conductivity > bicarbonate > WS potassium > available phosphorous > cation exchange capacity > available potassium > available nitrogen > saturation > organic carbon > pH > LST.

Table 3 LST estimated by split window approach at respective locations

Sampling number	Longitude-Latitude	LST (°K)	Name of block	Name of village
1	76° 17′ 2.59″ E-29° 57′ 8.06″ N	315.183	Guhla	Kharkhada
2	76° 16′ 2.30″ E-30° 00′ 0.31″ N	317.338	Guhla	Sultania
3	76° 24′ 6.65″ E-30° 03′ 5.76″ N	316.042	Guhla	Bhagal
4	76° 22′ 5.77″ E-30° 05′ 2.27″ N	313.710	Guhla	Dhandhauta
5	76° 16′ 9.21″ E-30° 01′ 0.47″ N	316.920	Guhla	Kharodi
6	76° 22′ 5.07″ E-30° 04′ 3.13″ N	311.307	Guhla	Badsui
7	76° 14′ 1.86″ E-30° 01′ 4.51″ N	317.369	Guhla	Kheri Dabah
8	76° 16′ 4.71″ E-30° 01′ 3.19″ N	316.422	Guhla	Dusharpur
9	76° 23′ 8.87″ E-30° 06′ 6.04″ N	313.760	Guhla	Kasauli
10	76° 23′ 4.86″ E-29° 43′ 6.53″ N	312.085	Kaithal	Titram
11	76° 21′ 0.58″ E-29° 50′ 4.14″ N	312.011	Kaithal	Serta
12	76° 24′ 5.88″ E-29° 41′ 0.41″ N	312.655	Kaithal	Deobon
13	76° 23′ 1.66″ E-29° 46′ 2.58″ N	313.530	Kaithal	Chandana
14	76° 26′ 9.44″ E-29° 45′ 3.28″ N	314.414	Kaithal	Sega
15	76° 24′ 0.96″ E-29° 49′ 8.88″ N	313.218	Kaithal	Khurana
16	76° 26′ 9.24″ E-29° 46′ 4.53″ N	315.981	Kaithal	Narar
17	76° 30′ 3.02″ E-29° 50′ 4.40″ N	317.386	Kaithal	Tik
18	76° 32′ 0.61″ E-29° 48′ 2.04″ N	316.751	Kaithal	Naina
19	76° 27′ 2.67″ E-29° 49′ 5.43″ N	315.026	Kaithal	Jagdishpura
20	76° 30′ 2.43″ E-29° 47′ 0.00″ N	316.711	Kaithal	Mundri
21	76° 22′ 4.38″ E-29° 51′ 3.03″ N	311.887	Kaithal	Ujana

Table 4 Statistical analysis of data for the study area

Parameter	Minimum value	Maximum value	Mean	Std. deviation	Coefficient of variation (%)
LST (°K)	311.307	317.386	314.748	2.004	0.637
Saturation (%)	33.5	47.5	37.750	3.878	10.272
CEC (cmol kg^{-1})	8.73	15.26	10.869	1.855	17.069
OC (%)	0.36	0.49	0.448	0.040	8.849
Calc. carbonate (%)	0	7	1.247	1.688	135.385
Available N (kg ha^{-1})	155.16	258.86	197.631	23.627	11.955
Available P (kg ha^{-1})	14.4	32	23.339	4.199	17.993
Available K (kg ha^{-1})	178.1	302.3	245.372	32.265	13.149
ESP (%)	5.4	25.55	14.658	5.531	37.732
ECe (dSm^{-1})	1.25	6.67	3.663	1.277	34.858
pH	7.21	9.32	8.421	0.596	7.081
Carbonate (me l^{-1})	0	4.82	2.457	1.294	52.655
Bicarbonate (me l^{-1})	6	18	12.865	3.653	28.395
Chloride (me l^{-1})	2	42.25	12.739	9.066	71.171
WS sulphate (me l^{-1})	0.89	11.77	5.731	2.834	49.455
WS calcium (me l^{-1})	1	11.64	4.662	3.055	65.530
WS magnesium (me l^{-1})	1.72	13.55	6.049	3.426	56.644
WS sodium (me l^{-1})	5.92	39.5	24.101	8.483	35.197
WS potassium (me l^{-1})	1.29	2.79	1.885	0.422	22.365

Note cmol represents centi-mole, dSm^{-1} represent deci-Siemens per meter and me l^{-1} represents milli-equivalent per litre

3.2 Karl Pearson's Correlation Analysis of LST with All Measured Soil Chemical Parameters

From the correlation matrix (Table 5), it was concluded that LST has no significant correlation with any of the soil chemical parameter. Although, LST shows correlation coefficient, 0.47 with carbonate but this correlation cannot be considered for modeling purpose. Correlation of LST from highest to lowest can be arranged in following manner; carbonate > potassium > available potassium > organic carbon > calcium carbonate > available nitrogen > saturation, cation exchange capacity, exchangeable sodium percentage > pH, chloride, sulphate > sodium > magnesium > bicarbonate > electrical conductivity, calcium > available phosphorous.

3.3 Best Input Selection Using Gamma Test

Gamma test was performing with all soil parameters for the selection of best input structure. Total 44 inputs structures were prepared by applying proper combination technique (Table 6). The structure which shows lower values of absolute gamma and variance ratio, was considered as best input structure. Structure number "30" which has combination of available nitrogen, available potassium and carbonate shows lowest value of absolute gamma (0.011) and second lowest value of variance ratio (−0.043). So, combination "30" has been selected as best input combination among all. In structure number "22", which has combination of available nitrogen and water soluble magnesium, also shows lowest value of variance ratio (−0.745), but this structure has poor value of absolute gamma (0.186), so it was not considered for study. Structure number "39" with combinations of available nitrogen, available potassium, electrical conductivity and carbonate has highest values of absolute gamma (0.577) and variance ratio (2.307) and it has been found inferior than all the combinations (Table 6). Boxplot of LST, N, K and soil carbonate (Fig. 3), demonstrates that LST and carbonate have least variation and minimum, maximum, first quartile, median and third quartile values almost coincide, notwithstanding LST values are much higher than the carbonate values. N and K values shows some variation in comparison to LST and carbonate, with deviation in minimum, maximum, first quartile, median and third quartile values.

3.4 Model Generation Using Multiple Linear Regression Analysis and Residual Study

Multiple linear regression model (Eq. 8) shows that the value of intercept is very high (301.309) whereas partial regression coefficients of available nitrogen, available potassium and carbonate are found low as 0.024, 0.028 and 1.242, respectively.

Table 5 Correlation matrix of LST and all soil chemical parameters

	LST	Saturation	CEC	Organic carbon	$CaCO_3$	N	P	K	ESP	ECe	pH	CO_3^{2-}	HCO_3^-	Cl^-	SO_4^{2-}	Ca^{2+}	Mg^{2+}	Na^+	K^+
LST	1.00																		
Saturation	0.10	1.00																	
CEC	0.10	1.00	1.00																
Organic carbon	−0.13	0.25	0.26	1.00															
$CaCO_3$	0.12	−0.26	−0.27	−0.32	1.00														
N	0.11	0.24	0.26	0.81	−0.15	1.00													
P	0.01	0.56	0.55	0.60	−0.22	0.44	1.00												
K	0.29	0.47	0.47	0.08	0.09	0.17	0.25	1.00											
ESP	−0.10	−0.64	−0.66	−0.18	0.08	−0.42	−0.33	−0.52	1.00										
ECe	−0.02	0.12	0.11	−0.24	−0.08	−0.43	−0.22	−0.07	0.21	1.00									
pH	−0.08	−0.54	−0.56	−0.31	0.23	−0.55	−0.25	−0.46	0.90	0.38	1.00								
CO_3^{2-}	0.47	−0.32	−0.33	−0.33	0.54	−0.25	−0.11	−0.10	0.42	0.13	0.42	1.00							
HCO_3^-	−0.03	−0.26	−0.29	−0.35	0.17	−0.54	−0.09	−0.34	0.61	0.41	0.77	0.30	1.00						
Cl^-	−0.08	0.21	0.22	−0.03	−0.21	−0.18	−0.18	−0.03	−0.01	0.90	0.09	−0.07	0.02	1.00					
SO_4^{2-}	−0.08	0.18	0.17	−0.19	−0.30	−0.38	−0.08	0.04	0.05	0.86	0.23	−0.12	0.20	0.82	1.00				
Ca^{2+}	0.02	0.62	0.62	−0.02	−0.20	−0.08	0.11	0.19	−0.41	0.77	−0.21	−0.15	0.04	0.80	0.71	1.00			
Mg^{2+}	0.04	0.43	0.44	−0.15	−0.10	−0.11	−0.15	0.20	−0.44	0.70	−0.27	−0.23	−0.10	0.81	0.72	0.87	1.00		
Na^+	−0.07	−0.27	−0.28	−0.30	−0.02	−0.56	−0.35	−0.29	0.62	0.88	0.72	0.30	0.57	0.70	0.73	0.38	0.33	1.00	
K^+	−0.34	−0.45	−0.45	0.08	0.44	0.03	−0.27	−0.33	0.39	−0.20	0.34	0.09	0.08	−0.16	−0.40	−0.46	−0.39	0.01	1.00

Table 6 Gamma test results with all possible combinations of input structure

Str. No.	Input	$\lvert\Gamma\rvert$	Variance ratio	Str. No.	Input	$\lvert\Gamma\rvert$	Variance ratio
1	Saturation	0.281	1.123	23	K, ECe	0.045	0.178
2	CEC	0.224	0.897	24	K, CO_3^{2-}	0.438	1.752
3	Organic carbon	0.295	1.179	25	K, Mg^{2+}	0.293	1.171
4	$CaCO_3$	0.262	1.048	26	ECe, CO_3^{2-}	0.271	1.084
5	N	0.099	0.397	27	ECe, Mg^{2+}	0.270	1.079
6	P	0.429	1.715	28	CO_3^{2-}, Mg^{2+}	0.080	0.320
7	K	0.132	0.528	29	N, K, ECe	0.143	0.573
8	ESP	0.370	1.481	30	**N, K, CO_3^{2-}**	**0.011**	**−0.043**
9	ECe	0.084	0.335	31	N, K, Mg^{2+}	0.098	0.390
10	pH	0.216	0.863	32	K, ECe, CO_3^{2-}	0.358	1.432
11	CO_3^{2-}	0.156	0.622	33	K, ECe, Mg^{2+}	0.166	0.663
12	HCO_3^-	0.206	0.825	34	ECe, CO_3^{2-}, Mg^{2+}	0.238	0.950
13	Cl^-	0.424	1.695	35	N, ECe, CO_3^{2-}	0.094	0.378
14	SO_4^{2-}	0.313	1.251	36	N, CO_3^{2-}, Mg^{2+}	0.173	0.691
15	Ca^{2+}	0.399	1.600	37	K, CO_3^{2-}, Mg^{2+}	0.277	1.109
16	Mg^{2+}	0.174	0.695	38	N, ECe, Mg^{2+}	0.088	0.351
17	Na^+	0.301	1.206	39	N, K, ECe, CO_3^{2-}	0.577	2.307
18	K^+	0.290	1.161	40	N, K, ECe, Mg^{2+}	0.031	0.122
19	N, K	0.072	0.289	41	K, ECe, CO_3^{2-}, Mg^{2+}	0.198	0.791
20	N, ECe	0.124	0.494	42	N, ECe, CO_3^{2-}, Mg^{2+}	0.196	0.784
21	N, CO_3^{2-}	0.194	0.775	43	N, K, CO_3^{2-}, Mg^{2+}	0.282	1.129

(continued)

Table 6 (continued)

Str. No.	Input	|Γ|	Variance ratio	Str. No.	Input	|Γ|	Variance ratio
22	N, Mg^{2+}	0.186	−0.745	44	N, K, ECe, CO_3^{2-}, Mg^{2+}	0.229	0.917

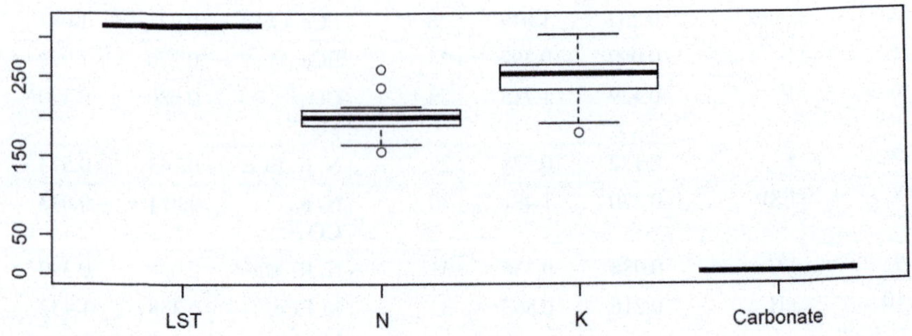

Fig. 3 Boxplot of land surface temperature, available nitrogen, available potassium and soil carbonate

Multiple linear regression model demonstrates that LST has less dependency on available nitrogen and available potassium in comparison to carbonate. Coefficient of multiple determination and multiple correlation coefficient between LST and input parameters were obtained as 0.367 and 0.606, respectively. RMSE was obtained as 2.276 °K.

$$LST = 301.309 + 0.024\,N + 0.028\,K + 1.242\,Carbonate \quad (8)$$

Now, for the population analysis a hypothesis was prepared whether multiple correlation coefficient is significant or not. Following null and alternative hypothesis were made for this:

H_0: All partial regression coefficients are zero or multiple correlation coefficient is not significant

H_1: Not all partial regression coefficients are zero or multiple correlation coefficient is significant

For testing the hypothesis, F-test was performed. Standard error, t-distribution value and probability values for intercept and three input variables have been given in Table 7. F-statistic was obtained as 3.291 on 3 and 17 degree of freedom. Final p-value was obtained as 0.04602 which is less than 0.05; therefore, null hypothesis was rejected at 5% level of significance i.e. multiple correlation coefficient is significant.

In this study, residual values associated with each LST observation, have also been studied (Table 8). In Fig. 8, scatter plot between residuals and remote sensing based LST has been plotted; this shows the increasing trend of residual over LST.

Table 7 F-test results

Coefficients	Estimate	Standard error	t-value	Pr(>|t|)
Intercept	301.30867	6.46735	46.589	<2e−16***
N	0.02433	0.02497	0.974	0.3436
K	0.02798	0.01780	1.572	0.1345
Carbonate	1.24205	0.45195	2.748	0.0137*

Note Level of significant codes: '***' 0.001, '*' 0.05

Minimum, first quartile, second quartile, third quartiles and maximum values of residuals distribution were obtained as −5.783, −1.492, −0.092, 1.7 and 3.876, respectively. The standardized residual is the ratio of residual and its standard deviation. The normal probability plot of residuals is a graphical representation in order to comparing a given data set with its normal distribution. We can use it with the standardized residual of the MLR model and see if the error term ϵ is actually follows normally distributed or not (Fig. 4). Scatter plot of scale location between square root of standardized residual versus fitted value of LST and standardized residual versus leverage value have also been plotted (Figs. 5 and 8, respectively). Normal Q-Q plot between standardized residual and theoretical quantities is very close to the 1:1 line (Fig. 6). Line of best fit for small intervals of fitted values of LST is very close to the "0" residual line (Fig. 7).

Table 8 Residual value associated with each observation of land surface temperature

Sampling number	LST (°K)	Residual (°K)	Sampling number	LST (°K)	Residual (°K)
1	315.183	−0.348	12	312.655	−0.76
2	317.338	3.876	13	313.53	0.195
3	316.042	0.693	14	314.414	−1.767
4	313.71	2.951	15	313.218	−0.505
5	316.92	2.198	16	315.981	2.792
6	311.307	−2.455	17	317.386	−1.492
7	317.369	3.493	18	316.751	−5.783
8	316.422	−0.092	19	315.026	0.126
9	313.76	1.7	20	316.711	−0.685
10	312.085	−2.772	21	311.887	−1.645
11	312.011	0.28			

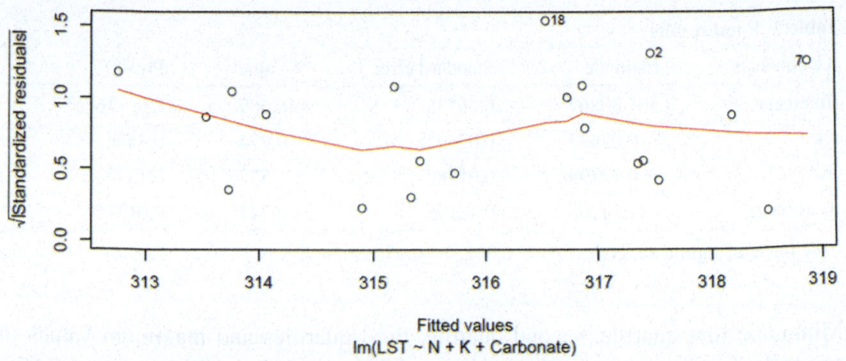

Fig. 4 Scale location between square root of standardized residual and fitted value of land surface temperature

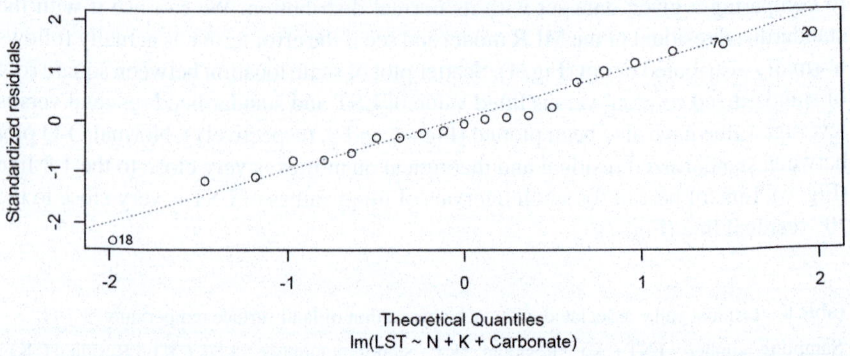

Fig. 5 Normal Q-Q plot between standardized residual and theoretical quantities

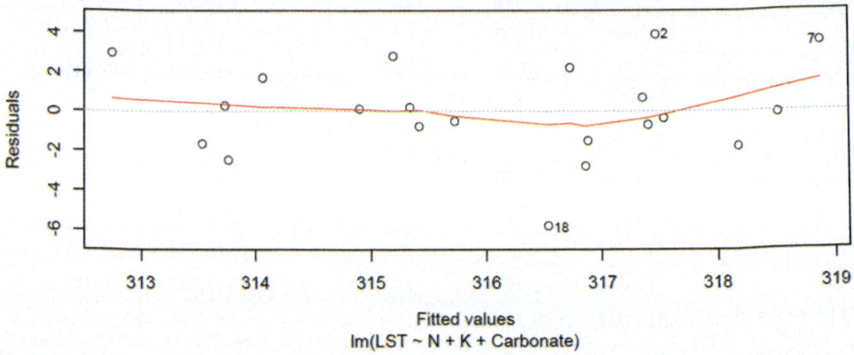

Fig. 6 Scatter plot between residuals and fitted values of land surface temperature

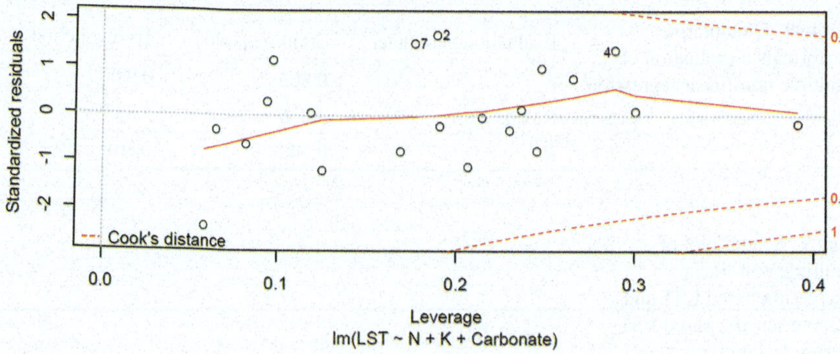

Fig. 7 Scatter plot between standardized residual and leverage value

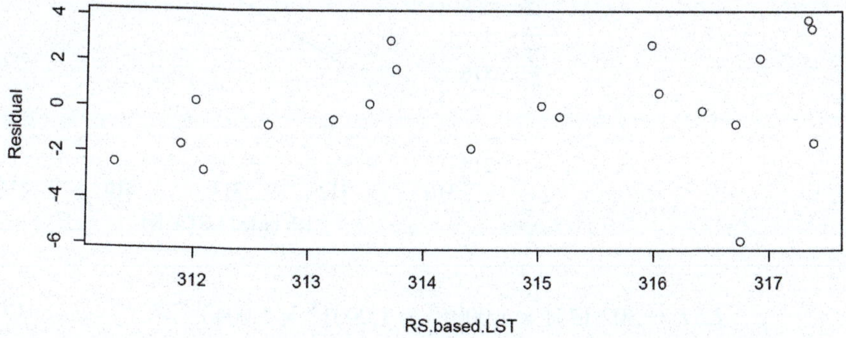

Fig. 8 Scatter plot between residuals and remote sensing based land surface temperature

3.5 *Model Generation Using Multiple Non-linear Regression Analysis*

LST dataset length was divided into two parts viz. training dataset and testing dataset. In this study 70% data (i.e. sampling number 1–15) have been used for training purpose and remaining 30% (i.e. sampling number 16–21) have been used form testing of multiple non-linear regression model (Kisi et al. 2012). In multiple non-linear regression model (Eq. 9), the value of parameter (α) is high (302.1731), whereas coefficients β_1 and β_2 are found with low values as 1.0001. Coefficient of carbonate (β_3) is comparatively higher (1.0043) than β_1 and β_2. All the coefficients demonstrate that LST is slight dependent on available nitrogen and available potassium in comparison to carbonate. During training period over fitting may occur, due to which testing results have been taken under consideration. The R^2 value (0.861) of model during testing period is acceptable, notwithstanding RMSE value is comparatively high (3.333 °K) (Table 9). Scatter diagram between remote sensing based LST and model generated LST, shows that the model results are over estimated (Fig. 9).

Table 9 Quantitative performance evaluation of multiple non-linear regression model

Evaluation parameter	Training period	Testing period
r	0.866	0.928
RMSE (°K)	3.835	3.333
R^2	0.749	0.861

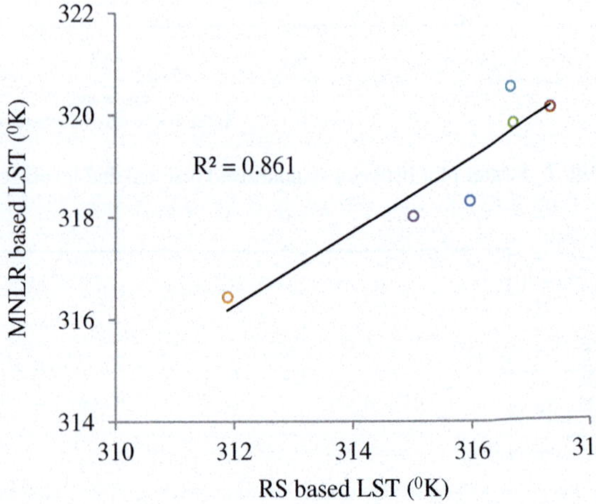

Fig. 9 Scatter plot of multiple non-linear regression based LST and remote sensing based LST during testing period

$$LST = 302.1731 \times 1.0001^{N} \times 1.0001^{K} \times 1.0043^{Carbonate} \quad (9)$$

Furthermore, this study were compared and influenced by the previous relevant work on this topic (Philip and De Vries 1957; Geiger 1961; DeVries 1963; Wierenga et al. 1969; Hanks et al. 1971; Hasfurther and Burman 1974; Khatry et al. 1978; DeBano and Conrad 1978; Lettau 1979; Meikle and Tredway 1979; Trangmar et al. 1985; Farouki 1986; Giovannini et al. 1990; Cambardella et al. 1994; Csiszar and Gutman 1999). Krarti et al. (1995) presented a parametric analysis in order to observe the impact of various properties of the soil such as evaporation from the soil surface, absorptivity and convective properties of soil. Convective properties of soil refer to the surface temperature, which occurs due to either temperature or density differences. This analysis indicated that simplified model predicts the surface temperature of soil within only 10% of the data of five locations. This data were obtained from measurement. Schaab et al. (1999) worked on 3-dimensional surface of either a plant canopy or a terrain and reported the effects on the amount of biogenically volatile organic compounds which were emitted. They have also developed a complex solar radiation model (SORAM) which was based on the GIS. The advantage of this model is that it is not limited to the certain test sites. Values of temperature were linked to the output of the given radiation and further, they have stated that this procedure enabled them to correct emission factors for light as well as temperature by using well known emission algorithms of Tingey et al.'s (1980) and Guenther et al.'s (1993).

Al-Temeemi and Harris (2001) predicted a soil profile of type of periodic variations in subsurface temperature. This variation was recorded with the soil depth and generation of the profile was based on equation for subterranean temperatures. These equations were based on the thermal and physical properties of the soil. At the end, these subsurface temperatures were compared with the ambient dry-bulb temperature. Mihalakakou (2002) stated that the soil surface temperature is a crucial factor in order to calculate the thermal performance of buildings in direct contact with the soil. He has also tested the efficiency of earth-to-air heat exchangers on the basis of his research. Gimeno-Garcia et al. (2004) created an artificial fire for assessing the surface temperature where was observed on spatial basis. This experiment was conducted in a field station on forest slope. Obtained results indicated that the spatial patterns of soil temperature really affect the crucial soil properties and depends on the fire severity. This fact could be contributed to change the spatial dynamics of soil nutrients and this research may help in the in the recovery of the burned vegetation. Fonseca et al. (2019) performed simulation trial for two week periods in the austral winter season. They have used surface temperature and relative humidity relationship by using a model called Weather Research and Forecasting (WRF). They have taken in-situ observations at three different sites in order to relate variables. They have also conducted sensitivity experiments to improve the observed measurements. In this sensitivity experiment surface albedo, moisture content and five tuneable parameters viz. soil porosity, soil suction, saturated permeability and b parameter were perturbed for Noah Land Surface Model. Wang et al. (2020) worked on effects on glomalin traits in Changchun city. For this purpose, 500 km^2 was selected in urban region and sampling was done in 281 random sites. To indicate the forest characteristics, 3 parameters were selected viz. plant diversity traits, tree size, and tree density, and 4 parameters were selected to indicate the soil properties viz. pH, bulk density, electrical conductivity, and temperature, and for indicating land use configuration, 4 characteristics were selected viz. road, green-space, building, and water. Association decoupling manifested that, in low-medium urbanization areas, soil properties explained 48–78% of inter-site variation. These all research indicates the importance of soil surface temperature on soil properties mainly chemical properties which motivated and helped us to work on this scientific research.

4 Conclusions

The purpose of this study is to check whether land surface temperature actually affects the soil chemical properties or not. MLR and MNLR based models have been generated with best selected inputs using gamma test, although all the variables have shown inferior correlation among one another. In multiple linear regression based model, multiple correlation coefficient was accepted at 5% level of significance and it has given very less values of residual, whereas due to high variation in the dataset, results of linear model are found inferior in comparison to non-linear model, notwithstanding multiple linear regression based model has lesser value of RMSE in comparison to multiple non-linear regression based model. This study will help us to determine fertility and alkalinity status of the soil. Furthermore, residual sodium carbonate can also be determined by some advancement of research in the areas which are actually suffering from salinity hazard. A promising correlation between remote sensing based LST and multiple non-linear regression based LST, provides a base for further research to determine the soil chemical parameters of difficult places where soil sampling is difficult or onerous, although multiple non-linear regression based model have shown over estimation which could be reduced by increasing the training data length. With this study, more research can also be performed to check the effect of surface temperature and soil chemical properties on ground water quality in semi-arid regions.

Acknowledgements Authors are very thankful to departmental staffs of soil science of Chaudary Charan Singh Haryana Agricultural University. They have provided necessary assistance in onerous work of soil sampling in one day and its analysis. We are also grateful for Govind Ballabh Pant University of Agriculture and Technology, Pantnagar for providing us computer laboratory for entire research work.

Conflicts of Interest None.

References

Agalbjorn S, Koncar N, Jones AJ (1997) A note on the gamma test. Neural Comput Appl 5:131–133. https://doi.org/10.1007/BF01413858

Al-Temeemi AA, Harris DJ (2001) The generation of subsurface temperature profiles for Kuwait. Energy Build 33(8):837–841

Anh DT, Tanim AH, Kushwaha DP, Pham QB, Bui VH (2023) Deep learning long short-term memory combined with discrete element method for porosity prediction in gravel-bed rivers. Int J Sedim Res 38(1):128–140. https://doi.org/10.1016/j.ijsrc.2022.08.001

Buchas GD (2001) Soil temperature regime. In: Smith KA, Mullins ED (eds) Soil and environmental analysis: physical methods. Marcel Dekker, New York, pp 539–594

Cambardella C, Moorman TB, Novak JM, Parkin TB, Karlen DL (1994) Field-scale variability of soil properties in Central Iowa soils. Soil Sci Soc Am J 58:1501–1511

Changa F, Heinemann PH (2018) Optimizing prediction of human assessments of dairy odors using input variable selection. Comput Electron Agric 150:402–410

Chatterjee RS, Singh N, Thapa S, Sharma D, Kumar D (2017) Retrieval of land surface temperature (LST) from landsat TM6 and TIRS data by single channel radiative transfer algorithm using satellite and ground-based inputs. Int J Appl Earth Observat Geo Inf. https://doi.org/10.1016/j.jag.2017.02.017

Csiszar I, Gutman G (1999) Mapping global land surface albedo from NOAA/AVHRR data. J Geophys Res 104:6215–6228

Davidson EA, Janssens IA (2006) Temperature sensitivity of soil carbon decomposition and feedbacks to climate change. Nature 440(7081):165–173

DeBano LF, Conrad CE (1978) The effect of fire on nutrients in a chaparral ecosystem. Ecology 59:489–497

DeVries DA (1963) Thermal properties of soils, physics of plant environment. In: Van Wijk WR (ed) North Holland Publishing Company, Amsterdam, pp 210–235

Dousset B, Gourmelon F (2003) Satellite multi-sensor data analysis of urban surface temperatures and land cover. Photog Rem Sens 58:43–54

Farouki OT (1986) Thermal properties of soil. Trans Tech Publications, Clausthal Zellerfeld, Germany

Fonseca R, Zorzano-Mier M, Azua-Bustos A, Gonzalez-Silva C, Martin-Torres J (2019) A surface temperature and moisture intercomparison study of the Weather Research and Forecasting model, in-situ measurements and satellite observations over the Atacama Desert. Clim Resilience Sustain 145(722):2202–2220

Gabriele C, Amanda LS, Ming P, Eric FW (2015) Creating consistent datasets by combining remotely-sensed data and land surface model estimates through Bayesian uncertainty post-processing: the case of land surface temperature from HIRS. Rem Sen Environ. https://doi.org/10.1016/j.rse.2015.09.010

Gadekar K, Pande CB, Rajesh J, Gorantiwar SD, Atre AA (2023) Estimation of land surface temperature and urban heat island by using google earth engine and remote sensing data. In: Pande CB, Moharir KN, Singh SK, Pham QB, Elbeltagi A (eds) Climate change impacts on natural resources, ecosystems and agricultural systems. Springer Climate. Springer, Cham. https://doi.org/10.1007/978-3-031-19059-9_14

Geiger R (1961) The climate near the ground. Harvard University Press, Cambridge, MA

Gimeno-Garcia E, Andreu V, Rubio JL (2004) Spatial patterns of soil temperatures during experimental fires. Geoderma 118(1–2):17–38

Giovannini G, Lucchesi S, Giachetti M (1990) Effect of heating on some chemical parameters related to soil fertility and plant growth. Soil Sci 149:344–350

Ghuman BS, Lal R (1982) Temperature regime of a tropical soil in relation to surface condition and air temperature and its Fourier analysis. Soil Sci 134:133–140

Guenther AB, Zimmerman PR, Harley PC, Monson RK, Fall R (1993) Isoprene and monoteme emission rate variability: model evaluation and sensitivity analyses. J Geophys Res 98(7):12609–12617

Gupta SC, Radke JK, Larson WE (1981) Predicting temperatures of bare and residue covered soils with and without a corn crop. Soil Sci Soc Am J 45:405–412

Hanks RJ, Austin DD, Ondreehen WT (1971) Soil temperature estimation by a numerical method. Soil Sci Am Proc 35:665–667

Hasfurther VR, Burman RD (1974) Soil temperature modeling using air temperature as a driving mechanism. Trans ASAE 17:78–81

Jebamalar AS, Raja S, Thambi A, Sunitha BSJ (2012) Prediction of annual and seasonal soil temperature variation using artificial neural network. Indian J Radio Space Phys 41(1):48–57

Jiang H, Eastman JR (2000) Application of fuzzy measures in multi-criteria evaluation in GIS. Int J Geogra Info Sys 14(2):173–184

Jones AJ, Margetts S, Durrant P (2002) The winGammaTM user guide. Cardiff U.K., University of Wales. http://users.cs.cf.ac.uk/O.F.Rana/Antonia.J.Jones/GammaArchive/Gamma%20Software/winGamma/winGammaManual2001.pdf

Kang S, Kim S, Oh S, Lee D (2000) Predicting spatial and temporal patterns of soil temperature based on topography, surface cover and air temperature. For Ecol Manage 36:173–184

Kaup C (2020) The optimum of heat recovery—determination of the optimal heat recovery based on a multiple non-linear regression model. J Build Eng. https://doi.org/10.1016/j.jobe.2020.101548

Khatry AK, Sodha MS, Malik MAS (1978) Periodic variation of ground temperature with depth. Sol Energy 20:425–427

Kisi O, Dailr AH, Cimen M, Shiri J (2012) Suspended sediment modeling using genetic programming and soft computing techniques. J Hydrol 450–451:48–58

Krarti M, Lopez-Alonzo C, Claridge DE, Kreider JF (1995) Analytical model to predict annual soil surface temperature variation. J Sol Energy Eng 117(2):91–99

Kumar M, Tripathi DK, Maitri V, Biswas V (2017) Impact of urbanisation on land surface temperature in Nagpur, Maharashtra. In: Sharma P, Rajput S (eds)

Kumar M, Kumari A, Kushwaha DP, Kumar P, Malik A, Ali R, Kuriqi A (2020) Estimation of daily stage–discharge relationship by using data-driven techniques of a perennial river, India. Sustainability 12:7877. https://doi.org/10.3390/su12197877

Kushwaha DP, Kumar D (2017a) Multilayer perceptron and suspended sediment modeling: a case study. Lambert Academic Publishing, Chisinau, Republic of Moldova

Kushwaha DP, Kumar D (2017b) Modeling suspended sediment concentration using multilayer feedforward artificial neural network at the outlet of the watershed. Int J Agric Eng 10(2):1–9. https://doi.org/10.15740/HAS/IJAE/10.2/1-9

Kushwaha DP, Kumar D (2017c) Suspended sediment modeling with continuously lagging input variables using artificial intelligence and physics based models. Int J Curr Microbiol Appl Sci 6(10):1386–1399. https://doi.org/10.20546/ijcmas.2017.610.164

Kushwaha DP, Kumar A (2021) Modeling of sediment yield and nutrient loss after application of pre-determined dose of top soil amendments. Pharma Innov J 10(4):1199–1206

Kushwaha DP, Singh VK, Tarate SB (2017a) Land surface temperature estimation using split window approach over US Nagar district of Uttarakhand state, India. Int J Agric Eng 10(2):354–359. https://doi.org/10.15740/HAS/IJAE/10.2/354-359

Kushwaha DP, Singh VK, Saran B (2017b) Daily pan evaporation estimation based on heuristic, regression and climate based techniques. In: Singh VK, Paras, Ramdayal (eds) Second all India seminar on advances in engineering and technology for sustainable development, The Institution of Engineers, Pantnagar, Uttarakhand, 25–26 November 2017a, pp 31–42

Kushwaha DP, Malik A, Singh SK (2019) Geographic information system for generating spatial pattern of natural streams: a case study in Nainital, India. Bull Env Pharmacol Life Sci 8:56–63

Kushwaha DP, Kumar A, Chaturvedi S (2021) Determining the effectiveness of carbon-based stabilizers blends in arresting soil erosion and elevating properties of Mollisols soils of North Western Himalayas. Environ Technol Innov 23:101768. https://doi.org/10.1016/j.eti.2021.101768

Lafdani EK, Moghaddam NA, Ahmadi A (2013) Daily suspended sediment load prediction using artificial neural networks and support vector machines. J Hydro 478:50–62. https://doi.org/10.1016/j.jhydrol.2012.11.048

Lahti M, Aphalo PJ, Finer L, Lehto T, Leinonen I, Mannerkoski H, Ryyppo L (2002) Soil temperature, gas exchange and nitrogen status of 5-year old Norway spruce seedlings. Tree Physio 22:1311–1316

Langholz H (1989) A simple model for predicting daily mean soil temperatures. J Agron Crop Sci 163:312–318

Lehnert M (2013) The soil temperature regime in the urban and sub-urban landscapes of olomoric Czech Republic. Morarian Geograph Rep 21(3):27–36

Lettau H (1979) Determination of the thermal diffusivity in the upper layers of a natural ground cover. Soil Sci 112:173–177

Lloyd J, Taylor JA (1994) On the temperature dependence of soil respiration. Funct Ecol 8:315–323

Mairizal AQ, Awad S, Priadi CR, Hartono DM, Moersidik SS, Tazerout M, Andres Y (2020) Experimental study on the effects of feedstock on the properties of biodiesel using multiple linear regressions. Renew Energy 145:375–381

Malczewski J (2006) GIS based multi-criteria decision analysis: a survey of literature. Int J Geogra Inf Sys 20(7):703–726

Malik A, Kumar A (2018) Comparison of soft-computing and statistical techniques in simulating daily river flow: a case study in India. J Soil Water Conserv 17(2):192–199

Malik A, Kumar A, Kushwaha DP, Kisi O, Salih SQ, Al-Ansari N, Yaseen ZM (2019) The Implementation of a hybrid model for hilly sub-watershed prioritization using morphometric variables: case study in India. Water 11(6):1138. https://doi.org/10.3390/w11061138

Mallick J, Kant Y, Bharath BD (2008) Estimation of land surface temperature over Delhi using Landsat-7 ETM Plus. J Ind Geophys Union 12(3):131–140

Manrique LA (1990) Estimating soil surface temperatures under different crop covers in Hawaii. Commun Soil Sci Plant Anal 21:2105–2117

Marill KA (2004) Advanced statistics: multiple linear regression. Acad Emerg Med 11(1):94–102. https://doi.org/10.1197/S1069-6563(03)00601-8

Meikle RW, Tredway TR (1979) A mathematical model for estimating soil temperatures. Soil Sci 128:226–242

Mihalakakou G (2002) On estimating soil surface temperature profiles. Energy Build 34(3):251–259

Moghaddamnia A, Ghafari M, Piri J, Han D (2009) Evaporation estimation using support vector machines technique. Int J Eng Appl Sci 5(7):415–423

Navale MM, Kashyap PS, Singh SK, Kushwaha DP, Kumar D, Kumar P (2018) Estimation of deterministic component of monthly rainfall time series: A case study for Pantnagar. Mausam 69(3):449–458. https://doi.org/10.54302/mausam.v69i3.338

Noori R, Karbassi AR, Moghaddamnia A, Han D, Zokaei-Ashtiani MH, Farokhnia A, Ghafari Gousheh M (2011) Assessment of input variables determination on the SVM model performance using PCA, Gamma test and forward selection techniques for monthly stream flow prediction. J Hydrol 401:177–189

Pahlavan-Rad MR, Dahmardeh K, Hadizadeh M, Keykha G, Mohammadnia N, Gangali M, Keikha M, Davatgar N, Brungard C (2020) Prediction of soil water infiltration using multiple linear regression and random forest in a dry flood plain, eastern Iran. CATENA 194:104715

Pajari B (1995) Soil respiration in poor upland site of Scots pine stand subjected to elevated temperatures and atmospheric carbon concentration. Plant Soil 168:563. https://doi.org/10.1007/BF00029369

Parton WJ (1984) Predicting soil temperatures in a shortgrass steppe. Soil Sci 138:93–101

Paul KI, Polglase PJ, Smethurst PJ, Anthony MO, Carlyle CJ, Khannaa PK (2004) Soil temperature under forests: a simple model for predicting soil temperature under a range of forest types. Agric for Meteorol 121:167–182

Peng F, Weng Q (2016) Consistent land surface temperature data generation from irregularly spaced Landsat imagery. Rem Sen Environ 184:175–187

Philip JR, De Vries DR (1957) Moisture movement in porous media under temperature gradients. Trans Am Geophys Union 38:222–232

Probert RJ (2000) The role of temperature in the regulation of seed dormancy and germination. In: Fenner M (ed) Seeds: the ecology of regeneration in plant communities. CABI Publishing, Wallingford, pp 261–292

Rastgou M, Bayat H, Mansoorizadeh M, Gregory AS (2020) Estimating the soil water retention curve: comparison of multiple nonlinear regression approach and random forest data mining technique. Comput Electron Agric 174:105502

Rahman KU, Pham QB, Jadoon KZ, Shahid M, Kushwaha DP, Duan Z, Mohammadi B, Khedher KM, Anh DT (2022) Comparison of machine learning and process-based SWAT model in simulating streamflow in the Upper Indus Basin. Appl Water Sci 12:178. https://doi.org/10.1007/s13201-022-01692-6

Rajesh J, Pande CB (2023) Estimation of land surface temperature for rahuri taluka, Ahmednagar District (MS, India), Using remote sensing data and algorithm. In: Pande CB, Moharir KN, Singh SK, Pham QB, Elbeltagi A (eds) Climate change impacts on natural resources, ecosystems and agricultural systems. Springer Climate. Springer, Cham. https://doi.org/10.1007/978-3-031-19059-9_24

Reimer A, Shaykewich CF (1980) Estimation of Manitoba soil temperatures from atmospheric meteorological measurements. Can J Soil Sci 60:299–309

Remesan R, Shamim MA, Han D (2008) Model input data selection using gamma test for daily solar radiation estimation. Hydrol Process 22:4301–4309

Repo TI, Leinonen AR, Finer L (2004) The effect of soil temperature on bid phenology, chlorophyll fluorescence, carbohydrate content and cold bardiness of Norway spruce seedlings. Physio Plant 121:93–100

Richards LA (1954) Diagnosis and improvement of saline and alkali soils. U. S. D. A. Hand Book No. 60. Oxford & IBH Publishing Co., New Delhi

Schaab G, Lenz R, Seufert G (1999) A temporal-spatial solar radiation model to improve scaling of biogenic emissions from a sparse Mediterranean pine/oak forest. Phys Chem Earth Part B 24(6):673–680

Singh JS, Gupta SR (1977) Plant decomposition and soil respiration and soil respiration in terrestrial ecosystems. Bot Rev 43:449–528

Singh VK, Prakash R, Paul R, Kumar S, Singh K (2017) Satyavan, evaluation of groundwater quality for irrigation in Gulha block of Kaithal district in Haryana. J Soil Salinity Water Q 9(2):241–248

Singh VK, Prakash R, Bhat MA, Deep G, Kumar S (2018a) Evaluation of groundwater quality for irrigation in Kaithal block (Kaithal District) Haryana. Int J Chem Stud 6(2):667–672

Singh VK, Singh BP, Kisi O, Kushwaha DP (2018b) Spatial and multi-depth temporal soil temperature assessment by assimilating satellite imagery, artificial intelligence and regression based models in arid area. Comput Electron Agric 150:205–219. https://doi.org/10.1016/j.compag.2018.04.019

Singh SK, Kashyap PS, Kushwaha DP, Tamta S (2020) Runoff and sediment reduction using hay mulch treatment at varying land slope and rainfall intensity under simulated rainfall condition. Int Arch Appl Sci Technol 11(3):144–155. https://doi.org/10.15515/iaast.0976-4828.11.3.144155

Sobrino JA, Coll C, Vicente C (1991) Atmospheric correction for land surface temperature using NOAA-11 AVHRR channels 4 and 5. Rem Sens Environ 38(1):19–34

Sobrino JA, Jimenez-Munoz JC, Paolini L (2004) Land surface temperature retrieval from LANDSAT TM 5. Rem Sens Environ 90:434–440

Sobrino JA, Jimenez-Munoz JC, Zarco-Tejada PJ, Sepulcre-Canto G, de Miguel E (2006) Land surface temperature derived from airborne hyperspectral scanner thermal infrared data. Remote Sens Environ 102:99–115

Sobrino JA, Jimenez-Munoz JC, Soria G, Ruescas AB, Danne O, Brockmann C, Ghent D, Remedios J, North P, Merchant C, Berger M, Mathieu PP, Gottsche FM (2016) Synergistic use of MERIS and AATSR as a proxy for estimating Land Surface Temperature from Sentinel-3 data. Rem Sen Environ 179:149–161

Stangierski J, Weiss D, Kaczmarek A (2019) Multiple regression models and Artificial Neural Network (ANN) as prediction tools of changes in overall quality during the storage of spreadable processed Gouda cheese. Eur Food Res Technol 245:2539–2547. https://doi.org/10.1007/s00217-019-03369-y

Stefansson A, Koncar N, Jones AJ (1997) A note on the gamma test. Neural Comput Appl 5:131–133

Subbaiah BV, Asija GL (1956) A rapid procedure for the estimation of available nitrogen in soil. Curr Sci 25:259

Tabari H, Marofi S, Sabziparvar A (2010) Estimation of daily pan evaporation using artificial neural network and multivariate non-linear regression. Irrigat Sci 28:399–406

Tabari H, Sabziparvar AA, Ahmadi M (2011) Comparison of artificial neural network and multivariate linear regression methods for estimation of daily soil temperature in an arid region. Meteorol Atmos Phys 110:135–142

Tamta S, Kumar A, Kushwaha DP (2023) Potential of roots and shoots of Napier grass for arresting soil erosion and runoff of mollisols soils of Himalayas. Int Soil and Water Conserv Res. https://doi.org/10.1016/j.iswcr.2023.02.001

Tarate SB, Singh VK, Kushwaha DP (2017) Assessment of meteorological drought for Parbhani district of Maharashtra, India. Int J Agric Eng 10(2):260–267. https://doi.org/10.15740/HAS/IJAE/10.2/1-9

Tingey DT, Manning M, Gmthaus LC, Bums WF (1980) Influence of light and temperature on monoterpene emission rates from slash pine. Plant Physiol 65:797–801

Trangmar BB, Yost RS, Uehara G (1985) Application of geostatistics to spatial studies of soil properties. Adv Agron 38:45–94

Tsui APM, Jones AJ, Guedes de Oliveria A (2002) The construction of smooth models using irregular embeddings determined by Gamma test analysis. Neural Comput Applic 10(4):318–329. https://doi.org/10.1007/s005210200004

Toselli M, Flore JA, Marogoni B, Masia A (1999) Effects of root-zone temperature on nitrogen accumulation by non-breeding apple trees. J Hort Sci Biotech 74:118–124
Varley J (1971) A textbook of soil chemical analysis. In: Hesse PR, Murray J (eds) London, p 520
Walkley A, Black IA (1934) An examination of Degtjareff method for determining soil organic matter and a proposed modification of the chromic acid titration method. Soil Sci 37:29–37
Wang Q, Zhang D, Zhou W, He X, Wang W (2020) Urbanization led to a decline in glomalin-soil-carbon sequestration and responsible factors examination in Changchun, Northeastern China. Urban for Urban Green 48:126506
Weih M, Karlson S (1999) The nitrogen economy of mountain birch seedlings. Implicat Winter Survival J Ecol 87:211–219
Weng Q, Lu D, Jacquelyn S (2004) Estimation of land surface temperature–vegetation abundance relationship for urban heat island studies. Rem Sens Environ 89:467–483
Wierenga JJ, Nielsen DR, Hagan RM (1969) Thermal properties of a soil based upon field and laboratory measurements. Soil Sci Soc Am Proc 33:354–360
Zhao S, Qin Q, Yang Y, Xiong YJ (2009) Comparison of two split-window methods for retrieving land surface temperature from MODIS data. J Earth Syst Sci 118(4):345–353. https://doi.org/10.1007/s12040-009-0027-4
Zheng X, Jiang Z, Ying Z, Song J, Chen W, Wang B (2020) Role of feedstock properties and hydrothermal carbonization conditions on fuel properties of sewage sludge-derived hydrochar using multiple linear regression technique. Fuel 271:117609

Assessment of Rainfall (R), Evapotranspiration (ET), and Crop Coefficient (K_c) Using Satellite Data

Susanta Das, Navneet Sharma, Puneet Sharma, and Nand Lal Kushwaha

Abstract Earth-observing satellites could provide regular estimates of rainfall (R), evapotranspiration, and crop coefficient on a global scale. Through active and passive remote sensing methods, it is possible to measure rainfall from space, expanding the scope of space and time parameters for precipitation data at global level. There are several space organizations around the world, that predict microwave-based rainfall with different special resolutions from the cloud and also show a good correlation with the ground station. Satellite products like Climate Hazards Group InfraRed Precipitation with Stations (CHIRPS), Global Precipitation Measurement (GPM), Tropical Rainfall Measuring Mission (TRMM), and CMORPH—CPC Morphing Technique, etc. are available. Similarly, the evapotranspiration and crop coefficient was estimated by using the radiation and water balance method from different satellite images like MODIS, LANDSAT, Sentinel-2, etc. The most popular method, SEBAL is used to estimate ET and K_c, where NDVI plays an important role. From a case study review, it was observed that estimation of R, ET, and K_c from satellite images has a good correlation with ground-based analysis. From the overall study, it can conclude that estimation of satellite-based R, ET, and K_c can provide spatial information, and also it will enhance the real-time monitoring and scheduling of the irrigation.

Keywords Remote sensing · Rainfall · Evapotranspiration · Crop coefficient · Satellite data

S. Das
North Eastern Space Application Centre, Department of Space, Government of India, Meghalaya 793103, India

N. Sharma
Department of Soil and Water Engineering, Punjab Agricultural University, Ludhiana, India

P. Sharma (✉)
Krishi Vigyan Kendra, Punjab Agricultural University, Pathankot, India
e-mail: sharmapuneet840@pau.edu

N. L. Kushwaha
Division of Agricultural Engineering, ICAR-Indian Agricultural Research Institute (IARI), New Delhi, India
e-mail: nand.kushwaha@icar.gov.in

C. B. Pande et al. (eds.), *Surface and Groundwater Resources Development and Management in Semi-arid Region*, Springer Hydrogeology,
https://doi.org/10.1007/978-3-031-29394-8_19

1 Introduction

The significance of precipitation for the hydrological cycle and the energy industry has been emphasised repeatedly (Faurès et al. 1995; Faridzad et al. 2018; Ebert et al. 2007). More precisely, precise observation of the rainfall as a process is essential for hydrological cycle modelling and the local, regional, and even global forecasting of extreme weather events (Bharti and Singh 2015; Beck et al. 2017). However, insufficient coverage of survey stations limits our understanding of this crucial process (Teng et al. 2020; Tuo et al. 2016). Due to technological issues or political concerns, the number of weather stations over an unit area is relatively less or of low significance in many countries of the world (Chen et al. 2017; Musie et al. 2019). In addition, the confined data sharing system and other factors, such as brief observation histories or poor data standards, limit the accessibility of the data from the current stations, making it difficult to use the observed data in hydro-meteorological research (Rahaman et al. 2022). Fortunately, the release of satellite-based precipitation databases has made it possible to complement gauge observations in data-limited regions such as desert, sparsely inhabited zones and alpine areas (Taylor et al. 2017; Mölg et al. 2014). Significant advancements in tropical and subtropical satellite precipitation estimation were accomplished with the actions of the Tropical Rainfall Measuring Mission (TRMM) satellite data (Sun et al. 2018). Since then, more highly-precise and extensive satellite precipitation datasets have been available. The continuation of TRMM is represented by TMPA (TRMM Multi-Satellite Precipitation Analysis) and GPM (Global Precipitation Measurement) (Hou et al. 2014). Infrared remote sensing technology is mainly used by PERSIANN (Precipitation Estimation from Remotely Sensed Information using Artificial Neural Networks) and CHIRP (Climate Hazards Infrared Precipitation) (Funk et al. 2014). There are also multisource datasets, like CMORPH (Climate Prediction Center Morphing Technique) and GSMaP (Global Satellite Mapping of Precipitation), among many others (Kubota et al. 2007).

The water and energy cycles frequently include land evapotranspiration (ET), which connects the surface to the atmosphere. For a better understanding of the interplay between hydrology and climate, accurate world-scale predictions of ET are essential. The FLUXNET project's local scale ET observations are provided (Baldocchi et al. 2001). Intensive world coverage by specified observations is not feasible, and current research is being done to determine how representative unit-scale in-situ data are for wider regions. In recent years, several alternative worldwide multiple years ET datasets are being produced to check this limitation. These datasets include estimates from satellites, land surface models that use forcing based on observations, re-analysis data results, projections based on empirical upscaling of detailed observations, and projections of the atmospheric water budget. The LandFlux-EVAL project (see http://www.iac.ethz.ch/url/research/LandFlux-EVAL) seeks to assess and contrast various ET datasets that are currently accessible. The project is a crucial part of the GEWEX LandFlux initiative, a programme of the GEWEX Radiation Panel that aims to provide a reliable and excellent worldwide ET dataset for climate studies. It is necessary to be aware of the uncertainties in the available ET products for many applications, including the assessment of climate change projections (McCabe et al., 2016). The IPCC Fourth Assessment Report's combined 11 atmospheres, ocean, and land GCMs are also examined (AR4). The sensible and latent heat fluxes in a subset of twelve satellite-based, LSM and re-analysis datasets are the subject of a complementary analysis by Jimenez et al. (2011) for three years (1993–1995).

The crop coefficient (K_c), which measures crop-specific water demand, is necessary to accurately predict the amount of irrigation needed for various crops in any given location (Tyagi et al. 2000). Plant characteristics known as crop coefficients are used to forecast evapotranspiration (ET). Simply put, the simplest crop coefficient, K_c, is the difference between the ET measured for the crop under study and the ET measured for the well-calibrated reference crop under the identical circumstances (Allen et al. 2011). The FAO-56 technique, based on combined effect of reference evapotranspiration ET_0 and crop coefficients, can be used to compute the crop coefficient (K_c). The single crop coefficient and the dual crop coefficient are techniques used to calculate crop evapotranspiration from remotely sensed observations in form of datasets. The impacts of crop transpiration and soil evaporation are merged into a single K_c in the single crop coefficient (K_c single) (Kamble et al. 2013). The soil evaporation coefficient (K_e) and the basal crop coefficient (K_{cb}) make up the dual crop coefficient (Er-Raki et al. 2007). For the study of vegetation and crops, MODIS vegetation indices like the Normalized Difference Vegetation Index (NDVI) and the Enhanced Vegetation Index (EVI) data are frequently utilised (Sharma et al. 2016). The NDVI can be used to estimate vegetation density (Singhal and Goyal 2012).

2 Descriptions of Different Models of Satellite-Based Rainfall Measurement

2.1 The Tropical Rainfall Measuring Mission

To better understand how rainfall is distributed and varies throughout the world as a component of the hydrological cycle in the contemporary climate system, TRMM was a satellite of research objective that was in action from years 1997–2015. TRMM is the source of important data on precipitation and the heat release it causes by covering the tropical and sub-tropical parts of the planet, which fuels the world's atmospheric circulation that determines climatic and weather conditions. To improve our knowledge of the dynamics between water vapour, clouds, and rainfall essential to controlling Earth's atmosphere, TRMM delivered significant precipitation information in cooperation with other satellites systems of NASA.

Since its debut in 1997, TRMM has offered vital precipitation observations in our planet's tropical and subtropical areas. The Precipitation Radar (PR) examined the rainfall column and offered fresh information on the structure and intensity of tropical storms. To determine the amount of water vapour, water from clouds, and precipitation intensity in the atmosphere, the TRMM Microwave Imager (TMI) analysed microwave radiation released by the Earth and its atmosphere. Numerous additional disciplines, including a wide range of social applications and tropical cyclone projections, numerical weather forecasting, and rainfall specific climatologies, have significantly benefited from TRMM precipitation measurements.

After the spaceship used up all of its fuel, TRMM was formally terminated on April 15, 2015. On June 15, 2015, TRMM was deactivated and re-entered Earth's atmosphere over the South Indian Ocean. TRMM delivered ground-breaking 3-D photographs of precipitation and storms for period of 17 observation-years despite being only intended to operate for three years. Additional microwave radiometer-equipped precipitation measurement satellites, including the GPM Core Observatory, have been sparked by TRMM.

3 Climate Data Record (CDR)

The rainfall projection from Remote Sensing data observation applying Artificial Neural Networks-Climate Data Record (CDR) is a data set that mainly depends on infrared data, and it was transformed from a challenging PERSIANN method on GridSat-B1 infrared satellite data. The CDR was modified using the Global Precipitation Climatology Project (GPCP) monthly product version 2.2 (GPCPv2.2). The data set was created at a spatial resolution of 0.25° in the latitude band 60S–60N from 1983 to the recent past (Ashouri et al. 2015) and was first made public on June 1, 2014. It is accessible on the Center for Hydrometeorology and Remote Sensing website at the University of California (https://chrsdata.eng.uci.edu/).

4 CHIRPS

The Climate Hazards Group Infrared Precipitation with Stations (CHIRPS) data set depends on previously applied strategies to "smart" projection methods and high resolution rainfall projections from long durations of observation, depend on infrared cold cloud duration (CCD) recordings. The data set was launched in the year 2015, and was formed at two spatial resolutions of 0.05° and 0.25° in the latitude band 50S–50N from the year 1981 to till date (Funk et al. 2015). The information was obtained from the website of the Climate Hazards Center (https://data.chc.ucsb.edu/products/CHIRPS-2.0/), University of California.

5 CMORPH

The CMORPH approach generates world rainfall studies at high time and space data in terms of resolutions by using precipitation estimates from microwave measurements made by low orbiter satellites. The NOAA file transfer protocol website (ftp:/ftp.cpc.ncep.noaa.gov/precip/CMORPHV1.0/) was used to download the dataset version used in this study (CMORPH IFLOODS V1.0 CRT), which was created at two spatial resolutions of 0.07° and 0.25° in the latitude band 60S-60N from 1998 to the end of the year 2019 (Xie et al. 1998).

6 GPM

The Integrated Multi-satellite Retrievals for GPM (IMERG) program of the GPM with objective to combine multiple types of satellite data, including microwave satellite information and infrared satellite record, station gauge record, and others. The GPM was developed as a continuation and improvement of the TRMM mission. In March 2019, the most recent version (GPM IMERG Final Precipitation L3 V06) was made available. The coverage spans the months of June 2000 through August 2020, with a latitude range of 90S to 90N and a spatial resolution of 0.10° (Huffman et al. 2019). NASA's Data and Information Services Center (DISC) has the dataset available at https://disc.gsfc.nasa.gov/datasets/GPM3IMERGDF06.

7 GSMaP

The Japan Aerospace Exploration Agency created the GSMaP, a GPM algorithm (JAXA). The use of numerous attributes acquired from the TRMM precipitation radar (TRMM PR) and GPM Dual-Frequency Precipitation Radar Ku Band is the key component of the GSMaP algorithm (GPM DPR Ku). It should be mentioned that the GSMaP V6 Gauge version was used in this analysis rather than the most recent version of the dataset (GSMaP V7) because of its brief period (2017–present) (Kubota et al. 2020). This version, which was made at two spatial resolutions of 0.10° and 0.25° in the latitude range 60S–60N from March 2000 to the present, was published in April 2016. The dataset was downloaded via the JAXA Earth Observation Research Center's transfer protocol website (ftp:/hokusai.eorc.jaxa.jp).

8 TMPA

The TMPA, is characterised by its intensive collection of high-quality microwave data and fill ins made up of microwave calibration infrared estimations. The dataset version used in this investigation (TMPA 3B42 daily V7) was published on May 15, 2016, and it was produced by the DISC of NASA (Huffman et al. 2016) with a spatial resolution of 0.25°. The data source is the DISC of NASA (https://disc.gsfc.nasa.gov/datasets/TRMM3B42_Daily7/), and the coverage spans 1998 through December 30, 2019, with a latitude range of 50S to 50N.

9 Descriptions of Different Models of Satellite-Based ET and K_c Measurement Models

9.1 MODIS ET

We synthesized the world yearly MOD16A3 (ET) parameters from the NASA Earth Observation System (EOS) program. The world level data informations were recorded by the Numerical Terradynamic Simulation Group (NTSG) at University of Montana (UMT) (available from http://www.ntsg.umt.edu/). The MOD16A3 product was obtained using the MOD17 program (Running et al. 2004; Zhao et al. 2005) and improvised MOD16 program (Mu et al. 2007, 2011, 2013), respectively. These informations were open for use from 2000 to till date at a spatial resolution of 1 km. The MOD17 algorithm is defined on basis of Monteith's radiation use efficiency procedure (1972). The program considers that vegetation productivity under well irrigated and soil health nutrition conditions is directly proportional to the amount of Absorbed Photosynthetically Active Radiation (APAR). A conversion efficiency parameter (ε) is applied to transform APAR to the projected productivity. The parameter ε varies for types of vegetations and climate conditions. To quantify MOD17 also projects daily leaf and fine root maintenance respiration, yearly growth respiration, and yearly maintenance respiration of live cells in woody tissue. The MOD17 operator procedure gives further description of the program and is open source at http://www.ntsg.umt.edu/files/modis/MOD17UsersGuide2015_v3.pdf. The MODIS NPP dataset has been validated and used in many studies (Turner et al. 2006; Zhao et al. 2005, 2006). The MOD 16 program is dependent on Mu et al.'s improvised ET program (2011), which based on Penman–Monteith equation. The ET comprises evaporation from wet and moist soil, rainwater intercepted by the canopy before coming into the ground, and transpiration from stomata on plant leaves and stems. The remotly sensed information from MODIS allocates input for surface biophysical variables affecting ET, including albedo, biome type and leaf area index (LAI). The detailed program of MOD16 is available at http://www.ntsg.umt.edu/project/modis/mod16.php. This global ET product was validated by Mu et al. (2011) using 46 eddy flux towers dataset and found that the dataset can be used to estimate actual ET with satisfactory accuracy in Asia. Previous studies in India used MOD16 ET to forecast actual ET (e.g. Shah and Mishra 2016). Due to the lack of ground observational recordings, the authors couldn't endorse these products.

9.2 SEBAL Based ET

The Surface Energy Balance Algorithm for Land (SEBAL) is an image processing algorithm that includes 25 computing levels to determine the actual (ET_{act}) and potential (ET_{pot}) evapotranspiration rates as well as other energy exchanges between the ground surface and surroundings. The primary input data for SEBAL are spectral radiance measurements in the spectrum's visible, near-infrared, and thermal infrared regions (Figs. 1 and 2).

For each pixel, SEBAL calculates a comprehensive radiation and energy balance and the resistances for momentum, heat, and water vapour transport. The resistances

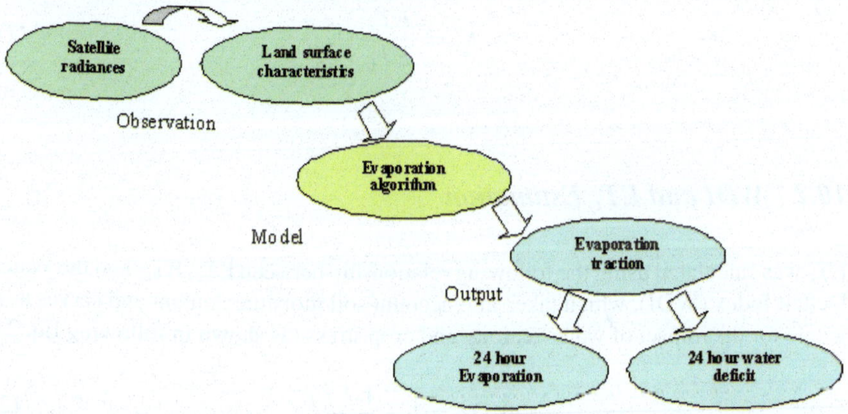

Fig. 1 Flow chart of the principal steps in SEBAL to derive instantaneous 24-h ET_{act} and ET_{pot} values (http://www.waterwatch.nl/tools0/sebal/sebal-a-scientific-description.html)

Fig. 2 Surface energy balance (Jaber and Pradhan 2016)

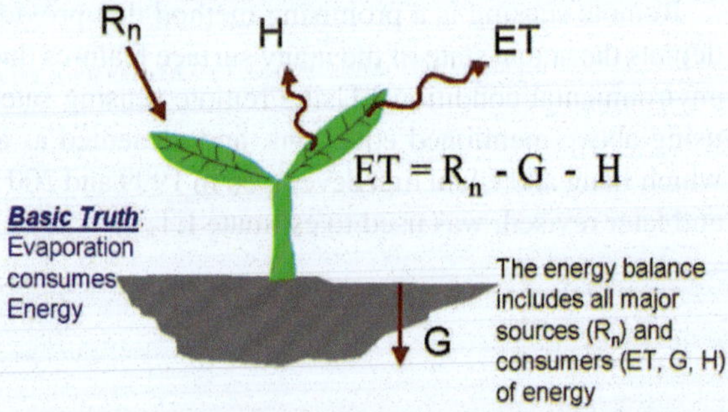

depend on state variables that change daily, including soil water potential (and consequently soil moisture), wind speed, and air temperature. Along with the satellite data, the SEBAL model requires daily weather parameters like wind speed, humidity, solar radiation, air temperature etc.

10 K_c Using Remote Sensing

10.1 Crop Coefficient (K_c) Estimation Using the Traditional Method

K_c is a parameter that displays the ratio of the (ET_c) and the (ET_o). A variety of factors influences it A variety of factors, including crop type, crop stage, and crop management influences it. Consequently, using the following equation, the K_c can be determined (as shown in following Eq. 1).

$$K_c = \frac{ET_c}{ET_0} \quad (1)$$

10.2 WDI and ET_c Estimation

ET_c was calculated using the following relationship between ET_c, ET_a, and the Water Deficit Index (WDI), which takes into account soil moisture content and serves as a gauge for the impact of water scarcity and crop stress (as shown in following Eq. 2).

$$WDI = 1 - \frac{ET_a}{ET_c} \quad (2)$$

where ET_a is the actual crop evapotranspiration (mm/day), ET_c is the crop evapotranspiration (mm/day), and WDI is the water deficit index (dimensionless).

Remote sensing is a promising method that provides important surface data and depicts the actual state of the many surface features due to heterogeneity and varying environmental conditions. Using remote sensing satellite data, ET_c was calculated using above mentioned equations. and presented as a map. The triangle approach, which Jiang and Islam first developed in 1999 and 2001 (Jiang and Islam 1999, 2001) and later revised, was used to estimate ET_a.

11 Relationship Between NDVI and K_c

Red and near-infrared bands from remote sensing data were used to calculate NDVI. It is based on the plant's reflectance in the red (R) and near-infrared (NIR) regions of the electromagnetic spectrum. This index of vegetation is provided by the following Eq. 3.

$$NDVI = \frac{NIR - R}{NIR + R} \tag{3}$$

where R is the reflectance in the red portion of the spectrum, and NIR is the reflectance in the near infrared portion. The scientists concurred that there is a strong correlation between the NDVI values and vegetation characteristics like net primary productivity, leaf area index, and gross primary productivity (Gamon et al. 1995). We must determine the K_c and NDVI values from the same pixel in order to simulate the K_c-NDVI relationship.

To obtain the intended relationship in the form (K_c NDVI $= a \times$ NDVI $\pm$ b), generated values are to be employed in a linear regression model. Another empirical approach used normalized NDVI (NDVIn) to estimate basal crop coefficient and actual crop evapotranspiration from the satellite data (Hunsaker et al. 2005).

$$K_{cb} = min[0.15; 0.176 + 1.325NDVIn - 1.466NDVI^2 + 1.146NDVIn^3 \tag{4}$$

where NDVIn is normalized NDVI as calculated in Eq. 3.

As defined in FAO56, when potential effects of water stress on ET_c are considered, actual ET_c ($ET_{c\,act}$) is computed as following Eq. 5.

$$ET_{c\,act} = (k_s k_{cb} + k_s)ET_0 \tag{5}$$

where K_{cb} is a crop transpiration coefficient (T_c), and K_e is a soil coefficient K_s is the water stress coefficient, ET_0 is the grass reference, and evaporation, evapotranspiration.

12 Case Study: Ludhina District of Punjab, India

The case study region is the Ludhiana district of Punjab, which is divided into four subdistricts: Jagraon, Samrala, Khanna, and Ludhiana. The study region is mainly dominated by the agricultural land, and followed by built-up and range land (Fig. 3). The land use land cover (LULC) map with major classes like water, forest, agriculture, built-up, base ground, and rangeland (dominated by grasses) were generated for year 2021 from sentinel 2.

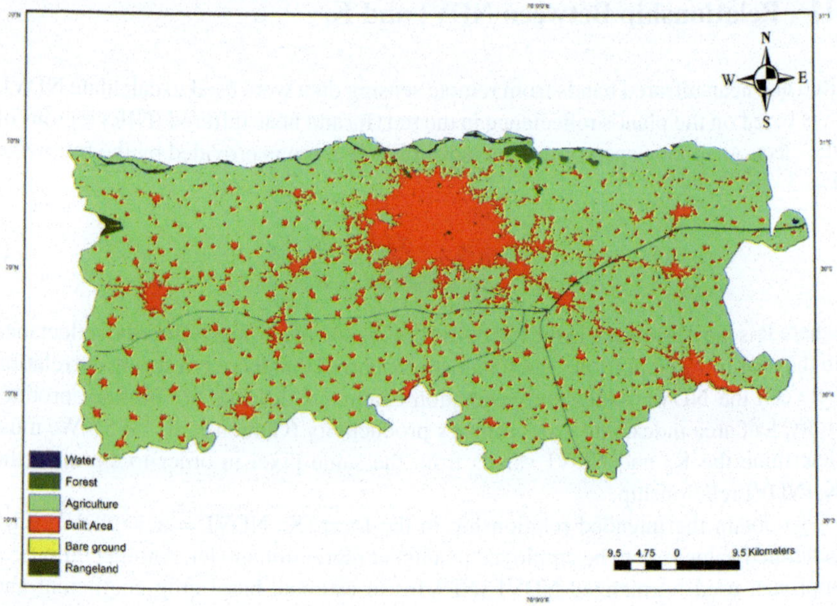

Fig. 3 Land use land cover of Ludhiana district of 2021

In the present study, spatially distribution of rainfall, ET, K_c were extracted from remote sensing data. For the study periods, the daily gridded rainfall data from CHIRPS with a high spatial resolution (0.25° × 0.25°) from 2000 to 2021 was acquired. This daily gridded values were further used in ArcGIS domain to calculated the mean and total annual rainfall. The highest mean annual rainfall was recorded in Samrala (832.87 mm), and the lowest was recorded in Jagraon (589.09 mm) (Fig. 4). A rainfall pattern was constructed using the point mean annual rainfall values that were extracted from the spatial map. According to the study, the sub-districts of Ludhiana and Samrala received the highest rainfall in 2012, while Khanna and Jagraon received the highest amounts in 2011 and 2010, respectively (Fig. 5).

Similar to rainfall, spatial form (0.25° × 0.25°) of evapotranspiration was acquired from Global annual MOD16A3 product from the NASA Earth Observation System (EOS) program (http://ntsg.umt.edu/project/mod16). The MOD16 ET datasets were estimated using improved ET algorithm discussed in Sect. 9.1. In the present study, mean MOD16 ET for the month of July (from the year 2000 to 2021), were acquired. The study shows that, the highest and lowest ET values in Jagraon and Ludhiana as 347.55, and 32.93 mm, respectively (Fig. 6). Figure 7 shows the actual evapotranspiration (AET) of Ludhiana district for the month of July 2021, where maximum area of the district having higher AET (>175 mm) except the city. As kharif maize is often planted in the study region and enters its development stage in July, when there is the greatest water intake, due to this region July month was chosen for AET values. Evaporation pattern were calculated from by extracting the pixel (point) values from

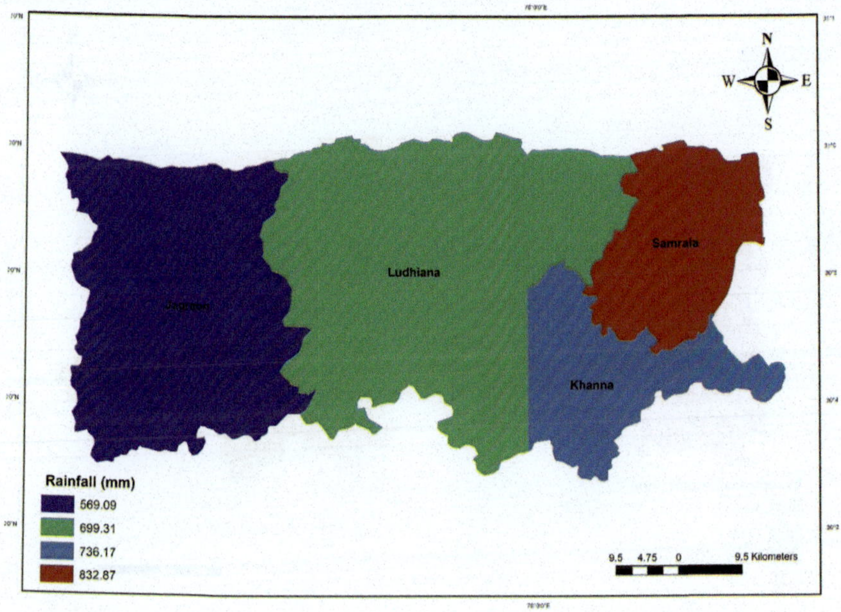

Fig. 4 Mean annual from 2000 to 2021 rainfall of Ludhiana district

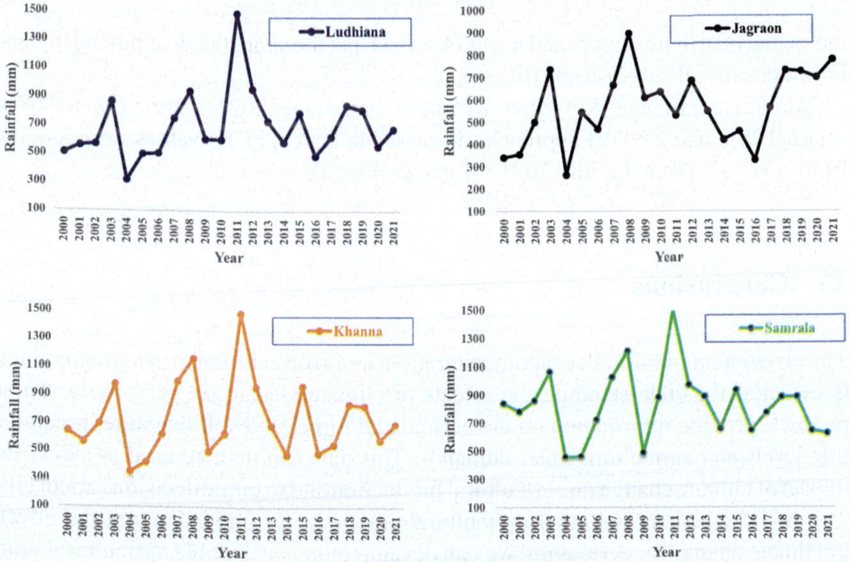

Fig. 5 Mean annual from 2000 to 2021 rainfall of sub-districts of Ludhiana district

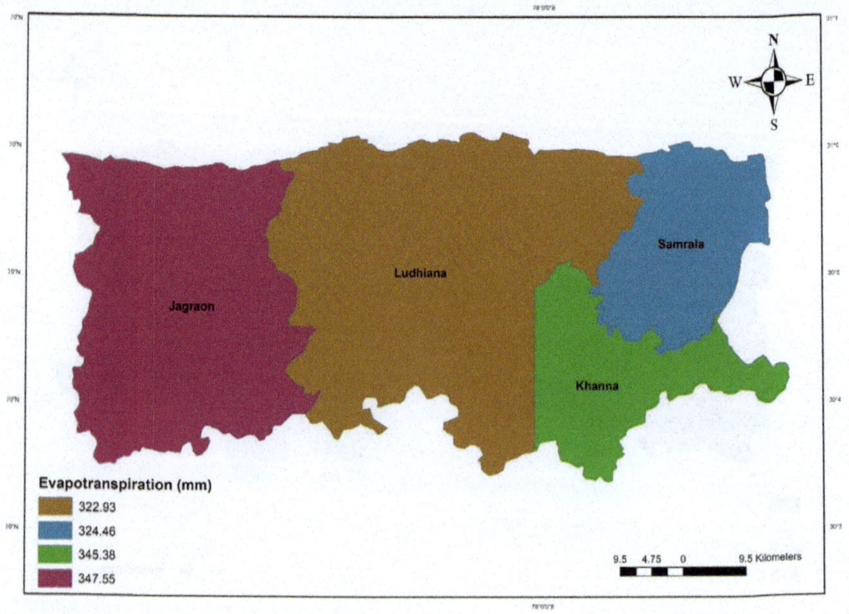

Fig. 6 Mean annual of July from 2000 to 2021 MODIS evapotranspiration of Ludhiana district

the spatial distribution map, and it was observed that the year 2006 was having highest ET values for all sub-districts (Fig. 8).

The crop coefficient (K_c) values for kharif maize (July 2021) were estimated from sentinel 2 by using NDVI approach (discussed in Sect. 11). K_c values were ranging from 0.31 to 1.06 in the July 2021 for maize (Fig. 9).

13 Conclusions

The assessment of rainfall, evapotranspiration, and crop coefficient satellite products is essential for understanding the effects of climate change on agriculture. These products provide information on the current and forecasted soil moisture, precipitation levels and agriculture water demands. This data can then be used to assess the impact of climate change on agricultural production and to guide decisions about crop selection, irrigation, and other agricultural practices. By understanding the effects of climate change on agriculture we can develop more sustainable agricultural practices and ensure the security of our food supply. Satellite-based products can help to improve irrigation and agricultural water management in a variety of ways of Ludhiana district of Punjab. They can provide detailed information on soil moisture, evapotranspiration, and crop health, which allow farmers to make more informed decisions about when and how much to irrigate their crops. They can also be used

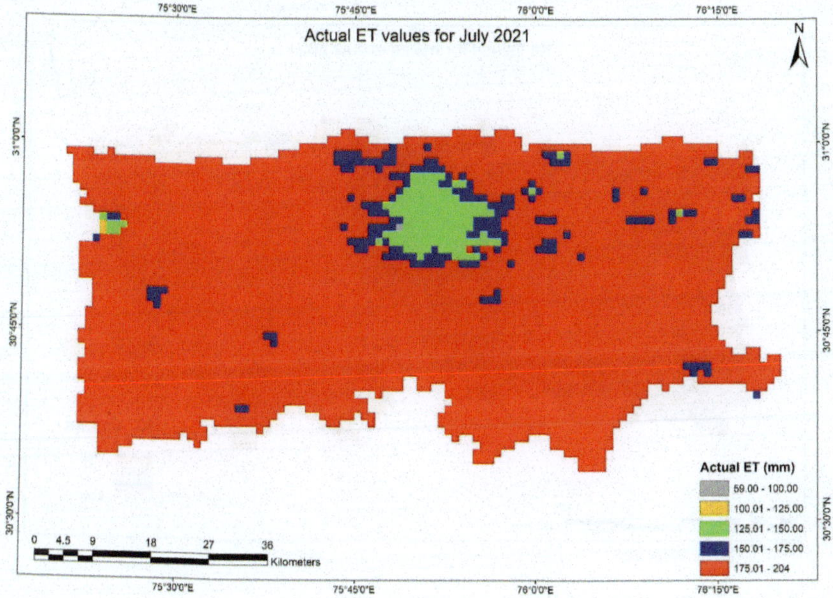

Fig. 7 Actual evapotranspiration for July 2021 of Ludhiana district

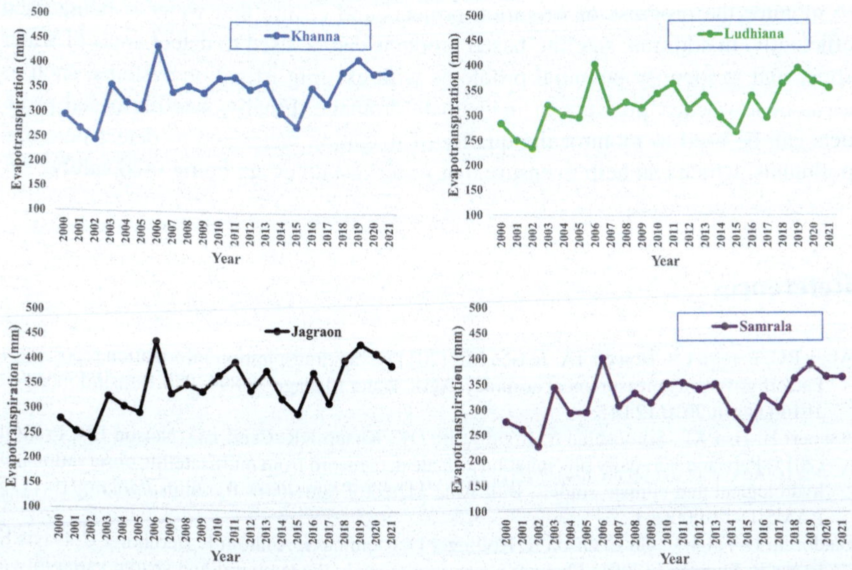

Fig. 8 Mean annual of from 2000 to 2021 Evapotranspiration of sub-districts of Ludhiana district

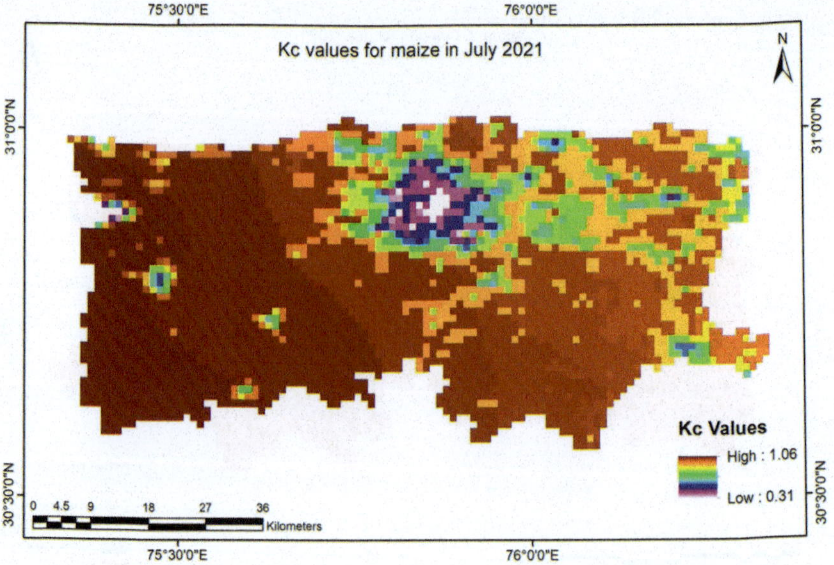

Fig. 9 Crop coefficient (K_C) of Ludhiana district for maize crop

to monitor the progress of irrigation projects and ensure that water is being used efficiently. In addition, satellite-based products can be used to detect areas of water stress and to identify potential problems with existing irrigation systems, such as areas of low water pressure or inadequate drainage. Finally, satellite-based products can be used to monitor the quality of irrigation water and to detect potential pollutants, which can help to ensure that water resources are being used safely.

References

Allen RG, Pereira LS, Howell TA, Jensen ME (2011) Evapotranspiration information reporting: I. Factors governing measurement accuracy. Agric Water Manage 98:899–920. https://doi.org/10.1016/j.agwat.2010.12.015

Ashouri H, Hsu KL, Sorooshian S, Braithwaite DK, Knapp KR, Cecil LD, Nelson BR, Prat OP (2015) Persiann-cdr: daily precipitation climate data record from multisatellite observations for hydrological and climate studies. Bull Ame Meteorol Soc 96:69–83. https://doi.org/10.1175/BAMS-D-13-00068.1

Baldocchi D, Falge E, Gu L, Olson R, Hollinger D, Running S, Anthoni P, Bernhofer C, Davis K, Evans R, Fuentes J (2001) Fluxnet: a new tool to study the temporal and spatial variability of ecosystem-scale carbon dioxide, water vapor, and energy flux densities. BullAm Meteorol Soc 82:2415–34. https://doi.org/10.1175/15200477(2001)082%3C2415:FANTTS%3E2.3.CO;2

Beck HE, Vergopolan N, Pan M, Levizzani V, van Dijk AIJM, Weedon GP, Brocca L, Pappenberger F, Huffman GJ, Wood EF (2017) Global-scale evaluation of 22 precipitation datasets using gauge observations and hydrological modeling. Hydrol Earth Syst Sci 21:6201–6217. https://doi.org/10.5194/hess-21-6201-2017

Bharti V, Singh C (2015) Evaluation of error in TRMM 3B42V7 precipitation estimates over the Himalayan region. J Geophys Res Atmos 120:12458–12473. https://doi.org/10.1002/2015JD023779

Boé J, Terray L (2008) Uncertainties in summer evapotranspiration changes over Europe and implications for regional climate change. Geophy Res Lett 35. https://doi.org/10.1029/2007GL032417

Chen C-J, Senarath SUS, Dima-West IM, Marcella MP (2017) Evaluation and restructuring of gridded precipitation data over the Greater Mekong Subregion. Int J Climatol 37:180–196. https://doi.org/10.1002/joc.4696

Ebert EE, Janowiak JE, Kidd C (2007) Comparison of near-real-time precipitation estimates from satellite observations and numerical models. Bull Am Meteorol Soc 88:47–64. https://doi.org/10.1175/BAMS-88-1-47

Er-Raki S, Chehbouni A, Guemouria N, Duchemin B, Ezzahar J, Hadria R (2007) Combining FAO-56 model and ground-based remote sensing to estimate water consumptions of wheat crops in a semi-arid region. Agric Water Manage 87:41–54. https://doi.org/10.1016/j.agwat.2006.02.004

Faridzad M, Yang T, Hsu K, Sorooshian S, Xiao C (2018) Rainfall frequency analysis for ungauged regions using remotely sensed precipitation information. J Hydrol 563:123–142. https://doi.org/10.1016/j.jhydrol.2018.05.071

Faurès J-M, Goodrich DC, Woolhiser DA, Sorooshian S (1995) Impact of small-scale spatial rainfall variability on runoff modeling. J Hydrol 173:309–326. https://doi.org/10.1016/0022-1694(95)02704-S

Funk CC, Peterson PJ, Landsfeld MF, Pedreros DH, Verdin JP, Rowland JD, Romero BE, Husak GJ, Michaelsen JC, Verdin AP (2014) A quasi-global precipitation time series for drought monitoring. US Geol Surv Data Ser 832:1–2. https://doi.org/10.3133/ds832

Gamon JA, Field CB, Goulden M, Griffn K, Hartley A, Joel G, Penuelas J, Valentini R (1995) Relationships between NDVI, canopy structure and photosynthesis in three Californian vegetation types. Ecol Appl 5:28–41

Hou AY, Kakar RK, Neeck S, Azarbarzin AA, Kummerow CD, Kojima M, Oki R, Nakamura K, Iguchi T (2014) The global precipitation measurement mission. Bull Am Meteorol Soc 95:701–722. https://doi.org/10.1175/BAMS-D-13-00164.1

Huffman G J, Bolvin D T, Nelkin E J, Stocker EF, Tan J (2011) V06 IMERG release notes. NASA/GSFC: Greenbelt, MD. https://gpm.nasa.gov/sites/default/files/2020-10/IMERG_V06_release_notes_201006_0.pdf

Hunsaker DJ, Pinter-Jr PJ, Kimball BA (2005) Wheat basal crop coefficients determined by normalized difference vegetation index. Irrig Sci 24:1–14

Jaber HS, Pradhan B (2016) Evaluation of SEBAL model for evapotranspiration mapping in Iraq using remote sensing and GIS

Jiang L, Islam S (1999) A methodology for estimation of surface evapotranspiration overlarge areas using remote sensing observations. Geophys Res Lett 26(17):2773–2776

Jiang L, Islam S (2001) Estimation of surface evaporation map over Southern Great Plains using remote sensing data. Water Resour Res 37(2):329–340

Jiménez C, Prigent C, Mueller B, Seneviratne SI, McCabe MF, Wood EF, Rossow WB, Balsamo G, Betts AK, Dirmeyer PA, Fisher JB (2011) Global intercomparison of 12 land surface heat flux estimates. J Geophy Res Atmos 116. https://doi.org/10.1029/2010JD014545

Kubota T, Shige S, Hashizume H, Aonashi K, Takahashi N, Seto S, Hirose M, Takayabu YN, Ushio T, Nakagawa K (2007) Global precipitation map using satellite-borne microwave radiometers by the GSMaP project: production and validation. IEEE Trans Geosci Remote Sens 45:2259–2275. https://doi.org/10.1109/TGRS.2007.895337

Mölg T, Maussion F, Scherer D (2014) Mid-latitude westerlies as a driver of glacier variability in monsoonal High Asia. Nat Clim Chang 4:68–73. https://doi.org/10.1038/nclimate2055

Mu Q, Heinsch FA, Zhao M, Running S W (2007 Dec 28) Development of a global evapotranspiration algorithm based on MODIS and global meteorology data. Remote Sens Environ 111(4):519–36

Mu Q, Zhao M, Running SW (2011) Improvements to a MODIS global terrestrial evapotranspiration algorithm. Remote Sens Environ 115:1781–800. https://doi.org/10.1016/j.rse.2011.02.019

Mu Q, Zhao M, Running SW (2013) MODIS global terrestrial evapotranspiration (ET) product (NASA MOD16A2/A3). Algorithm Theor Basis Doc Collect 5:600

Musie M, Sen S, Srivastava P (2019) Comparison and evaluation of gridded precipitation datasets for streamflow simulation in data scarce watersheds of Ethiopia. J Hydrol 579:124168. https://doi.org/10.1016/j.jhydrol.2019.124168

Rahman G, Rahman AU, Anwar MM, Dawood M, Miandad M (2022) Spatio-temporal analysis of climatic variability, trend detection, and drought assessment in Khyber Pakhtunkhwa, Pakistan. Arab J Geosci 15:81. https://doi.org/10.1007/s12517-021-09382-4

Running SW, Nemani RR, Heinsch FA, Zhao M, Reeves M, Hashimoto H (2004) A continuous satellite-derived measure of global terrestrial primary production. Bioscience 54:547–60. https://doi.org/10.1641/0006-3568(2004)054[0547:ACSMOG]2.0.CO;2

Seneviratne SI, Corti T, Davin E L, Hirschi M, Jaeger EB, Lehner I, Orlowsky B, Teuling A J (2010) Investigating soil moisture–climate interactions in a changing climate: a review. Earth-Sci Rev 99:125–61

Sharma G, Zaidi S, Goyal R (2016) Trend analysis of temporal variations in evi with respect to rainfall of jaipur district. Int J Res Eng Technol 5

Singhal V, Goyal R (2012) A methodology based on spatial distribution of parameters for understanding affect of rainfall and vegetation density on groundwater recharge. Eur J Sustain Devel 1:85

Sun Q, Miao C, Duan Q, Ashouri H, Sorooshian S, Hsu K-L (2018) A review of global precipitation data sets: data sources, estimation, and intercomparisons. Rev Geophys 56:79–107. https://doi.org/10.1002/2017RG000574

Taylor CM, Belušić D, Guichard F, Parker DJ, Vischel T, Bock O, Harris PP, Janicot S, Klein C, Panthou G (2017) Frequency of extreme Sahelian storms tripled since 1982 in satellite observations. Nature 544:475–478. https://doi.org/10.1038/nature22069

Teng J, Jakeman AJ, Vaze J, Croke BF, Dutta D, Kim SJ (2017) Flood inundation modelling: a review of methods, recent advances and uncertainty analysis. Environ Model Softw 90:201–16. https://doi.org/10.1016/j.envsoft.2017.01.006

Tuo Y, Duan Z, Disse M, Chiogna G (2016) Evaluation of precipitation input for SWAT modeling in Alpine catchment: a case study in the Adige river basin (Italy). Sci Total Environ 573:66–82. https://doi.org/10.1016/j.scitotenv.2016.08.034

Turner DP, Ritts WD, Cohen WB, Gower ST, Running SW, Zhao M, Costa MH, Kirschbaum AA, Ham JM, Saleska SR, Ahl DE (2006) Evaluation of MODIS NPP and GPP products across multiple biomes. Remote Sens Environ 102:282–92. https://doi.org/10.1016/j.rse.2006.02.017

Xie P, Joyce R, Wu S, Yoo S H, Yarosh Y, Sun F, Lin R (2017) Reprocessed, bias-corrected CMORPH global high-resolution precipitation estimates from 1998. J Hydrometeorol 18:1617–41. https://doi.org/10.1175/JHM-D-16-0168.1

Zhao M, Heinsch FA, Nemani RR, Running SW (2005) Improvements of the MODIS terrestrial gross and net primary production global data set. Remote Sens Environ 95:164–76. https://doi.org/10.1029/2004JG000004

Zhao M, Running SW, Nemani RR (2006) Sensitivity of Moderate Resolution Imaging Spectroradiometer (MODIS) terrestrial primary production to the accuracy of meteorological reanalyses. J Geophy Res Biogeosci 111(G1). https://doi.org/10.1029/2004JG000004

Study on Development of Design Rainfall for Stormwater Management System in an Urban Catchment

Manish Kumar Sinha, Klaus Baier, Rafig Azzam, Bhupesh Choudhary, and Mukesh Kumar Verma

Abstract Urban flooding, increase in storm runoff is caused due to development of urban infrastructure without paying much attention to changing rainfall patterns in an urban area. Design rainfall is one of the essential aspects while designing urban stormwater infrastructure. Many of the methods have been developed to prepare design rainfall from the shorter duration rainfall series. Annual maxima with extreme value distribution were very commonly used. This paper covers a study conducted in a pilot area i.e. Raipur City in India using 2D stormwater modelling for various design rainfall scenarios. Thirty-six years of shorter duration (15 min) rainfall has been collected to design the rainfall scenarios. Frequency analysis is performed using best fit distribution i.e. Gumbels Extreme Value distribution for the different return periods. For development of Intensity–Duration–Frequency (IDF) curves GEV distribution and Central Public Health and Environmental Engineering Organization (CPHEEO) methods were used. To create IDF curves shorter duration observed rainfall data 15, 30, 45 min, 1, 1.5, 2, 3, 4, 5, 6, 7, 8, 10, 11, 12, 16, 24 h were developed. Design storm intensity derived from IDF curves for different return periods estimates discharge and runoff to simulate the Storm water management model (SWMM).

Keywords Storm water management model (SWMM) · Rainfall-Runoff simulation · Intensity–Duration–Frequency Curve · Raipur city

M. K. Sinha (✉) · B. Choudhary
Environmental & Water Resources Engineering, UTD, Chhattisgarh Swami Vivekanand Technical University, Bhilai, India
e-mail: manishsinha200389@gamil.com

M. K. Sinha · K. Baier · R. Azzam
Department of Engineering Geology and Hydrogeology, RWTH Aachen University, Aachen, Germany

M. K. Verma
Department of Civil Engineering, National Institute of Technology Raipur, Raipur, India

C. B. Pande et al. (eds.), *Surface and Groundwater Resources Development and Management in Semi-arid Region*, Springer Hydrogeology,
https://doi.org/10.1007/978-3-031-29394-8_20

1 Introduction

The current hydraulic systems (urban wastewater drains) in developing nations like India are insufficient to quickly drain stormwater under harsh climatic conditions. It is caused not only by its capacity but also by the selection of design rainfall. This issue might be linked to changes in the area of urban growth as well as changes in the climate (Shrestha et al. 2014). Storm drains serve a vital function in carrying rainwater from metropolitan areas to an outlet point, which is usually a water body such as a river or lake. Since the drainage system in the nation is old and degraded, it lacks the ability to convey excess runoff caused by significant flooding events, and hence flash floods occur almost exclusively in short-duration, high-intensity rainstorm events (Schmitt et al. 2004; Sharma and Singh 2017). For water resource planning and management, it is necessary to estimate changes in catchment surface runoff due to rapid urbanization. Climate change-related rainfall uncertainty is difficult to connect to current changing surface runoff conditions (Sinha et al. 2019a). As a result of the short period of severe rainfall, significant flood peak flows from altered urban catchments are occurring. Short-duration rainfalls are required for runoff calculation, especially in metropolitan areas (Suriya and Mudgal 2012). However, in developing nations like India, short-term rainfall data is limited, and the data that is accessible is largely for daily rainfall data. In such instances, estimating design rainfall becomes an estimate, resulting in frequent drainage network failures and resultant floods (Ahmed et al. 2012). The hydraulic structure frequency analysis was performed to determine the design rainfall value. The frequency analysis was performed using the Intensity–Duration–Frequency (IDF) curve to graphically represent the projected rainfall intensities. The fact that IDFs built from historical climatic scenarios cannot be applicable for future climatic situations unless they are updated to reflect future climate trends has been acknowledged (Sinha et al. 2021, 2022). Before beginning any system performance study, it is necessary to assess the climate-induced change in the pattern, severity, and frequency of extreme rainfall events. Rainfall-Runoff simulation software is available in a variety of formats. The first computerized models of urban storm drainage were created in the late 1960s, and a variety of models have been used since then (Mitchell et al. 1966; Pohlert 2017; Zoppou 2001). Design models, flow prediction models, and planning models are the three types of models (Fewtrell et al. 2008; Rangari et al. 2015). Graphical User Interface (GUI) tools like SWMM, HEC-HMS, HEC-RAS, MIKE FLOOD, and others became available. Understanding the characteristics of rainfall and runoff in the urban area, as well as the consequences of heavy rainfall on the runoff of urban catchments and the many socio-economic elements of rainfall-runoff, is the major goal of rainfall-runoff modeling (Guhathakurta et al. 2011; Halefom et al. 2017).

In this study, the Gumbels extreme value (GEV) distribution and the CPHEEO manual on Rainfall Analysis have been used to prepare Intensity–Duration–Frequency curves of Raipur city. To select the optimal design storm events a rainfall-runoff modeling was done using stormwater management model of the Environmental Protection Agency (EPA) (SWMM) (Ahmed et al. 2012; Shrestha et al. 2017).

The SWMM is a dynamic rainfall-runoff simulation model based on conservation principles of momentum, mass, and energy. This model is used to build, analyze, and plan drainage systems, as well as to simulate the quality and amount of runoff in metropolitan regions (Rossman and Huber 2016).

2 Study Area and Data Used

The area chosen for the present study is a small region of Raipur city, the capital of Chhattisgarh. The geographic location of the study region is 21° 17′ N Latitude, 81° 77′ E Longitude. The location map is shown in Fig. 1. The region of research corresponds to the metropolitan catchment basin that drains the southwest portion of the city of Raipur. It enters the Kharun River from the Buddha Talab region, which implies' talab' in Hindi. The urban basin has a total area of 13.58 km^2 with an area of about 3.30 km north–south and an area of about 5.70 km east–west (about 13,58 km^2). Its highest height is roughly 254.9 m (m) high, the minimum height is 204.5 m above sea level (Sinha et al. 2016). The Kharun River is a major tributary of the Seonath River, which begins in the hamlet of Petechua in the Balod block of Durg district and flows for 164 kms before joining the Seonath River. The Kharun Basin has a dry sub-humid climate (Sinha and Rajput 2020). The region is flat and has a slight slope to the west-southwest (less than 1%). The average annual rainfall is 1200 mm from 1980–2016 and the mean temperature is 46.4 °C. Raipur is one of Chhattisgarh's most rapidly expanding cities.

In the research area, the stormwater that generates runoff follows the open drains. In Raipur, there is a lot of building and development going on. As a result, impermeable land has increased rapidly in recent years. As a result, more and more runoff enter open drains, causing overflow in many areas, and the hydraulic structure is no longer suitable to drain rainwater due to changing rainfall patterns. It is caused not only by its capacity but also by the choice of design rainfall. As a result, there is a need to study the optimal storm event selection for stormwater management. In this study 1980–2016 (37 years) of 17 continuous duration i.e. 15, 30, 45, 60, 90 min, 1, 2, 3, 4, 5, 6, 7, 8, 10, 12, 16, 24 h of rainfall data are used to prepare IDF curve are used and another dataset should be shown in Table 1.

3 Methodology

This paper follows methodology in three parts:

I. To check the fitness of rainfall data using the various distribution.
II. Developing Intensity–duration–frequency (IDF) curves for the available data best-fitted distribution and CPHEEO manual on rainfall analysis stormwater water management.
III. Selection of optimal storm events by using rain rainfall-runoff model.

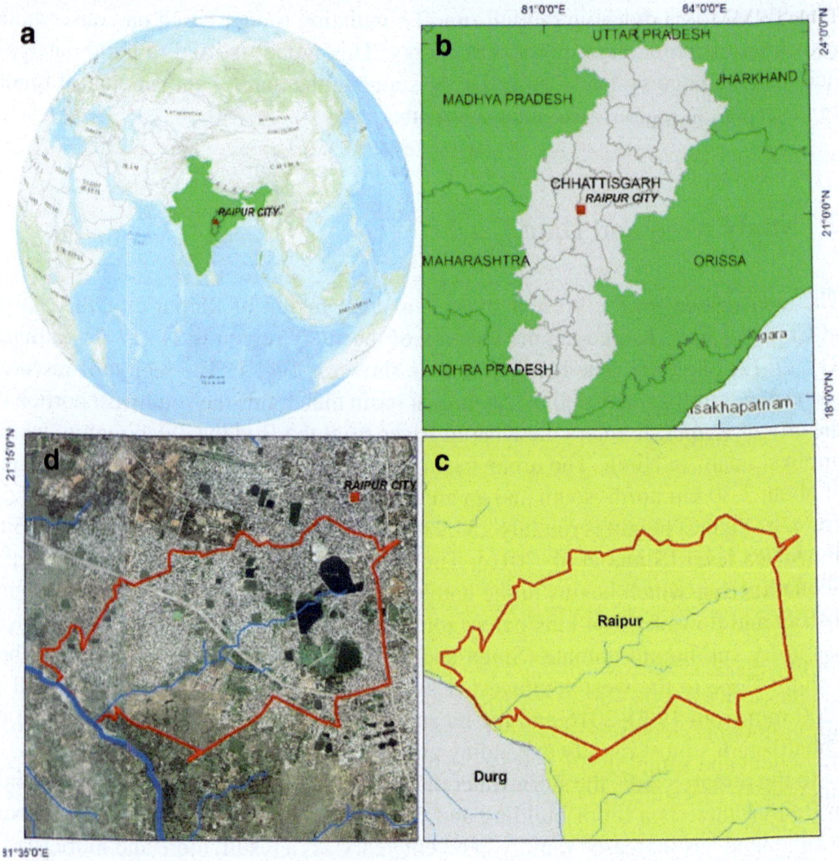

Fig.1 Location map of the study area (modified after Sinha et al. (2019b))

Simulation of rainfall-runoff model of the chosen study area is carried out by well-known software stormwater management model (SWMM). The model should be set up by digital elevation model (DEM) of 12.5 m resolution, topographical map, and land use—land cover map and drainage area network details collected from Raipur Municipal Cooperation (RMC). The catchment area should be divided into sub-catchment using the ArcGIS tool. The time of concentration should be found out by rainfall–runoff hydrograph and design rainfall intensity are obtained from IDF curves.

Table 1 Data used in this study

SNo	Data	Use	Source
1	Rainfall (1980–2016)	15 min shorter rainfall database used for preparing IDF curves	Indian Meteorological Department (IMD)
2	Digital Elevation Model	12.5 m resolution data along with field DGPS survey used for model preparation	ALOS-PALSAR (asf.alaska.edu/#/?dataset=SEASAT)
3	Drainage network	Point-line-vector data representing hydrological condition of the area	Raipur Municipal Corporation
4	Runoff coefficient and Hortons's parameter	Used in Rainfall-Runoff Model	From literature survey and manual of PCSWMM

4 Check the Fitness of Rainfall Data with Various Distribution

For this purpose, we have applied Goodness–of–fit test on various distributions. The goodness-of-fit test is used to test if sample data fits a distribution from a certain population, goodness-of-fit of a probability distribution can be tested by comparing the theoretical and sample values of the relative frequency or the cumulative frequency function. In this study three testes were applied Chi-Square, Kolmogorov–Smirnov, and Anderson–darling test to check the test fitness of all the distribution, Gumbel distribution was found the most suitable from all. All test is non-parametric test that is used to find out how the observed value of a given rainfall is significantly different from the expected value. The null and the alternative hypotheses are: H_0 is the data follow the Gumbel distribution and H_A is vice versa. The hypothesis regarding the distributional form is rejected at the chosen significance level (α) if the test statistic, is greater than the critical value obtained from a table. The fixed values of α (0.2, 0.1, and 0.05) are used to evaluate the H_0 at various significance levels.

5 Preparation of Intensity–Duration–Frequency (IDF) Curves

For this purpose, we have taken the annual maximum daily data from 37 years (1980–2016) 15-min. rainfall record collected from the Indian Meteorological Department (IMD). For the preparation of Intensity–Duration–Frequency (IDF) curves Gumbel distribution and CPHEEO Manual method are used. Gumbel distribution is provided a better fit than the other distribution based on the goodness of fit test. The Intensity–Duration–Frequency (IDF) curves have been generated for 2, 5, 10, 25, 50, 100 yr return period, and the time period is divided into three part one is 1980–2016 that is P1, second is 1980–1993 that is P2 and third one is 1994–2016 that is P3 for considering non-stationarity from change point (Sinha et al. 2021, 2022). From these IDF curves we can derive the design rainfall intensity for all return periods mentioned above.

6 Simulation Rainfall-Runoff Model

The SWMM model is developed for the study area shown in Figs. 2 and 3. Figure 2 shows pre-development-scenario model of the year 1965 in which total area divided into 3 sub-catchment and Fig. 3 shows post-development-scenario model of the year 2018 in which the total area is divided into 51-subcatchment by considering topography maps, sewer lines, and digital elevation models (DEMs). Area, slope, imperviousness, manning's coefficient, the infiltration rate is provided as an input in the SWMM model to define runoff. IDF data used in rain gauge layer to derive the design intensity of rainfall for different return periods and simulate the SWMM model for different storm duration and plot Rainfall-Runoff hydrograph to found out the time of concentration and runoff condition.

Critical storm or design storm period are essential component in storm water management. development of design storm event by using IDF curve. From the simulation of storm water management model shown in Fig. 2 derived rainfall—runoff plot. By using this plot, we can find out time of concentration using this formula:

$$t_p = 0.6 * t_c$$

where, t_p = Lag time (min); t_c = Time of concentration (min).

7 Result and Discussion

K-S test, chi-square, and Anderson darling goodness of fit test are used to determine the data's fitness for various distribution at the significance level of 0.20, 0.10, and 0.05. The result shows that the Gumbel distribution is best fit for the rainfall data than the other distribution. But some of the data are not fit. For example: K-S test has failed the 15 min rainfall data at 0.2 significance level and Chi-Square test and Anderson darling tests were failed at 0.20, 0.10 and 0.05 significance level. The 30 min rainfall data failed to fit in Gumbel as Chi Square test for 0.20 and 0.10 significance and Anderson darling test for 0.20 significance level. The 12 h rainfall is also not fit the Chi Square at 0.20 significance level. As a result, the fitness of 15, 30 min, and 12

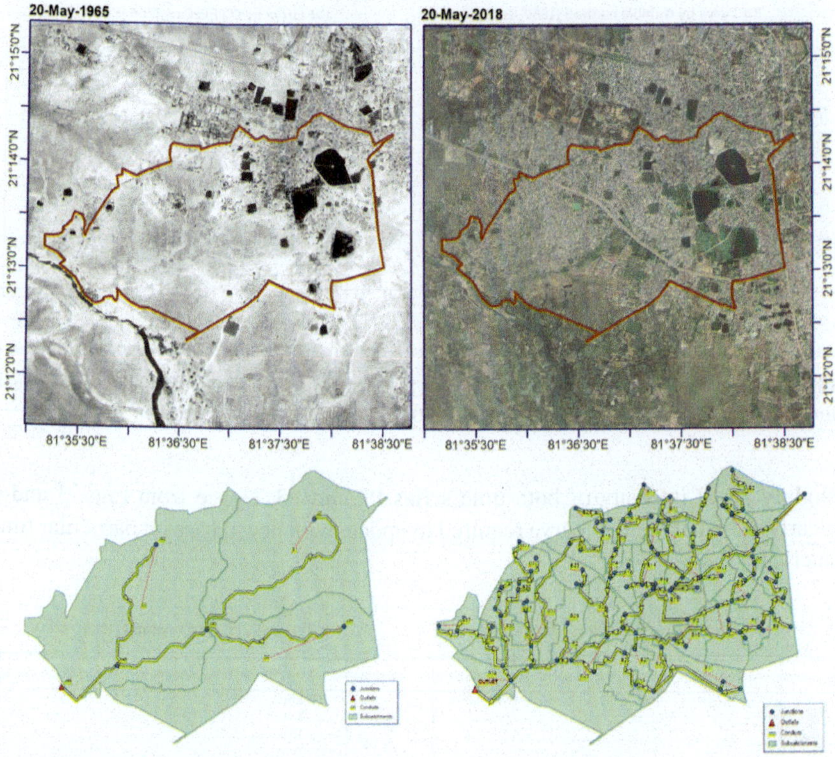

Fig. 2 Pre and Post development-scenario Model

h is insignificant for Gumbel distribution frequency analysis, rest all durations were fitted in GEV.

The Gumbel distribution result from the goodness of fit test shows the projected intensities of annual maximum precipitation for 15, 30, 45 min, 1, 1.5, 2, 3, 4, 5, 6, 7, 8, 10, 11, 12, 16, and 24 h duration for 2, 5, 10, 25, 50, and 100-year return periods. Figure 3 shows that the higher the return period, the greater the intensity of rainfall, which gradually decreases as the duration of the return period grows. Hence, a very precise records of shorter duration rainfall events (such as 1 h, 30, 15 min, and so on) is necessary in order to monitor the expected anomaly of high-intensity rainfall at the appropriate return time.

Figure 4 shows the IDF curves for the time series P2 and P3 by Gumbels method and CPHEEO manual method. The range of projected intensity for time series P2 is 45 to 96 mm/hr in Gumbel distribution method and 97–107 mm/hr in CPHEEO manual method for 2, 5, 10, 25, 50 and 100 years return period. And The range of projected intensity for time series P3 is 51–212 mm/hr in Gumbel distribution method and 91–107 mm/hr in CPHEEO manual method for 2, 5, 10, 25, 50 and 100 years return period. Which indicate that the Gumbel distribution method shows higher intensity in time series P3 as compared to P2 and CPHHEO method shows

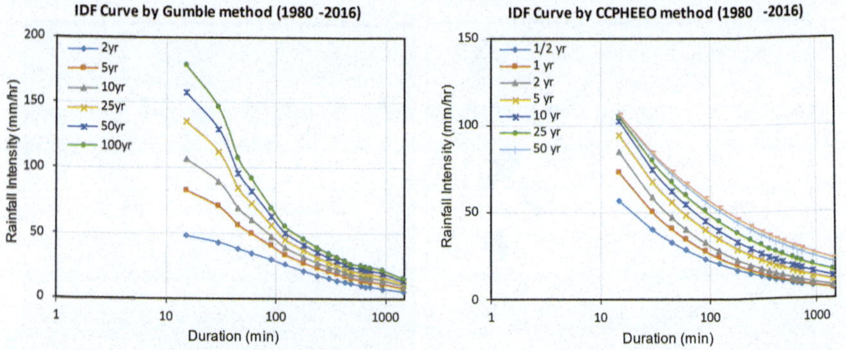

Fig. 3 IDF curves of Raipur city derived using Gumbel distribution and CPHEEO manual method for time series P1 (1980–2016): (Data-base:1980–2016:15 min storm interval)

slightly equal intensity of both time series P2 and P3. Hence from Figs. 3 and 4 clearly seen that the IDF curve required to update with new dataset at particular time interval.

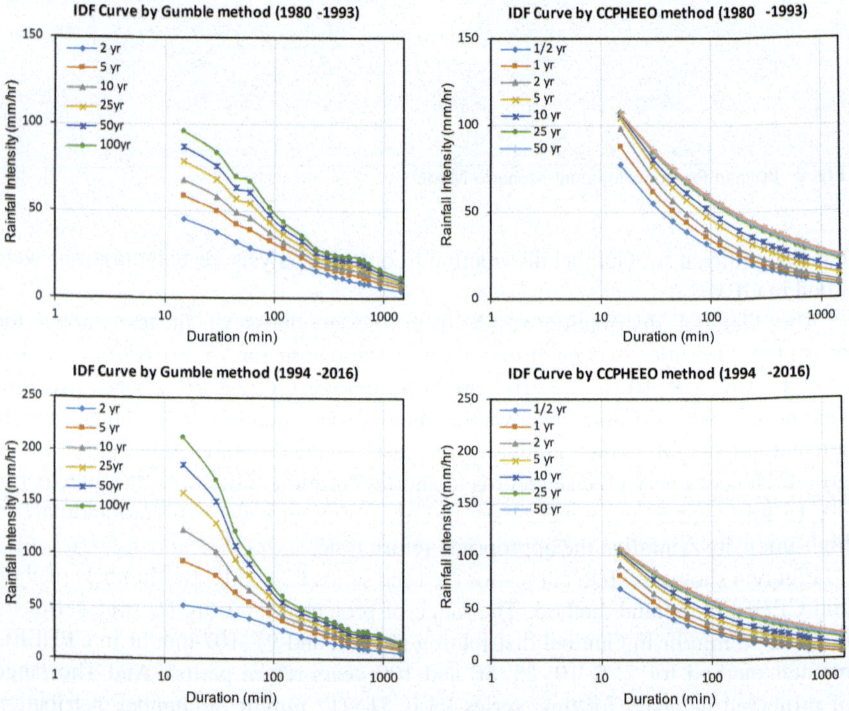

Fig. 4 IDF curves of Raipur city derived using Gumbel and CPHEEO for time series P2 (1980–1993) and P3 (1994–2016)

Figure 5 shows change in runoff in pre and post development scenarios for 10 year and 50-year return period. In pre-development scenario the runoff value is lesser than the post development scenarios i.e. Runoff value in pre-development scenario is 1.62 m^3/sec and in post development scenario is 45 m^3/sec for 10-year return period and runoff value in pre-development scenario is 1.23 m^3/sec and in post development scenario is 49 m^3/sec for 50-year return period. That shows the peak is higher after the urbanization which places new demands and challenges in storm water management. After the simulation of rainfall—runoff model with various return period junction j45, j44. j43, j30 are flooded node shown in below Fig. 7. And the flood inundation map shown in Fig. 6 for 10-year return period.

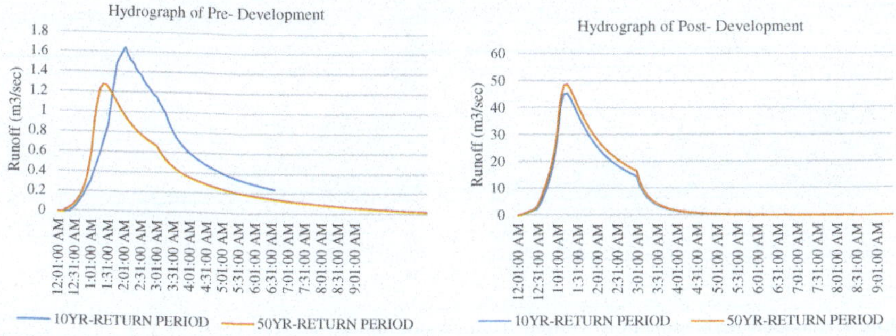

Fig. 5 Change in runoff in pre and post development scenario

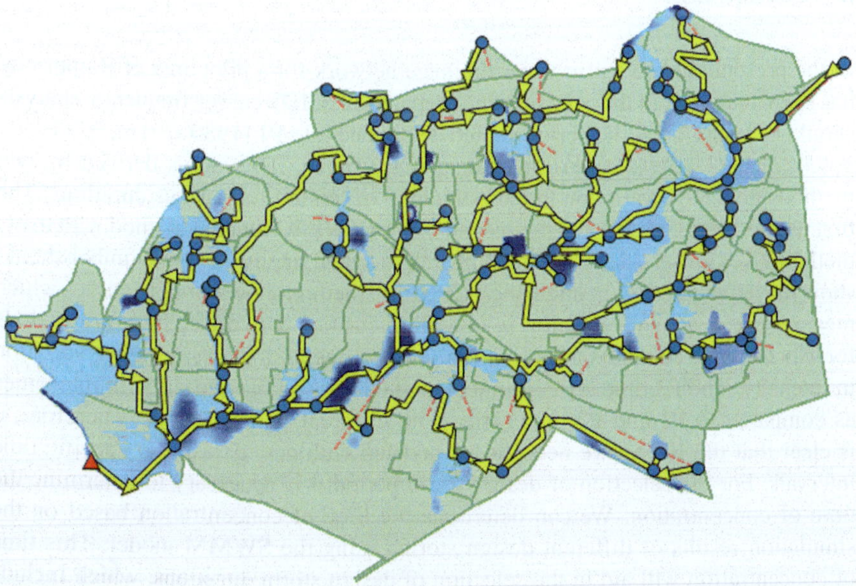

Fig. 6 Flood inundation map of the study area indicating flooding area-duration-depth

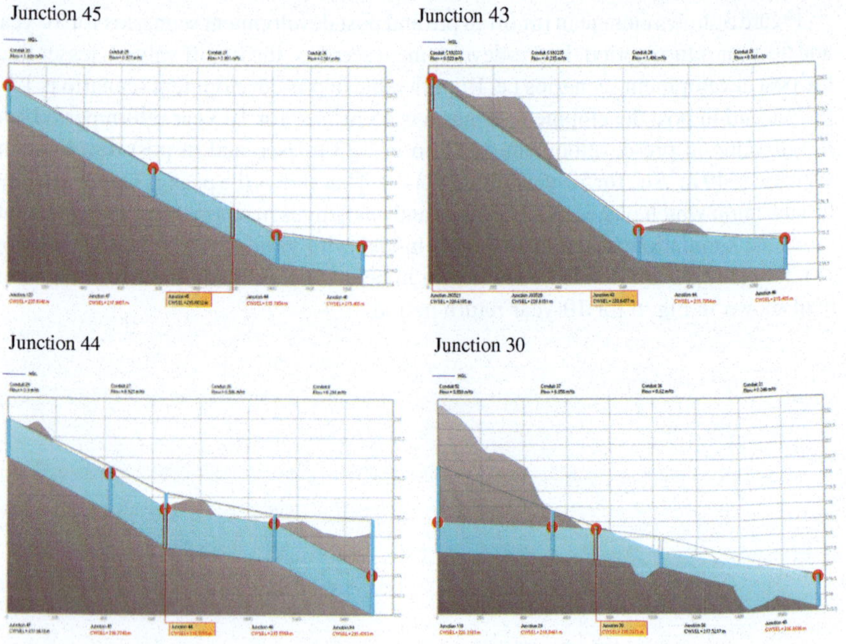

Fig. 7 Flooded node (Junction) Output from design rainfall derived from 10 year return period IDF

8 Conclusion

In the present study a storm water drainage network for a pilot area of Raipur city has been analyzed. In this study for the preparation of IDF curves frequency analysis should be done. The time period should be divided into P1 (1980–2016), P2 (1980–1993), and P3 (1994–2016) for making IDF Curves. IDF curves derived by two methods give large variation in the estimation of rainfall for various durations. The first method was Gumbel and the second one is CPHEEO manual method. CPHEEO method gives closer results when compared to actual intensities but Gumbel should show the large variation in intensities. CPHEEO method shows smooth curves which means there is a small variation in rainfall intensities with different return periods for P1, P2, and P3 time series, but the Gumbel method shows a higher variation in intensity. The P3 time series Gumbel method shows large variation in intensities as compared to P1 and P2 time series with different return periods. Therefore, it is clear that the IDF curve needs to be updated with new datasets at specific time intervals. For the selection of design storm period it is necessary to determine the time of concentration. We can determine the time of concentration based on the simulation results of different design storms using the SWMM model. This time of concentration will aid in the selection of design storm durations, which include 2, 5, 10, 50 year return periods with 1 and 3 h durations in pre and post duration.

This research demonstrates the design storm value, which will be helpful hydraulic structure design and to identify the flood prone area and reduce the risk of flooding.

Acknowledgements The authors would like to thank the Deutscher Akademischer Austauschdienst- German Academic Exchange Service (DAAD), as the research material is based upon work within the program "A New Passage to India". One of the authors would like to acknowledge the help in field data collection provided by LIH workgroup.

References

Ahmed Z, Rao D, Reddy K, Raj E (2012) Rainfall intensity variation for observed data and derived data—a case study of imphal. Arpn J Eng Appl Sci 7
Fewtrell T, Bates PD, Horritt M, Hunter N (2008) Evaluating the effect of scale in flood inundation modelling in urban environments. Hydrol Process Int J 22:5107–5118
Guhathakurta P, Sreejith OP, Menon PA (2011) Impact of climate change on extreme rainfall events and flood risk in India. J Earth Syst Sci 120:359–373
Halefom A, Sisay E, Khare D, Singh L, Worku T (2017) Hydrological modeling of urban catchment using semi-distributed model. Model Earth Syst Environ 3:683–692
Mitchell JM, Dzezerdzeeskii B, Flohn H, Hofmeyer WL, Lamb HH, Rao KN et al (1966) Climatic change. WMO Tech. Note 79, WMO No. 195. TP-100. Geneva 79
Pohlert T (2017) Non-parametric trend tests and change-point detection. CC BY-ND. 4
Rangari VA, Patel AK, Umamahesh NV (2015) Review of urban stormwater models. In: HYDRO 2015 international, 20th international conference on hydraulics, water resources and river engineering, IIT Roorkee
Rossman LA, Huber WC (2016) Storm water management model reference manual volume I–hydrology (Revised). USEPA. National Risk Management Laboratory
Schmitt TG, Thomas M, Ettrich N (2004) Analysis and modeling of flooding in urban drainage systems. J Hydrol 299:300–311
Sharma S, Singh PK (2017) Long term spatiotemporal variability in rainfall trends over the State of Jharkhand, India. Climate 5:18
Shrestha S, Khatiwada M, Babel MS, Parajuli K (2014) Impact of climate change on river flow and hydropower production in Kulekhani hydropower project of Nepal. Environ Process 1:231–250
Shrestha A, Babel M, Weesakul S, Vojinovic Z (2017) Developing Intensity–Duration–Frequency (IDF) curves under climate change uncertainty: the case of Bangkok, Thailand. Water 9:145
Sinha MK, Verma MK, Ahmad I, Baier K, Jha R, Azzam R (2016) Assessment of groundwater vulnerability using modified DRASTIC model in Kharun Basin, Chhattisgarh, India. Arab J Geosci 9:98
Sinha MK, Rajput P (2020) Geospatial evaluation of drought resilience in Sub-basins of Mahanadi river in India. Water Supply
Sinha MK, Baier K, Azzam R, Verma MK, Kumar S (2022) Impacts of climate variability on urban rainfall extremes using statistical analysis of climatic variables for change detection and trend analysis. In: Kumar P, Nigam GK, Sinha MK, Singh A (eds) Water resources management and sustainability. Springer Singapore, Singapore, pp 333–387
Sinha MK, Baier K, Azzam R, Verma MK, Jha R (2021) Analysis of intensity–duration–frequency and depth–duration–frequency curve projections under climate variability. Clim Change Impacts Water Resour Hydraul Water Resour Coast Eng 407–421
Sinha MK, Baghel T, Baier K, Verma MK, Jha R, Azzam R (2019a) Impact of urbanization on surface runoff characteristics at catchment scale, pp 31–42

Sinha MK, Baier K, Azzam R, Baghel T, Verma MK (2019b) Semi-distributed modelling of stormwater drains using integrated hydrodynamic EPA-SWM model, pp 557–567
Suriya S, Mudgal BV (2012) Impact of urbanization on flooding: the Thirusoolam sub watershed—a case study. J Hydrol 412–413:210–219
Zoppou C (2001) Review of urban storm water models. Environ Model Softw 16:195–231

Evaluation of Effective Criteria on Determination of Capable Areas to Construction of Underground Dam in Shahrekord Watershed, Iran

Sayed Naeim Emami, Saleh Yousefi, and Mohammad Nekoeimehr

Abstract In this study, in order to help in water supply, the effective criteria on the locating of the construction underground dam in Shahrekord watershed is investigated in three steps. In the first step of the removing criteria (slope, geology and land use), zero and one values were labeled to the suitable and unsuitable areas respectively using Boolean logic and suitable areas for the underground dam is determined. In the second step, the axes that do not stop the Qanat furnace and are suitable from the aspect of length and surface of the reservoir were selected, which ultimately identified 15 axes. In the third step, the proposed locations were prioritized using the analytic hierarchy process. For this purpose, four criteria including water status, reservoir, dam axis and socio-economic factors were used. Therefore, the indices with high suitability are in accordance with more suitable axes for the construction of underground dams in the studied area. Five scenarios were considered for the prioritization. The results of this study showed that the best axes for the construction of an underground dam are located in the alluvial river bed with high subsurface flow, areas with large capacity of the reservoir or high runoff volume and also in areas with high permeability and low slope. The proposed method in this study is able to accurately determine suitable area for construction of underground dams through precise and detailed studies and considering factors affecting underground dams.

Keywords Criteria · Susceptible area · Underground dam · Multi-criteria decision making

---

S. N. Emami
Soil Conservation and Watershed Management Research Department, Chaharmahal and Bakhtiari Agricultural and Natural Resources Research and Education Center, AREEO, P.O. BOX: 415, Shahr-e Kord, Iran
e-mail: emami1348@yahoo.com

S. Yousefi (✉) · M. Nekoeimehr
Soil Conservation and Watershed Management Research Department, Chaharmahal and Bakhtiari Agricultural and Natural Resources Research and Education Center, AREEO, Shahr-e Kord, Iran
e-mail: s.yousefi@areeo.ac.ir

C. B. Pande et al. (eds.), *Surface and Groundwater Resources Development and Management in Semi-arid Region*, Springer Hydrogeology,
https://doi.org/10.1007/978-3-031-29394-8_21

1 Introduction

By suppling ~36% of drinking water and ~42% of agricultural water, groundwater is a key freshwater resource globally (Wang et al. 2022). Iran's water crisis is entering a new paradigm where its impacts are becoming visible in the daily lives of millions of people (Kadam et al. 2012; Yousefian et al. 2022). The average annual precipitation in Iran is less than one third of the world average annual rainfall (Panda et al. 2007). According to the distribution of time and location of rainfall and runoff in the country and the lack of adequate moisture and permanent surface flows, the main sources of water supply are aquifers (Kharazi et al. 2019). The increasing importance of water supply causes that use of groundwater has been a great importance in comparison to surface water as a sustainable source because these resources are lower subject to drought and dehydration than other sources (Panda et al. 2007; Rahmati et al. 2016). In order to use groundwater resources and availability of water in all seasons, the underground dam is an appropriate solution in some specific areas (Yazdandoost 2016). An underground dam is a designed structure that obstructs and control flow of groundwater and stores water below the ground surface (Sehat et al. 2013). A subsurface dam is constructed below ground level and arrests the flow in a natural aquifer whereas sand-storage dam impounds water in sediments caused to accumulate by the dam itself (Talebi et al. 2019). The most important problem in the development and establishment of underground dams is the complexity of determining the appropriate areas for the construction of the dam. These problems are due to many criteria and factors including physical and socio-economic criteria that they are effective in locating (Chezgi et al. 2016). As a rule, underground water resources are affected by changes in climatic and hydrological condition (Naghibi and Pourghasemi 2015). Despite the indiscriminate extraction of underground water resources, they still play a vital role in water supply. Currently, about 33% of the world's people suffer from water shortage (Mehrabi et al. 2013).

Therefore, the development of a decision support system is essential in order to determine the criteria and identify the appropriate sites for the construction of the underground dams (Talebi et al. 2019). Sub-surface dams are obstacles to block subsurface water flow and these construction cause water to be stored in local aquifers or transferred to adjacent aquifers (Chezgi et al. 2016; Ebrahimi et al. 2021). The underground dams, as barriers, control the underlying water flow in the alluvial basin. These barriers can be physical or hydraulic. Hydraulic barriers are usually established adjacent to saltwater aquifers near the sea to protect freshwater aquifers (Sehat et al. 2013; Chezgi et al. 2016). The general difference between the underground and surface dams is that surface barriers are constructed obstacles in the rivers and theirs purpose is storing upstream surface water and collecting them in an open reservoir, while the underground dam is designed to store water at the lower levels of the ground and in the form of obstacles, it performs the collection of water and diversion of it to arbitrary points (Chezgi et al. 2016; Kharazi et al. 2019). This research has organized for determining and prioritizing factors affecting the location of underground dams construction using remote sensing techniques and preparing a multivariable decision-making system (DSS) in the Shahrekord Watershed.

2 Materials and Methods

This research consists of three stages:

(a) Identification and selection of locations with acceptable conditions for the construction of an underground dam: For this purpose, the data extracted from the basic maps, findings in this field and also expert opinions were used. Suitable areas include the alluvial bed of rivers that have slopes of up to 6% and there are no agricultural, industrial, residential or aqueduct furnaces there. In addition to the above conditions, these alluvial beds should not conform to linear structures such as faults. In this research, Boolean logic was used to remove inappropriate points.

(b) Identification of suitable straits in these areas: After preparing the base maps of geology, slope and land use using Boolean logic, the areas with a value of 1 are determined except for the areas prone to underground dams. At this stage, the remaining axes should be evaluated from three aspects. In addition to having the smallest width for a low structural workload, the desired axis in each area must also have a suitable tank volume.

(c) Evaluating the points relative to each other and prioritizing them for the construction of the underground dam: After specifying the axes in the second stage of locating, in this stage, using the MADM method and based on the decision-making process, suitable points for the construction of the underground dam are prioritized.

2.1 Study Area

Shahrekord watershed is located in north east of Chaharmahal va Bakhtiari province and in geographical area of Shahr-e-kord city (Yousefi et al. 2020). Also this watershed is located between 50°20′42.2″ to 51°10′19.76″ eastern longitude and 32°07′31.07″ to 32°33′13.28″ northern latitude. The location of the study area is illustrated in Fig. 1. The cities of Shahr-e-kord, Hafshchan, Nafech, Qian, Soureshjan, Harooni, Taqanak and Farokh-shahr are located in this area and have a population of 248,219 people. The area is 1454.8 km^2. The average annual rainfall is 330 mm and 442 mm at Shahr-e-kord and Morghmalek Stations, respectively. This basin is located in Sanandaj-Sirjan zone based on structural divisions and the atmospheric rainfall in the basin is mainly influenced by low-pressure Mediterranean systems (Emami et al. 2020).

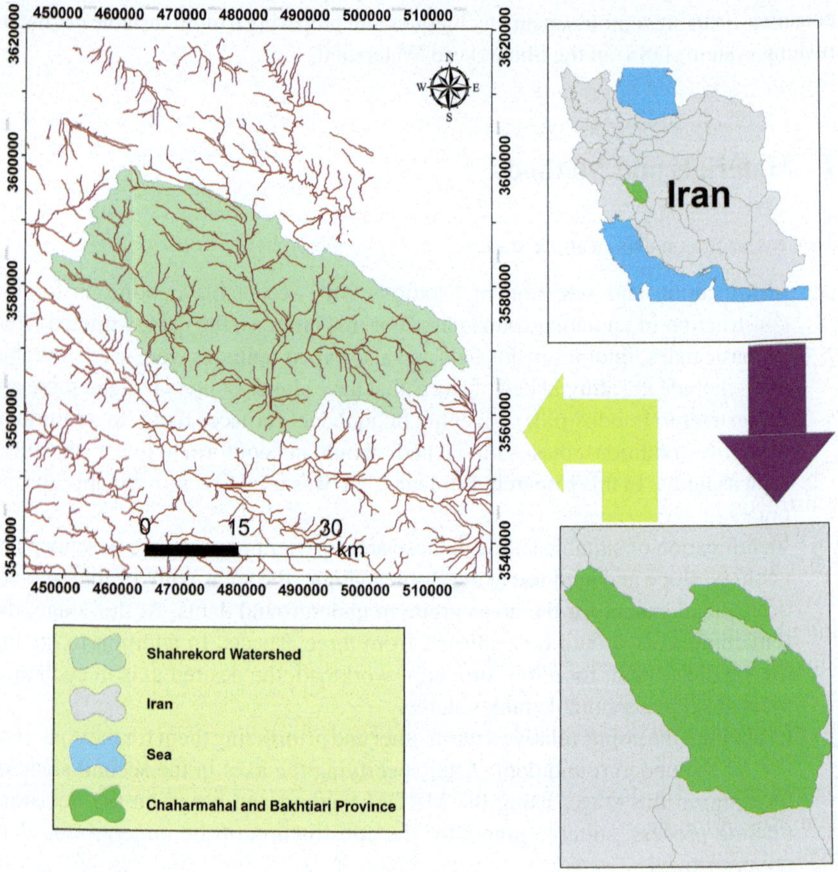

Fig. 1 Geographical location of the study area in Shahrekord, Chaharmahal va Bakhtiari province

2.2 *Identification and Selection of Appropriate Areas for the Construction of Underground Dams*

In order to accelerate the decision-making process, as well as avoiding the accumulation of excessive information for a surveyed problem, it is first necessary to eliminate inappropriate points by considering a number of key criteria and factors (Mallick et al. 2018). Suitable areas are including alluvial beds in rivers with maximum slopes 6% and in this area there are no agricultural, industrial, residential or Qanat furnace (Chezgi et al. 2016). In addition to the above mentioned conditions, these alluvial beds should not be in line with linear structures such as faults. The proper slope of the waterway should not be more than 5% to make it possible to create a suitable reservoir for subsurface water by constructing an underground dam (Mohammady et al. 2019; Ebrahimi et al. 2021). In this research, Boolean logic was used to remove inappropriate points. Boolean logic is derived from the name of the mathematician

(Mahdavi et al. 2017), in which weighing the units in each layer of information is based on the score of zero and one (Pourghasemi et al. 2012). The Boolean model has two operator of AND Boolean and OR Boolean, which based on theories, set of AND operator, the subscriber and operator of OR, extracts the set of communities (Mahdavi et al. 2017; Ebrahimi et al. 2021). According to the purpose of the research, AND operator was used to select suitable areas for all parameters. Geological, slopes and land-use maps were prepared using Bolin logic and were given value 1 to the areas prone to constructing a dam and value 0 to the restricted areas for construction.

2.3 Determination of Initial Proper Area for Construction Underground Dam

The desired axis in each range, in addition to having the lowest width for a small construction workload, should also have the appropriate reservoir volume. The presence of water resources in these areas should be investigated (Yazdandoost 2016). After determining these areas, using field observations, straits in these areas were investigated. If the length of the axis be less, the volume of structural work will reduce and implementation operation progresses rapidly (Ebrahimi et al. 2021). The most suitable axes are the ones that, in addition to the short length, have a large surface expansion of the reservoir at the upper hand axis. Qanats must also be considered not to be destroyed (Rahmati et al. 2019). An area with a distance of 100 m from the Qanats axis was defined, because the axis of underground barrier should not cut the Qanat furnace. The areas inside this area are unsuitable for creating an underground dam. Regarding these factors and field observations in these axes, 15 proper axes were identified.

2.4 Prioritization of Axes for the Construction of Underground Dams

After determining the axes in the second step of locating suitable site, at this stage, using the MADM method and based on the decision flow, the appropriate points for the construction of the underground dam are prioritized (Sehat et al. 2013; Talebi et al. 2019). The overall structure of decision-making criteria is presented in four levels. Level one in a hierarchy shows the target and the second level shows the criteria for achieving the desired target. At the third level (if this level exists), described the basic sub-criteria that effect on selection and finally, at the last level is mentioned options or solutions (Zucca et al. 2008). After determining the axes in the first stage of the locating suitable site and determining the criteria and their value, at this stage, using the MADM method and based on the AHP decision method, the proper points are prioritized for the construction of the underground dam (Kumar et al. 2017).

The required data in the GIS environment is provided in order to prioritize the 15 specified axes for the existing indicators in the lower part of the decision tree, which are the information layers for each axis. Subsequently, the fit index was calculated for these indices and criteria and using the formula for each axis by using the determined relative values in the binary method (Ebrahimi et al. 2021) (Fig. 2 and Table 1).

After binary comparing of the criteria and determining their relative importance, the suitability index for each criterion or option is achieved by combining the relative importance of the existing criteria (formula 1).

$$S_I = R_I A_1 \times \sum_{i=1}^{m} R_I \cdot B_i \times \mathrm{R}_I \cdot KB_i + R_I \cdot A_2 \times \sum_{y=1}^{L} R_I C_y + \ldots R_I A_N \times \sum_{z=1}^{j} R_I D_z \times R_I KD_Z \quad (1)$$

S_I = suitability index, N = Number of the major criteria of A, R_IA_N … R_IA_1, R_IA_2 = The relative importance of the criteria of A_1, A_2, …, A_N, L, j = the number of sub criteria related to the main criteria of A_1, A_2, …, A_N, R_IB, R_IC, R_ID = The relative importance of the criteria B, C and D, which are related to the main criteria A1, A2, …, and A_N. R_IKD and R_IKB, R_IKC = The relative importance of the indicators for sub criteria of B, C and D associated with the main criteria A1, A2, … A_N.

The mentioned formula will be correct If the decision sequence has been lower or higher level. The Inconsistency rate for the validity of the results of the AHP method should be less than 0.1; otherwise the comments should be modified.

3 Results and Discussion

3.1 *Detection and Selection of Appropriate Areas for Underground Dams*

Geologically, quaternary formations have the value of one and suitable for the construction of underground dam (Fig. 3). The slopes of 4–2 and 6–4 are suitable for the construction of underground dam in the slope map (Fig. 4). Poor, medium and good rangelands and garden lands are suitable for underground dam (Fig. 5). After combining these maps with Boolean logic, areas with a numerical value 1 are the appropriate area for construction underground dams (Fig. 6).

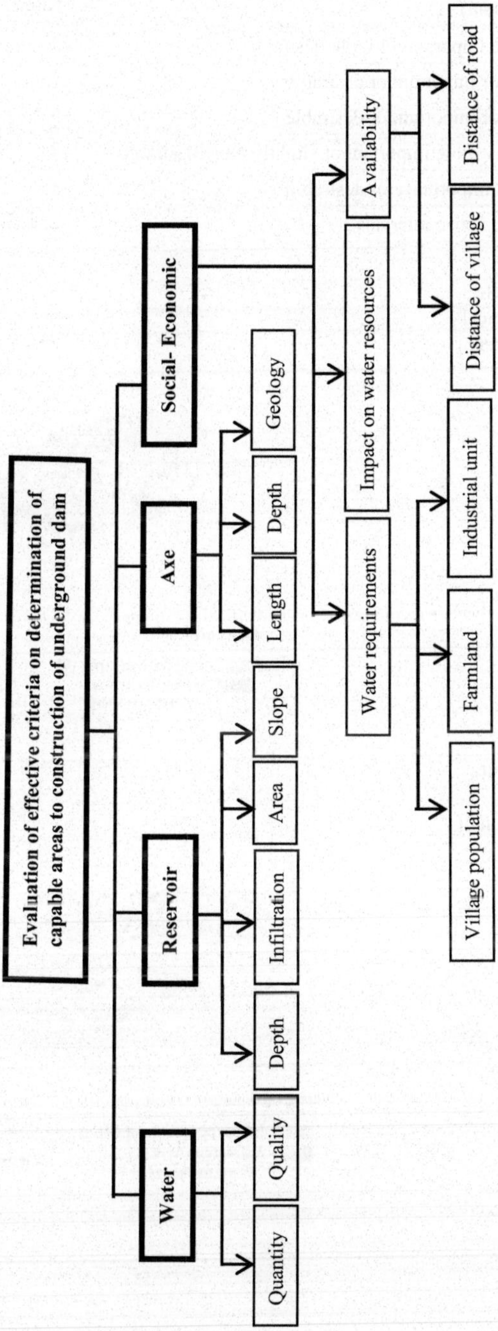

Fig. 2 The structure of decision making criteria in four levels to prioritize the construction of an underground dam (adapted from Chezgi et al. 2016)

Table 1 Quantitative value of judgments of the AHP model

Preferences (oral judgment)	Numerical value
Fully preferable or quite important or quite desirable	9
Very strong preference or important or desirable	7
Very strong preference or important or desirable	5
Slightly better or slightly more important or slightly more desirable	3
The equal preference or importance or desirability	1
Preferences between the above intervals	2, 4, 6, 8

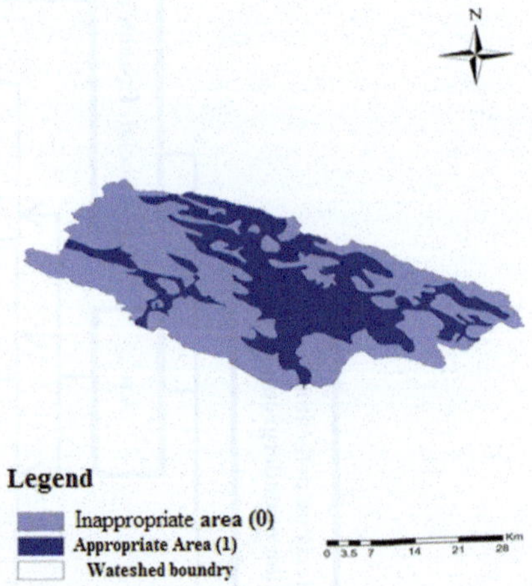

Fig. 3 Map of appropriate and inappropriate geological ranges for the construction of underground dam

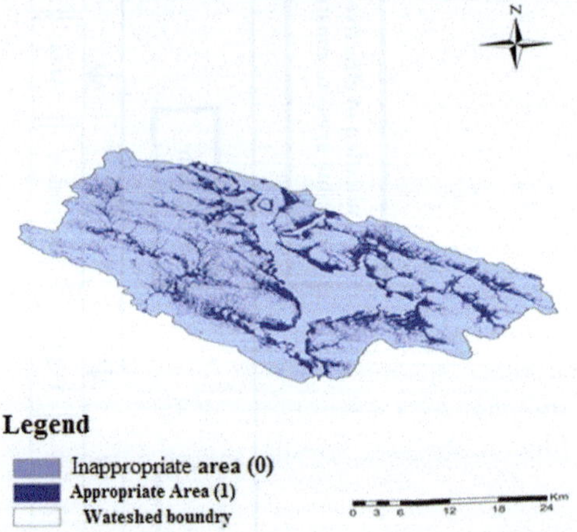

Fig. 4 Map of appropriate and inappropriate slope ranges for the construction of underground dam

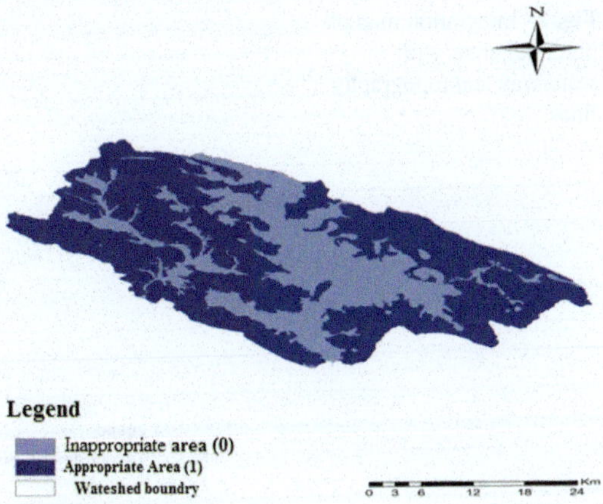

Fig. 5 Map of appropriate and inappropriate land-use ranges for the construction of underground dam

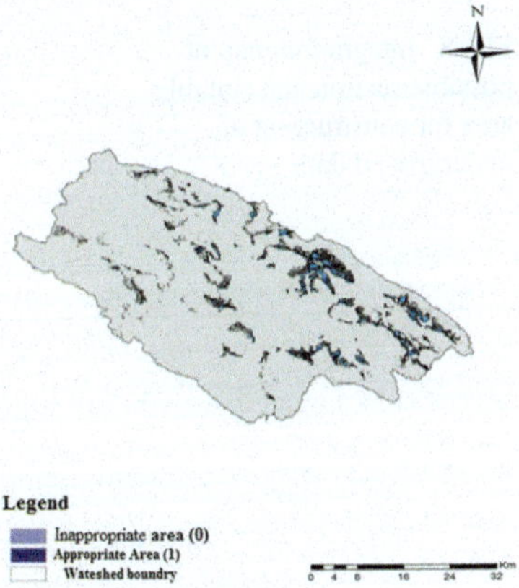

Fig. 6 Map of appropriate area for the construction of underground dam using removing criteria in Boolean method

3.2 *Identification of Appropriate Locations Within These Ranges*

After designing the map of suitable ranges for construction the underground dam by Boolean method, these ranges were mapped with a topography map 1: 25,000 and a drainage map derived from the altitude map was adapted and appropriate gaps were identified for the construction of underground dam (Figs. 7 and 8).

Axes of 52,8 due to low width of waterway, axes of 24, 15 and 62 due to lack of suitable reservoir and axes of 73, 71, 65 and 59 due to unsuitable foundation are removed. This assumption is considered that qanats should not be exposed to destruction or their water discharge would not be dropped sharply. Therefore, at this stage of finding potential, areas with Qanat furnaces are considered as unsuitable areas for the construction of underground dam. Figure 9 shows the distribution of Qanats in the region. By integrating the maps of the Qanats axis of the study area and

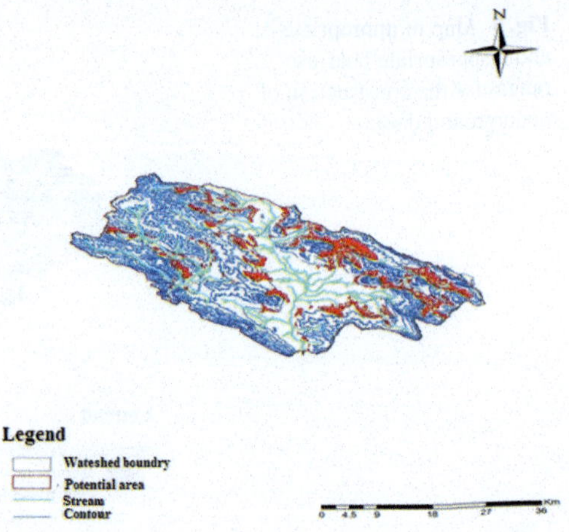

Fig. 7 Integration map of potential areas with waterway and topography lines

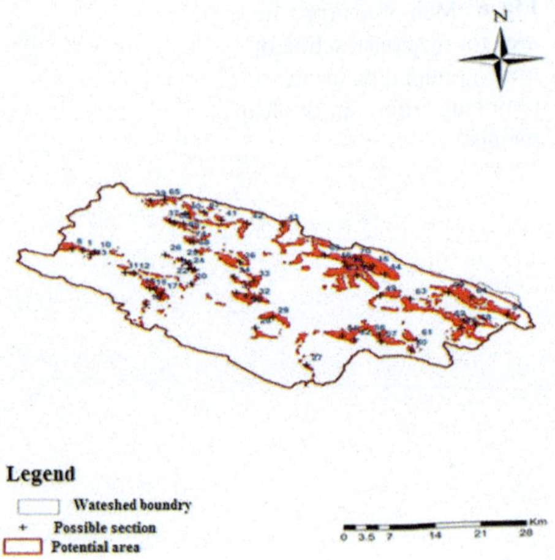

Fig. 8 Integration map of possible section and suitable area for construction of underground dam

the map of possible gaps of underground dam construction (Figs. 9 and 10), a total of 15 axes were identified for the construction of underground dam in ShahreKord and Morghmalek watersheds (Fig. 11).

3.3 Prioritizations of the Axes for the Construction of the Underground Dam

The suitability index is calculated for each of the main and secondary criteria in each branch of the decision tree and finally, after merging and aggregating them, is shown as a final number. If this amount be high, desired axis has more value for the construction of the underground dam (Tables 2, 3, 4, 5, 6, 7, 8, 9, 10 and 11).

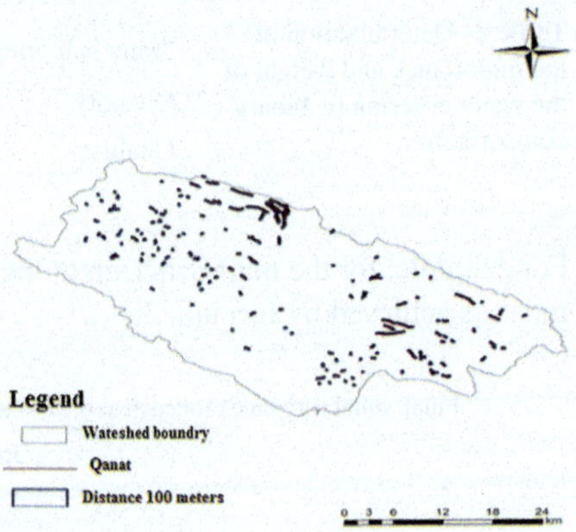

Fig. 9 Distribution map of Qanats in ShahreKord and Morghmolk watershed

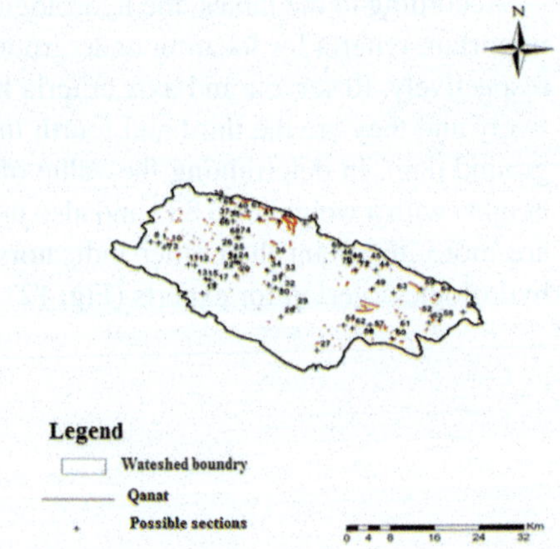

Fig. 10 The map of integrating the Qanats axis and possible sections of the underground dam

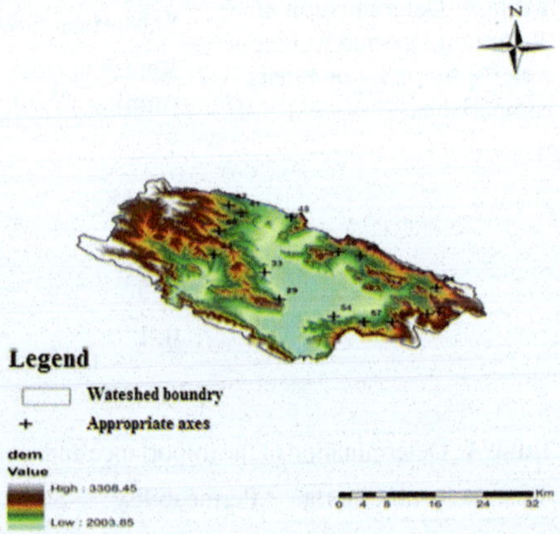

Fig. 11 Appropriate axes for building underground dams in ShaheKord and Morghmolak watersheds

Table 2 Determination of the importance and weight of the water criterion by binary comparisons

Water sub-criteria	Quantity	Quality	Relative importance
Quantity	1	5	0.83
Quality	1.5	1	0.17

For example, for the main criterion of the reservoir, the calculation of the suitability index is achieved by formula 2:

$$\begin{aligned}\text{Final suitability index for each axis} &= \text{Water suitability index} + \text{Axis suitability index}\\ &+ \text{Reservoir suitability index}\\ &+ \text{Socio-Economic suitability index}\end{aligned} \tag{2}$$

According to the tables, the hydrologic and socio-economic criteria are the most important criteria for locating underground dam with the weights of 0.51 and 0.30, respectively. Reservoir and axis criteria have the weight of 0.13 and 0.058, respectively and they are the third and fourth important criteria for prioritizing the underground dam. In determining the value of the indicators, the water quantity (runoff height) with a weight of 0.833 and also the depth of alluvium with a weight of 0.743 are more important than other indicators that this indicates the importance of the hydrologic criterion for experts (Fig. 12).

Table 3 Determination of the importance and weight of water sub-criteria by binary comparisons

Subsurface flow		Water quality	
Runoff height (mm)	Relative importance	Quality	Relative importance
15<	0.033	C_2–S_1	0.8
10–15	0.063	C_3–S_1	0.2
5–10	0.129		
1–5	0.261		
0–1	0.513		

Table 4 Determination of the importance and weight of the reservoir criteria by binary comparisons

Reservoir sub-criteria	Permeability	Slope	Surface	Depth	Relative importance
Permeability	1	3	5	5	0.56
Slope	1.3	1	3	3	0.25
Surface	1.5	1.3	1	1	0.095
Depth	1.5	1.3	1	1	0.095

Table 5 Determination of the importance and weight of reservoir sub-criteria by binary comparisons

Depth of alluvium		Slope		Surface		Permeability	
Depth (m)	RI	Slope (%)	RI	Reservoir surface (ha)	RI	Surface permeability (cm\h)	RI
0–20	0.46	0–2	0.56	40<	0.51	0.01–0.1	0.03
20–40	0.33	2–4	0.26	30–40	0.26	0.03–0.5	0.05
40–60	0.15	4–6	0.12	20–30	0.13	0.25–0.5	0.085
60<	0.06	6<	0.06	10–20	0.06	0.8–2.0	0.15
				10>	0.03	1.3–7.6	0.25
						2.5–25.0	0.44

Table 6 Determination of the importance and weight of the axis criterion by binary comparisons

Axis sub-criteria	Depth	Length	Foundation	Relative importance
Depth	1	5	9	0.74
Length	1.5	1	4	0.19
Foundation	1.9	1.4	1	0.06

Table 7 Determination of the importance and weight of axis sub-criteria by binary comparisons

Depth of alluvium		Axis length		Axis foundations	
Depth (m)	RI	Length (m)	RI	Lithology	RI
0–10	0.57	0–20	0.51	Group 1	0.34
10–20	0.27	20–40	0.26	Group 2	0.24
20–30	0.13	40–60	0.13	Group 3	0.15
30<	0.04	60–80	0.06	Group 4	0.09
		80<	0.03	Group 5	0.06
				Group 6	0.04
				Group 7	0.02

Table 8 Determination of the importance and weight of the socio-economic criterion by binary comparisons

Socio-economic sub-criteria	Influence on water resources	Water requirement	Availability	RI
Influence on water resources	1	5	7	0.73
Water requirement	1.5	1	3	0.19
Availability	1.7	1.3	1	0.08

Table 9 Determination of the importance and weight of influence on water resources and water requirement sub-criteria by binary comparisons

Reducing the water resources of the subterranean sub-basin Qanat		Water requirement		Rural population		The area of agricultural land		Industrial units	
Influence on water resources (%)	RI	Type of requirement	RI	Population (thousands of people)	RI	Area (ha)	RI	Existence of unit	RI
0–25	0.51	Need drinking water	0.73	>4	0.51	>3000	0.51	Available	0.9
25–50	0.26	Need agricultural water	0.19	3–4	0.26	2000–3000	0.26	Not available	0.1
50–75	0.13	Need industrial water	0.8	2–3	0.13	1000–2000	0.13		
75–90	0.06			1–2	0.06	100–1000	0.06		
90<	0.03			<1	0.03	<100	0.03		

Table 10 Determination of the importance and weight of access to the underground dam with binary comparisons

Access criteria	Distance from the village	Distance from the road	Relative importance
Distance from the village	1	7	0.88
Distance from the road	1.7	1	0.12

Table 11 Determination of the importance and weight of access to dam sub-criteria by binary comparisons

Distance from the village		Distance from the road	
Distance (m)	RI	Distance (m)	RI
<1000	0.56	<1000	0.51
1000–2000	0.26	1000–2000	0.26
2000–3000	0.12	2000–3000	0.13
3000<	0.05	3000–4000	0.06
		4000<	0.03

Fig. 12 Prioritization of the proper axis of underground dam in the first scenario

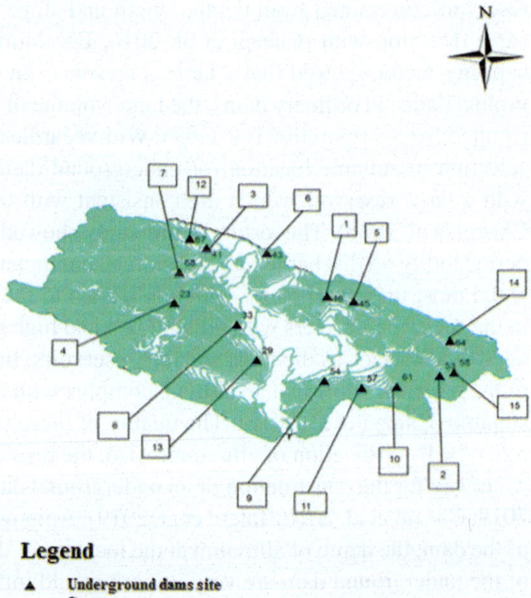

4 Conclusions

The aim of this study was to investigate the effective criteria on determining the proper location of construction underground dam in Shahrekord watershed of Chaharmahal and Bakhtiari province. The results showed that the quantity of water is more important than its quality, because in the absence or lack of subsurface flow, the underground reservoir is not fully dewatered and will be faced with many problems, such as supplying water right which is consistent with results of (Talebi et al. 2019) about collecting water using an underground dam with emphasis on water quantity in African countries. If the waterway has been a higher subsurface flow, it has more relative importance than other waterways. The results of this study showed that the most suitable waterways for the construction of underground dams are those with 3rd and 4th rank, which is consistent with the results of (Chezgi et al. 2016).

Due to the formation of the underground reservoir in the underground and between the pores and alluvial sediments, it is very difficult to obtain the volume of the reservoir, which is considered as disadvantages of underground dam. The volume of the reservoir determined from depth, length and slope indices that these results are in same direction with (Chezgi et al. 2016; Ebrahimi et al. 2021) results. Reservoir capacity factor showed that a large reservoir is an necessary factor. Unlike underground dams, in ordinary dams, the large volume of the reservoir due to water losses resulted from evaporation is a defect. With regardless of other effective factors in the selection of suitable location for underground dam's construction are in upstream with a large reservoir, which are consistent with the results of (Sehat et al. 2013; Chezgi et al. 2016). The results of this study showed that due to the high influence of permeability on discharge rate and storage coefficient of dam reservoir, the reservoir is the most important factor which is related to the results of (Kharazi et al. 2019) in the selection of rivers with alluvial bed and high permeability, low slope, suitable depth and surface as the most suitable reservoirs. Investigations related to the depth of the reservoir show that an alluvial complex with a thickness of about 10 m is very suitable (Chezgi et al. 2016). The quality of these waters is very suitable due to the role of self-purification of alluvium. Also, the results show that the best slope of the waterway for the construction of an underground dam is less than 5% (Kharazi et al. 2019; Chezgi et al. 2016; Talebi et al. 2019). Among the indicators related to the axis of the dam, the depth of alluvium at the location of the axis and the length of the axis of the underground dam are very important and influential from an economic point of view. This criterion due to limit in drilling depth in comparison to axis length, is first and foremost priority that is same with the results of (Talebi et al. 2019) that notified the most important limitation in the construction of the underground dam is the consideration of the axis depth. The results of the water requirement showed that the supply of drinking water needed by the villagers is more important than the agricultural sector's needs, which is in same direction with the results of (Yazdandoost 2016; Ebrahimi et al. 2021). The index of distance from the village is more important than the distance from the road. Presence of the road at the site of the underground dam increases the access speed and reduces the cost of construction and maintenance of the structure. The axes near the road have more priority than other axes, which are in same direction with the results of (Chezgi et al. 2016; Talebi et al. 2019). By investigating the effective factors on construction of structures related to water storage, including underground dams in arid and semi-arid regions, achieving the purpose of comprehensive management of watersheds and sustainable development will be provided.

Acknowledgements This study was funded by Agriculture and Natural resources research and education center of chaharmahal and Bakhtiari Province as part of the National Research Project No. 24-42-29-009-000208.

References

Chezgi J, Pourghasemi HR, Naghibi SA, Moradi HR, Kheirkhah Zarkesh M (2016) Assessment of a spatial multi-criteria evaluation to site selection underground dams in the Alborz Province, Iran. Geocarto Int 31:628–646

Ebrahimi J, Moradi HR, Chezgi J (2021) Prioritizing suitable locations for underground dam construction in south-east of Bushehr Province. Environ Earth Sci 80:1–16

Emami SN, Yousefi S, Pourghasemi HR, Tavangar S, Santosh M (2020) A comparative study on machine learning modeling for mass movement susceptibility mapping (a case study of Iran). Bull Eng Geol Environ 1–18. https://doi.org/10.1007/s10064-020-01915-7

Kadam AK, Kale SS, Pande NN, Pawar NJ, Sankhua RN (2012) Identifying potential rainwater harvesting sites of a semi-arid, basaltic region of Western India, using SCS-CN method. Water Resour Manage 26:2537–2554. https://doi.org/10.1007/s11269-012-0031-3

Kharazi P, Yazdani MR, Khazealpour P (2019) Suitable identification of underground dam locations, using decision-making methods in a semi-arid region of Iranian Semnan Plain. Groundw Sustain Dev 9:100240

Kumar S, Srivastava PK, Snehmani (2017) GIS-based MCDA–AHP modelling for avalanche susceptibility mapping of Nubra valley region, Indian Himalaya. Geocarto Int 32:1254–1267. https://doi.org/10.1080/10106049.2016.1206626

Mahdavi A, Ghasemi M, Jafarzadeh A (2017) Determination of suitable areas for reforestation and afforestation with indigenous species. Casp J Environ Sci 15:29–46. https://doi.org/10.22124/cjes.2017.2215

Mallick J, Singh RK, AlAwadh MA, Islam S, Khan RA, Qureshi MN (2018) GIS-based landslide susceptibility evaluation using fuzzy-AHP multi-criteria decision-making techniques in the Abha Watershed, Saudi Arabia. Environ Earth Sci 77:276. https://doi.org/10.1007/s12665-018-7451-1

Mehrabi H, Zeinivand H, Hadidi M (2013) Site selection for groundwater artificial recharge in Silakhor rangelands using GIS technique. J Rangeland Sci 2:687–695

Mohammady M, Pourghasemi HR, Amiri M (2019) Land subsidence susceptibility assessment using random forest machine learning algorithm. Environ Earth Sci 78:503. https://doi.org/10.1007/s12665-019-8518-3

Naghibi SA, Pourghasemi HR (2015) A comparative assessment between three machine learning models and their performance comparison by bivariate and multivariate statistical methods in groundwater potential mapping. Water Resour Manage 29:5217–5236. https://doi.org/10.1007/s11269-015-1114-8

Panda DK, Mishra A, Jena SK, James BK, Kumar A (2007) The influence of drought and anthropogenic effects on groundwater levels in Orissa, India. J Hydrol 343:140–153

Pourghasemi HR, Pradhan B, Gokceoglu C (2012) Application of fuzzy logic and analytical hierarchy process (AHP) to landslide susceptibility mapping at Haraz watershed, Iran. Nat Hazards 63:965–996

Rahmati O, Golkarian A, Biggs T, Keesstra S, Mohammadi F, Daliakopoulos IN (2019) Land subsidence hazard modeling: machine learning to identify predictors and the role of human activities. J Environ Manage 236:466–480. https://doi.org/10.1016/j.jenvman.2019.02.020

Rahmati O, Pourghasemi HR, Melesse AM (2016) Application of GIS-based data driven random forest and maximum entropy models for groundwater potential mapping: a case study at Mehran Region, Iran. CATENA 137:360–372. https://doi.org/10.1016/j.catena.2015.10.010

Sehat M, Kamanbedast AA, Asadilour M (2013) Zonning underground dams using GIS in Halayjan valley, Izeh-Iran. Tech J Eng Appl Sci 3:3752–3756

Talebi A, Zahedi E, Hassan MA, Lesani MT (2019) Locating suitable sites for the construction of underground dams using the subsurface flow simulation (SWAT model) and analytical network process (ANP)(case study: Daroongar watershed, Iran). Sustain Water Res Manage 5:1369–1378

Wang X, Chen Y, Fang G, Li Z, Liu Y (2022) The growing water crisis in Central Asia and the driving forces behind it. J Clean Prod 378:134574

Yazdandoost F (2016) Dams, drought and water shortage in today's Iran. Iran Stud 49:1017–1028

Yousefi S, Pourghasemi HR, Emami SN, Rahmati O, Tavangar S, Pouyan S, Tiefenbacher JP, Shamsoddini S, Nekoeimehr M (2020) Assessing the susceptibility of schools to flood events in Iran. Sci Rep 10. https://doi.org/10.1038/s41598-020-75291-3

Yousefian E, Faghihi A, Daneshfard K (2022) Designing a model of integrated policy for water governance in Iran. Iran J Manage Sci 16

Zucca A, Sharifi AM, Fabbri AG (2008) Application of spatial multi-criteria analysis to site selection for a local park: a case study in the Bergamo Province, Italy. J Environ Manage 88:752–769. https://doi.org/10.1016/j.jenvman.2007.04.026

Evaluation of Multi Indicators for Groundwater Recharges Conditions of the Asna River Basin, Maharashtra, India with Integration of Fuzzy Logic and GIS Tools

Udaykumar Sahu, Vasant Wagh, Ajaykumar Kadam, Dipak Panaskar, and Satyajit Gaikwad

Abstract The accurate representation of groundwater recharge conditions performs a crucial role in hard rock terrain; where, incidence, flow and occurrence of subsurface water have limitations. This results into a domestic water scarcity, mainly in the summer season period of the Asna River basin, Maharashtra, India. The present work has been carried out in the Asna river basin using Fuzzy logic and GIS techniques to identify the potential groundwater recharge zones. In view of this, various multi-indicators (thematic layers) like Drainage, Lineaments, Geomorphology, Slope, Geomorphology, Rainfall and Water Table Fluctuation etc., have been used to delineate the best appropriate sites for artificial subsurface water rejuvenation of the Asna River basin. The thematic indicators were generated from inventory maps and database, satellite imageries, and field surveys for identification of the best appropriate sites for groundwater recharge. These layers were classified, weighted, and combined using geographic information system (GIS) techniques and fuzzy logic (FL) method. This method was undertaken to get the weights to be assigned to numerous feature covers and these feature were coupled with geospatial devices to define groundwater recharge potential zones. The groundwater potential recharge map was classified into three classes; Low, Moderate and High. The results of fuzzy logic techniques were validated with the groundwater fluctuation data. The outcomes

---

U. Sahu
Department of Geology, Toshniwal Arts, Commerce and Science college, Sengaon, Hingoli, Maharashtra, India

V. Wagh (✉) · D. Panaskar
School of Earth Sciences, Swami Ramanand Teerth Marathwada University, Nanded, Maharashtra, India
e-mail: wagh.vasant@gmail.com

A. Kadam
Department of Environmental Sciences, Savitribai Phule Pune University, Pune, Maharashtra, India

S. Gaikwad
Department of Geology, Savitribai Phule Pune University, Pune, Maharashtra, India

C. B. Pande et al. (eds.), *Surface and Groundwater Resources Development and Management in Semi-arid Region*, Springer Hydrogeology,
https://doi.org/10.1007/978-3-031-29394-8_22

of the study will be helpful for precise demarcation of subsurface recharge arrangements and inventive controlling strategy leading to sustainable development. Also, this study will be helpful to researchers, engineers and stakeholders for development plan of Asna river basin.

Keywords Fuzzy logic · GIS · Groundwater potential · Asna river · India

1 Introduction

Groundwater resources are the precious and regular fresh water reserves of the country and help to sustain life. It is a supply of water to local communities, commercial and agricultural activities, and individual households as a common property resource (Shiferaw et al. 2008; Wagh et al. 2016). The groundwater scarcity problem is the destabilizing effects by climatic changes in the arid and semi-arid regions. The lack of awareness and participation of local communities in water conservation that results exhibits mismanagement (Kumar et al. 2013). The rural drinking water supply is a primary goal of every country in the world beginning with small scale (household) to (nation) large scale. In rural areas, growing demand for drinking water and agricultural usage has led to the high groundwater extractions (Kadam et al. 2022). In the world, more than a billion individuals use consumption water and about 280 million ha farming parcels are used for subsurface water (Richey et al. 2015; Siebert et al. 2013). In the South Asian countries like India, Pakistan and Bangladesh is about 85% groundwater exploiters is mainly used for agriculture relatively 40% rest of the world (Razzaq et al. 2022). The water shortage leads to reduce the industrial, urban and agricultural growth in various parts of the India (Black and Talbot, 2005; Tiwari et al. 2014). An annually quantified subsurface water resource revived in India is 433 km^3 and a net gain of 399 km^3 (Chatterjee and Purohit, 2009). The appraisal of groundwater is typically to understand the groundwater storing and accessibility, but also consider other constituents like hydrogeological and socioeconomic factors (Seiler and Gat 2007).

In Maharashtra state, hard rock basalt is dominant, depleting water levels due to heavy pumping in the agricultural fields and reduces the recharge leads to degradation, excessive water extraction and unscientific sand mining too major hurdles in groundwater recharge. Nevertheless, hard basaltic terrain show demanding conflict in subsurface water storage and have been projected to agrarian scarcities (Thomas et al. 2017; Wagh et al. 2019). Recently, Marathwada semi-arid regions have caused severe precipitation of different tehsils (blocks) influenced terrible and rain-fed undeveloped areas (Budhiraja and Kulkarni 2015; Sahu et al. 2018; 2022). The agriculturalists have unobstructed irrigation pumping and have been scarce subsurface water stores. Early, Theis (1940) suggested that the effects of pumping well bound on subsurface water release and more revealing to revive inside aquifers displays those ideas of safe yields. The depiction of aquifer yields and sources limited to subsurface geology included joints, fractures and weathered sections (alternates) limited to deepening. The elementary ground station well data is consisting of well locality, depth of well, water level (pre & post-monsoon), variation and well exterior. It also got exact capability limits of discrete well such as drawdown, recovery and sequence of geology with distance below ground.

The morphometric study has been carried out through extent of linear, aerial, relief, and slope of the basin (Magesh et al. 2011; Pande and Moharir 2017). The parameters like drainage pattern and density, stream order and frequency, bifurcation and elongation ratio etc., are the primary elements to evaluate watershed characteristics (Sahu et al. 2018). In many study, it is proved that GIS is an essential tool for morphometric analysis from different data sources (Thomas et al. 2012; Patel and Dholakia 2010). Using groundwater recharge potential mapping is a scientific tool for systematic planning and development of the water resources (Díaz-Alcaide and Martínez-Santos 2019; Pande et al. 2020; Kadam et al. 2023). For evaluating problem-solving methodologies, fuzzy logic approaches integrated geospatial data; while, uncertain and partial data were graphical vague (Zadeh, 1965; Nguyen and Walker 2000). FL mechanisms have been reviewed by scientific communities in order to solve the problems with high accuracy and smoother control (Cagman et al. 2011; Liu and Pedrycz, 2009). Fuzzy Logic sets have a degree of truth that can be applied in a variety of areas like atmospheric transmission or groundwater management (Bardossy et al. 1995; Guan and Aral 2005). Fuzzy logic is the selection method using fuzzy values by combining promising and disapproving methods simultaneously. The integration of all assigned FL values spatial maps functions under the fuzzy logic tools in the ArcGIS 10.6 environs. Therefore, the fuzzy membership function specified scale ranges from 1 (full membership) to 0 (full non-membership) (Bonham-Carter, 1994).

With considering the above factors, here we tried to identify the potential groundwater recharge zones within Asna river basin located in the Marathwada region of Maharashtra state. In this region, climate is typically dry with average rainfall 890.28 mm received from SW monsoon. As per GSI report (1997), 98% of area occupied by Deccan basalt and 2% is alluvium. Generally, in vesicular basalt has limited primary porosity; however, secondary porosity and permeability was developed by weathering, jointing, fracturing, etc. In the Asna river basin, the terrain slope observed at north to south and west to south directions and groundwater flow observed in north plateaus to southern plain areas. With low and uncertain rainfall scenario and agriculture is prime occupation with food grain and fodder crops in this region. Most of the rural population is depends on canal and groundwater resources for their domestic and agricultural needs. The present study is initiated with integration of Fuzzy logic and GIS techniques to identify the potential groundwater recharge zones in Asna river basin. The study will provide the baseline information to local water planners and young researchers for the sustainable management of water resources in the Asna watershed.

2 Study Area

The area comes under in the Hingoli and Nanded regions at southern part of Maharashtra state of India. The Asna river basin covers an area of about 1187km^2, which extent from19^{0}10′ to19^{0}32′ N latitudes and 77^{0}04′ to 77^{0}31′ E longitudes (Fig. 1). The origin of basin having elevation of 551 m above mean sea level and lowest elevation is 333 m (amsl). Elevation of area 218 m (amsl) down gradient of southern trends associated to dendritic and paralleled drainage pattern up to 7th order of stream. The river flow NW-ES course through physiographical undulant to plain and bedrock consist of Deccan basalt. The climate of the area is dry with mean yearly rainfall is 890.28 mm occur from the SW monsoon and tropical with hot summer and slight cooled with mean temperature from 12.7° to 41.7 °C. The 30% moisture upsurges in the morning time, high moistureduring the rainsperiods and summer low moisture occurs (CGWB 2013). Based on Geological Survey of India (GSI 1997) report, 98% of the study area covered by Deccan basalt and rest of 2% is alluvium. The vesicular basalt (30 to 60%) has limited primary porosity. However, the secondary porosity and permeability was generated due to weathering, jointing, shearing, fracturing, etc. (CGWB 2013). Generally the terrain slope observed at north to south and west to south directions in the basin. The groundwater flow mainly from the north plateaus to southern plain areas.

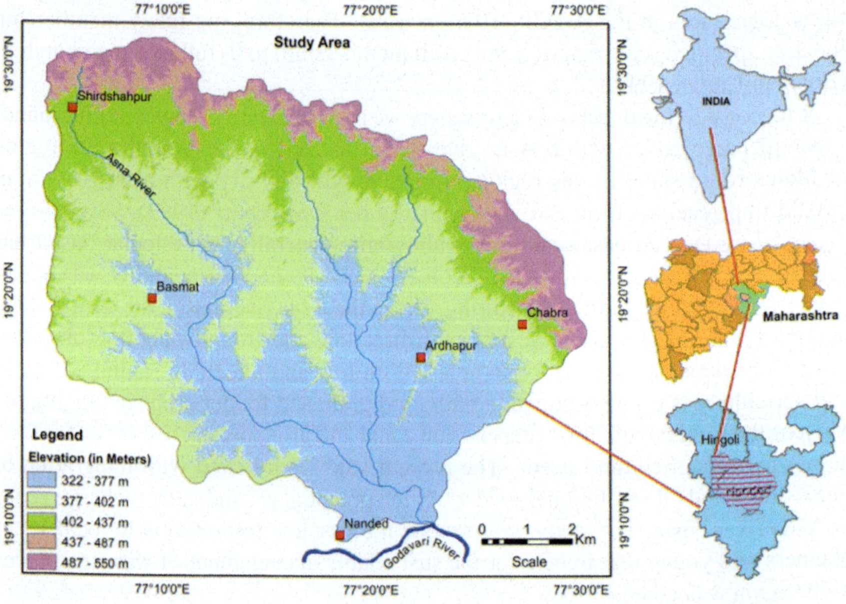

Fig. 1 Study area map

3 Methodology

In this study Survey of India (SOI) toposheets maps (56E/2, 3, 4, 7and 8) on 1:50,000 scales were geo-referenced and transformed in ArcGIS platform to demarcate the drainage divide boundary of the Asna River basin. The drainage layer of the study area was digitized directly from SOI toposheets. A detailed depiction of drainage pattern is dendritic to parallel with assigned drainage orders numbers and corresponding from field checks (Goodrich et al. 1997; Hilpert and Miller 2001). Drainage density (Dd) accurately measured like total length (km) of all the streams divided by the entire basin area (km^2) (Carlson et al. 1963). The method of lineaments was identified from satellite imageries (IRS P6 LISS IV) and support of ASTER (GDEM) data to refined polyline features of the lineaments within the study area using ArcGIS 10.6v software. The extracted and calculated lineament density is based on the polyline features of the lineaments per units area (number/km^2) (Hung et al. 2005). The extracted lines feature of the lineaments executed under the density extension tool to define the lineament density maps classified on the basis of natural breaks in ArcGIS. The higher lineament density areas show that significant prospect for groundwater zones. The terrain elevation attributes and slope percentage maps created from standard ASTER-GDEM (https://earthexplorer.usgs.gov/) with 30 m resolution of the Earth raster data used in the ArcGIS 10.6 v software. Remote sensing visual interpretation signatures based on the acquired satellite imagery data in reference of SOI topographic map features. The characteristic of landform boundaries clearly observed on the satellite imagery with help of image elements like colours/tones, textures, etc. The interpreted geomorphic map verified for ground truth in field checks in the account to relief and terrain features. These geomorphic units were classified into shallow, moderate and deep supported on the well inventory data and profile section cutting observed in nalla/road cutting etc. The entire geo-database was converted into a digital format to create groundwater potential recharge zones. Drainage and lineament layer maps were prepared to density raster maps using a density tool and slope maps in percentage converted from ASTERGDEM data. The corresponding attributes data as rainfall, water table of wells, etc. The groundwater potential recharge zones were evaluated in Fuzzy logic analysis tool into a GIS environment. Using of the fuzzy analysis tool that helped to separate the maps with new class features. The fuzzy map sub-classes were assigned fuzzy membership values based on their influence and significant characteristics on groundwater recharge (Fig. 2). The Table 1 represents various thematic layers with classification, FL values and area %.

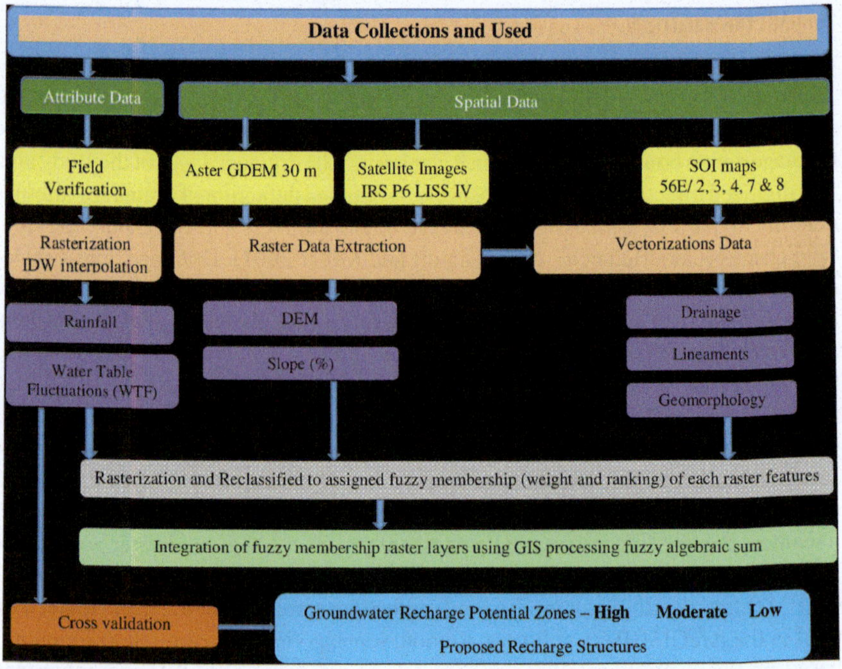

Fig. 2 Flow chart of the methodology

4 Result and Discussion

4.1 Drainage and Drainage Density

The surface natural drainage divides were extracted by identifying drainage basin boundaries and streams (drainage) outlet can be delineated. The infiltration of surface water flow from river and drainage systems to the contribute groundwater recharge. In our study, drainage divide assigned the Strahler- drainage order has to join each others up to the 7th higher order. First order drainages are most abundant and highest total length (1192.60 km) and the last seventh drainage length (7.94 km) found within the watershed area (Horton, 1945; Schumm, 1956; Rinaldo et al.1993). Drainage density (Dd) illustrates the compactness of the streams that presents quantitative measures of the total length of drainage of the basin (Horton, 1932; Strahler, 1964). It also explained the correlation with landform dissection, climate and vegetation, soil and water infiltration efficiency, surface runoff to end products erosion and sedimentation in the watershed (Farhan and Anaba 2016).The drainage density output is prepared from the drainage layer in GIS environment. Four drainage density (Dd) categories were produced and assigned Fuzzy Logic (FL) values to each (Fig. 3) namely, very low (265.43 km^2), low Dd (408.25 km^2), moderate Dd (341.71 km^2)and high Dd (171.60 km^2).Most of the areas covered with very low and low drainage density

Table 1 The thematic layers with fuzzy logic values (After Sahu et al. 2022; Arulbalaji et al. 2019)

Thematic layer	Features	Class	FL value	Area in km^2	Area in %
Drainage density	4.25–2.58 (km/km^2)	High	0.20	171.61	14.45
	2.58–1.88 (km/km^2)	Moderate	0.60	341.71	28.78
	1.88–1.26 (km/km^2)	Low	0.80	408.25	34.39
	1.26–0 (km/km^2)	Very low	1.00	265.43	22.38
Lineament density	1.74–0.82 (km/km^2)	High	1	278.46	23.45
	0.82–0.38 (km/km^2)	Moderate	0.75	525.42	44.28
	0.38–0.00 (km/km^2)	Low	0.25	383.12	32.27
Geomorphology	Plateau undissected (PLU)	–	0.50	17.5	1.47
	Plateau highly dissected (PLH)	–	0.25	78.53	6.61
	Plateau moderately dissected (PLM)	–	0.80	87.1	7.86
	Valley (V)	–	0.9	37.56	3.01
	Plateau weathered canal command (PLWC)	–	1	759.06	63.94
	Plateau weathered (PLW)	–	0.80	202.51	16.99
	Mesa/Butte (M/B)	–	0.05	1.5	0.2
	Flood Plain (FP)	–	1	3.24	0.38
Slope (%)	0–1.55%	Plain	1	798.79	67.29
	1.59–4.56%	Gentle	0.97	326.45	27.5
	4.56–9.69%	Moderate	0.95	0.12	0.03
	9.69–16.07%	High	0.45	37.44	3.15
	16.07–29.07%	Steep	0.25	24.2	2.03
Rainfall	<800 mm		0.1	99.46	8.37
	800–900 mm	–	0.16	266.32	22.43
	900–1000 mm	–	0.25	505.73	42.63
	1000–1200 mm	–	0.35	212.36	17.89
	>1200 mm	–	1.00	103.13	8.68

have significant groundwater storage, which comprised plain and flat terrain and less runoff zones. Therefore, high drainage density area exhibits the low permeability and is limited to groundwater.

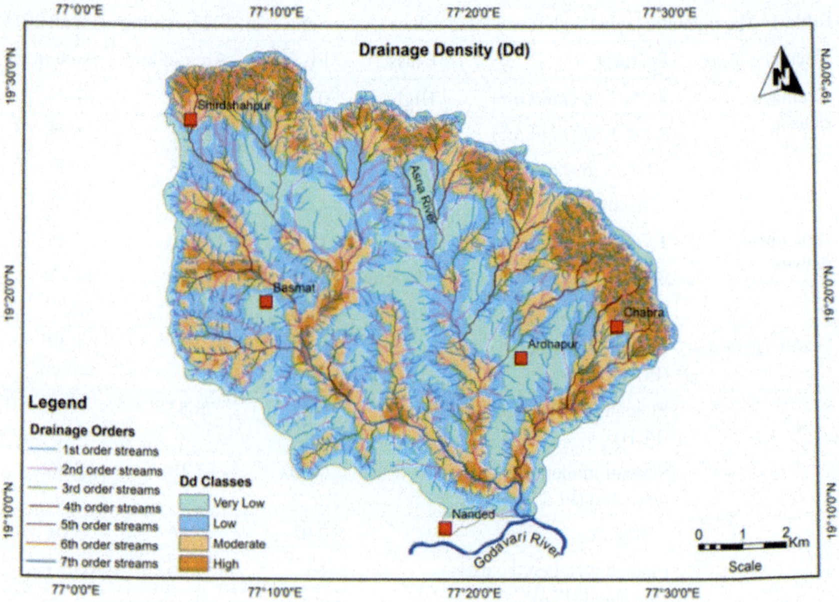

Fig. 3 Drainage (D) and Drainage Density (Dd) map of the study area

4.2 Lineaments & Lineaments Density (Ld)

Lineaments are linear structures on surface the earth that have increased permeability factor, which is extensively covered by the earth (Suganthi et al. 2013; Ni et al. 2016).This is indication of the subsurface geological structures, fractures, faults or joints associated with the porosity and permeability of water cropping pattern and low lying topography etc., (Masoud et al. 2022; Rajaveni et al. 2017; Pradhan and Youssef 2010). Densely fractured and weathered basaltic rocks exhibit good groundwater recharge zones. The lineament map was prepared from satellite images. Lineaments were calculated for using lineaments density map using the density technique in ArcGIS mode. In this study, Ld classified into 3 classes like high Ld (278.46 km^2), moderate Ld (525.42 km^2) and low Ld (383.12 km^2) (Fig. 4). Lineaments are the major factor in groundwater recharge. The high Ld area about 23.45% area covered in the form of small patches with gentle and plain slopes within plateau weathered landform, denoted with FL is 1 value. The Low Ld areas continuously covered the edge basin boundaries and center of the study area.

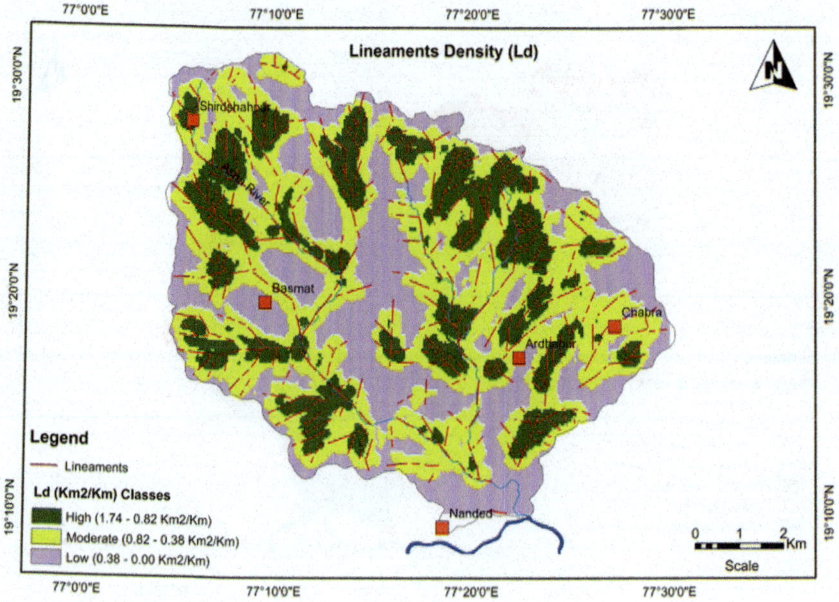

Fig. 4 Lineament density (Ld) mapof the study area

4.3 Geomorphology

Geomorphologyis important terms for infiltration and groundwater recharge into the earth (Etikala et al. 2019). This study shows that the surface produced three types of landforms like denudational, erosional and depositional. The denudational landforms include Plateau Undissected (PLU), Plateau Highly Dissected (PLH), Plateau Moderately Dissected (PLM), Plateau Weathered Canal Command (PLWC), Valley (V), Plateau Weathered Canal Command (PLWC), Plateau Weathered (PLW), Messa/Butte (M/B), and Depositional Landform is Flood Plain (FP) (Fig. 5). The groundwater augmented under the plateau canal command areas compared to the non-canal command area with high fuzzy value (1) assigned the PLWC unit. Messa/Butte (M/B) isolated flat topped hills of basalt with steps to moderate slopes, high runoff zone and not suitable for groundwater recharge potential. It is occupied by the 1.54 km^2 (0.2%), it is not significant for groundwater recharge unit due to high runoff zone therefore it provided the fuzzy value (0.05). A flood plain (FP) is the portion of a river valley adjacent to the river bed, which consists of loose deposited during the overflow of river water discharge. It is a low elevation of 342 to 320 m (amsl), completely flat area.

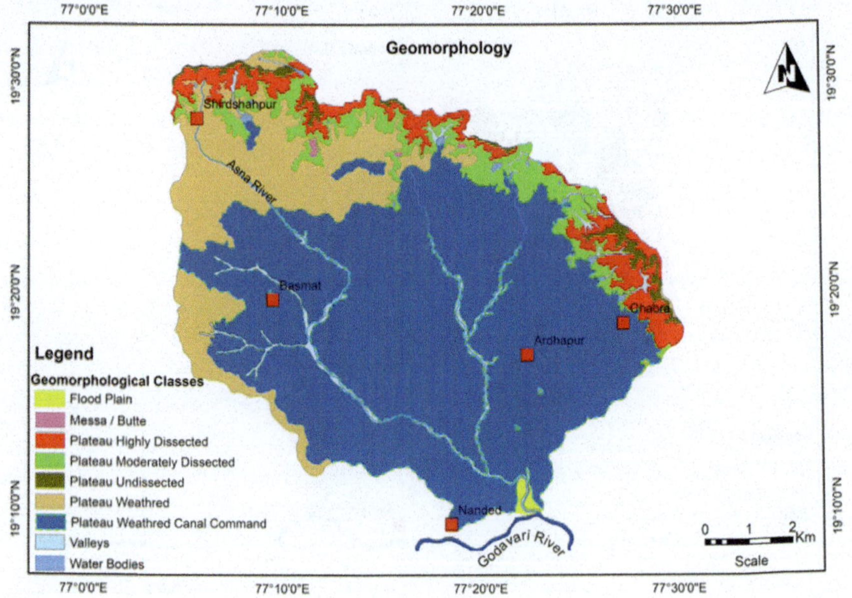

Fig. 5 Geomorphology map of the study area

4.4 *Slope*

The 90% part of the study area is covered by gentle to plain slopes, including the northern side stretched near the western margins of watershed, elevated terrain or plateaus sloping downwards in the southern part. A maximum terrain elevation is 551 m (amsl) to minimum 333 m (amsl) and relief of 218 m (amsl) respectively. Slope map prepared from ASTER GDEM elevation data of 30 m resolution under processed into ArcGIS. The analyzedwatershed surface slopes in to percentage (%) classesranges from (0% to > 15%). Therefore, the slope mapdivided five classes (Fig. 6), (i) plain slope preference for the flat and almost zero gradient surfaces covering the 798.79 Km^2 (67.29%). It is assigned great fuzzy value (1) to successful groundwater recharge. (ii) Secondly, dominantly gentle slope covered by a total area 326.45 km^2 (27.5%) and comprises a rough surface to flat topography sloping towards the south of the watershed. Whichever denote the good prospects of groundwater with a fuzzy value (0.80). (iii) Generally, the terrain of this area is an undulating surface area that covers much less area 0.12 km^2 (0.03%) in this watershed. The fuzzy logic membership is specifying (0.70) and it's moderately suitable for groundwater recharge potential. (iv) High slopes are mainly conjugated to form high drainage density values, symbolize the high peak with rugged terrain. It contributes to high runoff surface, less infiltration and low groundwater recharge practices. (v) Steep slope class accounts for a small area 24.2 km^2 (2.03%) part of the watershed. It

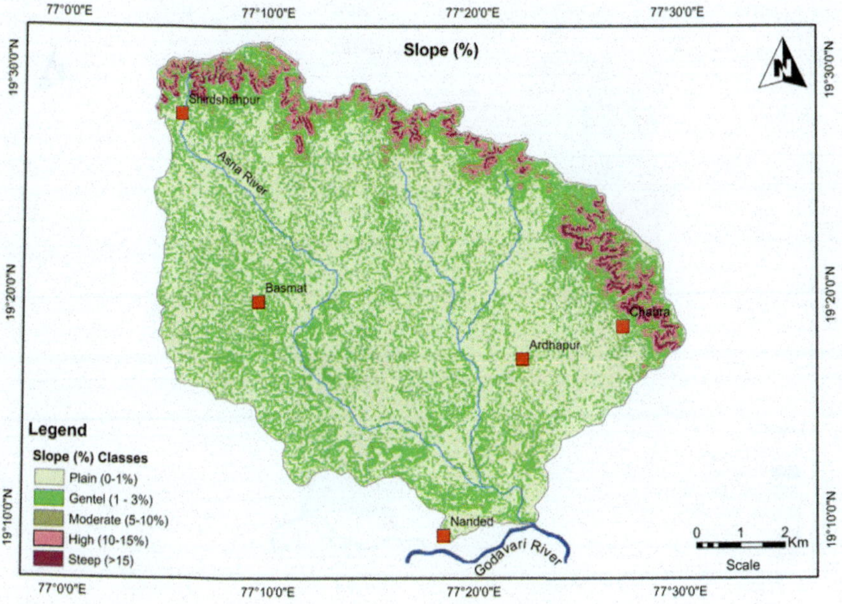

Fig. 6 Slope (%) map of the study area

mainly occurs in the PLH and PLU landforms, flowing 1st and 2nd order drainages because of greater runoff to low fuzzy value (0.30) provided.

4.5 Rainfall

Rainfall is the principal source of water in arid and semiarid environment, which has significance in hydrological cycle (Tapiador et al. 2012; Keune and Miralles 2019). The study area consist of hard basaltic terrain, this rainfall factor was influenced in the groundwater recharge process. An interpolation method is used to create gridded rainfall data that depict the rainfall distribution at altered time interims (Giambelluca et al. 2013). A rainfall distribution map was prepared for the study area to comprehend the groundwater recharge (Fig. 7).

4.6 Water Table Fluctuation (WTF)

The water table fluctuation (WTF) specifies that transient water-level rises are directly related to renew water reaching the water table (Nimmer et al. 2010). The water table fluctuation from 0.5 m to as high as 40 m based on well record data. The pre and post-monsoon water table data were computed in ArcGIS platform to produce a water

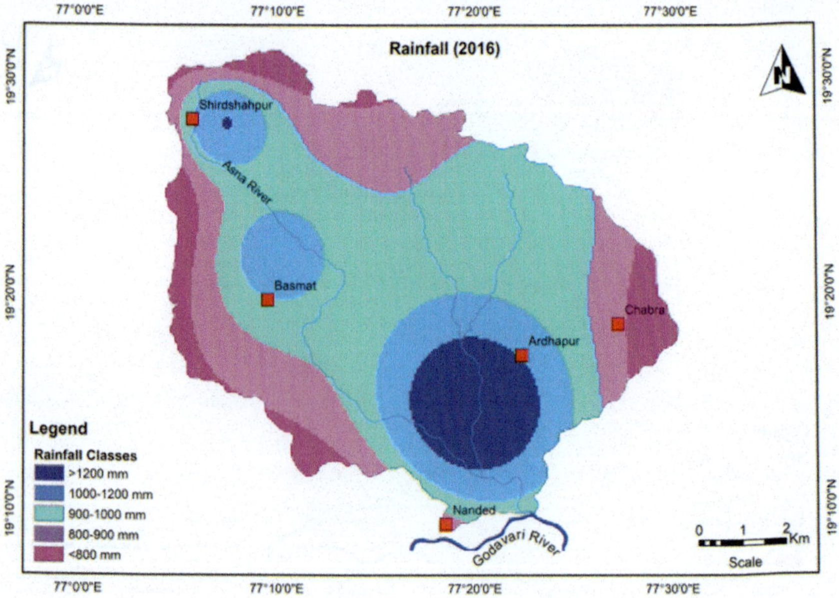

Fig. 7 Rainfall distribution map of study area

table fluctuation map, which classified into three categories like shallow, moderate and deep (Fig. 8). Based on the priority excellent recharge potential zone that assigned the fuzzy logic (0.80) for shallow water table fluctuation (WTF) zone, aerial extended about 570.54km^2 (48.06%) in the north to east trajectory in the Asna watershed. The moderate fluctuation portion perspective northwest and south down slope of the watershed. It is indicating moderate prospects of groundwater recharge, which is exposed near about 500.38km^2 (42.15%) area obtained the fuzzy logic (0.60). The negative water-level fluctuations shows in deep water-level zones covered the 116.08 km^2 (9.77%) and were distributed in the form of pockets in southwest side of the study area. Low (0.40) fuzzy logic value designated for the deep WTF zone.

4.7 *Groundwater Potential Recharge Zones (GWPRZ)*

Fuzzy analysis was performed to identify the groundwater potential recharge zones (GWPRZ) in the study area (Fig. 9). In this study area, high groundwater potential recharge zones are mainly observed in western and southern parts and vicinity near to Basmat town due to low slope of surface associated with high drainage density and rainfall with low groundwater fluctuation. The moderate groundwater potential recharge zones are occurred in patches where gentle and moderate slopes with low to moderate drainage density. Moreover, precipitation in this zone is low to moderate and groundwater fluctuation is shallow to moderate. However, majority part of the

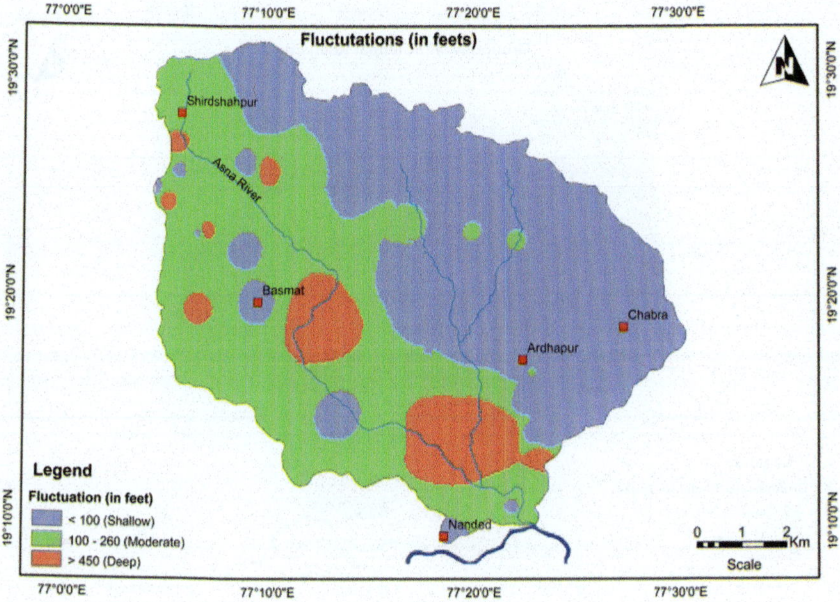

Fig. 8 Fluctuation map of study area

study area are classified as low groundwater recharge potential due to steep to gentle slopes, high to low drainage density and groundwater fluctuation encountered as moderate to deep. Therefore, this zone is classified as low groundwater recharge potential.

5 Conclusion

The present study has been carried out with integration of Fuzzy logic and GIS techniques to identify the potential groundwater recharge zones in the Asna river basin, Maharashtra. In view of this, thematic layers such as drainage, lineaments, geomorphology, slope, geology, etc., were considered to delineate the best appropriate sites for artificial subsurface water rejuvenation. The weight assigned to drainage and lineament density, geomorphology, slope, precipitation and WTF, which are significant parameter for the controlling the groundwater. The groundwater potential recharge map was generated by using three classifications like Low, Moderate and High. The results of fuzzy logic techniques were validated with the groundwater fluctuation data. In Asna river basin, high groundwater potential recharge zones are mainly encountered in western and southern parts and vicinity of Basmat region due to low slope, high drainage density, rainfall and low groundwater fluctuation. Moreover, moderate groundwater potential recharge zones are observed in patches; where, gentle and moderate slopes with low to moderate drainage density. However,

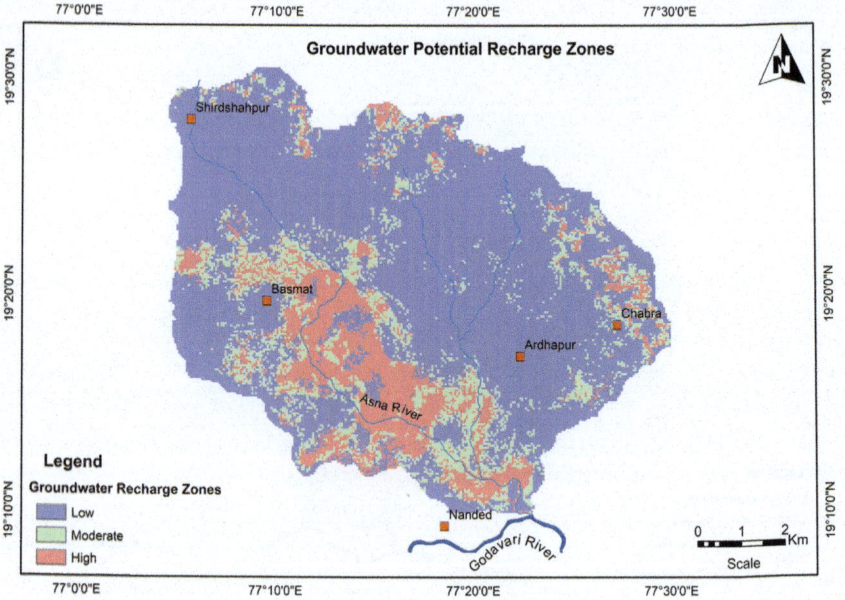

Fig. 9 Groundwater potential recharge zones of the study area

most part of the basin characterized as low recharge potential due to elevated plateaus comprises with massive hard basalt and high to moderate drainage density because groundwater fluctuation as moderate to deep. The outcomes of the study will be useful for the precise demarcation of groundwater potential recharge structures of the watershed. Also, provide the baseline information to new researchers, stakeholders and policy makers to develop a sustainable water management plan in the Asna river basin.

References

Arulbalaji P, Padmalal D, Sreelash K (2019) GIS and AHP techniques based delineation of groundwater potential zones: a case study from southern Western Ghats India. Sci Rep 9(1):1–17

Black M, Talbot R (2005) Water, a matter of life and health: water supply and sanitation in village India. Oxford University Press

Budhiraja RB, Kulkarni SA (2015) Water sector regulatoty reforms in Maharashtra: a step forward. NDCWWC J (A half Yearly J New Delhi, Centre of WWC) 4(2):5–10.

Bardossy A, Duckstein L, Bogardi I (1995) Fuzzy rule-based classification of atmospheric circulation patterns. Int J Climatol 15(10):1087–1097

Bonham-Carter GF, Bonham-Carter G (1994) Geographic information systems for geoscientists: modelling with GIS (no 13). Elsevier

Carlson CW (1963) Drainage density and stream flow: US Geol Survey Prof Paper 422

Cagman N, Enginoglu S, Citak F (2011) Fuzzy soft set theory and its applications. Iranian J Fuzzy Syst 8(3):137–147

Chatterjee R, Purohit RR (2009) Estimation of replenishable groundwater resources of India and their status of utilization. Curr Sci 1581–1591

Central Ground Water Board (CGWB) (2013) Nanded District, Maharashtra Profile, 1804/DBR/2013

Díaz-Alcaide S, Martínez-Santos P (2019) Advances in groundwater potential mapping. Hydrogeol J 27(7):2307–2324

Etikala B, Golla V, Li P, Renati S (2019) Deciphering groundwater potential zones using MIF technique and GIS: a study from Tirupati area, Chittoor District, Andhra Pradesh, India. HydroResearch 1:1–7

Farhan Y, Anaba O (2016) A remote sensing and GIS approach for prioritization of WadiShueib mini-watersheds (Central Jordan) based on morphometric and soil erosion susceptibility analysis. J Geogr Inf Syst 8(1):1–19

Geological Survey of India (GSI) (1997) Quadrangle map 56E series

Giambelluca TW, Chen Q, Frazier AG, Price JP, Chen YL, Chu PS, …, Delparte DM (2013) Online rainfall atlas of Hawai ‘i. Bull Am Meteor Soc 94(3):313–316

Goodrich LV, Milenkovic L, Higgins KM, Scott MP (1997) Altered neural cell fates and medulloblastoma in mouse patched mutants. Science 277(5329):1109–1113

Ground water year book of Maharashtra and union territory of Dadra and Nagar haveli (year 2019–2020) (aap 2020–2021)

Guan J, Aral MM (2005) Remediation system design with multiple uncertain parameters using fuzzy sets and genetic algorithm. J Hydrol Eng 10(5):386–394

Hilpert M, Miller CT (2001) Pore-morphology-based simulation of drainage in totally wetting porous media. Adv Water Resour 24(3–4):243–255

Hung LQ, Batelaan O, De Smedt F (2005) Lineament extraction and analysis, comparison of LANDSAT ETM and ASTER imagery. Case study: Suoimuoi tropical karst catchment, Vietnam. In: Remote sensing for environmental monitoring, GIS applications, and geology, vol 5983. SPIE, pp 182–193

Horton RE (1932) Drainage-basin characteristics. Trans Am Geophys Union 13(1):350–361

Horton RE (1945) Erosional development of streams and their drainage basins; hydrophysical approach to quantitative morphology. Geol Soc Am Bull 56(3):275–370

Kadam A, Wagh V, Jacobs J, Patil S, Pawar N, Umrikar B, ..., Kumar S (2022) Integrated approach for the evaluation of groundwater quality through hydro geochemistry and human health risk from Shivganga river basin, Pune, Maharashtra, India. Environ Sci Pollut Res 29(3):4311–4333

Kadam AK, Patil SN, Gaikwad SK et al (2023) (2023) Demarcation of subsurface water storage potential zone and identification of artificial recharge site in Vel River watershed of western India: integrated geospatial and hydrogeological modeling approach. Model Earth Syst Environ. https://doi.org/10.1007/s40808-022-01656-4

Keune J, Miralles DG (2019) A precipitation recycling network to assess freshwater vulnerability: challenging the watershed convention. Water Resour Res 55(11):9947–9961

Kumar M, Patle GT, Khanna M, Dagar JC (2013) Climate smart water conservation management technologies. Int J Water Resour Environ Eng 5(10):556–572

Liu X, Pedrycz W (2009) Axiomatic fuzzy set theory and its applications, vol 244. Springer, Heidelberg

Magesh NS, Chandrasekar N, Soundranayagam JP (2011) Morphometric evaluation of Papanasam and Manimuthar watersheds, parts of western Ghats, Tirunelveli district, Tamil Nadu, India: a GIS approach. Environ Earth Sci 64(2):373–381

Masoud AM, Pham QB, Alezabawy AK, El-Magd SAA (2022) Efficiency of geospatial technology and multi-criteria decision analysis for groundwater potential mapping in a Semi-Arid region. Water 14(6):882

Monitoring and Evaluation of Artificial Recharge of Groundwater programs/Schemes/project in Rainfed Area Authority, Planning Commission, Government of India (2011)

Ni C, Zhang S, Liu C, Yan Y, Li Y (2016) Lineament length and density analyses based on the segment tracing algorithm: a case study of the gaosong field in gejiu tin mine, China. Math Problems Eng 2016
Nimmer M, Thompson A, Misra D (2010) Modeling water table mounding and contaminant transport beneath storm-water infiltration basins. J Hydrol Eng 15(12):963–973
Pande CB, Moharir K (2017) GIS based quantitative morphometric analysis and its consequences: a case study from Shanur River Basin Maharashtra India. Appl Water Sci 7(2):861–871
Pande CB, Moharir KN, Singh SK, Varade AM (2020) An integrated approach to delineate the groundwater potential zones in Devdari watershed area of Akola district, Maharashtra, Central India. Environ Dev Sustain 22(5):4867–4887
Patel DP, Dholakia MB (2010) Identifying probable submergence area of Surat city using digital elevation model and geographical information system. World Appl Sci J 9(4):461–466
Pradhan B, Youssef AM (2010) Manifestation of remote sensing data and GIS on landslide hazard analysis using spatial-based statistical models. Arab J Geosci 3(3):319–326
Rajaveni SP, Brindha K, Elango L (2017) Geological and geomorphological controls on groundwater occurrence in a hard rock region. Appl Water Sci 7(3):1377–1389
Razzaq SM, Jiang A, Jafarpour B (2022) Latent-space inversion (LSI): a deep learning framework for inverse mapping of subsurface flow data. Comput Geosci 26(1):71–99
Richey AS, Thomas BF, Lo MH, Reager JT, Famiglietti JS, Voss K, ..., Rodell M (2015) Quantifying renewable groundwater stress with GRACE. Water Resour Res 51(7):5217–5238
Rinaldo A, Rodriguez-Iturbe I, Rigon R, Ijjasz-Vasquez E, Bras RL (1993) Self-organized fractal river networks. Phys Rev Lett 70(6):822
Sahu U, Panaskar D, Wagh V, Mukate S (2018) An extraction, analysis, and prioritization of Asna river sub-basins, based on geomorphometric parameters using geospatial tools. Arab J Geosci 11(17):1–15
Sahu U, Wagh V, Mukate S, Kadam A, Patil S (2022) Applications of geospatial analysis and analytical hierarchy process to identify the groundwater recharge potential zones and suitable recharge structures in the Ajani-Jhiri watershed of north Maharashtra, India. Groundw Sustain Dev 17:100733
Schumm SA (1956) Evolution of drainage systems and slopes in badlands at Perth Amboy, New Jersey. Geol Soc Am Bull 67(5):597–646.3
Seiler KP, Gat JR (2007) Groundwater recharge from run-off, infiltration and percolation, vol 55. Springer Science & Business Media
Shiferaw B, Reddy VR, Wani SP (2008) Watershed externalities, shifting cropping patterns and groundwater depletion in Indian semi-arid villages: the effect of alternative water pricing policies. Ecol Econ 67(2):327–340
Siebert S, Henrich V, Frenken K, Burke J (2013) Update of the digital global map of irrigation areas to version 5. Rheinische Friedrich-Wilhelms-Universität, Bonn, Germany and Food and Agriculture Organization of the United Nations, Rome, Italy
Strahler AN (1964) Quantitative geomorphology of drainage basin and channel networks. Handbook of Applied Hydrology
Suganthi S, Elango L, Subramanian SK (2013) Groundwater potential zonation by Remote Sensing and GIS techniques and its relation to the Groundwater level in the Coastal part of the Arani and Koratalai River Basin Southern India. Earth Sci Res J 17(2):87–95
Tapiador FJ, Turk FJ, Petersen W, Hou AY, García-Ortega E, Machado LA, ..., De Castro M (2012) Global precipitation measurement: Methods, datasets and applications. Atmosph Res 104:70–97
Tiwari AK, Singh PK, Mahato MK (2014) GIS-based evaluation of water quality index of ground water resources in West Bokaro Coalfield India. Curr World Environ 9(3):843
Thomas J, Joseph S, Thrivikramji KP, Abe G, Kannan N (2012) Morphometrical analysis of two tropical mountain river basins of contrasting environmental settings, the southern Western Ghats India. Environ Earth Sci 66(8):2353–2366
Thomas, R., Duraisamy, V., Scodanibbio, L., Hoffman, T., & Misquitta, K. (2017). Vulnerability to groundwater drought in semi-arid areas of western Ahmednagar District, India.

Theis CV (1940) The source of water derived from wells. Civ Eng 10(5):277–280

Wagh VM, Panaskar DB, Varade AM, Mukate SV, Gaikwad SK, Pawar RS, ..., Aamalawar ML (2016) Major ion chemistry and quality assessment of the groundwater resources of Nanded tehsil, a part of southeast Deccan Volcanic Province, Maharashtra, India. Environ Earth Sci 75(21):1–26

Wagh VM, Panaskar DB, Jacobs JA, Mukate SV, Muley AA, Kadam AK (2019) Influence of hydrogeochemical processes on groundwater quality through geostatistical techniques in Kadava River basin Western India. Arabian J Geosci 12(1):1–25

Zadeh L (1965) Fuzzy sets. Inf Control 8(3):338–353. https://earthexplorer.usgs.gov/